A Solar Car Primer

Eric Forsta Thacher

A Solar Car Primer

A Guide to the Design and Construction of Solar-Powered Racing Vehicles

Springer

Eric Forsta Thacher
Potsdam
New York
USA

ISBN 978-3-319-17493-8 ISBN 978-3-319-17494-5 (eBook)
DOI 10.1007/978-3-319-17494-5

Library of Congress Control Number: 2015936143

Springer Cham Heidelberg New York Dordrecht London

Printed on acid-free paper

Springer International Publishing AG Switzerland is part of Springer Science+Business Media (www.springer.com)

To Clarkson University's Solar Racing Teams for all their hours of loving labor, and to Susan for her endless patience and for all those chocolate chip cookies.

Preface

Environmental Purpose

In 1990, after the first Sunrayce, the *Courier-Observer,* a newspaper serving the Potsdam area where I live, commented in an editorial that "…we don't expect solar cars to become the wave of the future…" This opinion may be taken as typical of many. However, there is another view. This view explains, in part, the motivation for writing this book.

The future is upon us. Each day our transportation system dumps millions of tons of pollutants into the atmosphere. This poisoning, even if it were the only environmental poisoning taking place (which it is not, of course), must be viewed as a crisis in the history of the earth. Furthermore, each day the millions of gallons of fuel consumed is subtracted from the remaining, decidedly finite, supply, some of which comes to us from overseas.

We must make a radical shift away from the combustion of carbon-based fuels as the energy source of our transportation system to a pollution-free transportation system with an energy source that can sustain that system indefinitely. This must be done quickly compared to the scale of historical time. To think otherwise is, as a teacher I know is fond of saying, to "live in la-la land."

What solutions are available that meet, or nearly meet, the criteria just set out? Only one: solar energy, in some form. Near-term examples are fuel cells using solar-generated hydrogen and air to power electric vehicles, electric vehicles driven directly or indirectly (by solar-charged batteries) by solar cell arrays, or vehicles driven directly by the combustion of solar-generated hydrogen with air. In the hydrogen-powered cases, the gas would be generated by the hydrolysis of water using the electric current produced by a fixed solar cell array, or by wind turbines driving electric generators.

None of these methods perfectly meet the criteria or is free of problems: the combustion of hydrogen with atmospheric air generates nitrous oxides; disposal of some battery types is a solid waste problem; the fabrication of solar cells creates wastes; large-scale solar power production requires large land areas. However, it is important to have a diversity of approaches under development. Otherwise we cannot hope to find the best approach, or combination of approaches. The history of this country is full of examples of technology rejected by our industrial establishment or our government without adequate examination, only to surface elsewhere. Perhaps, in the case of automobiles, opposition to progress would be less if gasoline prices were more than $ 5.00 per gallon (as in many countries as of this writing[1]), a price that better reflects the avoided cost of this fuel.

The potential for a net reduction in vehicular atmospheric pollution even using ordinary electric vehicles recharged by the utility grid is great. Of course, as long as the utility grid is supplied by energy from carbon-based fuels, use of electric vehicles will neither eliminate pollution entirely nor will it eliminate the consumption of these fuels. Bentley et al. (1992) studied these problems. And there is an additional risk: the reliability of the vehicle fleet would then depend upon the availability of the utility grid. Directly or indirectly solar-powered automobiles do not share this problem.

Driven by the urgent need to meet the environmental crisis, our transportation system is now changing, and it must continue to change at an increasing rate. Lightweight, hybrid-powered, very efficient but hydrocarbon-burning cars have entered the market, as have electric vehicles. Hamilton (1989) is an early general study of electric and hybrid vehicles. The "hypercar" concept of the Rocky Mountain Institute (Lovins et al. 1993, Moore and Lovins 1995) is a more recent example.

Educational Purpose

The remaining part of the motivation for this book is the improvement building solar cars brings to the education of the student engineers who build them. To design, build, test, and race a solar-powered car is experiential learning of the most effective sort. College graduates who have participated in such projects have helped to solve a real, complex, engineering problem. As such, they are prime prospective employees. And not just the engineers: the projects include marketing, fundraising, project planning and management, and public relations. Students from the Business School can be involved as well.

As of this writing (2013), 11 biennial American Solar Challenge races have been staged (the first five were named "Sunrayce"). During the period 1989–2012, there were 18 American Tour de Sol[2] races and several overseas solar-electric vehicle races, including the biennial World Solar Challenge[3] in Australia. Entrants in these events were teams from colleges, high schools, automobile companies, and individuals. It is notable that high school teams typically represented a substantial proportion of the American Tour de Sol entrants and that an annual international race just for high schools, the Solar Car Challenge,[4] was created in 1993.

---

[1] See, for example, MyTravelCost.com, for a list of worldwide gasoline prices.

[2] The final American Tour de Sol, the eighteenth, was run in 2006.

[3] See http://www.worldsolarchallenge.org.

[4] See http://www.solarcarchallenge.org—a site rich with information.

This activity implies a strong interest in solar-electric and electric vehicle racing at colleges and also at secondary and post-secondary schools. However, solar car design, construction, and racing, demand technical knowledge that may not be readily available to private individuals, secondary school students, and in an integrated form, to college students.

Books have been written at the professional level on the aerodynamics of vehicles, on suspension design, on solar cells, and on composite materials—on every component used in a solar car. The book aims to provide a primer on those subjects as they apply to the design of solar-electric cars. The book emphasizes the integration and application of fundamental knowledge and skills to the process of creating a low-energy-consumption vehicle.

Background Assumed

The readers of this book are likely to be of diverse technical backgrounds. Therefore, emphasis has been placed on physical explanations. Mathematical relationships have, when possible, been accompanied or replaced by graphs. However, I have assumed a high-school-level knowledge of algebra, trigonometry (elementary calculus will be helpful, but is not vital), physics (statics, dynamics, thermodynamics), and electrical circuits. Knowledge of basic fabrication techniques and terminology also has been presumed.

Use of this Book

I have defined "courses" to include both conventional design courses and project-based learning communities (PBLC). The former tend to produce only paper designs[5] and center around the familiar classroom environment: lecturing, homework, etc. They generally last a semester and include students of one grade level. The latter run more or less continuously, are likely to include students of various grade levels and educational specialties, and are very fruitful when built around a large, multidisciplinary project such as a solar car.

The features of *A Solar Car Primer* already discussed make it useful for both conventional design courses and PBLCs. The book contains a discussion of the relevant physics, which is keyed to the project to motivate the discussion and a detailed example based on an actual solar car. The book is modular, so advanced students can skip the basic theory and apply the example directly to their project. Beginners can do the reverse. The test, energy management, and fund raising and public relations chapters should be especially useful to PBLCs.

---

[5] Although there is a trend toward more product realization: multi-semester sequences in which prototypes are designed, built, and tested.

Acknowledgments

In Chaps. 9, 10, and 11, I have drawn upon some of the designs made, and some of the manufacturing methods used, by Clarkson University's 1995 and 1999 Sunrayce teams. I would like to thank those teams and to acknowledge certain members of those teams whose work contributed, directly or indirectly, to this book. These are: Vivek Sarin, Leslie Ann Hummel, Mathieu Gratton, Renay Bouchard Gratton, Jules Leduc, Dan Retajczyk, Tim St. Hilaire, Nate Ryder, Craig Evans, Scott Leonard, Rob Preuss, Matt Duquette, Lou Fasulo, Allison Green, Brian Lisiecki, and Dan Lyskawa.

The cost information in Chap. 14 is based in large part on information submitted by many of the US teams which have participated since 2010 in the American Solar Challenge and the Formula Sun Grand Prix.[6] Most of these teams were in the midst of designing, building, or testing new solar racing cars. So I am especially grateful that the following team members took the time to submit cost information: Wilkins White, Oregon State University; Logan Wells, University of Kentucky; Amy Sunderlin, Southern Illinois University Edwardsville; Min Ju Lee, University of California Berkeley; and Eric Hausman, University of Michigan.

Most of the photographs of solar cars were taken by Brian Lisiecki, Mike Griffin, Director of News Services at Clarkson University, and Jules Leduc. The photographs in Chap. 12 were taken by Mike Griffin and the staff of the Advanced Aerodynamic Laboratory in Ottawa, Canada. Photographs not otherwise credited were taken by the author.

Many thanks to Chuck Goodwin, Glen Botto, Kelly Eston-Thacher, Nancy Boyer-Rechlin, Drs. Dana Barry, Norbert Ackerman, Jim Carroll, and Ken Visser for reading drafts, or portions, of this book and for their helpful comments there on. Any errors remaining are my own.

I acknowledge with gratitude Dr. Russ Read, who was a co-advisor for Clarkson's 1990 and 1993 Solar Car Teams, Dr. Tom Ortmeyer, who was a co-advisor for the 1990 team, and Dr. Larry Compeau, who was a co-advisor for the 1995 and 1997 teams, Mr. Mike Griffin, and Ms. Karen St.Hilaire, who were staunch supporters of the project from its inception.

References

Bentley, J. M., Teagan, P., Walls, D., Bailes, E., Parish, T., 1992, "The Impact of Electric Vehicles on CO_2 Emissions," EGG-EP-10296, DE93 003719, U.S. Government Printing Service, Washington, DC.

Hamilton, W., 1989, "Electric and Hybrid Vehicles," DOE/ID–10252, DE90 002218, U.S. Government Printing Service, Washington, DC.

Lovins, A. B., Barnett, J. W., and Lovins, L. H., 1993, "Supercars, the Coming Light-Vehicle Revolution," Rocky Mountain Institute, Snowmass, Colorado.

Moore, T. C., and Lovins, A. B., 1995, "Vehicle Design Strategies to Meet and Exceed PNGV Goals," SAE Paper No. 951906, Society of Automotive Engineers, Warrendale, Pennsylvania.

[6] See Chapter 16.

Contents

List of Figures

Chapter 1
Introduction

1.1 Solar Racing

Solar car racing helps to push the development of automotive technology in new directions because it is free from the constraints by which automobile manufacturers regard themselves bound. For example, solar racers have demonstrated that it is possible to do much with far less energy than that which is lavished on commercial automobiles. Solar racing, as do other racing venues, provides a means for manufacturers to test and showcase their products. It puts a lot of young, flexible, intelligent minds to work on these difficult problems, minds that do not know what they cannot do.

1.2 Organization

Figure 1.1 shows a solar car in operation. The interactions of the vehicle with its environment (the road, the atmosphere, and the sun) must be understood physically and modeled mathematically in order to understand how to select the characteristics of the vehicle (its shape and weight, for example) to produce a certain speed and range under the design conditions. Aerodynamic drag, gravity, and rolling resistance produce the energy demand that must be met to travel at a given speed. The book begins with a discussion of this demand.

The conversion of the sun's radiant energy into mechanical energy delivered to the driving wheel, or wheels, is the supply that meets this demand. The discussion of this topic, following that of the energy demand, proceeds in the order of the conversion chain: solar energy into electrical energy by means of the photovoltaic effect, electrical energy into chemical energy in batteries, electrical energy from the batteries or from the photovoltaic array into mechanical energy by an electric motor, and delivery of the mechanical energy to the wheel or wheels by a transmission.

After covering the background material, the book presents chapters that are devoted to an example of the design, manufacture, and testing of a solar-racing car.

E. F. Thacher, *A Solar Car Primer*, DOI 10.1007/978-3-319-17494-5_1

Fig. 1.1 A solar car in action

Following these, a chapter on energy management explains a method for maximizing the average speed of a solar car during a race. Energy management strategy answers the question: at what speed should the car be moving at this moment to maximize the average speed over some planning period?

Fund raising and public relations, the topics of the penultimate chapter, are strongly coupled; both sustain the design and construction of the car. Projects at academic institutions will also find that fund raising and public relations are tools for institutionalization—the permanent integration of the project into the academic program of the school.

Chapter 15, *A Solar Car-Based Learning Community,* discusses some implications for engineering and business education of learning communities driven by a large, multidisciplinary project.

1.3 Characteristics of Design

This section introduces the design process using ideas from Thacher (1995). The introduction is intended for readers who are new to design work. Often such persons find the process circular and therefore frustrating. To some extent this is a result of our educational process that emphasizes the solution of artificially well-defined problems that involve only one area of knowledge. Such a problem would be to find the average speed required to traverse a given distance in a specified time, a well-defined problem in physics, only. In design, the practitioner must invent the problem; very little is given, and there is no single, right answer—no certainty.

Engineering design is the choice-making process used to evolve a set of instructions for making an artifact or developing a system to meet a need. It begins with broad concepts and continues in the direction of ever-increasing detail. The process is iterative because some early decisions must be made with incomplete knowledge.

Here is an example of the design process applied to an artifact more familiar than a solar car.

Suppose a man hires an architect to design a house. The man has purchased a lot of a certain size, which has a view to the east and is located in a certain climate region. He wants the house to have low heating costs. His children are no longer living at home. He and his wife have avocations they want to pursue at home. They are fond of light and color and interestingly-shaped rooms. And they have an idea of how much money they can spend on the house.

The architect must transform this small collection of artistic feelings, facts, and requirements into a set of construction plans for a house. The need is poorly defined relative to the construction plans. What does "low heating costs" mean in terms of insulation in the walls? How is a liking for light, color, and interestingly-shaped rooms to be realized by the size and orientation of the windows and the room layout? Where and how should the house be placed on the lot? How do the features of the house relate to its cost? The number of house configurations that could answer these questions is very large; each of them is a possible "right answer." But there is no unique right answer.

The architect decides to enhance the view to the east by placing a large window in the east wall of the house. However, large windows, even modern two-pane designs, have large heat losses, compared to the insulated walls. So the requirement to enhance the view conflicts with the requirement for low heating costs. It may even conflict with the budget; large windows are expensive. The customer's problem statement is not self-consistent because the requirements conflict. Hence, no solution can satisfy all of the requirements.

The architect and the customer meet to clarify the problem statement and to discuss the solutions. The architect shows the customer some preliminary sketches. Viewing these proposed solutions begins to clarify for the customer what he means by, say, interestingly shaped rooms. The architect's insight into the customer's desires is thereby also improved. Suggesting solutions is a way of understanding the design problem.

Designing the house requires the architect to apply knowledge drawn from several fields. For example, he must understand how to control the climate in the building, how to provide a proper structure, how to control the project's cost, what laws and standards must be followed—besides how to shape the building and arrange its interior space to please the customer. Design requires synthesis of knowledge from diverse fields.

The architect may make some preliminary drawings and suggest the customer that they be given to builders interested in bidding for the construction of the house to collect suggestions from them for making the house more buildable. Design cannot be separated from manufacturing; the way something is to be built influences how it is designed, and vice versa.

The final solution to the design problem is a set of instructions for building the house: the construction drawings. This is a communication packed with information for the builder. Along the way, the architect has communicated orally, and probably also in writing, with his customer. The customer, the architect, and the builder have collaborated. They must communicate in order to collaborate; communication is essential to design.

Note that the design problem confronting the architect is ill-posed. The problem is poorly defined compared to the information necessary to solve it (the construction drawings). It contains requirements that conflict with each other. The architect's, the customer's, and the builder's understanding of the problem depends upon the solutions that have been proposed and the way the building is to be manufactured. There is a range of possible solutions, but no unique solution.

The essence of design problems is their ill-posedness. Engineers, scientists, entrepreneurs, writers, artists—all professionals—are confronted with such problems. Solving ill-posed problems is the quintessential characteristic of all professional work; their solution is a demanding art.

The primary objective of education follows directly from this seminal characteristic: students must be prepared to solve ill-posed problems. The design process must be the unifying theme in education, at all levels. Alfred North Whitehead wrote (Whitehead 1929), "Education is the acquisition of the art of the utilization of knowledge."

Solution Process The design process has identifiable parts that are revealed by the example above. The problem-solving team (the customer, the architect, and the builder, in the example) first clarifies the problem statement, then creates and chooses between alternate solution concepts, moves recursively between stages of the solution as revision requires, and finally produces a document that expresses the solution, i.e., the artifact or system, in sufficient detail to allow it to be built.

The solar racer design example is intended to be used as a model. Therefore, it is organized into these same stages: specification-writing (clarification), concept generation, concept selection, and detailed design.

References

Thacher, E. F. (1995). Teaching design. In *Engineering Technology Education, Principles of Engineering*, The University of the State of New York, The State Education Department, Albany, New York.

Whitehead, A. N. (1929). The aims of education. New York: The MacMillan Company (The New American Library).

Chapter 2
Interactions with the Atmosphere and Road

2.1 Introduction

The interactions of a car with its environment—gravity, the atmosphere, and the road surface—create forces which act on the car, usually opposing its motion. This chapter shows how these forces are related to the characteristics of the car under the designer's control, such as its shape and weight, and to the effort required to move it: the tractive force. A magnitude will be calculated to give the reader an idea of the importance of each interaction. The calculations will use the characteristics of actual solar racing cars to make the numbers realistic.

No attempt is made to present an exhaustive treatment of each interaction. However, the most important features have been presented. Additional details will be found in Chaps. 17, 18, 19 and 20. Readers who wish to pursue topics in even greater depth may consult the references at the end of the chapter.

Cruise Condition This chapter concentrates on the interactions going on most of the time, which will be called the *cruise condition*. The cruise condition plays the strongest role in setting the energy consumption of the car, which determines the range. In this condition, the car moves straight ahead or turns through a large-enough radius or at a slow-enough angular rate, such that inertia-related forces transverse to the car's direction of motion are relatively small. Also, the wind relative to the car, which interacts with the shape of the car to create aerodynamic forces, blows from nearly directly ahead. The aerodynamic side forces are therefore relatively small. Thus, in the cruise condition all inertial and aerodynamic forces acting transverse to the direction of motion, and their associated moments, are neglected.

Intermittent Conditions However, important interactions are associated with events that happen intermittently, such as side gusts and emergency maneuvers. The transverse forces generated in these events strongly influence the controllability, stability, and structural design of a solar-electric vehicle.

When passed by a large vehicle, such as a truck, or when a wind gust blows suddenly from the side, a car experiences wind forces which tend to blow it sideways and rotate it about the vertical axis through its center of gravity. Solar racing cars

E. F. Thacher, *A Solar Car Primer*, DOI 10.1007/978-3-319-17494-5_2

tend to be light and are thus more sensitive to side gusts than conventional vehicles. Chapter 21, *Stability Calculations,* presents a method for predicting the effect of a side gust.

A solar car maneuvering in an emergency may be required to accelerate or brake while turning. In this situation, the car should remain controllable[1] and hence must not skid nor roll over, and its structure must withstand the moments and forces developed by the maneuver. Chapter 21 presents an analysis of the stability of the vehicle.

2.2 Equivalent Interactions

The forces on the car are distributed over the car or some portion of it. Gravity acts on the entire mass of the car. The friction force of the road on the tires acts over the area of the tire in contact with the road. In general, each force tends to both translate the car in and rotate the car about, at least one coordinate direction. To conveniently model the dynamics and energetics of the vehicle, we replace each distributed force by an equivalent isolated force and its associated moment. That is, the isolated force and moment have the same translational and rotational effects as the distributed force which they replace. Each of the equivalent forces acts at a convenient point, such as the center of gravity. Each of the equivalent moments acts about a convenient axis (usually a coordinate axis). Henceforward when the terms "force" and "moment" are used, it will usually be the equivalent, isolated forces and moments that are meant.

2.3 Coordinate Systems

Figure 2.1 shows a set of coordinate axes attached to the center of gravity of a solar-electric car so that the axes always point in the same directions relative to the car. The positive direction of each axis is shown. A force acting in the positive direction of each axis is defined as positive. A semicircular arrow about each coordinate axis shows the positive direction of the moments about those axes.

Aerodynamic forces arise from the motion of the air relative to the car. Thus, it is natural when discussing this relative motion to think of the car as stationary with axes attached to it. This is exactly the situation when the aerodynamic forces are measured in a wind tunnel, for instance. When it is more convenient, we will revert to thinking of the car as moving with respect to a coordinate system fixed to the earth, such as at the starting line at the Indianapolis Motor Speedway.

[1] Within specified design limits; absolute stability cannot be achieved.

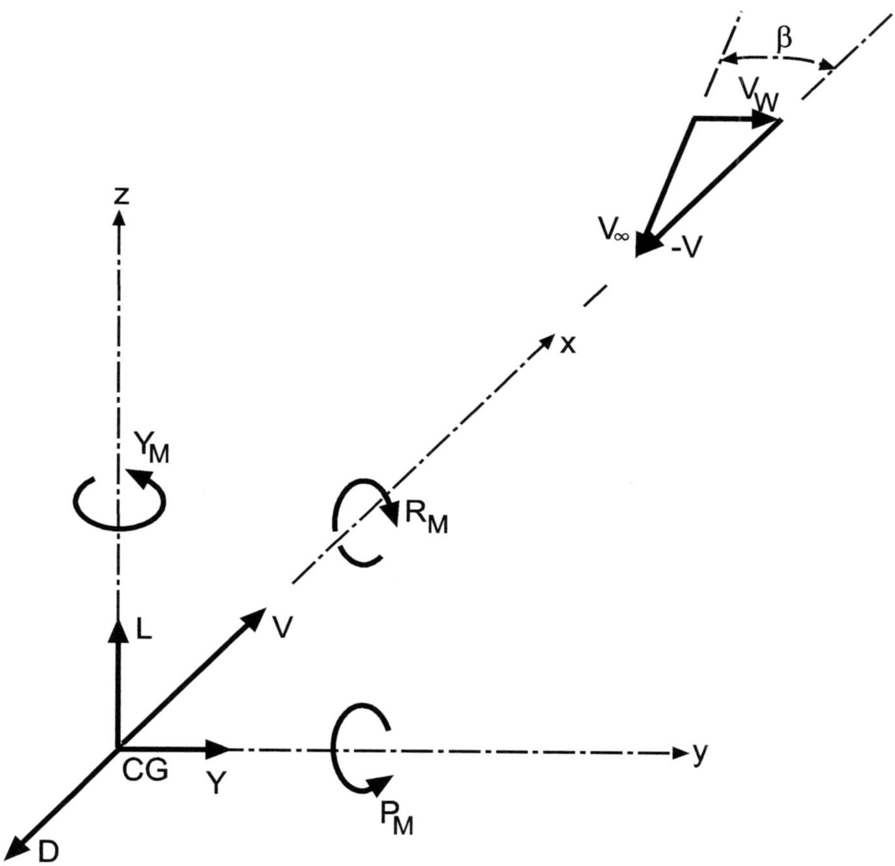

Fig. 2.1 Coordinate axes

2.4 Aerodynamic Interactions

We will select the point of action of aerodynamic forces as the center of gravity of the vehicle. Wind tunnel measurements of the moments of these forces are by convention often referenced to a point on the centerline of the car and halfway down the wheelbase. However, it is convenient when discussing their effects on motion to reference these moments to the center of gravity, as for the forces.

Figure 2.1 shows three aerodynamic forces and three aerodynamic moments, a force and moment for each coordinate axis, each named to suggest how it tends to affect the car's motion. The force acting along the x-axis is called *drag (D),* that acting along the y-axis is called *side force (Y),* and that acting along the z-axis is called *lift (L)*. The moment about the x-axis is called *roll* (R_M), that about the y-axis is called *pitch* (P_M), and that about the z-axis is called *yaw* (Y_M). In general, each of the forces and moments can be positive or negative.

Figure 2.1 also shows a vector representing the result of subtracting the car's velocity *(V)* from the true wind vector. This result is the *relative wind*, V_R, the motion of the air relative to the car, but sufficiently far upstream of the car so that it is undisturbed by the shape of the car. The relative wind blows from a *yaw angle*, β, measured from the x-axis and positive in the direction of positive Y_M.

In the cruise condition, the side force, yawing moment, rolling moment, and yaw angle are zero. Pitch, drag, and lift remain. How these arise from interactions between the flow field relative to the car and the car's shape, attitude, and internal flow passages will now be discussed.

Drag Experiment This section elaborates a bit on an example in Sherman (1990). Suppose you stir a mixture of small pepper grains and water in a white cup (so you can see the grains and thus visualize the flow) and then remove the spoon. The whirling motion of the mixture persists but eventually slows to a stop. The persistence depends upon the fluid's *momentum,* which in turn depends on both the mass of the fluid and its rotational speed. One would expect the liquid metal mercury in an identical cup to whirl for a longer time than water.

The mixture does not whirl forever but comes to rest because the friction force caused by the *viscosity* of the fluid opposes the rotation. Viscosity measures a fluid's resistance to flowing relative to itself, just like your hands resist being rubbed against each other. As in that case, the friction force is tangent to the flow. All fluids have viscosity; in some, such as air, it is small and in others, such as honey, it is large. (Try the experiment with a cup of honey. The pepper will stay on the honey's surface, but it will still help to visualize the flow. The friction force could rotate a light cup in this case. Would this show that the force is tangential to the cup's inner surface?)

If you observe the pepper, you will see (especially if you have been able to impart mostly circular motion to the mixture) that the grains near the inner surface of the cup slow down first. It turns out that the mixture actually contacting the cup's surface is at a speed of zero, which is called the *no slip* condition. So the rotational speed of the mixture is zero at the cup surface but increases toward the center. (You may observe other motions as well.) Because the friction force is created when the fluid resists flowing relative to itself, the speed difference (or *gradient*) must be present to give the friction force.

The foregoing discussion will be of use in understanding the friction drag on a car moving through air.

Parked Car A car parked along a road in still air, like the pepper grains in the cup or a fish motionless in a pond, is immersed in a fluid: the atmosphere. This mixture of gases (about 75 % nitrogen and 25 % oxygen) presses on every part of the outside (and inside) of the car. This pressure force distribution is called *static* because the atmosphere is not moving relative to the car at any point on it.

The static pressure is not uniformly distributed over the body of the fish, being greater underneath it because of the greater depth. This is also true for the parked car. However, the density of water is about 850 times greater than that of air at standard conditions (temperature 298.15 K, pressure 101.325 kPa). Consequently, the maximum pressure difference across the car is on the order of 0.0002 atm. So, differences in height between parts of the car may be neglected and the static pressure distribution taken as uniform over the car. Thus, the net static pressure force on the car is zero.

Moving Car As the car moves down the road, air flows over the surface of the car. This relative motion[2] changes the pressure distribution such that a net pressure force is created that opposes the car's motion. The external flow also applies a retarding tangential friction force to the car's surface, as in the stirred-cup experiment. Also, air flows through the car for ventilation. The net pressure loss in internal passages, caused by friction and the losses in ducting bends, dampers, and other components, also exerts a retarding force. It can be as much as 8–10 % of the total.

The total of the external and internal retarding forces we call *drag*. The magnitude of the drag is expressed by

$$D = c_{\mathrm{D}} A_D q. \tag{2.1}$$

The drag coefficient, c_{D}, a dimensionless quantity, characterizes the drag of the car and changes with the flow, in general. The *dynamic pressure* of the relative air speed far from the car, q, is given by

$$q = \frac{1}{2}\rho V_{\mathrm{R}}^2. \tag{2.2}$$

The dynamic pressure is the pressure increase above the ambient static pressure that would occur if the flow were brought to a halt with no losses (stagnate) against a surface. The air density (ρ) may be computed from the ideal gas equation

$$\rho = \frac{p}{R_{\mathrm{A}} T}. \tag{2.3}$$

The gas constant for air (R_{A}) is 0.287 kJ/kg·K. At standard temperature and pressure, Eq. (2.3) gives an air density of 1.184 kg/m^3. Note that for the same pressure, the drag is lower if the air is hotter and higher if the air is cooler.

In order to give units of force, the dynamic pressure must be multiplied by an area. By convention, the area used is the profile area (A_D), the area blocked out by the car when viewed from straight ahead. The product $c_{\mathrm{D}}A_D$ is called the *drag area*. Chapter 17 presents a means of estimating the drag area of a candidate body shape. Measurement of the drag area of a scale model or full-scale vehicle in a wind tunnel or by coast-down testing will be discussed in Chap. 12, *Testing*.

The drag coefficient incorporates all of the opposing drag force components mentioned: friction (c_{F}), pressure (c_S), and ventilation (c_{V}). Referring each component to $A_D q$ gives

$$c_{\mathrm{D}} = c_{\mathrm{F}} + c_S + c_{\mathrm{V}}. \tag{2.4}$$

We shall now explain in more detail why these components arise.

[2] It is the relative motion that counts; you could also blow on a stationary car and create drag, as in a wind tunnel.

2.5 Friction Drag

Boundary Layer Suppose that the relative airflow approaching the car is smooth and at zero yaw angle ($\beta = 0$ in Fig. 2.1). Like the water–pepper mixture in the cup, the air at the car's surface moves at zero speed relative to that surface. However, air farther from the surface moves nearer to the relative speed of the surrounding air, as shown in Fig. 2.2 (in which the n-axis is the local vertical). The air layer over which the local relative flow speed changes from zero to 99 % of that of the surrounding air is defined as the *boundary layer*.

The boundary layer thickens as the distance from the front of the car increases. A velocity gradient now exists in a viscous fluid. Hence, the air applies a retarding frictional force tangent to the surface of the car. As the car increases speed, the gradient becomes steeper, and the friction force at the surface increases. Figure 2.2 shows the gradient at the surface as the slope ($\Delta V_R/\Delta n$) of the tangent to the velocity distribution at that point. The symbol τ_0 represents the friction force per unit surface area. The *streamlines* shown in Fig. 2.2 are imaginary lines tangent to the local flow velocity.

Viscosity As we expect from the cup experiment, the proportionality factor between friction force and the velocity gradient is the viscosity of the air (μ).[3] If the car were moving through water, the viscosity of which is about 48 times that of air at 25 °C, the frictional drag would be much larger at a given speed (remember the honey). Near atmospheric pressure, the viscosity of air shows a weak tendency to

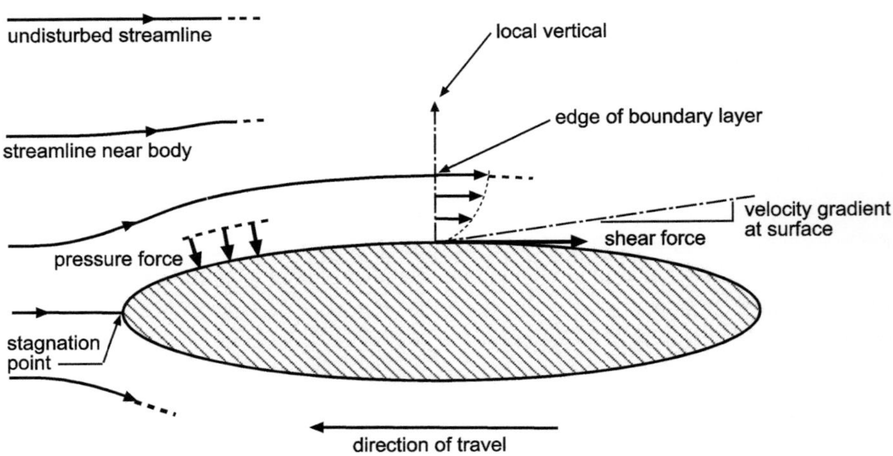

Fig. 2.2 Boundary layer and aerodynamic forces

[3] Many fluids obey this relation between the surface shear force and the velocity gradient, air and water, for instance. Such fluids are called newtonian, after Sir Isaac Newton, who first proposed this linear model.

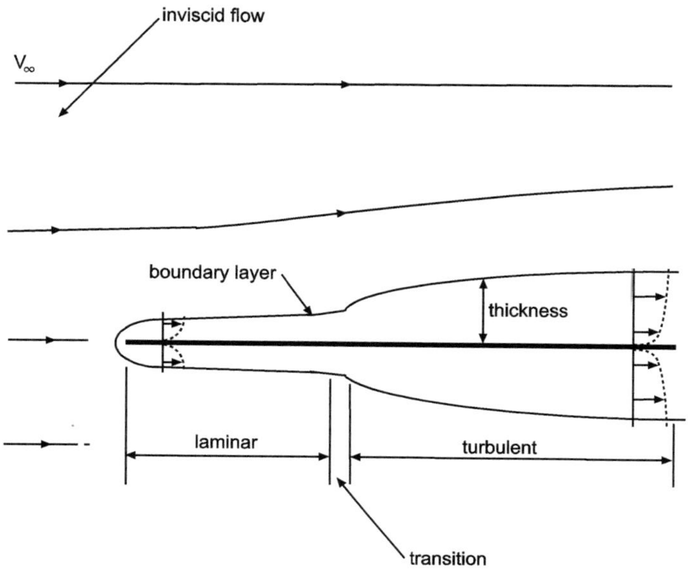

Fig. 2.3 Laminar and turbulent boundary layers

increase with pressure and a strong tendency to increase with temperature. Thus, the frictional drag increases with increasing temperature.

Is this the whole story of friction drag? No, unfortunately; we are headed for more trouble as we increase speed.

Laminar and Turbulent Flow Figure 2.3 shows a smooth, flat plate traversed by initially smooth air flow.

The flow contacts the plate, and the boundary layer forms. The flow in the boundary layer at this early stage is still smooth. We can visualize this as very thin air layers, *lamina,* moving relative to each other with no velocity components transverse to their motion. The flow is called *laminar* because of this characteristic. A particle of air striking, say, a small bump (no actual surface can be exactly smooth) may deflect up a bit, but the viscous friction of the other particles drags it back into line, keeping the flow laminar.

An impulsive force was applied between the fluid and the surface when the element glanced off the little bump and thereby gained momentum away from the surface. Further on, at the next bump (or other disturbance, maybe even loud rock music), the thickening of the boundary layer has magnified the destabilizing impulse force relative to the stabilizing viscous force because the velocity gradient is reduced. So, as the boundary layer thickens, small waves appear in it. The waves grow into chaotic eddies and the boundary layer makes a transition to *turbulence,* accompanied by additional thickening.

Besides the torque need to overcome the viscous friction, the engine of the car must now exert extra torque on the driving wheel or wheels to cause the eddies to circulate. Consequently, the friction drag in turbulent flow is higher than in laminar flow.

Reynolds Number The foregoing discussion implies that a number correlated with the ratio of the impulsive to viscous forces in the boundary layer would also correlate strongly with the transition to turbulence. Because force is proportional to the rate of change of the momentum, we expect the impulsive force will be correlated with the momentum flow rate of the air external to the boundary layer. This is ρV_R^2, when expressed as force per unit area perpendicular to the flow, or $\rho V_R^2 A_{flow}$ in force units, where A_{flow} is a conveniently chosen reference area perpendicular to the flow. The friction drag per unit surface area is proportional to $\mu V_R/\ell$, where ℓ is the local boundary layer thickness and A_{fric} is a reference area on the car's surface. Since $\ell A_{flow}/A_{fric}$ has units of length and A_{flow} and A_{fric} are arbitrary, the ratio of interest is:

$$\mathrm{Re}_\lambda = \frac{\rho V_R \lambda}{\mu}, \tag{2.5}$$

where λ stands for a conveniently chosen reference length. Equation (2.5) defines the *Reynolds number*. For the present discussion, we choose the distance x from the nose of the car to a point in the boundary layer measured along the surface as the characteristic length because the thickening of the boundary layer depends on the distance from the nose.[4] The number, now called the *local* Reynolds number because it depends upon the location, is:

$$\mathrm{Re}_x = \frac{\rho V_R x}{\mu}. \tag{2.6}$$

The local Reynolds number at which the transition to turbulence begins on the surface is called the *critical* local Reynolds number. This number is usually found by experiment. The transition to turbulence is affected by the roughness of the surface: the rougher the plate, the lower the critical Reynolds number at which it begins. On the other hand, as Eq. 2.6 implies, for a given fluid and surface, the critical Reynolds number will be reached at a shorter distance from the nose when the flow is faster.

Thickness Compared to the characteristic dimension of the body in the flow direction, say the length of the plate in Fig. 2.3, the boundary layer is quite thin, even in turbulent flow. Suppose the length of the plate were 2 m and the latter portion of its surface were in turbulent flow, as shown, the boundary layer thickness would be only of the order of 4 cm at the trailing edge. (Its dimensions have been exaggerated in the figures.) Nevertheless, all of the viscous interaction of the airflow with a body takes place in the boundary layer. Compared to the boundary layer, the flow external to this layer may be treated as if it had no viscosity.

Total Friction Drag Because the local frictional force discussed above is expressed as a force per unit area, the total frictional force on the car is proportional to the surface area of the car. The larger this area, the larger the force will be. However, the

[4] The Reynolds number is important in other contexts. So, other reference lengths more appropriate for the context are defined for these cases.

line of action of the drag force is antiparallel to the direction of motion. Therefore, the friction force at a particular location contributes to the drag in proportion as the surface upon which it acts is parallel to the direction of motion. The flow over the upstream face of a rear-view mirror is nearly perpendicular to the direction of motion and therefore contributes little to the total friction drag force, for example. It contributes to the pressure drag, however, as the following discussion demonstrates.

2.6 Pressure Drag

Frictionless Flow We now return to the effect of shape on pressure. Figure 2.4 shows a cross-sectional view of the steady flow of air over two smooth cylinders. Both are very long compared to their diameters, so the complicating effect of flow near their ends may be neglected. Consider first the flow over the upper cylinder, for which we imagine the viscosity of the air to be zero, so that the flow is frictionless.[5]

Since a streamline is an imaginary line tangent to the local flow velocity, if the flow is undisturbed, all the streamlines are parallel and flat. Note that, by definition, flow cannot cross a streamline. Now we can imagine that the upper half of the cylinder is in a channel.[6] The upper "wall" of this channel is a surface formed by the streamlines of air far enough from the cylinder to be undisturbed by its presence. These bounding streamlines, taken together, could be called a *stream surface.* The lower wall is formed by the stream surface that hits the front of the cylinder and then follows its surface.

Bernoulli's Equation Consider any streamline between two vertical planes, such as those marked 1 and 2 in the figure. The flow is steady, there is no friction and, we assume, no heating of the air. Then it is true that for any two points along the streamline

$$\frac{p_1}{\rho_1}+\frac{V_1^2}{2}+gZ_1-\left(\frac{p_2}{\rho_2}+\frac{V_2^2}{2}+gZ_2\right)=0. \tag{2.7}$$

Equation (2.7), called *Bernoulli's equation,* shows that

$$\frac{p}{\rho}+\frac{V^2}{2}+gZ=const. \tag{2.8}$$

along a streamline for the conditions assumed. Now, to unclutter things even more, we observe that the gravitational potential energy term, gZ, may be neglected for height changes on the order of the height of an automobile or truck. Also, if the flow were incompressible, the density, ρ, would be constant and then the quantity

[5] This apparently oversimplified scenario will still yield valid insights, believe it or not.

[6] Actually, we could equally well imagine the entire cylinder in a channel, but the drawing of a half cylinder takes up less space.

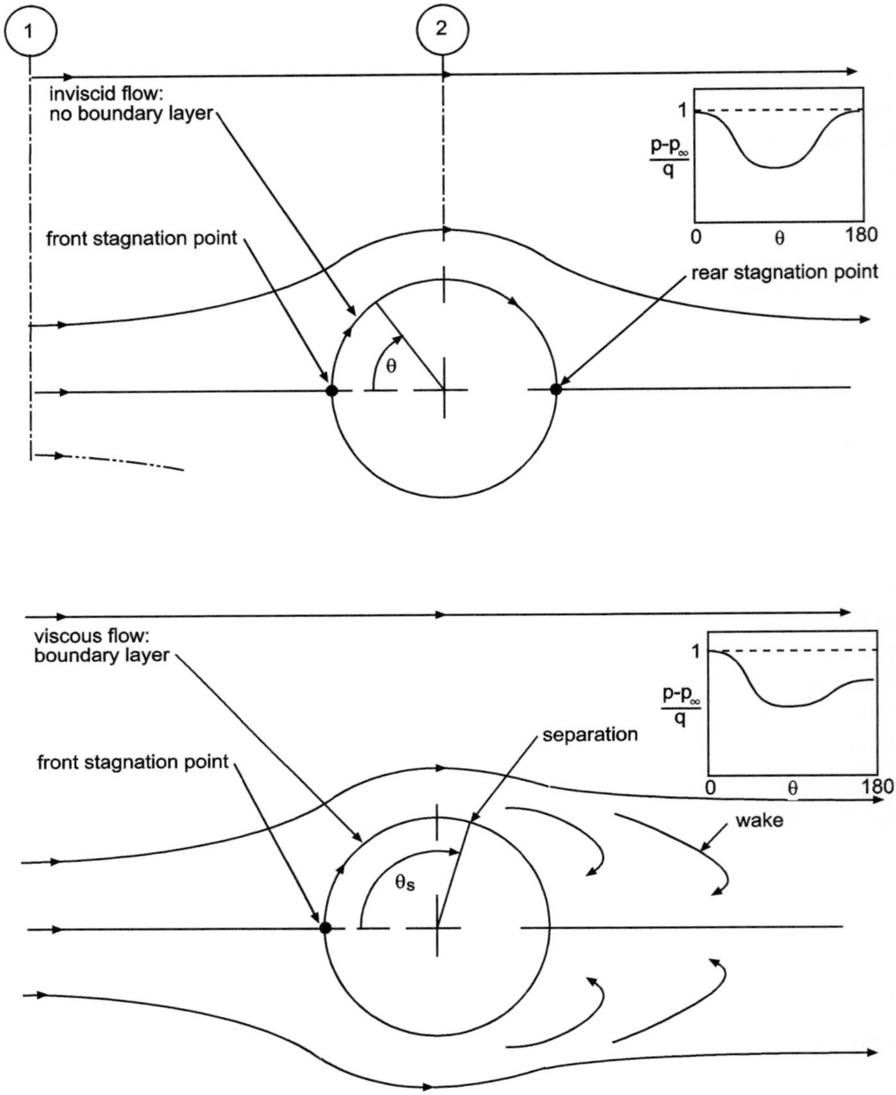

Fig. 2.4 Origin of pressure drag

$p+\rho V^2/2$, the *total pressure*, p_0, would be constant along a streamline. The pressure changes typical of external air flows produce only small changes in density. We will model such flows as incompressible.

Using our simple model, let us investigate the static pressure distribution on the upper cylinder of Fig. 2.4. Upstream of the disturbance of the cylinder, the velocity is uniform (all streamlines flat, parallel, and evenly spaced). Therefore, the pressure is uniform in the flow. At the forward-most point ($\theta=0°$ in the figure), the air speed is momentarily zero, so it is a stagnation point (even though it is really a line) and $p=p_0$. The flow then turns and moves up, tangent to the cylinder's surface. Hence, because the mass flow rate is steady and the density cannot change, the air speed increases to a maximum, and, as required by Bernoulli's equation, the static pressure decreases to a minimum as the point of minimum channel cross-sectional area at the top of the cylinder ($\theta=90°$). Beyond this point, the air speed decreases and the static pressure increases. At the downstream location opposite to the front stagnation point ($\theta=180°$), the tangential air speed component becomes zero. A rear stagnation point forms at which $p=p_0$ once again.

Figure 2.4 shows the pressure variation around the cylinder. Clearly the shape of the cylinder strongly influences the pressure distribution over it. But notice, there is no net pressure change across it in the flow direction for the ideal, frictionless conditions assumed. Therefore, there is no pressure drag.

Flow with Friction Pressure drag on objects immersed in a real, viscous fluid arises because of *boundary layer separation.* Consider the lower cylinder of Fig. 2.4, which is immersed in a real, viscous, approximately incompressible fluid such as air. A boundary layer now forms on the cylinder. Bernoulli's equation is invalid inside the boundary layer. But because, as we observed earlier, the flow external to the boundary layer is approximately frictionless and the boundary layer is quite thin, the pressure imposed on the boundary layer approximately obeys Bernoulli's law. The pressure increase on the downstream surface of the lower cylinder of Fig. 2.4 opposes the flow in the layer. The more sharply the surface curves down, the more rapid will be the opposing pressure increase predicted by Bernoulli's equation. At some position angle, this causes the velocity gradient at the surface to be zero.[7] At that point, the main flow ceases to follow the curved surface, and the boundary layer is said to *separate* from that surface. The flow then forms a turbulent wake, as shown. This causes the air pressure on the rear surface downstream of the separation area to drop below that near the front stagnation point and perhaps even below that of the ambient air. There is now a pressure force difference, high in front, low in back. This net opposing force is called *pressure drag* (or sometimes *profile drag*). Pressure or profile drag is reduced by making the shape less blunt.

Streamlining Figure 2.5 shows a cylindrically shaped body and a streamlined air-foil-shaped body, both with circular cross sections and having the same profile area. The drawing of the cylindrically shaped body shows that separation can occur at locations upstream of the trailing surface, such as at the forward corners of the box shape. Downstream of these locations, the flow may reattach to the car and some pressure loss be recovered.[8] Compare the cylindrical shape to the streamlined shape which minimizes separation and thus pressure drag, by avoiding rapid changes in the slope of its surface.

Qualitative pressure and friction force profiles for the two shapes are shown at the bottom of the figure (tear drop: dashed line). These curves were constructed using the exchange of pressure and velocity expressed in the Bernoulli equation.

---

[7] There will even be back flow downstream of the separation point.

[8] These local zones of separated flow are called separation bubbles.

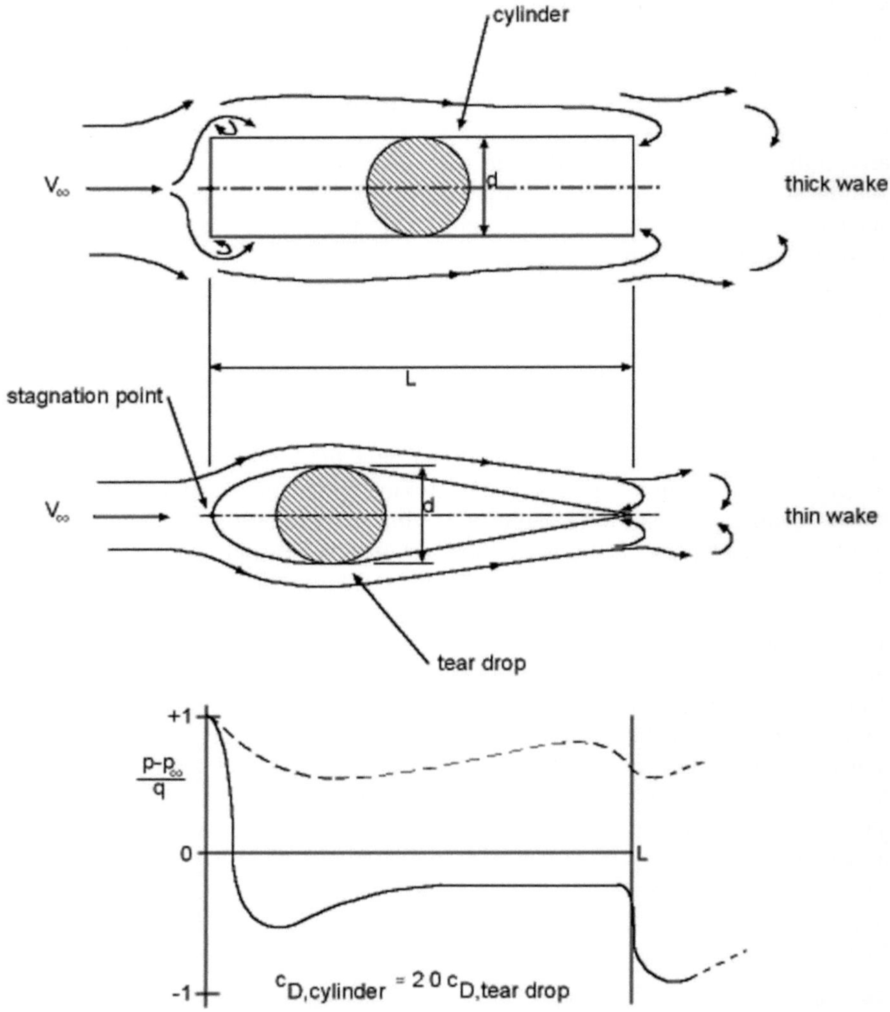

Fig. 2.5 Effect of streamlining

The velocity gradient at the surface in separated zones is small and therefore so is the shear force. However, streamlining may add to the external area that contributes to friction drag because, by filling in the areas having more abrupt slope changes, streamlining increases the component of surface area parallel to the direction of motion. Thus, there would seem to be a trade-off between pressure and friction drag. But separation usually dominates the drag from external flow (Hucho 1983). Hence, any gains in friction drag from streamlining are usually outweighed by the reduction in pressure drag. Nevertheless, one should bear in mind that solar racing cars are radically streamlined to give very low total drag. Friction drag is therefore of greater relative importance than in conventional vehicles. Pay close attention to details such as surface finish; in solar racing cars, the drag is in the details.

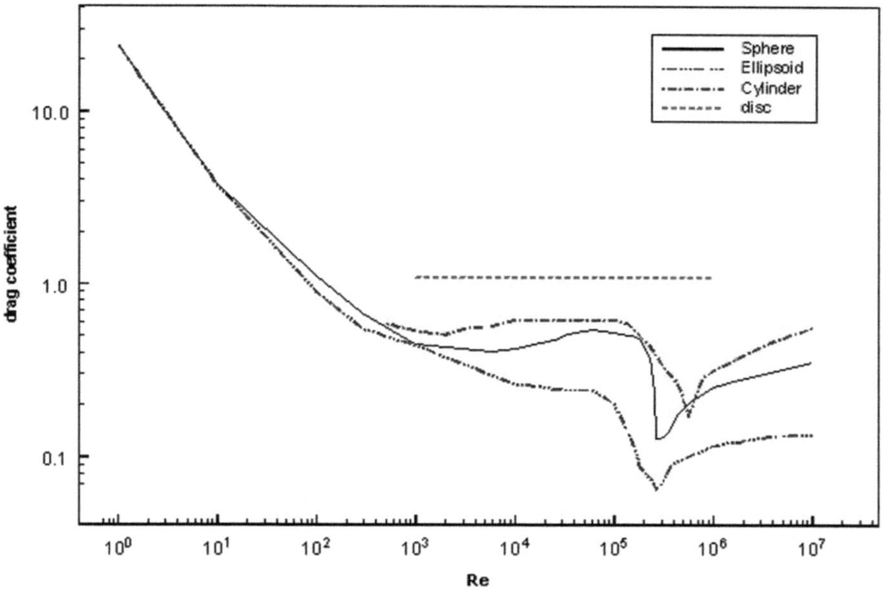

Fig. 2.6 Drag coefficients of four shapes. (Adapted from White 1986)

Total Pressure Drag The total pressure drag may be found in principle by adding the components of the local normal pressure force (pressure times area) in the direction of motion. Applying this rule to the shapes of Fig. 2.5, we see that the retarding and pushing components of the pressure force tend to be concentrated at the front and rear of the car (except for the "separation bubbles" mentioned in footnote 8).

Effect of Turbulence Turbulence in the boundary layer can reduce the total drag on a body, even though it increases the friction drag in non-separated portions of the boundary layer. Figure 2.6 illustrates this effect for the total drag of a disk, cylinder, ellipsoid, and sphere in cross flow. The curves are similar, except for that of the disc. In laminar creeping flow only friction is present, then laminar separation occurs, and the drag coefficient becomes nearly constant. Observe that the disk's drag is almost completely determined by separation because it has little surface area parallel to the flow, so that at most Reynolds numbers viscous boundary layer effects are negligible. However, boundary layers can develop on the other three objects. Turbulence begins on these shapes at about a Reynolds number of 10^5, where the drag coefficient drops. The increased momentum in the boundary layer causes the ring-shaped locus of separated flow to be blown farther downstream. Thus, the region of low pressure on the downstream side of the cylinder is smaller and the drag force is lower. After the transition to turbulence is complete, the drag coefficient slowly increases with the Reynolds number because of friction.

The drag-reducing effect of turbulence is employed in the manufacture of golf balls. The dimples on the surface of a golf ball tend to "trip" the flow to turbulence, thus reducing drag and increasing the distance the ball carries. This technique of introducing local roughness to trip the flow has been employed on solar racing cars to reduce the drag of the cockpit canopy.

Local wakes may be formed when the shape of the car forces flows of differing speeds to join, imparting a whirling *(vortex)* motion to the resulting flow. The low pressure in the vortex results in an additional drag. The section on lift explains how merging flows of differing speeds can be created.

2.7 Estimating Drag

The drag area (and drag coefficient) of a proposed vehicle may be estimated by summing the drag areas of the components that make up the vehicle's shape. The free air drag coefficients of the shape components must be known and corrections for ground effect and interference must be applied. This *drag build-up* method is capable of estimates that agree within ± 10 % with the drag areas measured in wind tunnel tests. Chapter 17 explains this method and Chap. 9, *Solar Racer: Concept Generation and Selection,* presents a detailed application of it.

2.8 Ventilation Drag

Ventilation of the cockpit forces out the hot, stale air, helping to keep the cockpit comfortable. Outside air supplied to the battery compartment flushes out gases evolved by the battery during charging. These might accumulate in the battery compartment in explosive amounts if it were sealed. Race regulations (Chap. 16) require a certain fan-forced airflow through the battery compartment whenever the battery is electrically connected to the solar car.

Sample System Figure 2.7 shows a ventilation system layout schematically. The air enters in the front of the car through a low-loss (round-edged) inlet. It flows through the cockpit (simplified to a box-shaped volume), the battery box, the fan, a diffuser, and then leaves the car. The diffuser increases the pressure of the flow before returning it to the outside. Heat inputs from the sun, instruments, and the driver, and heat losses to the outside are shown. These cannot be neglected as they were in deriving Eq. (2.7).

Drag Sources The internal flow drag arises in a manner similar to that from external flow. Consider Fig. 2.7. The internal flow applies viscous shear forces to the inside surfaces of the air ducts. The rougher the surface, the greater the force. The pressure increases as the flow velocity decreases in the sudden expansion when the air is discharged into the cockpit. Separation occurs, as shown, dissipating some energy and thereby reducing the pressure increase. This causes drag, in effect, by increasing the net pressure drop in the system. Separation in the sudden contraction at the discharge from the cockpit increases the pressure drop caused by the area decrease. Separation also can occur in the diffuser if its included angle is too large, reducing the pressure recovery as in the sudden expansion. The pressure loss in the

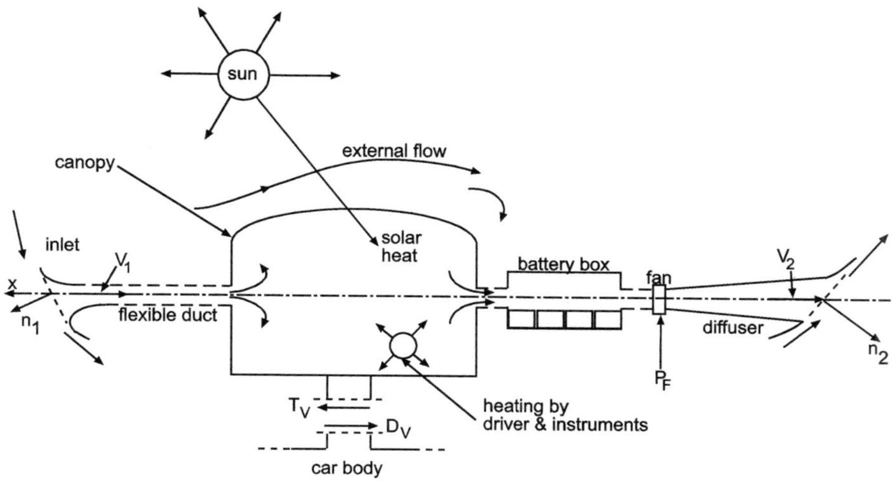

Fig. 2.7 A ventilation system

duct is reduced by the pressure increase across the fan. Note that race rules require the fan to be running when the battery is connected to the main bus.

Drag Force The locations where the system is attached to the car are symbolized by the single pylon. It has been cut to show the x-direction tractive force (T_V) exerted at the support. To find the drag, D_V, at a steady speed, sum the forces in the x-direction on the free body consisting of the ventilation system and the air in it and, following Newton's second law, equate this sum to the momentum change of the air between the inlet and outlet. Then note that $D_V = -T_V$. The drag on the ventilation system is transmitted through the pylon to the body of the car. Consequently, the motor must supply additional tractive force equal to D_V. If m_1 represents the mass flow rate through the ventilation system, the drag then is:

$$D_V = (\bar{p}_{G1}A_1 - \bar{p}_{G2}A_2)_X + \dot{m}_1(V_1 - V_2)_X. \tag{2.9}$$

In general, the pressure over the inlet and outlet openings will not be uniform. Hence, the average gauge pressures over the inlet and outlet, $\bar{p}_{G1}$ and $\bar{p}_{G2}$, respectively, must be estimated from information about the flow around the car. The "x" subscript denotes components in the x-direction.

The ventilation drag coefficient is computed from the drag force as:

$$c_V = \frac{D_V}{qA_D}. \tag{2.10}$$

A method for estimating the ventilation flow rate and for sizing the fans used in the system is explained in Chap. 18. An application of this method and of Eqs. 2.9 and 2.10, to estimate the drag caused by the ventilation flow, is presented in Chap. 10, *Solar Racer: Detailed Design.*

2.9 Lift

Lift is directed perpendicular to the *x, y* plane, the plane of the car's motion. As mentioned above, the net lift force is not necessarily upward. The lift coefficient, defined as

$$c_L = \frac{L}{qA_D}, \tag{2.11}$$

may therefore be positive or negative.

Ground Effect Figure 2.8 shows three views of a simplified solar car body. It has a symmetric airfoil shape longitudinally, is elliptic in cross section, and in plan view rounded in front and square in back. The angle between the horizontal and the chord line connecting the nose and the tail of the car we shall call the *pitch angle,* positive when up. Figure 2.8 shows the car at zero pitch. Underneath the drawing is a graph showing qualitatively the pressure distribution around the car.

The plot below the drawing shows that even though the pitch is zero and the shape is symmetric about the chord line, the pressure distribution on the underside of the car (dashed line) differs from that on the top (solid line). This is because of the proximity of the road surface. The variable cross-sectional area channel formed by the road and the car's underbody alters the speed and pressure distribution as required by the Bernoulli equation. The arrows indicate the direction and magnitude of the lift, with the large arrows showing that the lift causes a net upward pitching moment about the center of gravity (L_{CG}).

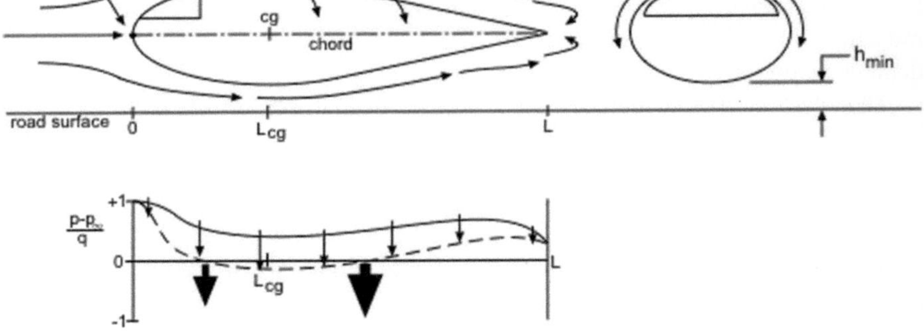

Fig. 2.8 Ground effect

If the pitch angle were made positive, the lift might become net positive because the stagnation point would shift underneath the car. If the pitch angle were made negative, the stagnation point could shift to the top of the car. A net negative pitching moment may occur.

Note in Fig. 2.8 that because of the pressure difference between the top and bottom of the car, flow from one side of the car can curl around and join the flow on the other side. This is called *upwash* or *downwash,* as appropriate. If the joining flows have different speeds, vortices may form and trail downwind. As explained in the pressure drag section, this is a source of drag. Upwash or downwash is associated with the lift distribution even when the net lift is nearly zero. However, the drag induced by each lift-induced vortex always adds to the total. Some lift is inescapable. The key to reducing vortex-induced drag is to smooth the joining of the flows. For example, use fillets at the junction of the body and fairings used to streamline exposed wheels. Chapter 17 gives rules for selecting the fillet radius and length.

Optimum Ground Clearance Morelli (1983) conducted wind tunnel tests on model shapes configured for low drag near the ground. His results showed that as the ratio h_{MIN}/W between the minimum height above the ground and the width of the car increased from the minimum value tested, near 0.05, to values of 1.0 or greater, the drag decreased, in some cases passed through a minimum that was less than the free-air value, and then approached free-air values as the influence of the tunnel floor on the pressure distribution diminished. These results may be contrasted with those Morelli presented for teardrop shapes which exhibited drag coefficients several times those of the special shapes at the same ground clearance ratios.

Study the "Shark" curves in Fig. 17.8. The lowest drag coefficient Morelli reported was about 0.045 for a 6.9% camber ratio (b_{max}/W) at a dimensionless ground clearance (h_{min}/W) of about 0.15. (This is the lowest of the three Shark's shapes tested.) For a 2-m wide car, this implies an actual minimum ground clearance of about 0.3 m. The teardrop shape's drag is not comparable to the Sharks until a clearance ratio of about 0.7. This corresponds to a clearance of 1.4 m for a 2-m wide car. Thus, the Morelli shape can have low drag without having to reduce roll stability by raising the clearance ratio. This is not true for the teardrop.

2.10 Example 2.1

Suppose a solar car (pitch angle −0.5°, profile area 1.46 m^2) is traveling at 96 kph (60 mph) in the cruise condition in still air at standard temperature and pressure. Under these conditions, c_D, and c_L are, respectively, 0.095 and 0.22. What are the drag and lift forces?

Solution The air density is:

$$\rho = \frac{101.325\dfrac{\text{kN}}{\text{m}^2}}{\left(0.287\dfrac{\text{kN}\cdot m}{\text{kg}\cdot K}\right)(298.15K)} = 1.184\frac{\text{kg}}{\text{m}^3}.$$

Dynamic pressure-profile area product (speed in m/s) then is:

$$q = \frac{1}{2}\left(1.184\frac{\text{kg}}{\text{m}^3}\right)\left(26.39\frac{\text{m}}{\text{s}}\right)^2 (1.46\text{m}^2) = 607.6\text{N}.$$

And the drag force becomes

$$D = (0.095)(607.6\text{N}) = 57.7\text{N},$$

directed opposite to the direction of motion. The lift is:

$$L = (0.22)(607.6\text{N}) = 133.7\text{N},$$

directed vertically up.

2.11 Pitch

The pitching moment, P_{M}, arises from the tendency of the lift distribution to rotate the car about the y-axis. The net lifting force may be nearly zero. But there still may be a lift-caused pitching moment.

The pitching moment coefficient is defined similarly to those of lift and drag, except that a length scale must be introduced into the denominator to non-dimensionalize the quotient. It is:

$$c_{P_{\text{M}}} = \frac{P_{\text{M}}}{qA_D L_{\text{W}}}. \tag{2.12}$$

2.12 Example 2.2

A wheel base length of 3.073 m will be used as the length scale in this example. For the conditions of Example 2.1, the pitching moment coefficient, $c_{P_{\text{M}}}$, is -0.196 about the y-axis through the center of gravity. Find the pitching moment.

Solution Using Eq. (2.13) gives

$$P_{\text{M}} = (-0.196)(607.6\text{N})(3.073\text{m}) = -366\text{N}\cdot\text{m}.$$

The negative sign denotes pitch down.

Pitch angle changes of a fraction of a degree regularly occur in the cruise condition. But in an emergency situation, such as a rear tire blow out in a three-wheeled car while cornering, a pitch up condition of two or three degrees might occur. At highway speeds, this can cause a positive pitching moment in lightweight, airfoil-shaped vehicles sufficient to momentarily sharply reduce the normal force on one or both front wheel contact patches. During that moment, the driver has little or no steering control.[9]

2.13 Road and Gravity Interactions

Figure 2.9 shows a three-wheeled car moving up a straight grade in the cruise condition. The forces acting parallel or antiparallel to its motion are drag *(D)*, rolling resistances (R_1, R_2, and R_3; numbered clockwise viewed from above starting with the left front wheel), the *x*-component of the weight (W_x), and the tractive force *(T)*. The tractive force is the force propelling the car up the hill. If the speed is steady, it equals the sum of the forces resisting motion. The forces acting normal to the direction of motion are lift *(L)*, the *z*-component of the weight (W_z), and the wheel reactions (N_1, N_2 and N_3). The pitching moment (P_M), shown positive, acts about the *y*-axis.

2.14 Gravity

The components of the weight of the car of Fig. 2.9 are

$$\begin{cases} W_x = W\sin\alpha \\ W_z = W\cos\alpha \end{cases}, \tag{2.13}$$

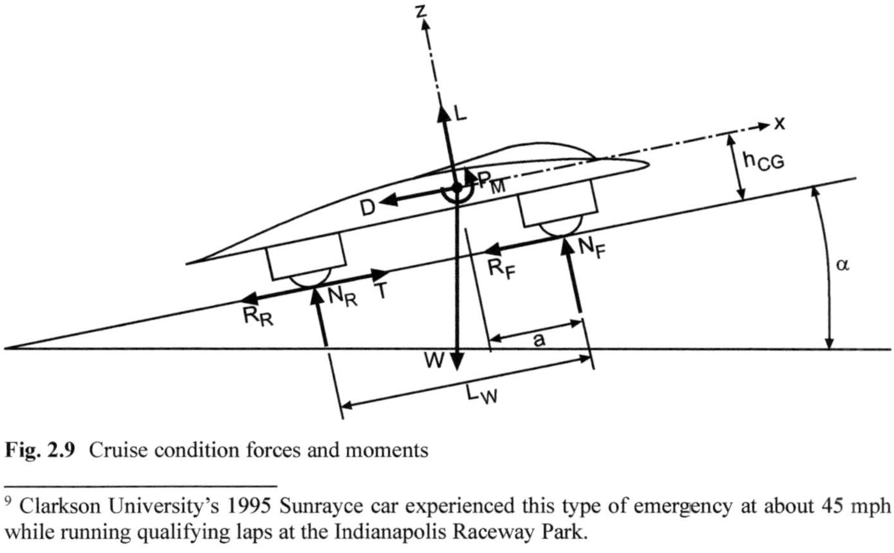

Fig. 2.9 Cruise condition forces and moments

[9] Clarkson University's 1995 Sunrayce car experienced this type of emergency at about 45 mph while running qualifying laps at the Indianapolis Raceway Park.

where W is the weight. Weight is the largest force on the car, in the cruise condition. W_x is the component of weight that directly opposes the motion when going up a hill (α positive) or aids the motion when descending a hill (α negative). The rolling resistance force is directly proportional to the vertical component of the weight W_Z, plus any up or down lift, as explained in the next section.

2.15 Example 2.3

Suppose the car of Fig. 2.9 weighs 3558.4 N and that the grade is 10 %[10] (α of 5.71°). Find W_x and W_z.

Solution The component of weight opposing the motion up the hill would be:

$$W_x = (3558.4\text{N})\sin 5.71^{\circ} = 354.0\text{N}.$$

This force is a bit more than six times the drag force calculated in Example 2.1 for horizontal travel at 96 kph. The drag coefficient used in that example is typical of high-performance solar cars as reported by Storey et al. (1994). However, doubling of the drag coefficient, or halving of the grade, would still give an opposing weight component much larger than the drag in the example. The normal weight component is:

$$W_z = (3558.4\text{N})\cos 5.71^{\circ} = 3540.7\text{N}.$$

Range and Stability Reducing the weight improves the range of the vehicle by reducing the energy needed to overcome rolling resistance, accelerate, and climb hills. On the other hand, the stability of the car is also reduced because a lighter car is more sensitive to side wind gusts and side forces caused by turning.

2.16 Rolling Resistance

Dynamic Resistance Figure 2.10 shows a wheel in contact with the ground and moving straight ahead at speed V, like those of the car in Fig. 2.9. The tire deforms under load so that it contacts the road over an elliptically shaped[11] area called the *contact patch*. The tire is not sliding over the road, so there is no relative motion between the contact patch and the road's surface.

---

[10] In the first US cross-country solar car race, the solar cars were required to show that they could climb a 10 % (rise/run times 100) grade.

[11] This deformed shape is typical of the small, high pressure, rounded-cross section tires usually used by solar racers.

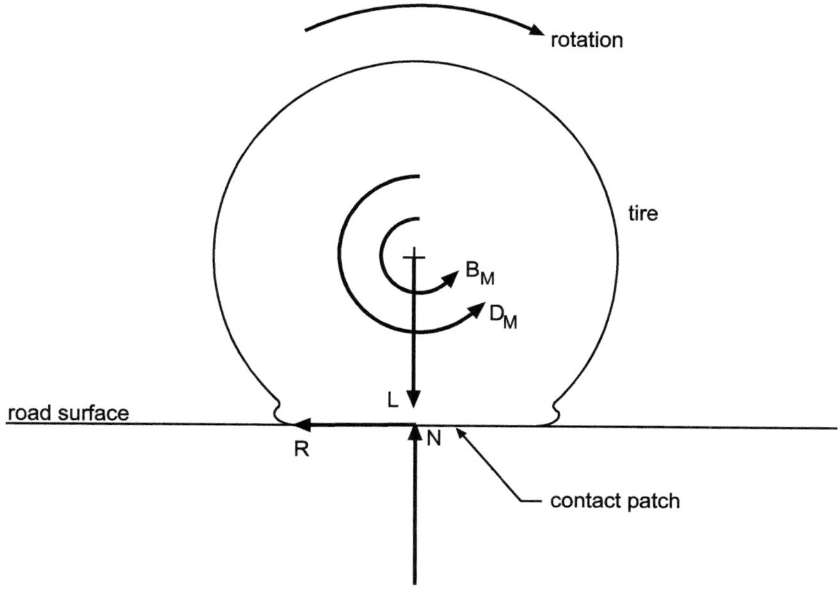

Fig. 2.10 Forces and moments on a wheel

We assume that the drag on the wheel caused by the flow over the car is accounted for in vehicle drag and focus on the drag resisting the wheel's rotation around its axis. A moment, D_M, about the wheel's axis opposing the wheel's rotation is imposed by the drag force created by the rotation of the wheel. The drag force is distributed over the wheel and depends upon the presence or absence of a wheel housing, the spacing between the wheel and its housing, and the flow field about the car. Another opposing moment, B_M, comes from friction in the wheel bearings. The moments may be represented by a moment-equivalent force applied to the contact patch acting on a moment arm attached to the axis of the wheel. We will call this force the *dynamic rolling resistance*.

Static Resistance As the tire rolls, the flexing of the tire needed to form the patch absorbs energy. The portion of the tire "flowing" into the leading edge of the patch is compressed, increasing the pressure on the leading edge above that caused by the weight. The portion of the tire leaving the contact patch expands, and most of the energy stored in the tire by the compression is recovered. However, the deformation is inelastic, so a portion of this energy is lost, manifesting itself as heating of the tire. Consequently, the contact pressure of the material leaving the contact patch is less than that entering the contact patch. The x-direction component of the net compressing force, F_C, on the material in the patch is called the *static rolling resistance.*

Total Resistance The *total rolling resistance, R,* is the sum of the static and moment-equivalent forces. The rolling resistance coefficient is defined as

Table 2.1 Rolling resistance coefficients

μ_1	μ_2 (h/km)	Source	Remark
0.00252	$3.14(10^{-5})$	Kyle 1990	Moulton 17×1 1/4 bicycle tire at 100 psig, $N = 100$ lbf; rotating drum test
0.0075–0.3	$3.11(10^{-5})$–0.0011	Steeds 1960	Typical μ_1 for automobile tires: lower value smooth pavement, higher unpaved

$$\mu = \frac{R}{N}. \tag{2.14}$$

We expect a dependence of μ on V through the dynamic component of resistance. Kyle (1990) and Steeds (1960) present models in which this coefficient is a linear function of V. SAE (1997) presents it as a function of V^2. The former model is adopted herein because it agrees with the test data of Kyle on small wheels typical of those used on solar racing cars. Also, it allows the dynamic coefficient to be easily separated from the drag area in coast-down testing, since otherwise both the dynamic rolling resistance and the drag would be functions of V^2.

The rolling resistance coefficient model adopted herein is:

$$\mu = \mu_1 + \mu_2 V. \tag{2.15}$$

When modeling the motion of the vehicle up a grade, the total load on the wheels is reduced to W_z, neglecting lift. Rolling resistance coefficients of tires are shown in Table 2.1.

2.17 Example 2.4

Suppose the vehicle of Fig. 2.9 were traveling 40 kph up the 10 % grade of Example 2.3. If it is equipped with the Moulton bicycle tires used in the drum test reported by Kyle, what would the total rolling resistance be?

Solution Using W_Z from Example 2.3 and Eqs. (2.14) and (2.15),

$$\mu = 0.00252 + (0.0000314\text{kph}^{-1})(40\text{kph}) = 0.00378,$$

$$R = (0.00378)(3540.7\text{N}) = 13.4\text{N},$$

where the lift has been neglected because of the low speed. The contact patch force accounts for 66 % of R.

Chapter 12, *Testing,* describes a method of measuring rolling resistance. Chapter 20 presents a method for estimating rolling resistance from tire properties during the design.

2.18 Tractive Force

The tractive force is the propulsive force equivalent to the torque delivered by the drive (the motor, controller, and transmission) to the driven wheels, or wheel. (Solar racing cars often have only one driven wheel. This reduces the weight, complexity, frictional losses, and cost of the car.) Suppose the radius of a driven wheel is r_W, and the flattening of the loaded tire is ignored. The equivalent tractive force is:

$$T = \frac{\tau}{r_W}. \tag{2.16}$$

2.19 Force Balance

At steady conditions (speed, weather, and road conditions independent of time), there is no acceleration, and so the sum of the forces in each coordinate direction must equal zero. And recall that in the cruise condition there is no force in the y-direction and consequently also no yawing moment. The force balance in the x-direction is:

$$T = D + W_x + R. \tag{2.17}$$

R is the total rolling resistance, related to the total normal force by Eq. (2.14).

The drag, D, and the rolling resistance, R, have no z- nor y-components and oppose the motion of the car.[12] A gravity force component opposes the motion when the car travels up a hill and aids the motion when the car travels down a hill. The lift force has no component along the x-axis and so neither opposes nor aids the motion directly. (However, it adds to or subtracts from the net downward force on the car and thus changes the rolling resistance.)

2.20 Example 2.5

Suppose the car in Fig. 2.9 is traveling at a steady 40 kph up the grade. Find the tractive force, wheel torque, and power required to climb the grade.

Solution The tractive force to maintain this speed is the sum of the opposing forces given in Eq. (2.17). The magnitude of the drag, D, may be found from the result of Example 2.1 (assuming c_D is constant) by a ratio:

$$D = (57.7\text{N})\left(\frac{40}{96}\right)^2 = 10.0\text{N}.$$

[12] Unless the relative wind blows from behind. This is an unusual situation and will be ignored.

The rolling resistance was found in Example 2.4. So the tractive force is (again neglecting lift):

$$T = 10.0\text{N} + 354.1\text{N} + 13.4\text{N} = 377.5\text{N}.$$

The torque required to climb the grade at 40 kph, assuming the 17.5-in wheel diameter of the Moulton wheels (and ignoring the flattening of the wheel at the contact patch), is:

$$\tau = (377.5\text{N})\left(\frac{17.5\text{in}}{2}\right)\left(0.0254\frac{\text{m}}{\text{in}}\right) = 83.9\text{N}\cdot\text{m}.$$

The total power required at the driven wheel(s) is the tractive force times the speed (m/s), or 4,194.4 W, about 5.6 hp.

2.21 Acceleration

Suppose the vehicle is changing cruising speed. Typical solar car speed changes are gradual; the y-axis moment sum and certainly the z-axis force sum will remain zero. In this case, the resultant of the x-forces equals the *effective mass* (M_e) times the acceleration (a).[13] So the tractive force must now equal

$$T = M_e a + D + R + W_x. \tag{2.18}$$

The tractive work done by the car in traveling a short distance ΔS over which T may be assumed constant is $T(\Delta S)$. Thus, T may be thought of as the tractive work per unit distance.

In order that the reader may begin to develop some intuition about the tractive force, we will study Eq. (2.18). We will assume that the drag coefficient is a constant, which is only approximately true.

Dimension Independence Non-dimensionalizing Eq. (2.18) will make our conclusions independent of, say, the weight of a particular car. We will therefore not calculate T, but T/W (T^*). The speed V will be replaced by V/V_D (V^*). The speed scale V_D used to non-dimensionalize the speed will be called the *drag speed.* This is the speed at which the drag force is equal to the weight. (Imagine the car in a stable, nose-first free fall at constant drag coefficient through a uniform atmosphere. The drag speed is the terminal velocity that would be predicted by Eq. (2.17).) It is:

[13] The effective mass is larger than the actual mass and accounts for the need to accelerate masses that translate as part of the body of the car, but also rotate about their own axes, such as the wheels. See Chapter 22.

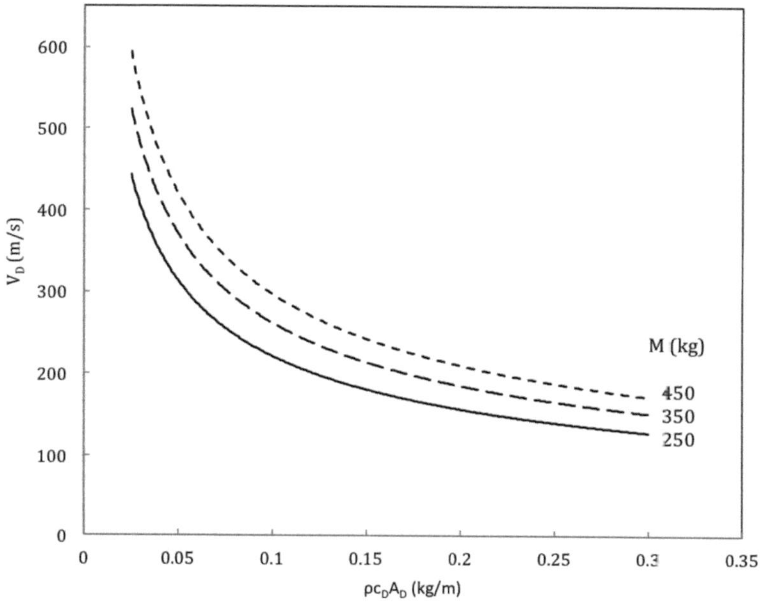

Fig. 2.11 Drag speed

$$V_D = \sqrt{\frac{2W}{c_D A_D \rho}}. \tag{2.19}$$

Using values from Chap. 8 gives V_D=237.8 m/s, or 856 kph (532 mph). Thus, if the car were moving at 88 kph (55 mph), V^* would be 0.103. Figure 2.11 shows how the drag speed is influenced by the vehicle's mass, drag area, and the air density.

The ratio a/g_0 (a^*) will be the nondimensional acceleration. The scale, g_0, is the standard acceleration of gravity (9.807 m/s^2), which we will assume to be uniform over the Earth. Using the models of the forces *R, D,* and W_X previously given and non-dimensionalizing as above, transforms Eq. (2.18) to

$$T^* = \frac{M_e}{M} a^* + D^* + R^* + W_x^*, \tag{2.20}$$

where, if the wind blows from ahead (a numerically negative wind speed),

$$D^* = (V^* - V_W^*)^2, \quad R^* = \mu_1 + \mu_2^* V^*,$$
$$\mu_2^* = \mu_2 V_D, \qquad W_x^* = \sin\alpha.$$

2.22 Steady Motion Studies

No Grade nor Wind The specification in Chap. 8 requires $\mu_1 = 0.004$ and $\mu_2 = 0.0001$ s/m, or 2.78(10^{-5}) h/km. Using these values, and the other specified parameters given above, let us examine T^*, D^*, and R^* as a function of V^*, with accelaration, no headwind, zero lift, and a horizontal road. Figure 2.12 shows the results of this study. Its quadratic speed dependence causes the drag to be less than the rolling resistance until V^* is approximately 0.075. This corresponds to 64.2 kph (about 40 mph) for the car of Chap. 8. At average speeds characteristic of stop-and-go city traffic, the drag is less than half the rolling resistance. On the other hand, at freeway speeds of 105 kph (65 mph) the drag is more than twice the rolling resistance. The exact crossover speed depends upon the drag speed of an individual car and its rolling resistance coefficients.

Grade Although it has been assumed here for convenience, a horizontal road is rare in reality. Imagine the car is climbing steadily up a 10 % grade. The tractive force is increased by 0.099, or the *sin* of the grade angle of 5.71°. This increase in the tractive force is about twice the largest magnitude shown in Fig. 2.12. A 3 % grade would increase the tractive force by 0.03. Figure 2.13 shows the tractive force required to climb various grades between 0 and 10 % at different speeds.

Also shown in Fig. 2.13 are curves of constant dimensionless power, p^*. This is the power that must be delivered to the driving wheel(s), not the power delivered to

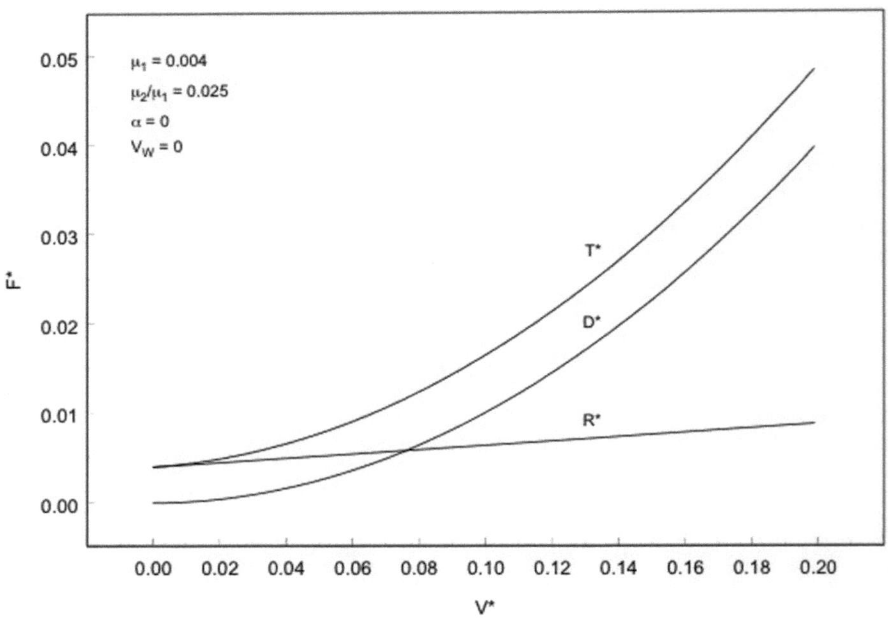

Fig. 2.12 Tractive force components

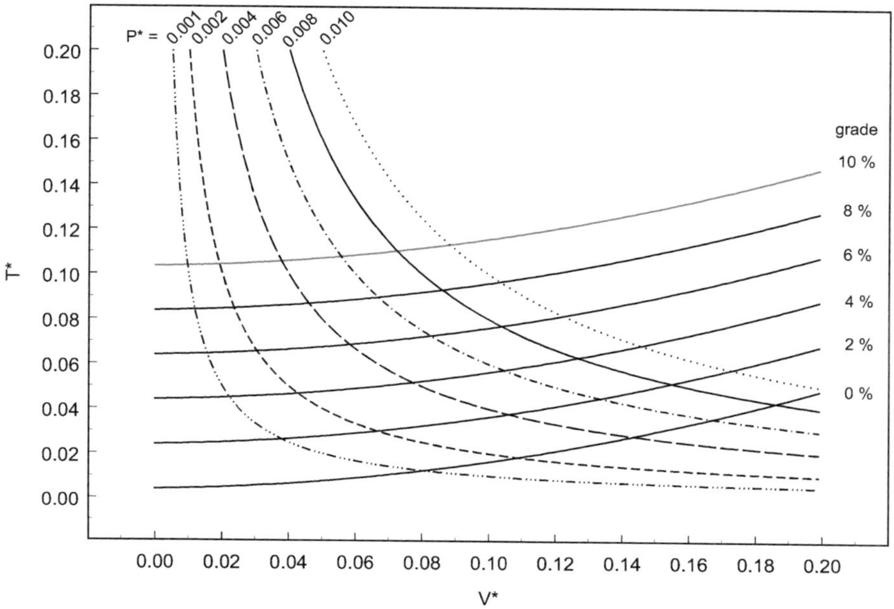

Fig. 2.13 Climbing at steady speed

the motor. The power required to travel at speed V is TV. The dimensionless power is TV/WV_D, symbolized as T^*V^*. The intersection of a power curve with the tractive force curve for a hill gives the climbing speed (found on the abscissa) and the tractive force (found on the ordinate) required to make that speed.

Solar Zone Where do solar racing cars built to Sunrayce specifications fall on the power curves of Fig. 2.13? These cars, because of the solar array and battery limitations imposed by the rules in Chap. 16, can deliver a maximum dimensionless power of roughly 0.006 in full sun (1000 W/m^2)—assuming a drive efficiency of 80%). To do so, they must discharge their batteries at a high rate (say 30 or 40 A). (These rates cannot be sustained for more than a short time without greatly depleting the battery's charge.) If only the solar array is used for power, the cars can deliver roughly 0.001 in full sun. Figure 2.13 shows that the power range 0.001–0.006 means a speed range of about 0.01–0.055 on a 10% grade and 0.08–0.17 on a horizontal road. For the car under study, these dimensionless speeds translate to 9–51 kph (about 6–32 mph) and 60–146 kph (about 37–91 mph), respectively.

Strategic Decision What goes up must come down, at least eventually. So it is possible to recover some of the energy expended to overcome gravity when climbing a hill, as gravity helps to accelerate the car on the following downhill grade.[14] The

[14] Only some of the energy expended is recoverable because of the losses in the battery and the drive. This will be discussed in more detail in Chap. 4, *Storing Energy*, and Chap. 5, *Electric Motor Drives*.

racing team must decide whether to return a portion of the energy to the battery through regeneration (running the motor as a generator) or to use it all to gain speed.

2.23 Wind and Drag

Headwind Suppose that the headwind of 8 kph specified in Chap. 8 is blowing. Because of the high drag speed of the car under study, the dimensionless relative wind speed is increased by only 0.0093. Nevertheless, for a horizontal road, Fig. 2.13 shows that at 88 kph (55 mph), the drag, and therefore the tractive force, is increased by about 19%.

Wind Averaging As Kurtz (1980) pointed out, real vehicles seldom operate exclusively at zero yaw; there is usually a crosswind. Hence, if a single drag coefficient is to be used in design, it must be the wind-averaged value. But to calculate this value, the relative wind speed and direction and the car's speed and course must be known at any moment, and the drag coefficient must be known as a function of the yaw angle. The design weather and route supply the former, but wind tunnel tests on models, at least, or coast-down tests with the full-scale car provide the best measurement of the latter.

Drag and Yaw Figure 2.14[15] shows the drag coefficient as a function of the yaw angle for a passenger car typical of those tested by Kurtz (1980) and a properly designed solar racer similar to the *Spirit of Biel* reported by Storey et al. (1994). The drag of the passenger car increases with yaw, whereas the drag of the solar racer has the opposite tendency. Therefore, the wind-averaged drag of the solar racer will be less, and the wind-averaged drag of the passenger car will be greater than its drag coefficient at zero yaw.

The yaw angle and speed of the relative wind over the solar racer at any moment depend on the racer's course and speed and the wind's speed and direction. Kurtz (1980) reported wind-weighting factors for estimating the effective drag coefficient of passenger cars from their drag coefficients at zero yaw angle for several standard driving cycles.[16] Kurtz's factors will not apply to streamlined solar cars with falling drag coefficients.

Illustration Figure 2.15 was prepared to illustrate the error in using the zero-yaw drag coefficient for a solar car assumed to have a falling drag-yaw characteristic similar that illustrated in Fig. 2.14.[17] The car was assumed to be driving due west at

[15] The curves were normalized to their respective drag coefficients at zero yaw because the drag coefficient of the typical passenger car over the yaw angle range displayed was three to five times that of the solar racer.

[16] The results showed that the weighting factor could vary from about 1.04 (a 4% correction) to more than 1.4 (a 40% correction), depending upon the driving cycle and the vehicle's drag characteristics.

[17] Other parameters: M_e, 338 kg; μ_1, 0.004; μ_2, 0.0001 s/m, A_D, 1.45 m^2.

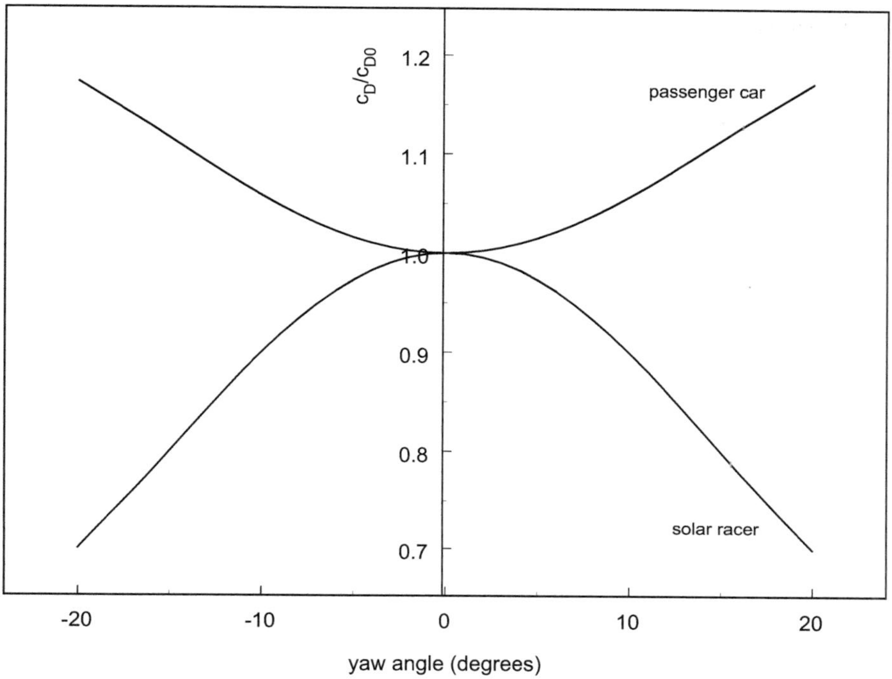

Fig. 2.14 c_D/c_{D0} as a function of yaw

various constant speeds on a level road under standard conditions. The wind speed and direction were taken as 8 kph and 225°, respectively. The Reynolds number dependency of the drag coefficient of the smooth, streamlined solar car was taken to be similar in shape to that of the 2:1 ellipsoid of Fig. 2.6.

Operating Zone The speed in the study ranged from 1.0 to 88 kph. Observe that even at low speeds the resulting Reynolds number placed the car after the laminar–turbulent transition. And if there had been no wind, the Reynolds number at low speeds would have been about 10^5, just before the transition to turbulence. During periods of acceleration from rest, the car would traverse the drag coefficient curve to the turbulent region in seconds. This transition shifts to somewhat lower Reynolds numbers as the vehicle's basic shape becomes more streamlined, as Fig. 2.6 implies.

The foregoing discussion supports this design thumb rule: A solar car may be assumed always to operate in the turbulent, separated-flow region. Notice that the zero-yaw drag coefficient in this region increases relatively slowly with the Reynolds number because of increasing skin friction drag. Remember, however, that the thumb rule applies to the car as a whole. The local flow over upstream areas of the car, such as a canopy, may still be laminar separated, or, near the nose, just laminar.

Force Error Figure 2.15 shows the percent relative error in the calculated tractive force compared to the yawed case as a function of the Reynolds number. A maximum error of about 3 % appears near the point at which the speed and yaw angle curves

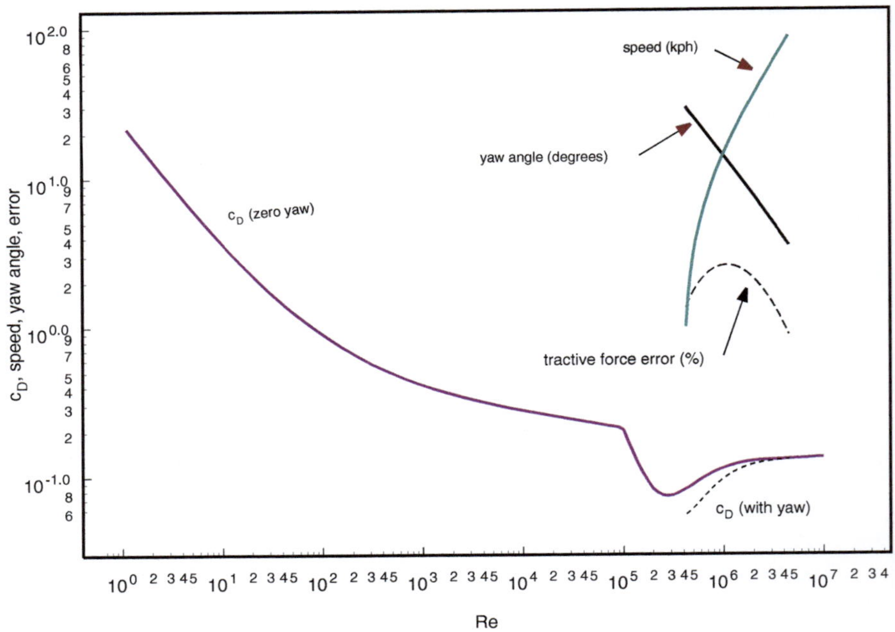

Fig. 2.15 Error from neglect of yaw

cross. At low speeds, the yaw angle is at its largest, but the quadratic dependence of the drag on the relative wind speed forces the drag to be small compared to the rolling resistance. Hence, the error is small. But as the speed increases, the importance of the drag exceeds that of the rolling resistance, as Fig. 2.12 shows. But the yaw angle is decreasing. These competing influences cause the error to pass through a maximum and decrease until it is again about 1 % at 88 kph vehicle speed. Higher wind speeds, or wind directions further south, increase the error. For example, doubling of the error is possible at wind speeds of 16 kph (10 mph).

2.24 Unsteady Motion Study

Steady conditions are uncommon (though simpler to analyze). The car is usually driving into an upgrade or a downgrade or into some other condition affecting its motion. Thus, the speed and torque are usually changing as functions of time. Equation (2.20) can be used to find the speed as a function of time. The time dependence of the torque (tractive force) must be known, however.

Little Taps If the acceleration were constant, the speed at the end of a time interval would be:

$$V_2 = V_1 + a_1(t_1 - t_2). \tag{2.21}$$

We stipulate that the time interval be small enough so that the acceleration is approximately constant over the interval: The car moves in little steps, as if tapped by a hammer. Then we can compute the speed at the end of each time interval from the speed at the end of the previous time interval using Eq. (2.20). In this way, we march forward in time.

2.25 Example 2.6

Calculate the velocity as a function of time for a car accelerating from 40 to 88.5 kph with the tractive force set equal to that required to sustain 88.5 kph and also four times that value. The road is horizontal.

Solution The tractive force corresponding to 88.5 kph is:

$$T = \frac{1}{2} c_D A_D \rho V_2^2 + (\mu_1 + \mu_2 V_2) W = (0.5)(0.11\mathrm{m}^2)\left(1.18\frac{\mathrm{kg}}{\mathrm{m}^3}\right)\left(24.58\frac{\mathrm{m}}{\mathrm{s}}\right)^2$$
$$+\left(0.004 + 0.0001\frac{s}{\mathrm{m}} \times 24.58\frac{\mathrm{m}}{\mathrm{s}}\right)(3315\mathrm{N}) = 58.57\mathrm{N}.$$

At the instant before the car begins to accelerate,

$$D_0 = (0.5)(0.11m^2)\left(1.18\frac{kg}{m}\right)\left(11.11\frac{m}{\sec}\right)^2 = 8.012N,$$

$$R_0 = \left(0.004 + 0.0001\frac{\sec}{m} \times 11.11\frac{m}{\sec}\right)(3315\,N) = 16.943N.$$

Therefore, taking $M_e/M = 1.05$, the acceleration is:

$$a_0 = \frac{T - D_0 - R_0}{M_e} = \frac{58.57\mathrm{N} - 8.012\mathrm{N} - 16.943\mathrm{N}}{354.9\mathrm{kg}} = 0.0947\frac{\mathrm{m}}{\mathrm{s}^2},$$

and the speed after a time step of 1.0 s is:

$$V_1 = V_0 + a_0 \Delta t = 11.11\frac{\mathrm{m}}{\mathrm{s}} + \left(0.0947\frac{\mathrm{m}}{\mathrm{s}^2}\right)(1.0s) = 11.205\frac{\mathrm{m}}{\mathrm{s}}.$$

Table 2.2 gives the results for the first 3 s. As illustrated above, the acceleration used to predict the next speed is found from the previous speed. The power delivered to the wheel, the product of the tractive force and the speed, is shown in the last column.

Figure 2.16 shows curves of velocity as a function of time for the required speed increase on a horizontal road with no headwind. The curve resulting from applying four times the first tractive force is also shown for comparison.

Table 2.2 Acceleration under constant tractive force

t(s)	*V* (m/s)	*D* (N)	*R* (N)	*a* (m/s²)	*P* (W)
0	11.11	8.012	16.943	0.0947	277.3
1	11.205	8.160	16.974	0.0942	656.3
2	11.299	8.299	17.006	0.0937	661.8
3	11.393	8.437	17.037	0.0933	667.3

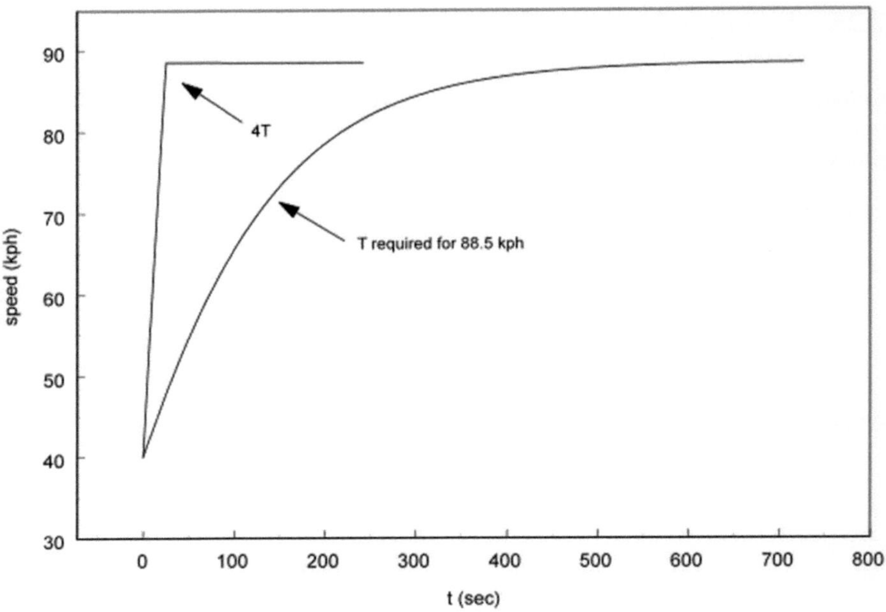

Fig. 2.16 Accelerated motion

References

Gillespie, T. D. (1992). *Fundamentals of vehicle dynamics*. Pennsylvania: Society of Automotive Engineers.

Hucho, W.-H. (1978). The aerodynamic drag of cars, current understanding, unresolved problems, and future prospects. In G. Sovran, T. Morel, & W. T. Mason (Eds.), *Aerodynamic drag mechanisms of bluff bodies and road vehicles* (p. 1). New York: Plenum Press.

Hucho, W.-H. (1987a). Aerodynamics of passenger cars. In W.-H. Hucho (Ed.), *Aerodynamics of road vehicles* (p. 106). London: Butterworth and Co. Ltd.

Kurtz, D. W. (1980). *Aerodynamic design of electric and hybrid vehicles, a guidebook*, N81–12943 (NASA-CR-163744 Jet Propulsion Lab) NTIS 1980.

Kyle, C. R. (1990). The sunraycer: Wheels, tires, and brakes. In P. MacCready et al. (Eds.), *Sunraycer case history*. Society of Automotive Engineers, lecture 3–3.

Morelli, A. (1983). Aerodynamic basic bodies suitable for automobile applications. *Int. J. of Vehicle Design*, Technological Advances in Vehicle Design Series, SP3. In M. A. Dorgham (Ed.), *Impact of aerodynamics on vehicle design* (p. 70). United Kingdom: Interscience Enterprises Ltd., La Motte Chambers.

SAE. (1997) Road load measurement and dynamometer simulation using coastdown techniques. SAE J1263 Feb. 96, *1997 Society of Automotive Engineers Handbook* (Vol. 2, p. 26.531).
Sherman, F. S. (1990). *Viscous flow*. New York: McGraw-Hill.
Steeds, W. (1960). *Mechanics of road vehicles*. London: Iliffe and Sons, Ltd.
Storey, J. W. V., Schinckel, A. E. T., & Kyle, C. R. (1993). *Solar racing cars*. Canberra: Australian Government Publishing Service.
White, F. M. (1986). *Fluid mechanics*. New York: McGraw-Hill Book Company.

Chapter 3
Interaction with the Sun

3.1 Introduction

This chapter focuses on three topics. First, how to estimate the rate at which solar energy strikes a planar surface, such as a solar cell, oriented at some position with respect to the sun, at a given time and location on the earth is explained. Then, a method for calculating the solar energy transmitted through a glazing is discussed. The final section explains how the solar energy irradiating a solar cell is converted into electric energy.

3.2 The Solar Source

Solar Spectrum Some of the energy released by the nuclear fusion reactions within the sun is broadcast as electromagnetic radiation, Q_S. Most of this radiation is distributed over wavelengths ranging from the ultraviolet to the infrared. Textbooks on solar energy, Hsieh (1986) and Duffie and Beckman (1991) for example, give details about the distribution of this radiation over wavelength, its *spectral distribution*, in space and at the earth's surface. Herein, we shall usually be concerned only with a broad division of the solar spectrum at the earth's surface into *visible* and *infrared* regions. The former we define as wavelengths extending from 0.25 to 1.0 μ ([μ], a micron, short for micrometer, is a millionth of a meter) and the latter as the wavelengths above 1.0 μ.

Solar Constant The sun emits energy at a rate of about $3.826(10^{26})$ W. The average intensity (W/m^2) of this radiation on a sphere having a radius equal to the average radius of the earth's orbit ($1.496(10^8)$ km, one *astronomical unit* or AU) is called the *solar constant*, G_{sc}. The solar "constant" is not actually constant; it changes over the approximately 30-day rotation period of the sun. Its variation is greatest during the peak of the 11-year sunspot cycle and decreases as the sunspot cycle

E. F. Thacher, *A Solar Car Primer*, DOI 10.1007/978-3-319-17494-5_3

approaches its nadir. We shall use the value recommended by Smith and West (1983): 1371 ± 5 W/m^2.

Perpendicular Plate Irradiance We are interested in knowing the solar radiation striking a surface. Therefore, we begin with a flat plate in space, perpendicular to the sun's radiation and following the earth's orbit. The intensity, G_{0n} ("n" stands for "normal to the plate," i.e., perpendicular to it), on this plate will differ from the solar constant, except close to the spring and autumn equinoxes. The earth's orbit is somewhat elliptical. Hence, the earth–sun distance varies over the year, causing changes in the intensity of the radiation. Figure 3.1 shows G_{0n} as a function of the

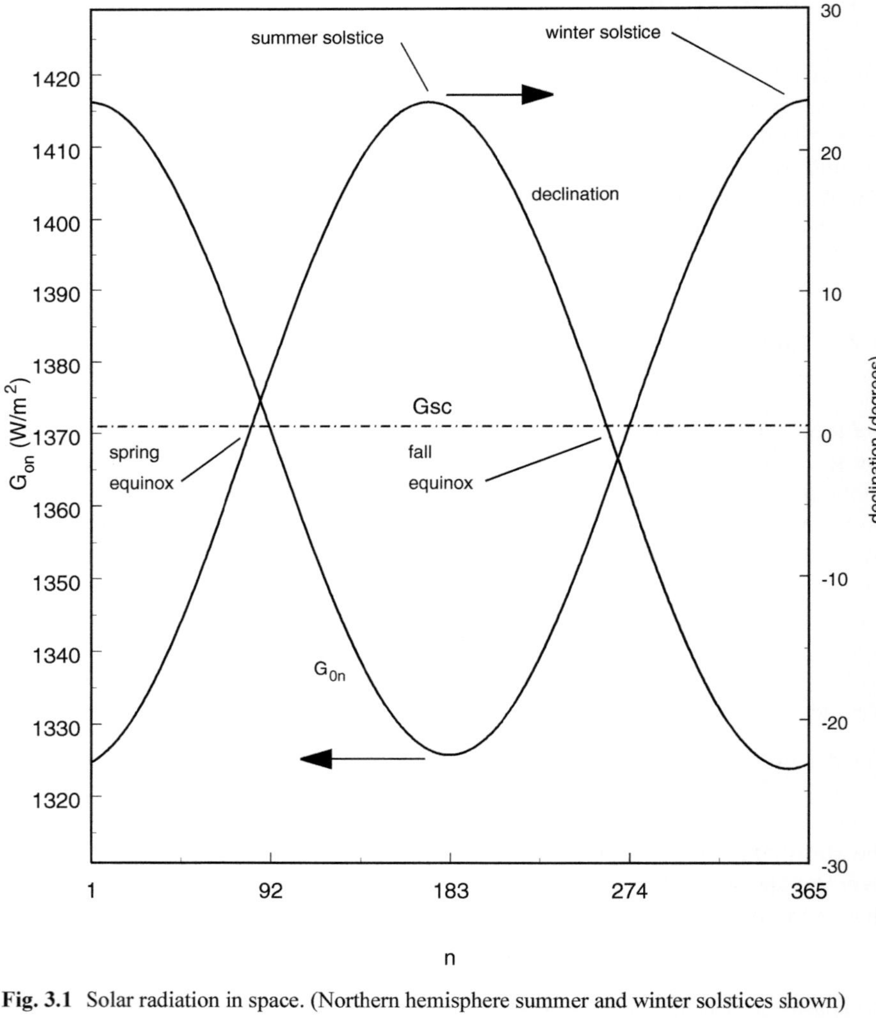

Fig. 3.1 Solar radiation in space. (Northern hemisphere summer and winter solstices shown)

serial day of the year. The declination and G_{0n} curves are based on formulas in Cooper (1969) and Duffie and Beckman (1991), respectively.

Note that G_{0n} is smallest (about 1325 W/m^2) when it is summer in the earth's northern hemisphere, but largest (about 1416 W/m^2) in the northern hemisphere's winter. This is counterintuitive. The earth's axis of rotation is inclined to the plane of its orbit by about 23.45°, the *declination angle* (δ), also shown in Fig. 3.1. In the northern hemisphere's winter, this tilt is away from the sun, thus causing this hemisphere to intercept less solar energy (see the discussion below of the radiation on a tilted plate). At the same time the southern hemisphere intercepts more radiation and experiences summer. The region near the equator is less affected.

Other sources of variation in G_{0n}, such as sunspot activity, cause variations more than an order of magnitude less than those discussed above.

Even though they are actually continuous functions of time, Fig. 3.1 shows that it is sufficiently accurate to consider G_{0n} and δ as constants for a particular day.

Tilted Plate Irradiance The solar radiation intensity at a particular moment and the total amount of solar energy available over a particular interval on a locally horizontal plate (one tangent to the surface of the earth), or on a locally tilted plate, depend upon the latitude, some other angles, and the time of year. Suppose that on a certain day a flat plate of area A (m^2) is located at some *latitude* (L_a, positive north of the equator) and longitude (L_o, positive east of the zeroth meridian which passes through Greenwich, England) on the earth, is inclined at a *tilt angle* (β) above the local horizontal plane, and is turned through a *surface azimuth angle*, (γ, positive when west of south) degrees. The plate is shown in Fig. 3.2.

For now, we continue to neglect the atmosphere and assume that the sun is a point source of radiation, so that its rays are approximately parallel.[1] A line drawn perpendicular to the plate's surface would make an *angle of incidence* (θ) with the incoming solar rays. The area of the plate intercepting solar radiation would be the plate area times the cosine of the angle of incidence. So the rate at which solar energy strikes the plate would be

$$Q_{\text{tilted}} = G_{0n} A_{\text{P}} = G_{0n} A \cos\theta. \tag{3.1}$$

A method for calculating the angle of incidence from information about the location and orientation of the plate and the time of day is explained later.

3.3 Solar Time

The position of the sun relative to the plate depends upon the time of day. This time of the day is the *solar time*, neither the standard nor daylight saving time kept in the time zone where the plate resides. A time zone is centered on a standard meridian, and the standard time of the zone is the time at this meridian. Table 3.1 gives the standard meridians of all the time zones in the contiguous USA.

[1] The sun's rays diverge from parallelism by only about 16 min of arc. They may be assumed parallel for our design purposes.

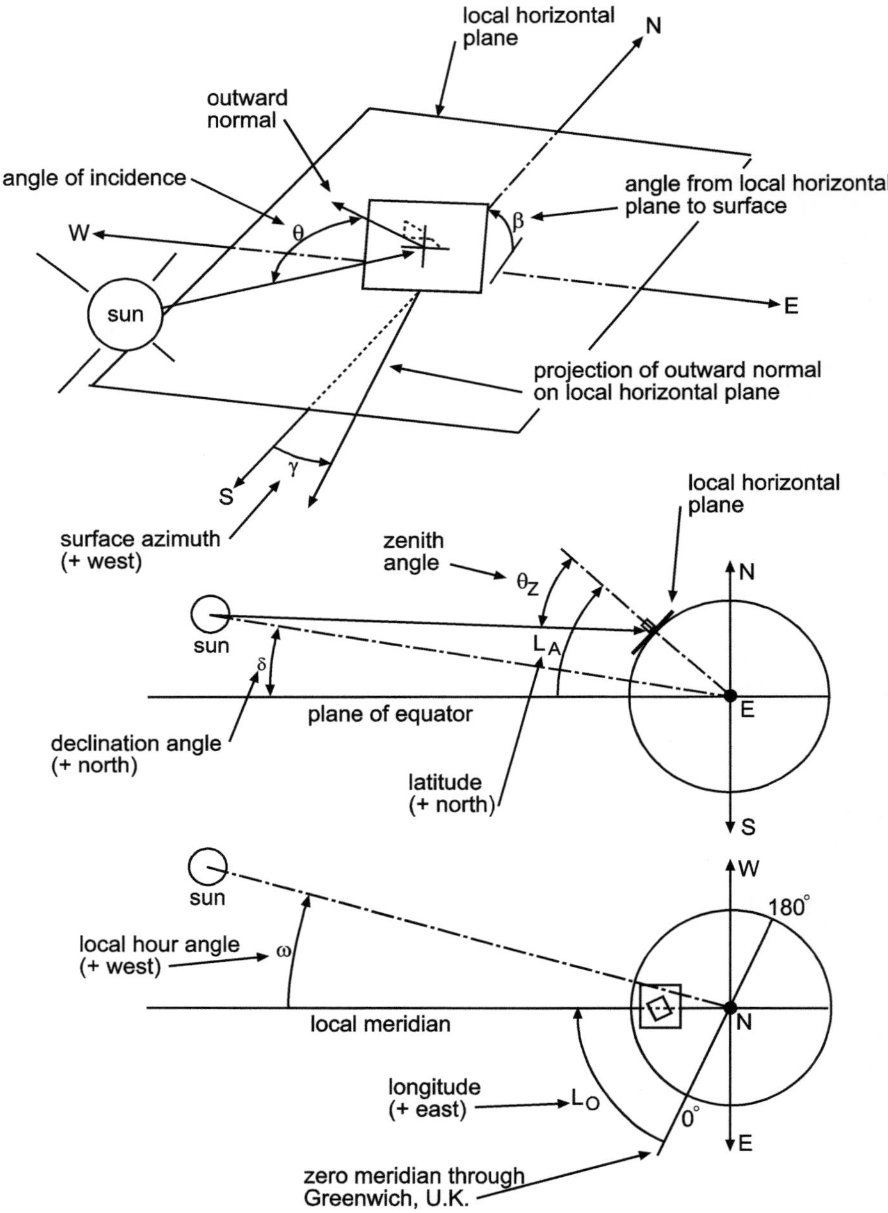

Fig. 3.2 Orientation of a flat plate

When the sun is directly over the meridian passing through the plate, it is noon solar time at the plate. The sun moves westward at 15° of longitude per hour (1° every 4 min), which is about the width of a typical time zone (or would be, politics aside). Therefore, solar time could be about half an hour different from the local

Table 3.1 Standard meridians

Meridian (W)	Time zone
75	Eastern
90	Central
105	Mountain
120	Pacific

standard time, if the plate were located near one of the time zone's borders. The solar time (t_{solar}) at a particular location may be found by applying a correction to the standard time (t_{std}) at that location.

$$t_{\text{solar}} = t_{\text{std}} + 4(L_{\text{std}} - L_{\text{local}}) + e. \quad (3.2)$$

The two correction terms are in minutes; the longitudes are expressed in degrees. If the plate were located at the standard meridian (L_{std}), the solar and local standard times would be within a few minutes of each other. This residual error, e, is caused by small variations in the earth's rotational speed. Spencer (1971) gives the residual error as shown in Fig. 3.3.

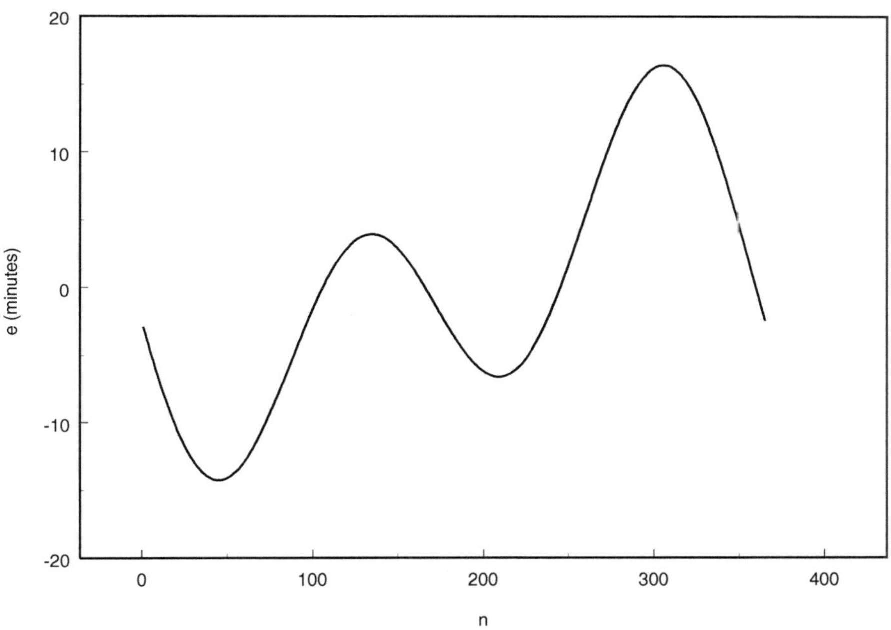

Fig. 3.3 Residual error

3.4 Angle of Incidence

Once the solar time, the latitude, and orientation of the plate are known, the angle of incidence may be calculated using a relation given by Benford and Bock (1939):

$$\begin{aligned}\cos\theta &= \sin\delta\sin L_a\cos\beta - \sin\delta\cos L_a\sin\beta\cos\gamma \\ &+ \cos\delta\cos L_a\cos\beta\cos\omega + \cos\delta\sin L_a\sin\beta\cos\gamma\cos\omega \\ &+ \cos\delta\sin\beta\sin\gamma\sin\omega.\end{aligned} \tag{3.3}$$

where ω, the *local hour angle* of the plate, is the angle between the longitude of the sun and the longitude of the meridian passing through the plate, caused by the apparent motion of the sun at 15°/h. It is measured from the local meridian with morning (east) negative and afternoon (west) positive. Note that if the plate is horizontal ($\beta = 0$ and $\theta = \theta_Z$), the equation simplifies considerably

$$\cos\theta_Z = \sin\delta\sin L_a + \cos\delta\cos L_a\cos\omega. \tag{3.4}$$

3.5 Example 3.1

Suppose a solar cell is lying on flat ground at Potsdam, New York, about 44° 56′ N latitude and 74° 51′ W longitude, at noon eastern standard time on June 15. The cell is 4 cm long and 2 cm wide. Neglecting the atmosphere, at what rate does the cell intercept solar energy?

Solution Equation (3.2) gives a negligible correction, and the longitude is equal to the standard meridian (Table 3.1): the solar time is approximately noon. The area of the cell is 8 cm² or 8(10⁻⁴) m². Using Eq. (3.4) we find $\theta = 21.7°$ and therefore the projected area is, from Eq. (3.1),

$$A_P = 8(10^{-4})\text{m}^2\cos(21.7°) = 7.433(10^{-4})\text{m}^2.$$

From Fig. 3.1, $G_{0n} = 1328$ W/m². The rate of energy intercept is therefore

$$Q = G_{0n}A_P = 1328\frac{\text{W}}{\text{m}^2} \times 7.433(10^{-4})\text{m}^2 = 0.99\text{W}.$$

Now suppose that the cell is pointed at the sun, that is, the angle of incidence is set to zero. (This requires[2] $\beta = L_a - \gamma$.) Then $A_P = A$ and

$$Q = 1328\frac{\text{W}}{\text{m}^2} \times 8(10^{-4})\text{m}^2 = 1.06\text{W}.$$

[2] This may be shown for the conditions of the example by applying the relations for the sin and cos of the difference of two angles to Eq. 3.8

As Example 3.1 shows, the smaller the angle of incidence, the greater will be the rate of solar energy interception. Solar racing teams have taken advantage of this in various ways. The direction of the 1990 Sunrayce was southeast to northwest (Disney World, near Orlando, Florida, to Warren, Michigan, near Detroit). Western Washington University made the average angle of incidence on their flat, fixed-position array small by tilting it toward the east. Their car had a cockpit at each end. At noon each day they turned the car around so that the array pointed west and drove from the other cockpit. Storey et al. (1994) presented a picture of this car.

3.6 Sunset Hour Angle and Day Length

The sunset hour angle and day length are frequently used in solar calculations. The sunset hour angle can be obtained from Eq. (3.4). At sunset, the angle of incidence on a horizontal surface (equal to the zenith angle) will be zero. Imposing this condition on Eq. (3.4) gives

$$\cos\omega_S = -\tan L_a \tan\delta. \tag{3.5}$$

For a clear horizon, which implies a day symmetric to solar noon, the sun will move through an angle of $2\omega_s°$ during the day at a rate of 15°/h. The number of hours in a day is therefore

$$N = \frac{2}{15}\omega_S. \tag{3.6}$$

3.7 Daily and Hourly Total Radiation

The daily total radiation on a horizontal surface will be of use later in this chapter. It is given by

$$H_0 = \frac{86,400}{\pi} G_{0n}\left(\cos L_a \cos\delta \sin\omega_S + \frac{\pi\omega_S}{180}\sin L_a \sin\delta\right). \tag{3.7}$$

In this equation, ω_S is the sunset hour angle in degrees from Eq. (3.5) and G_{0n} (W/m^2) is found from Fig. 3.1. H_0 is in J/m^2 for the day. The total radiation over a shorter period during a day may be computed from Eq. (3.8):

$$I_0 = \frac{43,200}{\pi} G_{0n}\left[\cos L_a \cos\delta\left(\sin\omega_2 - \sin\omega_1\right) + \frac{\omega_2 - \omega_1}{180}\sin \mathrm{L}_a \sin\delta\right]. \tag{3.8}$$

In this equation, ω_2 corresponds to the later time. I_0 is in J/m^2.

Equations (3.7) and (3.8) are developed in Duffie and Beckman (1991) (and in other similar texts).

3.8 Effects of the Atmosphere

Beam and Diffuse Solar radiation is either transmitted directly (*beam* radiation) or transmitted after scattering by the gases, clouds, and dust particles in the atmosphere (*diffuse* radiation). On its way through the atmosphere, most of the radiation below about 0.25 μ and above about 2.5 μ is absorbed. The substances mainly responsible for absorption are ozone (O_3) in the ultraviolet, and oxygen (O_2), carbon dioxide (CO_2), and water vapor (H_2O) at longer wavelengths.

Diffuse Components On a clear day, the diffuse radiation on a surface may be thought of as the sum of four parts. The first is radiation scattered strongly in the forward direction such that it appears to come from a ring of sky immediately around the solar disc: the *circumsolar* diffuse radiation. Next is uniform radiation from the sky dome: the *isotropic* diffuse radiation. The third component is the brightening observed near the horizon, especially on clear days: the *horizon brightening* diffuse radiation. A fourth term, *ground* diffuse, must be added to account for radiation reflected onto the surface from surrounding objects, such as the ground, if the surface is tilted up, or buildings.

3.9 Hourly Beam and Diffuse Radiation

Motivation The solar cell in Example 3.1 would receive its solar energy from all of the paths discussed above, rather than just directly, as it would if it were in space. The interaction of solar radiation with solar cell glazing depends upon the direction of the radiation and upon whether the radiation is beam or diffuse. Also, the forms of the correction factors for converting radiation data for a horizontal surface to equivalent values on a tilted surface depend on whether the radiation is beam or diffuse. For these reasons we must be able to separate the global radiation striking a surface into beam and diffuse components.

Methods Two methods will be presented. The first method will be most useful during design, or when assessing the typical performance of a solar car over a route during a particular time of the year. What is needed for these tasks is solar radiation data that are typical of the location and period of the analysis. We will call this the *design method,* which is explained in detail in this chapter. During racing, an estimate of the solar energy available ahead of a solar car is essential to maximizing its average speed. The actual weather at a particular moment will deviate randomly from the long-term, typical values used in the design method. The second method accounts approximately for this deviation. We will call this the *racing method*; it will be covered in Chap. 13.

Table 3.2 Typical day (Klein 1977)

Month	Typical day (date)	Find "*n*" on typical day "i"	n	Declination
January	17	i	17	−20.9
February	16	31+i	47	−13.0
March	16	59+i	75	−2.4
April	15	90+i	105	9.4
May	15	120+i	135	18.8
June	11	151+i	162	23.1
July	17	181+i	198	21.2
August	16	212+i	228	13.5
September	15	243+i	258	2.2
October	15	273+i	288	−9.6
November	14	304+i	318	−18.9
December	10	334+i	344	−23.0

3.10 Design Method

The design method has two parts. The first part is a means of estimating the hourly radiation on a surface for a typical day from monthly average values. The second method develops the hourly information from so-called typical meteorological year data.

Average Day Calculations Sizing calculations in the early stages of design may require solar radiation values, representative of a certain month, and which can be quickly obtained. The datum meeting this need is the daily *global* (beam plus diffuse) solar energy intensity on a horizontal surface averaged over the month. This value ($\bar{H}$, kJ/m²·day) is called the *average daily horizontal radiation*. The average daily horizontal radiation for each month at many locations in the USA is reported in Knapp et al. (1980), and may also be found in solar energy engineering texts, such as Hsieh (1986) and Duffie and Beckman (1991). These latter sources also give this data at locations around the world.

Table 3.2 gives the date of the "average day" recommended by Klein (1977) for each month of the year. The average day is the day having a daily extraterrestrial horizontal radiation closest to the average daily extraterrestrial horizontal radiation for the month ($H_0 \approx \bar{H}_0$).

3.11 Example 3.2

What is the total energy received by the horizontal solar cell in Example 3.1, including the effect of the atmosphere, for the typical day of June?

Solution Potsdam is not one of the locations tabulated in Knapp et al. (1980). Massena, which is about 20 miles north of Potsdam, is the nearest tabulated location.

Assuming Massena's data applies to Potsdam, we find $\bar{H}$ =20,190 kJ/(m² day). The total energy received on the average day is

$$Q = \bar{H}A_p = 20{,}190\frac{\text{kJ}}{\text{m}^2 \times \text{day}} \times 7.433(10^{-4})\text{m}^2 = 16.15\frac{\text{kJ}}{\text{day}}.$$

Let us compare this value to the daily total received by the cell when no atmosphere is present. This will be the product H_0A_p, because $H_0 \approx \bar{H}_0$ on the average day, by definition. From Table 3.2: n = 162, δ = 23.1°. From Fig. 3.1: G_{0n} = 1329 W/m². Equation (3.5) gives ω_S = 115.3°. Finally, from Eq. (3.7), H_0 = 41.88 MJ/m². Then, as above, Q = 31.13 kJ/day.

Clearness Index and Monthly Diffuse The ratio Q/Q_0 is 0.48 for the case in the preceding example. This ratio, called the monthly average clearness index, is given more simply as

$$\bar{K}_T = \frac{\bar{H}}{\bar{H}_0} \tag{3.9}$$

because the cell area cancels. It correlates strongly with the monthly average diffuse fraction, $\bar{H}_D / \bar{H}$. Figure 3.4, based on a statistical study reported by Erbs et al. (1982), shows this correlation. In it, ω_S is the sunset hour angle on the average day and the correlation is restricted to monthly average clearness indices of $0.3 \le \bar{K} \le 0.8$. $\bar{K}_T$ is tabulated in Knapp et al. (1980)

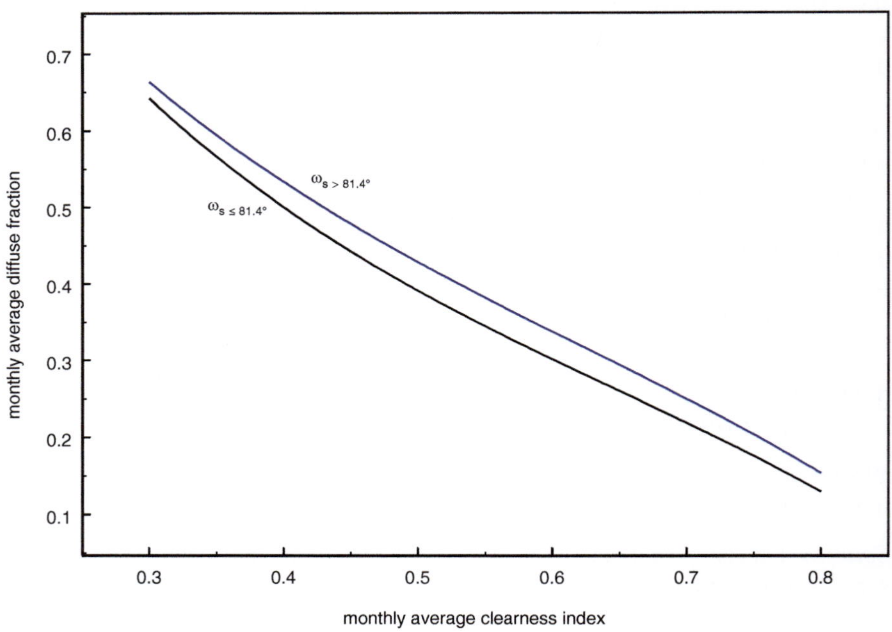

Fig. 3.4 Monthly average diffuse fraction

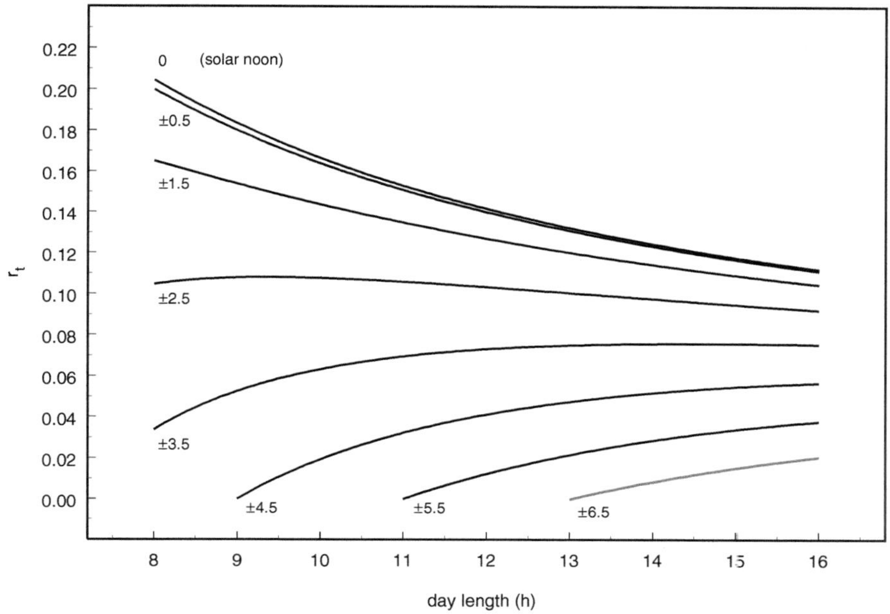

Fig. 3.5 Average hourly total radiation fraction

Hourly from Daily Liu and Jordan (1960) and Collares-Pereira and Rabl (1979) studied the distribution of the hourly global radiation and the hourly diffuse radiation over the hours of a day. They presented their results in terms of the ratios

$$r_{\mathrm{t}} = \frac{I}{\overline{H}}, \qquad r_{\mathrm{d}} = \frac{I_{\mathrm{D}}}{\overline{H}_{\mathrm{D}}} \tag{3.10}$$

where I is the total global radiation on a horizontal surface during an hour and I_D is the total diffuse radiation on a horizontal surface during an hour. Figure 3.5 shows r_t and Fig. 3.6 shows r_d at the midpoint of the hour. These figures are based on averages taken over many years and are intended to be used to find hourly average radiation. If used for individual days they work best if the day is clear, according to Duffie and Beckman (1991).

3.12 Example 3.3

Develop a table of the beam and diffuse solar energy intercepted each hour (solar time) of the typical June day by a horizontal solar cell at Massena, NY. The calculation will be done at the midpoint of each hour. We will display the procedure for the hour solar noon ± 0.5 h.

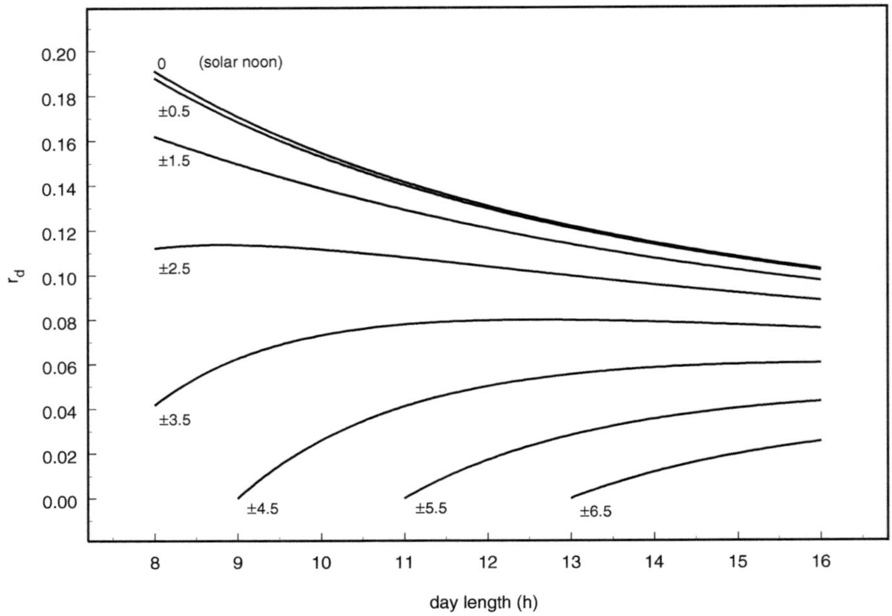

Fig. 3.6 Average hourly diffuse fraction

Solution From Example 3.2, L_a=44° 56′ N, L_o=74° 51′ W and $\bar{H}$=20,190 kJ/m². Knapp, et al. (1980) gives $\bar{K}_T$=0.481. (If $\bar{K}_T$ is not available, then calculate H_0 from Eq. (3.7) and find $\bar{K}_T$ from its definition, Eq. (3.9).) Using $\bar{K}_T$ in Fig. 3.4 gives $\bar{H}_d / \bar{H}$=0.448. Multiplying this ratio by $\bar{H}$ gives $\bar{H}_d$=9038 kJ/m². On the typical day, June 11, ω_S=115.3°. For the current hour, ω=7.5°. From Fig. 3.5 and the definition of r_t, $I=r_t\bar{H}$=(0.116)(20,190)=2313 kJ/m². The average total 9038 kJ/m². On the typical day, June 11, ω_S=115.3°. For the current hour, ω=7.5°. From Fig. 3.5 and the definition of r_t, $I=r_t\bar{H}$=(0.116)(20,190 kJ/m²)=2313 kJ/m². The average total rate of intercept during the hour is

$$\bar{Q} = \frac{2.323(10^6)\frac{\text{J}}{\text{m}^2}}{3600\,\text{s}} \times 7.433(10^{-4})\text{m}^2 = 0.48\,\text{W}.$$

From Fig. 3.6 and the definition of r_d, $I_d = r_b\bar{H}_d$=(0.105)(9038)=952 kJ/m². Finally, $I_b = I - I_d$=1361 kJ/m². Because the day length from Eq. 3.6 is 15.3 h, the table below shows the rest of the results from solar noon to ±7.5 h.

Hour (h)	r_d	I_d (kJ/m²)	r_t	I (kJ/m²)	I_b (kJ/m²)	$\bar{Q}$(W)
0.5	0.105	952.0	0.115	2313.0	1361.0	0.48
1.5	0.100	907.0	0.107	2166.0	1259.0	0.45
2.5	0.091	819.0	0.094	1892.0	1073.0	0.39
3.5	0.077	695.0	0.076	1528.0	833.0	0.32
4.5	0.060	543.0	0.055	1120.0	577.0	0.23
5.5	0.041	374.0	0.035	714.0	340.0	0.15
6.5	0.022	198.0	0.017	347.0	149.0	0.07
7.5	0.003	29.0	0.002	46.0	17.0	0.01

3.13 Beam and Diffuse on a Tilted Surface

Choice of Model Most radiation data is for a horizontal surface and therefore must be corrected for the orientation of the surface of interest. To do this, we neglect the circumsolar and horizon-brightening effects and regard the diffuse radiation from the sky as uniform in all directions. We further assume that the ground reflects a fraction ρ_G of the total horizontal radiation uniformly in all directions, and that we can ignore the atmosphere when correcting the beam radiation. This is the *isotropic diffuse* model reported by Liu and Jordan (1963). It is the most conservative of the models available in that it under-predicts the radiation by several percent.[3] However, it is the simplest to use.

Isotropic Diffuse Model When a surface, such as a solar cell, is tilted above the horizontal, it "views" both the sky and the ground. The radiation it intercepts thus is the sum of a beam radiation contribution, a sky-diffuse contribution, and a ground-reflected diffuse contribution. For a particular hour

$$I_T = I_{BT} + I_{DT} + I_{GT}. \tag{3.11}$$

The intercepted beam radiation, I_{BT}, is given by

$$I_{BT} = I_B R_B \tag{3.12}$$

where R_B is the beam tilt correction factor, the ratio of the radiation on a tilted surface to that on a horizontal surface at the same location and time. If the plate were horizontal ($\beta = 0$), the line perpendicular to its surface would be pointing at the local zenith, and θ would equal θ_Z. Thus

$$Q_{horizontal} = G_{0n} A \cos\theta_Z. \tag{3.13}$$

[3] See the comparison of several models in Duffie and Beckman (1991).

Then the beam tilt correction factor would be

$$\frac{Q_{\text{tilted}}}{Q_{\text{horizontal}}} = \frac{\cos\theta}{\cos\theta_Z} \equiv R_B. \tag{3.14}$$

I_{DT}, the intercepted sky-diffuse radiation, is

$$I_{DT} = I_D \frac{1+\cos\beta}{2}. \tag{3.15}$$

The factor containing cosβ is the fraction of the surface's view that is sky. I_{GD}, the ground-diffuse term, is

$$I_{GT} = I\rho_G \frac{1-\cos\beta}{2}. \tag{3.16}$$

The product $I\rho_G$ represents the radiation reflected uniformly in all directions by the ground. The last factor is the fraction of the surface's view occupied by ground. Because the surface can view only sky and ground, this is the fraction occupied by sky subtracted from 1. Applying the forgoing gives

$$I_T = I_B R_B + I_D \frac{1+\cos\beta}{2} + I\rho_G \frac{1-\cos\beta}{2}. \tag{3.17}$$

To find G_T at any moment we can use the same model but with I_B, I_D, and I_G replaced by G_B, G_D, and G_G.

3.14 Example 3.4

Calculate the average rate of solar energy interception by the cell of Example 3.3 for each hour of the average June day if the cell is tilted at an angle of 45°, pointed south, and the ground reflectivity is 0.2. The details of this calculation will be displayed for solar noon ±0.5 h. The beam and diffuse solar radiation components on a horizontal surface were found in Example 3.3.

Solution The information found in the first step of that example is used in Eqs. (3.3) and (3.4), along with the tilt angle above, to give $\cos\theta = 0.911$ and $\cos\theta_Z = 0.923$. From the definition of the beam correction factor, $R_B = 0.911/0.923 = 0.988$. Hence $I_{BT} = (0.988)(1361\ \text{kJ/m}^2) = 1345\ \text{kJ/m}^2$.

The collector views fractional amounts $(1+\cos 45°)/2 = 0.854$ of the sky and $1-0.854 = 0.146$ of the ground. Therefore, the sky-diffuse contribution is $(0.854)(952) = 813\ \text{kJ/m}^2$ and the ground-diffuse contribution is (0.146)(0.2)

(2314) = 68 kJ/m². The total amount intercepted during the hour is therefore 1345 + 813 + 68 = 2225 kJ/m². The average rate of intercept is

$$Q = \frac{(2.225\times10^{6}\,\mathrm{J})(7.443\times10^{-4}\,\mathrm{m}^{2})}{3600\,\mathrm{s}} = 0.46\,\mathrm{W}.$$

The table gives the results for the rest of the hour-pairs.

Note that the beam radiation intercepted by the plate for the hour-pairs after ±5.5 is zero. The local hour angle for these times exceeds 90°. The plate is pointed south and cannot "view" the sun at these hour angles. But why are the Q values less than those in Example 3.3?

n	R_B	I_{BT} (kJ/m²)	R_D	I_{DT} (kJ/m²)	R_G	I_{GT} (kJ/m²)	Q (W)
0.5	0.99	1345	0.85	813	0.03	68	0.46
1.5	0.97	1217	0.85	774	0.03	63	0.42
2.5	0.92	986	0.85	699	0.03	55	0.36
3.5	0.83	692	0.85	593	0.03	45	0.27
4.5	0.67	386	0.85	464	0.03	33	0.18
5.5	0.33	112	0.85	319	0.03	21	0.09
6.5	0	0	0.85	169	0.03	10	0.04
7.5	0	0	0.85	24	0.03	1	≈0

3.15 Hourly Typical Meteorological Year (TMY) Data

Definition A *typical meteorological year* (TMY) is one year of hourly weather and solar data that is representative of a weather station. Each datum is a measured, not an averaged, value, or if measurements are absent, it is generated from measured values. TMY (2008) reports the method used to generate the TMY's for 1020 weather stations; Wilcox and Marion (2008) did the same for the third set, named TMY3. Each TMY was constructed from 12 typical meteorological months chosen statistically from the weather data base for each station. In most cases this data base covered 23 years. TMY3 files are available for 1020 stations distributed over the continental USA, Alaska, Hawaii, Peurto Rico, Guam, and the US Virgin Islands.

Each TMY data set has 8760 hourly weather observations or ersatz constructions of different weather variables, including the dry bulb temperature, cloud cover, global solar radiation on a horizontal surface, and direct (beam) solar radiation on a horizontal surface.

TMY (1981) reports a study in which earlier TMY data for Madison, Wisconsin, were evaluated by simulating the annual performance of a solar heating system using both the hourly data for each of the years 1953–1974 and the TMY data. The results using TMY data were all within one standard deviation of the means of the results using the 22 years of data.

Habte et al. (2014) reports that "gridded" TMY data is available on the same 10-km by 10-km grid used for the National Solar Radiation Database.

Full TMY data sets may be obtained free from the National Renewable Energy Laboratory, Golden, Colorado. More conveniently sized TMY data sets with fewer than 24 variables may be obtained from the Solar Energy Laboratory, University of Wisconsin, Madison.

Beam and Diffuse Separation The radiation on a tilted surface for any hour in a TMY may be found from Eq. (3.17), once the total radiation has been separated into beam and diffuse components. These components may be found using statistical relations based on the hourly clearness index, k_T, defined as

$$k_T = \frac{I}{I_0}. \tag{3.18}$$

I_0 is calculated from Eq. (3.8). Correlations between I_D/I and k_T published by Orgill and Hollands (1977) and other workers (see Duffie and Beckman 1991) are in close agreement. The Orgill and Hollands correlation is plotted in Fig. 3.7.

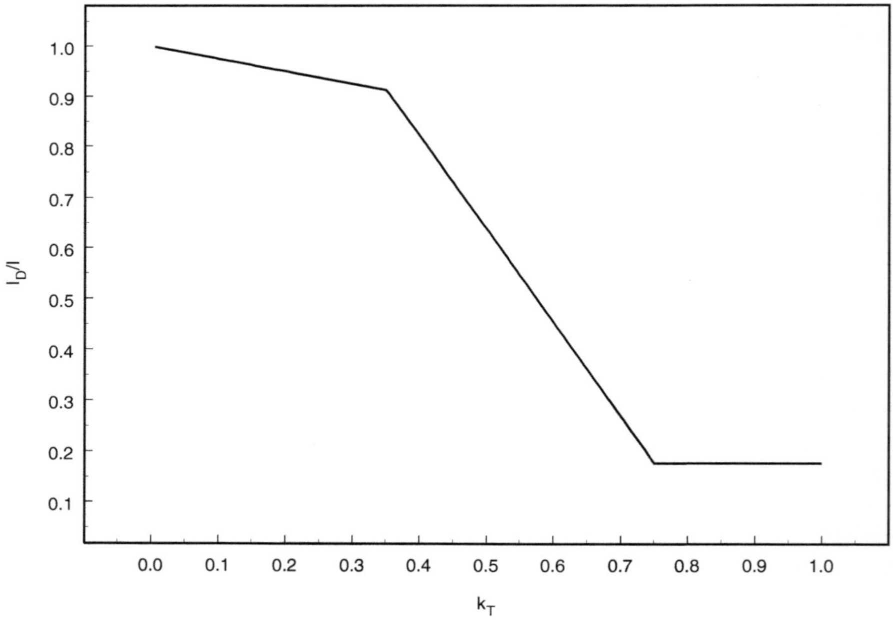

Fig. 3.7 The Orgill and Hollands (1977) correlation

3.16 Example 3.5

The TMY3 data file for the Massena, NY, airport shows a total, horizontal radiation of 750 W·h/m², or 2700 kJ/m², during the hour ending at 1 pm EST on June 11. Calculate the total horizontal solar energy and its beam and diffuse components and the total radiation intercepted by the solar cell of Example 3.4. Compare the results to those obtained in that example.

Solution The sunset hour angle and other relevant information have been found in Example 3.4. The beginning and ending hour angles for the interval of the calculation are $\omega_1 = 0^\circ$ and $\omega_2 = 15^\circ$. From Fig. 3.1, $G_{0n} = 1329$ W/m², and thus from Eq. 3.8, $I_0 = 4044$ kJ/m². Therefore $k_T = 2700/4404 = 0.613$. Entering Fig. 3.7 with this value gives $I_D/I = 0.44$. Consequently, $I_D = (0.44)(2700) = 1188$ kJ/m² and $I_{DT} = (0.854)(1188) = 1014$ kJ/m².

Then, $I_{BT} = (0.988)(2700–1014) = 1666$ kJ/m². Finally, the ground-reflected component is $(0.029)(2700) = 79$ kJ/m². The total on the tilted surface is $I_T = 1666 + 1014 + 79 = 2759$ kJ/m².

The average rate, calculated as in Example 3.4, is 0.57 W.

The following table contrasts results of Examples 3.4 and 3.5.

Example	I (kJ/m2)	I_{BT} (kJ/m2)	I_{DT} (kJ/m2)	I_{GT} (kJ/m2)	Q (W)
3.4	2313	1345	813	67	0.46
3.5	2700	1666	1014	79	0.57

The results are greater than those from Example 3.4. Q is 24 % greater. The results of Example 3.4 are based on long-term averages while those of Example 3.5 are derived from actual data, for the most part. If the TMY and long-term average data were used to find the total energy supplied to the cell over a longer period, such as a year, we would expect the results to be closer.

3.17 Transmission Through Glazing

Motivation We wish to estimate the solar heating of the cockpit and the electric power produced by the solar cells. Therefore, we must learn how to calculate the amount of solar radiation that is transmitted through a transparent material, such as the cockpit windshield or the glazing on the solar cells. Many authors have presented general, detailed developments of this topic, among them Siegel and Howell (1981), Duffie and Beckman (1991), and Hsieh (1986).

We will begin by considering beam radiation and include diffuse radiation after the main physical ideas have been established. Finally, we will consider radiation through a glazing system onto a planar absorbing surface, such as a solar cell, or into an enclosure, such as a cockpit.

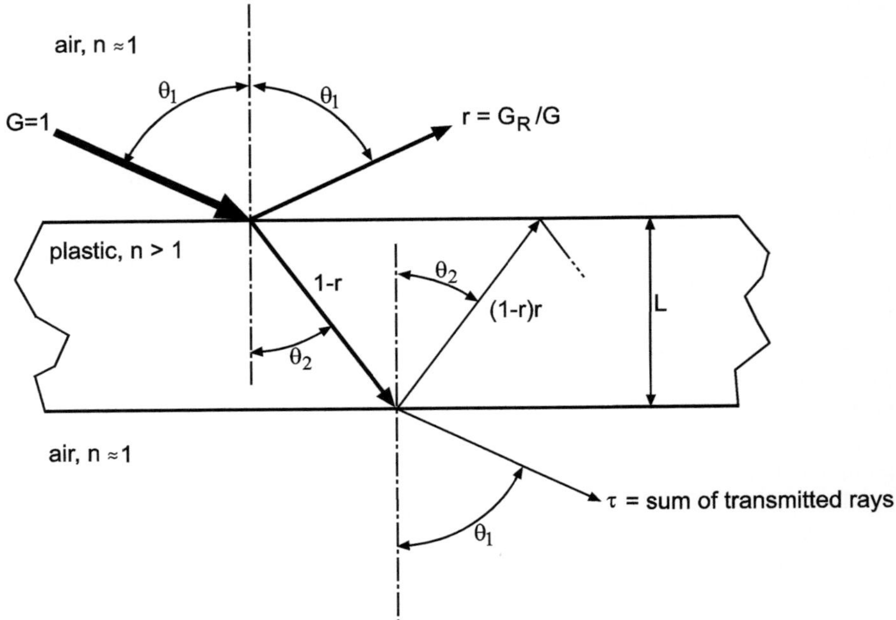

Fig. 3.8 Reflection, refraction, and transmission

Beam Radiation Figure 3.8 shows a ray of light traveling in air and then striking a glazing of thickness L at an angle of incidence θ_1. Some of the light is reflected, and some enters the glazing after being *refracted*; that is, bent toward the normal within the glazing. A portion of this light is absorbed or scattered from the beam by the glazing, some is reflected again at the lower face, and some is transmitted.

Index of Refraction The behavior of the solar radiation striking a glazing surface is a function of the *index of refraction* of the glazing material, the index of refraction of air in Fig. 3.8, and the angle of incidence. The index of refraction for non-electrically conducting materials, as glazings typically are, is:

$$n = \frac{c_0}{c} \tag{3.19}$$

where c is the speed of light in the material and c_0 is the speed of light in a vacuum. The speed of light in gases and solids is slower than that in a vacuum, so n is greater than one. The index of refraction is a function of wavelength. For example, the index of refraction of glass varies from 1.50 to 1.55 for wavelengths between 0.36 and 2.0 μm (Hodgman 1957). However, for design, we will ignore the wavelength dependence of n and instead use single values typical of the solar spectrum. Table 3.3 gives a table of these values for some transparent substances.

Table 3.3 Indexes of refraction

Material	N
Air	1.00
Water at 20 °C	1.33
Polycarbonate	1.60
Tedlar	1.45
Plexiglas	1.49
Glass (<0.01 % Fe_2O_3 "white")	1.53
Mylar	1.64

Snell's Law The wavefront in Fig. 3.8 is bent, refracted, toward the vertical, making an angle θ_2, smaller than θ_1, with it because n_2 is greater than n_1, i.e. c_2 is less than c_1. The second angle is called the *refraction angle* and can be predicted using *Snell's law*

$$n_1 \sin\theta_1 = n_2 \sin\theta_2. \tag{3.20}$$

Figure 3.9 shows how the refraction angle depends on n_2/n_1 and the incidence angle, according to Snell's law.

Reflection If the surface of the glazing is *specular* (mirror-like), it reflects a fraction r of the incoming radiation at an angle of reflection equal to the angle of incidence and contained in the plane established by the incoming ray and the normal to the surface, as shown in Fig. 3.8. Smooth glazings may usually be treated as approximately specular. In contrast, if a surface is *diffuse*, it reflects a fraction r of the incoming radiation uniformly over the hemisphere above it. The interior surfaces of a house are often covered with a dull-finish latex paint that is approximately diffuse. The reflection characteristic of most surfaces lies between specular and diffuse. That is, the reflected radiation may peak in one direction, but have components in all other directions too.

3.18 Transmittance

Reflection Loss As Fig. 3.8 implies, each time a ray encounters a change in the index of refraction, as it does at each glass–air interface, a portion of it is reflected, whether or not the ray is coming from the glazing or the air. Figure 3.8 shows that because of reflections, some of the radiation is not transmitted but leaves the glazing on the side on which it was originally incident.

Absorption Loss A portion of the radiation is removed from the beam by collisions with glazing material, and thus heats the glazing. Ignoring reflection losses, the

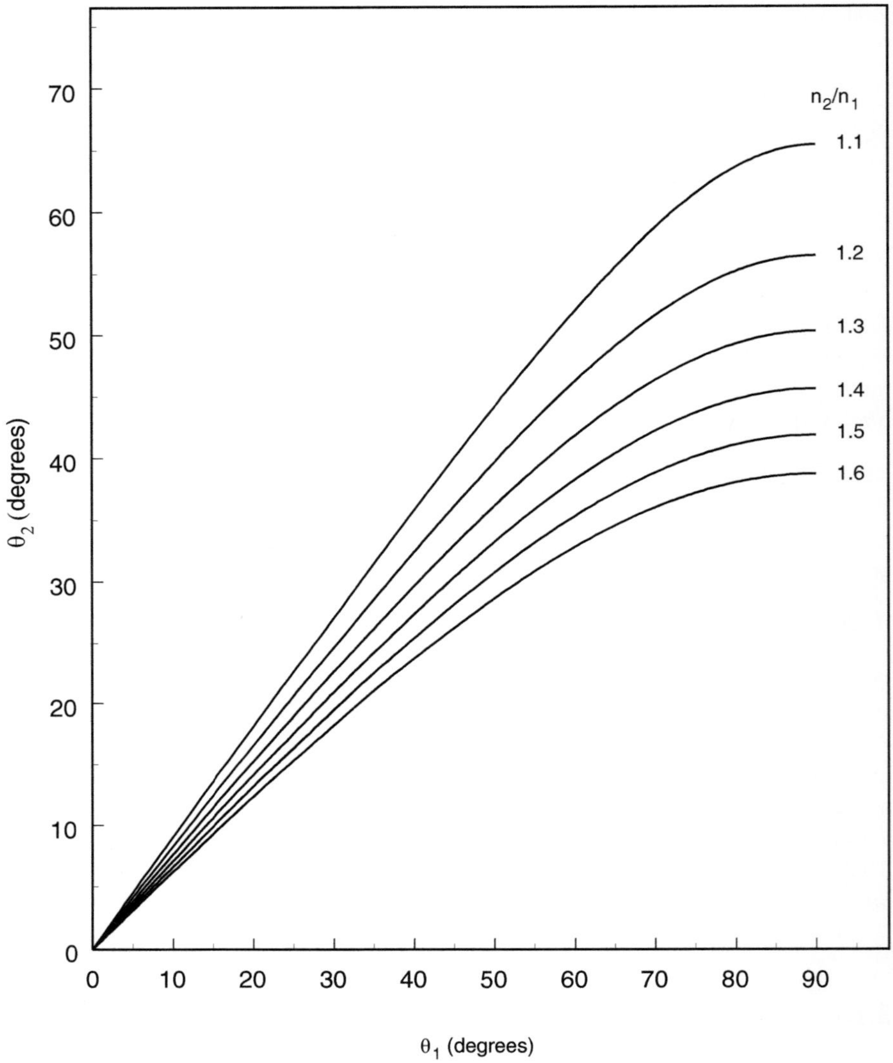

Fig. 3.9 Snell's law

fraction of the radiation entering the glazing layer that is not absorbed, the *absorption transmittance*, is[4]

$$\tau_a = e^{-\frac{KL}{\cos\theta_2}}. \tag{3.21}$$

[4] This relation is sometimes called Bouger's law.

Table 3.4 Extinction coefficients

Material	K (m-– 1)
Glass (< 0.01 % Fe_2O_3, "white")	~4
Glass (window, "green")	~30
Teflon	59
Glass (heat absorbing)	130–270
Tedlar	140
Polyethylene	165
Mylar	205

The *extinction coefficient*, K (m^{-1}), is a function of wavelength, and in general also of temperature and composition. The ratio $L/\cos\theta_2$ is the path length, S, of a ray through the glazing. As for the index of refraction, we will employ extinction coefficients averaged over the solar spectrum. Table 3.4 gives extinction coefficients for some transparent materials. This table shows that so-called "white" glass absorbs the least radiation. The thickness of an equivalent layer of polyethylene would have to be 4/165 times that of a layer of white glass to have the same optical thickness. On the other hand, high quality glass is more expensive, and glass is more brittle and more difficult to work with than plastic.

Transmittance Including Absorption The net transmittance through the glazing includes both reflection and absorption losses. Before showing how these effects depend on the angle of incidence, we will show a practical method for treating the transmission of diffuse radiation.

3.19 Diffuse Radiation

The material presented above is couched in terms of beam radiation. However, Brandemuehl and Beckman (1980) defined an *equivalent angle of incidence* for isotropic diffuse radiation as the angle of incidence of beam radiation having the same transmittance as isotropic diffuse radiation. They performed multiple calculations for the equivalent angle of incidence for one- and two-pane glazing systems with indices of refraction between 1.34 and 1.526 and KL products at or below 0.0524, tilted at various angles above the horizontal and irradiated diffusely. Figure 3.10 is a graph of the equations which best fit the equivalent angles of incidence for sky- and ground-diffuse radiation Brandemuehl and Beckman calculated.

3.20 Opaque Flat Plate Under a Cover System

Part of the radiation transmitted through the glazing over an opaque flat plate (which could be a solar cell) is absorbed in the plate and part is reflected to the glazing. Usually we are interested in the fraction of the radiation incident upon the glazing

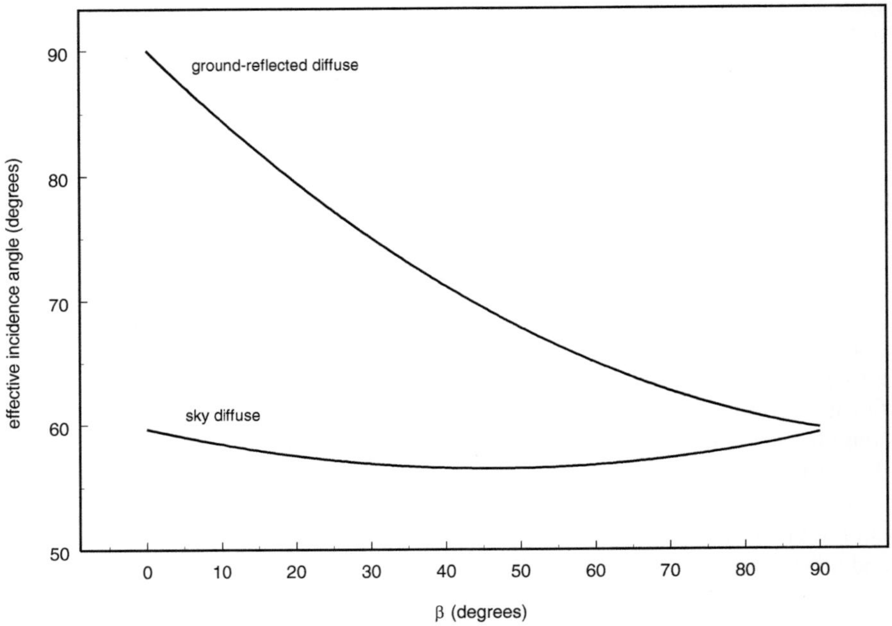

Fig. 3.10 Equivalent angles of incidence. (Brandemuel and Beckman 1980)

which is absorbed by the plate. This fraction, the *transmittance–absorptance product*, is given by Siegel and Howell (1981) as

$$(\tau\alpha) = \frac{\tau\sigma}{1-(1-\alpha)\rho_c} \tag{3.22}$$

where α is the absorptivity of the plate, τ is the transmittance of the glazing, and ρ_c is the reflectance of the glazing system for diffuse radiation from the bottom side. Figure 3.11 shows a plot of the transmittance–absorptance product for sets of glass covers ($n = 1.526$) over a plate with a diffuse absorptivity of 0.9 for solar radiation. The method of Whillier (1953) was used to account for two covers. Notice that at low angles of incidence, (τα) is approximately independent of the angle of incidence. Figure 3.12 shows the components of (τα). The reflectance is the dominant loss component at high incidence angles.

3.21 Glazed Enclosure

A glazed enclosure, such as a solar car cockpit, differs greatly from a glazed flat plate. The glazing may be curved, and the radiation transmitted through the glazing encounters surfaces of various orientations, shapes, and radiative properties,

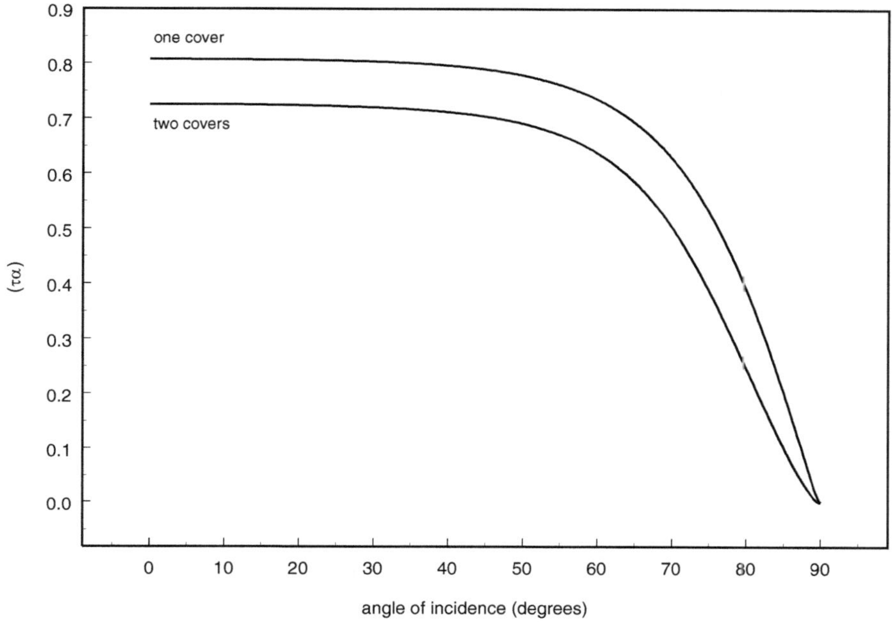

Fig. 3.11 Glazed flat plate transmittance-absorptance product

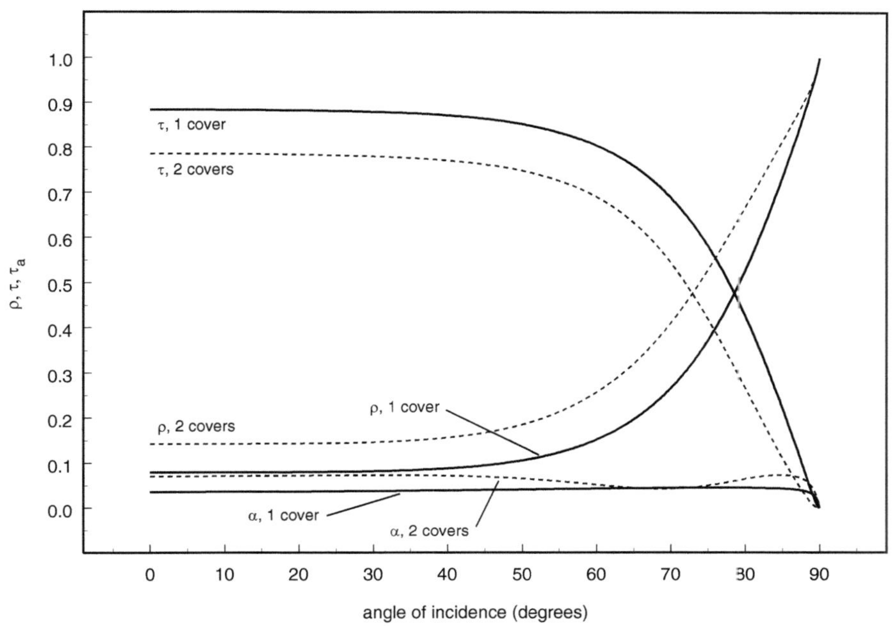

Fig. 3.12 Transmittance–absorptance product components

including the driver. Fortunately, our sole purpose in considering glazed enclosures is to estimate the heat gain in the cockpit from solar radiation. For an approach that avoids most of the complications but includes the major design elements, see Chap. 18.

3.22 Absorbed Solar Radiation

To calculate the absorbed solar energy, we must account for the angle of incidence, which is different for the beam, sky-diffuse, and ground-diffuse radiation components. Therefore, for each facet of the solar array glazing, each of the components appearing in Eq. 3.17 must be multiplied by a transmittance–absorptance product calculated for the applicable angle of incidence (using Fig. 3.10 to find the equivalent angles for the ground and sky diffuse) then the sum of these products is taken. For the *j*th array facet

$$G_{T_{abs\,j}} = [(\tau\alpha)_B G_{BT} + (\tau\alpha)_D G_{DT} + (\tau\alpha)_G G_{GT}]_j. \tag{3.23}$$

The subscripts B, D, and G refer to beam, sky-diffuse, and ground-diffuse radiation, as before. Note that

$$(\tau\alpha)_j = \left(\frac{G_{T_{abs}}}{G_T}\right)_j. \tag{3.24}$$

3.23 Example 3.6

Suppose the solar cell of Example 3.5 is irradiated such that $G_B = 650$ W/m^2 and $G_D = 125$ W/m^2. Imagine that the cell is covered with a single glazing with $L = 2.3$ mm, $K = 16.1$ m^{-1}, and $n = 1.526$ and that the cell's absorptivity is 0.9. Estimate the rate at which the cell absorbs energy just after solar noon.

Solution From Examples 3.3 to 3.5: $A_c = 8(10^{-4})$ m^2, $\theta = 24.4°$, $R_B = 0.988$, sky fraction viewed $(F_S) = 0.854$, ground fraction viewed $(F_G) = 0.145$, tilt angle $= 45°$, azimuth $= 0°$, and ground reflectivity $(\rho_G) = 0.2$.

Therefore, $G_{BT} = R_B G_B \approx (0.988)(650) = 642.2$ W/m^2, $G_{DT} = F_S G_D = (0.854)(125) = 106.8$ W/m^2, and $G_{GT} = F_G \rho_G G \approx (0.145)(0.2)(775) = 22.5$ W/m^2.

Fig. 3.10 gives $\theta_D \approx 57°$ and $\theta_G \approx 69°$. Entering Fig. 3.11 with θ and these equivalent angles of incidence gives $(\tau\alpha)_B \approx 0.81$, $(\tau\alpha)_D \approx 0.73$, and $(\tau\alpha)_G \approx 0.63$. Equation 3.23 then gives $G_{T_{abs}} = (0.81)(642.2) + (0.73)(106.8) + (0.63)(22.5) = 612.3$ W/m^2. Finally, $Q_{ABS} = A_C G_{T_{abs}} = 8(10^{-4})(612.3) = 0.50$ W.

3.24 Solar Cells

Introduction We have learned how to estimate the rate at which solar energy strikes the surface of a solar cell, having passed through glazing. In this section, we will learn how much of this intercepted energy the cell can convert into electric energy. We will also consider how a solar cell interacts with a resistive load. In Chap. 5, we will learn how solar cells interact with batteries and DC electric motors.

McCarney et al. (1987) is a good hands-on book on solar cells and Duffie and Beckman (1991) contains a chapter on practical ways of estimating solar cell performance. Hu and White (1983) is an advanced text on the subject. The material in Rauschenback (1980) is focused on the design of solar cells arrays for space applications. However, the book is very useful in gaining a theoretical and practical understanding of array design.

Production of Free Electrons Solar cells are thin, translucent wafers of special materials which, when exposed to sunlight and connected to a load such as an electric motor, produce a unidirectional electric current (DC or *direct current*). The material most commonly used to make solar cells is silicon, which is found in beach sand. How is the electric current produced and at what voltage?

Solar radiation may be thought of as traveling in small bits called *photons*. Photons move at light speed and their energy is directly proportional to the radiation's frequency; they could be thought of as "light bullets." When a bullet strikes something, say an apple on a tree, it may have enough energy to do the work necessary to break the stem of the apple and remove it from the tree. Suppose we think of the electrons in a silicon atom as the "apples." They are bound to the "tree," or nucleus, by a "stem" made of the electrostatic attraction between the positive nucleus and the negative electrons. If a photon with sufficient energy (high enough frequency) strikes an electron, the electron will be freed from the atom. The vacant electron orbit is called a *hole*.[5] The silicon atom then has a net positive charge equal to the magnitude of the electronic charge and is called an *ion*. The positive charge may be thought of as belonging to the hole.

The quantity of electrons liberated per second, the ionization rate, is proportional to the number of photons of energy sufficient to ionize the silicon absorbed by the cell per second. Thus, for a given cell, the ionization rate is proportional to the irradiation times the area of the cell.

Charge Separation A way of separating the charges must be supplied so that the electrons may be collected and forced to flow through an external circuit and do work before they return to the cell and recombine with the holes. Otherwise, the free electrons will wander until they find holes and fall in, or *recombine*, becoming bound to atoms again.

Charge separation is produced by placing *p-type* silicon, in which the majority of charge carriers are positively charged holes, on one side of a very thin junction

[5] The production of electron-hole pairs by radiation is called the photoelectric effect.

region and *n-type* silicon, in which the majority of charge carriers are negatively charged free electrons, on the other. These two kinds of silicon are produced by putting specially selected impurities into the pure silicon by a process called *doping*.

An electron-diffusion pressure exists on the *n*-type side of the junction because there are more electrons on that side than on the *p*-type side. The pressure causes electrons to diffuse across the junction into the *p*-type material. Similarly, holes diffuse across into the *p*-type material. The diffusion continues until a local net positive charge builds up on the *n*-type side and a local net negative charge builds up on the *p*-type side. This creates an electric potential barrier (voltage) which opposes the diffusion pressure of both species and eventually stops diffusion.

Suppose the *n*-type material is on the illuminated side, the top. Thin metal strips called *electrodes* are bonded to the top (they have to be sparse because this is the illuminated side) and the back of the cell is completely *metallized*, or covered by its electrode. Figure 3.13 illustrates this structure.

When the cell is illuminated, the free electrons in the *p*-type material are swept into the *n*-type material by the barrier potential and join any free electrons already on that side and flow to the top electrodes, which become negatively charged. Because the holes have positive charges, they are swept in the *opposite direction* by the barrier potential to the bottom electrode, which becomes positively charged. Thus the charges are separated. Just as in the case of the junction barrier potential, this migration continues until enough charge has been collected on the electrodes to prevent further migration (a very rapid process). Further ionizations are balanced by recombinations. Figure 3.13 shows this open circuit state.

Open Circuit Voltage The voltage thus produced across the electrodes is called the *open circuit voltage* (V_{oc}) because this is the voltage that would be measured by a voltmeter connected to the terminals of a cell when no load, such as a resistor, is connected. V_{oc} is the largest voltage that the cell can produce. V_{oc} depends upon the material of the cell and decreases as the cell's temperature increases. For silicon at 25 °C, V_{oc} is about 0.6 V, and decreases at about 0.002 V/°C as the cell's temperature increases (Hu and White 1983). V_{oc} increases nonlinearly with irradiation; a tenfold increase would increase it about 10 %.

Short Circuit Current If the positive and negative terminals of the cell are connected together, the short circuit current (I_{sc}), the largest current that the cell can produce, will flow. This current can be measured by connecting an ammeter across the terminals of the cell. I_{sc} is directly proportional to the intensity of the irradiation, or for a given irradiation it is directly proportional to the area of the cell. The proportionality constant for a unit area cell is called the *sensitivity* (S_{sc}, A/W). If you purchase a cell, the V_{OC} and I_{SC} for a cell temperature of 25 °C and irradiance (normal to the cell) of 1000 W/m^2 will usually be given on a data sheet accompanying the cell.

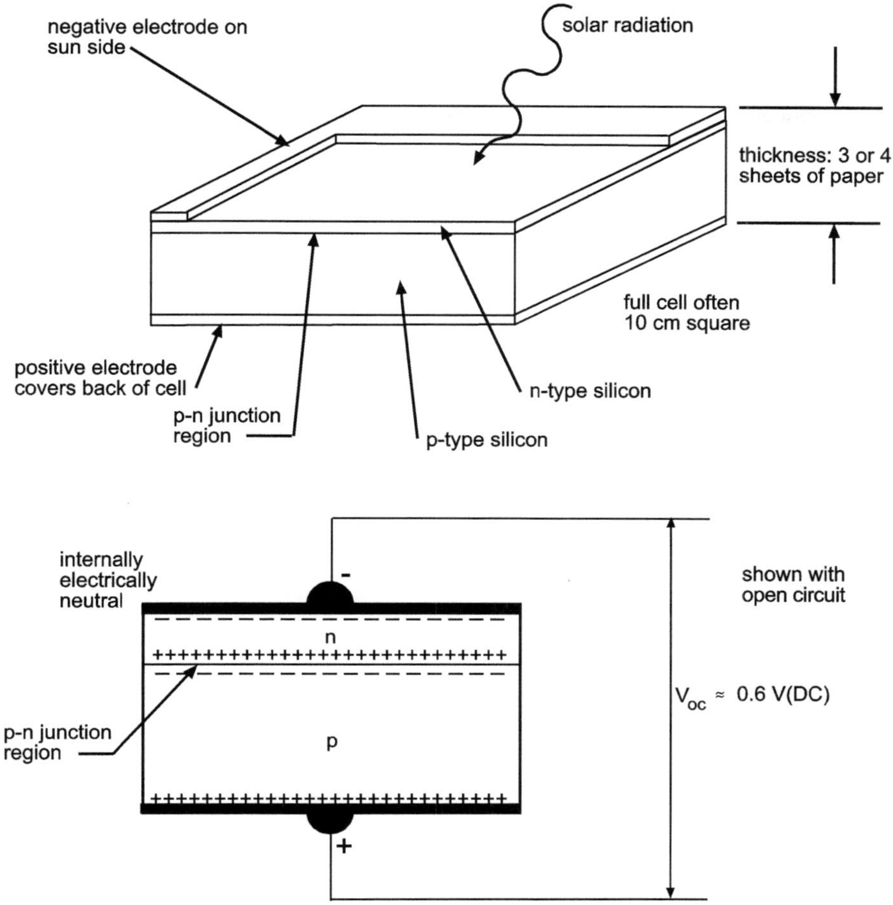

Fig. 3.13 A solar cell

3.25 Example 3.7

A solar cell with an area of 100 cm^2 has an I_{sc} of 3.0 A and a V_{oc} of 0.6 V at the standard rating conditions of 1000 W/m^2 solar irradiation and a cell temperature of 25 °C. Calculate the sensitivity of this cell and estimate I_{sc} under indoor lighting conditions of 300 W/m^2.

Solution The sensitivity is 3.0 A/(1000 W/m$^2 \times$ 0.01 m^2), or 0.3 A/W. The indoor lighting's spectral distribution is not the same as that of solar radiation, but we assume that the difference is not significant. Hence, I_{sc} is directly proportional to the radiation level. So

$$I_{SC} = 0.3\frac{\mathrm{A}}{\mathrm{W}} \times 300\frac{\mathrm{W}}{\mathrm{m}^2} \times 0.01\mathrm{m}^2 = 0.9\mathrm{A}$$

is our estimate.

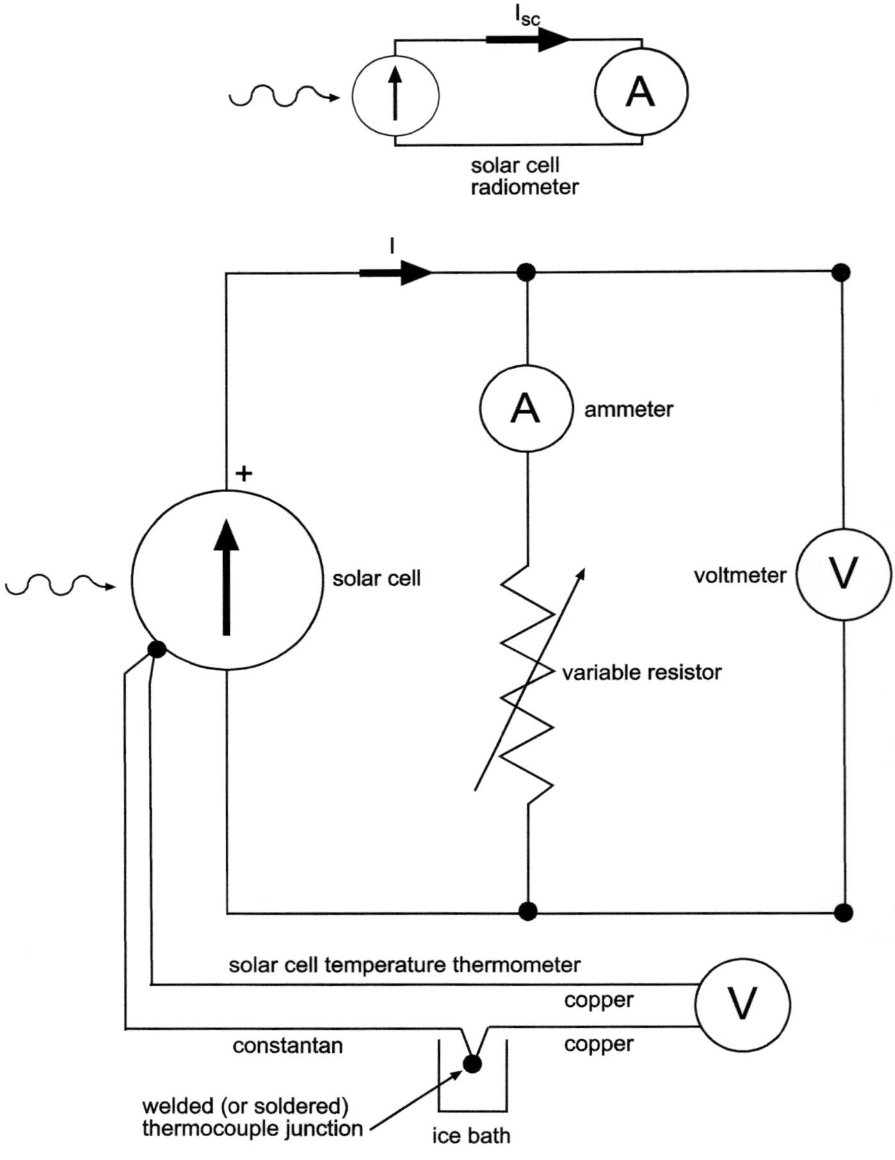

Fig. 3.14 *I–V* curve measurement

Current–Voltage Characteristic Suppose that we connect a variable resistor, a voltmeter, and an ammeter as shown in Fig. 3.14. Keeping the irradiance constant, we increase the resistance from a low value to a very high value, reading the ammeter and voltmeter after each change. We then plot the measurements with the current on the vertical axis and the voltage on the horizontal axis. The resulting curve, called

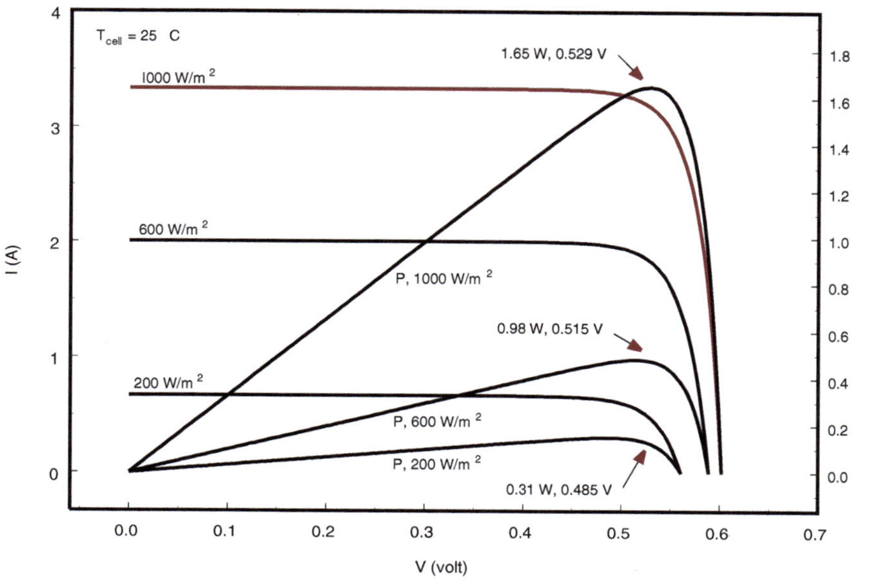

Fig. 3.15 Solar cell *I–V* curves

a *current–voltage characteristic*, or an *I–V characteristic*, appears in Fig. 3.15 for three different radiation levels. This information may also accompany a purchased cell.

Power–Voltage Characteristic The power (P) delivered to the resistor is equal to the product of the voltage and current. If we plot the power as a function of voltage we get peaked curves like those shown in Fig. 3.15. These are zero at both the short circuit and open circuit points and have a maximum at the knee of the corresponding *I–V* characteristic. The efficiency of the cell is

$$\eta_{\text{cell}} = \frac{P_{\text{cell}}}{GA_{\text{cell}}}. \tag{3.25}$$

The cell's efficiency is highest at the maximum power point. At standard rating conditions, the maximum power point of a silicon cell is at 0.53 V, approximately. However, note the reduction in V_{oc} and the shift in the maximum power (and efficiency) point toward lower voltages as the irradiation decreases.

Figure 3.16 shows a family of *I–V* characteristics each at a different cell temperature and irradiation of 1000 W/m_2. The reduction in V_{oc} with increasing cell temperature is evident. Note the decrease in efficiency caused by the increase in cell temperature.

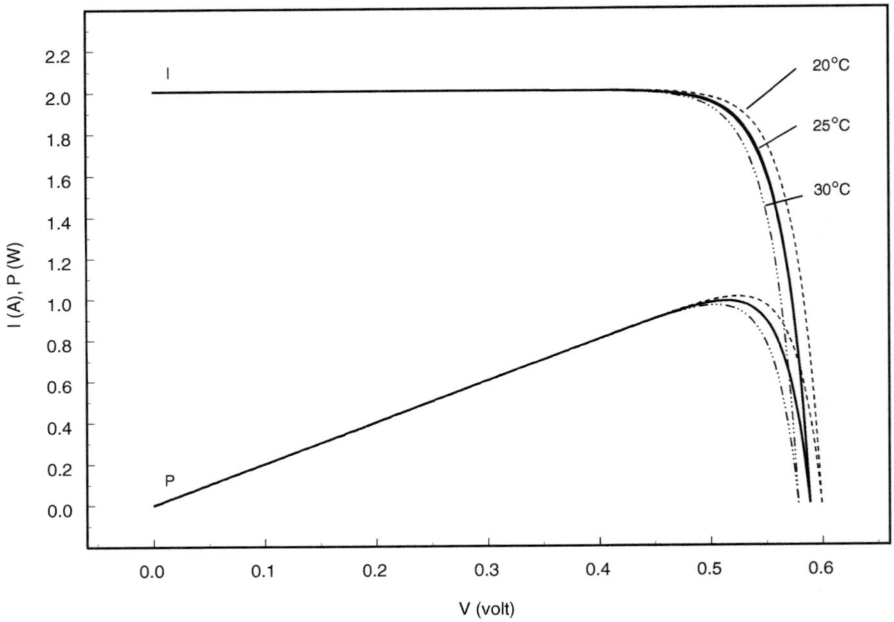

Fig. 3.16 Effect of cell temperature at constant irradiance

3.26 Solar Cell Arrays

Increasing Voltage Operating at a fraction of a volt will not do. We will need to have voltages on the order of 100 V. How can this be done? Suppose we measure the V_{oc}'s of two identical cells under the same conditions using the previously outlined method. Then, we connect the positive terminal of one cell to the negative terminal of the other cell. This circuit, shown in Fig. 3.17, is called a *series* connection because the current flows through each cell in sequence. Then, if we connect a voltmeter between the positive terminal of the second cell and the negative terminal of the first cell we read twice the V_{oc} of a single cell. Replacing the voltmeter with an ammeter to measure I_{sc}, we find it to be the same as for a single cell. The voltages of cells in series add, but the current through each cell is the same.

Increasing Current A 100-cm^2 cell produces about 3.0 A at standard conditions. This is respectable, but we will need more, perhaps four times that amount. How do we do it? Suppose instead of the series connection above, we connect the positive terminals of the cells together and the negative terminals of the cells together. This is called a *parallel* connection because the currents in each cell flow side-by-side. Now we measure I_{sc} by connecting an ammeter between the joined positive terminals and the joined negative terminals, as shown in Fig. 3.17. The meter reads twice the current of a single cell. But if V_{oc} is measured with the voltmeter it is the same as for a single cell. The currents of cells in parallel add, but the voltage across each cell is the same.

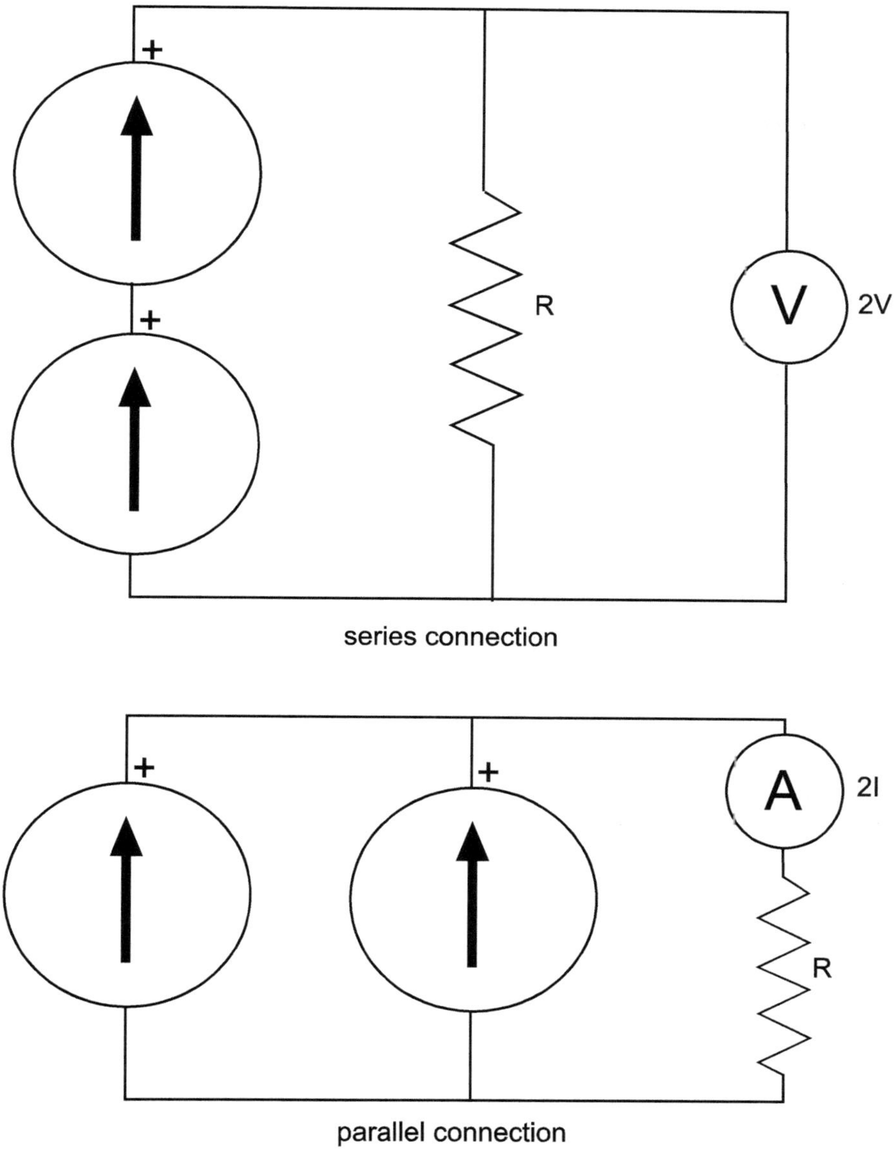

Fig. 3.17 Series and parallel connections

Increasing I and V, if we want more voltage and also more current, we can combine both series- and parallel-connected cells. For example, series strings could be connected in parallel, instead of just single cells. A solar cell circuit containing more than one cell is called an *array*.

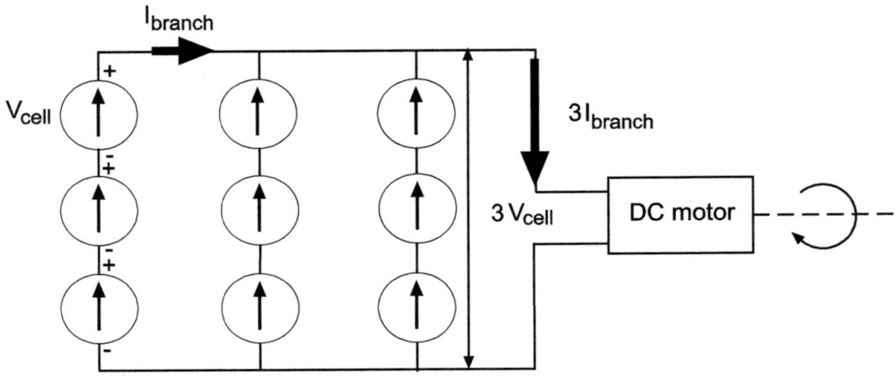

Fig. 3.18 A solar cell array

3.27 Example 3.8

It has been estimated that a small, DC electric motor with which we will power a model car requires 1.5 VDC at 0.3 A (0.45 W) to move the car at 1.0 m/s across a high school gym floor. The irradiance on the gym floor is 300 W/m^2. Small, 15-cm^2 silicon solar cells, purchased at a local electronic parts store and having the same S_{sc} as the cells of Example 3.7, will power the model car. We wish to design an array to power the car.

Solution Our cells can produce 0.6 V, maximum. So we must string some cells in series to get 1.5 V. This number in series will be greater than 1.5/0.6, or 2.5. Round this up to 3, because only whole cells are possible.

One cell will yield I_{sc}=(0.3 A/W)(300 W/m^2)(0.0015 m^2)=0.135 A., maximum, under the gym's lighting conditions. The string of three cells in series produces the current of one cell in the string. It will be necessary to connect some three-cell strings as parallel branches to get the current we need. The number of branches will be greater than 0.3 A/0.135 A/string=2.22. Round this up to three. There are three branches with three cells per branch, so nine cells are required. Figure 3.18 shows how the cells are connected.

Array I–V Characteristic An array may be regarded as one big cell with an *I–V* characteristic that is a composite of the *I–V* characteristics of the cells in it. In the case of the array of Example 3.8, this composite *I–V* characteristic has an I_{sc}=0.405 A and V_{oc}=1.7 V in the gym lighting, as shown in Fig. 3.19[6]. For comparison, the *I–V* curve of a single cell is also shown. The array's maximum power point voltage, V_m, is 1.46 V and the corresponding current, I_m, is 0.39 A.

[6] Note that the *V*oc is less than three times that for standard conditions. A logarithmic dependence on the irradiation causes the open circuit voltage to drop at low irradiances.

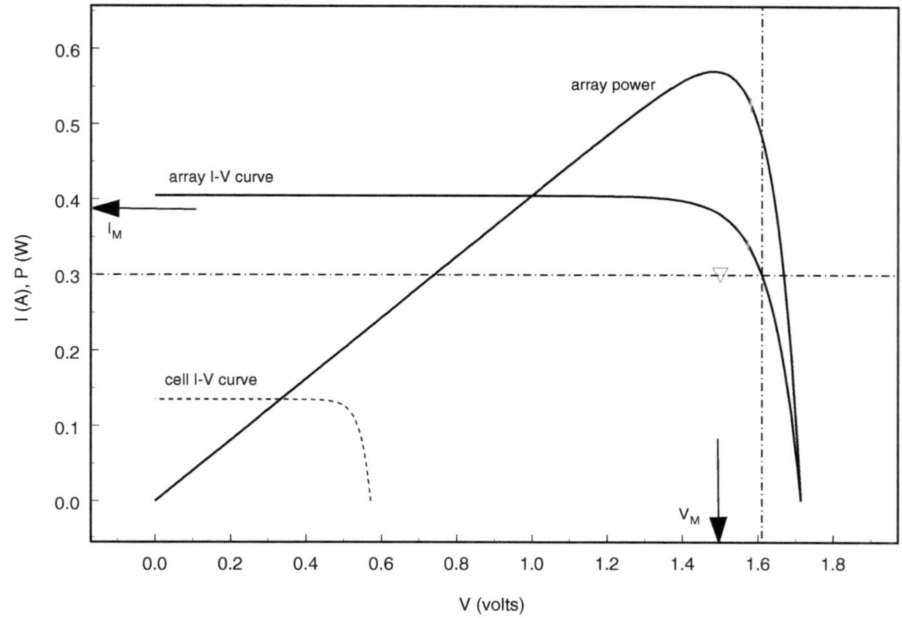

Fig. 3.19 Small array *I–V* curve in gym lighting

Observe that the desired motor operating point, 0.3 A at 1.5 V (marked by a triangle in the figure), does not exist on the array *I–V* curve. The torque (or tractive force) required by the motor to drive the model car depends only on the desired speed, 1.0 m/s in this case, once the car design and the operating conditions are fixed. The torque produced by a particular motor is directly proportional to the motor current. Therefore, the current to the model must be 0.3 A. However, as shown in the figure by the broken lines, at this current the array produces about 1.6 V, not 1.5 V. The equivalent resistance of the motor at its desired operating point is 5 Ω (1.5 V/0.3 A), whereas the array requires a load resistance of 5.33 Ω to deliver 0.3 A. If no other action were taken, the model's speed would change until the motor's operating point lay on the array's *I–V* curve. This would result in a small speed increase, if the motor's equivalent resistance were approximately constant. If a 0.33-Ω resistance were placed in series with the motor the operating point of this composite load would lie on the array's *I–V* curve at the intersection of the broken lines. The model would then operate at 1.0 m/s, as desired. Chapter 5, *Electric Motor Drives*, contains more discussion about this kind of interaction between the array and the drive of a solar-electric car.

The wires that connect the cells together in strings add resistance in series with the cells. This additional resistance is heated by the electric current passing through it and thus consumes power, reducing the array voltage at a given current, and thus power available to the load. The array *I–V* characteristic may be regarded as a composite of the *I–V* characteristics of its cells (as stated above), but the efficiency is reduced by the additional series resistance of the interconnecting wires. This interconnection loss was not included in the curves of Fig. 3.19.

In Example 3.8, since the terminal voltage is just above 1.5, we might have considered adding an additional cell in each series string to make sure that connection losses do not reduce the array's output voltage below 1.5 V. On the other hand, this might add $ 10 to the cost of the model solar car. This demonstrates again one of the generic characteristics of design, pointed out in Chap. 1: the requirements of a design are often conflicting. In this case, lower cost and increased performance cannot both be attained.

Array Diodes Suppose that one of the wires connecting a cell in an array string on a solar car were broken by a rock thrown up by the car ahead, opening the circuit. The string's output would immediately drop to zero. Is there a way of protecting the array against such happenings, or against lesser, but more likely, events such as the shading of a cell, or part of a string? There is, using devices called *diodes* which pass current in their *forward biased direction* with little resistance but interpose a very high resistance to the passage of current in their *reverse biased direction*. Figure 3.20 shows a string with a diode connected in parallel with each cell. The arrow in the diode symbol shows the forward direction. Ordinarily, the diode would be reverse biased by the cell's voltage. However, if the cell's circuit were to open, this bias would be lost. Current from the other cells in the string would then flow around the open circuit through the diode in the direction of the arrow, so it is called a *bypass diode*. The string would still function, but with its voltage reduced by about 0.5 V. Contrary to Fig. 3.20, bypass diodes are usually connected across several cells in a string. It is impractical to bypass every cell.

Diodes also keep string performance up when a cell, or cells, is shaded. Under these conditions, the maximum current of the string must be less than the short circuit current of the shaded cell, which of course could be quite small. As the voltage

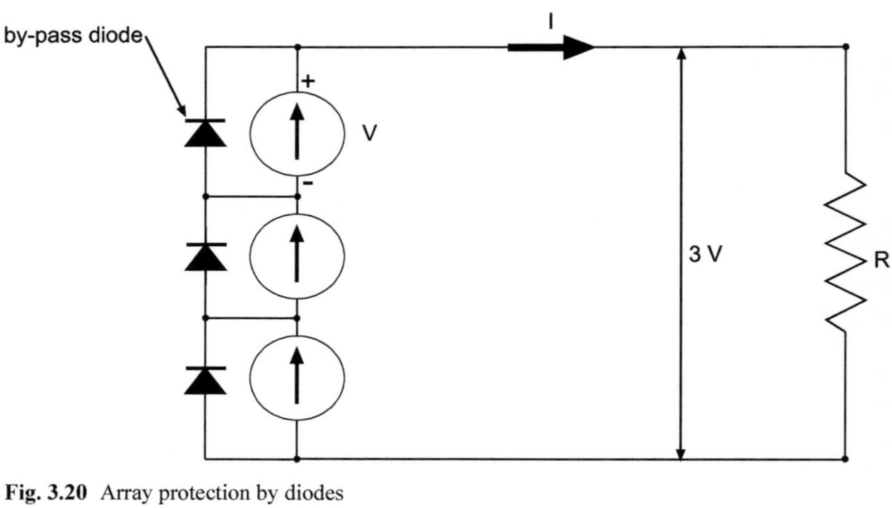

Fig. 3.20 Array protection by diodes

of the bypassed cell or substring drops at the lower illumination, the reverse bias on the diode drops, allowing current to flow around the shaded cell or substring.

Area Efficiency It will usually not be possible to completely cover the area allocated to the array by race rules with cells. This may be caused by spaces between the cells left for thermal expansion. Or, the need to lower aerodynamic drag may cause a non-rectangular array surface. Laying down strings of rectangular cells on a non-rectangular surface results in unused area. Lack of complete area coverage is expressed by an *area efficiency*

$$\eta_{\text{area}} = \frac{A_{\text{cells}}}{A_{\text{array}}}. \tag{3.26}$$

A_{cells} represents the actual cell area; A_{array} represents the area of the cells plus unused space. Clearly, it is desirable to have area efficiency close to 1.0; efficiencies greater than 0.95 are achievable through careful design.

Load Line A resistor operating at a constant temperature has a constant resistance, R. The quotient of I/V, where I is the current through the resistor and V is the applied DC voltage, is $1/R$. On I–V coordinates, $1/R$ is the slope of a straight line passing through the origin: the I–V characteristic of the resistor, or its *load line*.[7] Electric motors and storage batteries can also be represented by load lines. The interaction between the solar array, battery, and motor of a solar car will be discussed using these load lines in Chap. 5.

If the resistor is the load for a solar array, the operating point, the current and voltage, of the array-load system must lie on the array's I–V characteristic and also on the resistor's load line. Therefore, the intersection of the load line and the I–V characteristic establishes the operating point graphically. This is illustrated in Fig. 3.21, which shows three load lines: high load (low R), optimal load ($R = R_{\text{m}} = V_{\text{m}}/I_{\text{m}}$) and low load (high R). The product of the voltage and current supplied to each load, read from the intersections with I–V curve, show the power is less in each case than when the load resistance is R_{m}.

Changes in temperature, irradiation, or the load can move the operating point away from the maximum power point. What can be done about this?

3.28 Maximum Power Point Tracking

Suppose an electric circuit that keeps the array (or segment of the array to which it is connected) operating at the array's maximum power point was connected between the array and the load. Then, the array would always convert the available solar irradiance into the maximum possible power. Such a device is called a *maximum power point tracker* (MPPT).

[7] The term *load* usually means the current drawn by the device being powered, although it can also mean the power drawn.

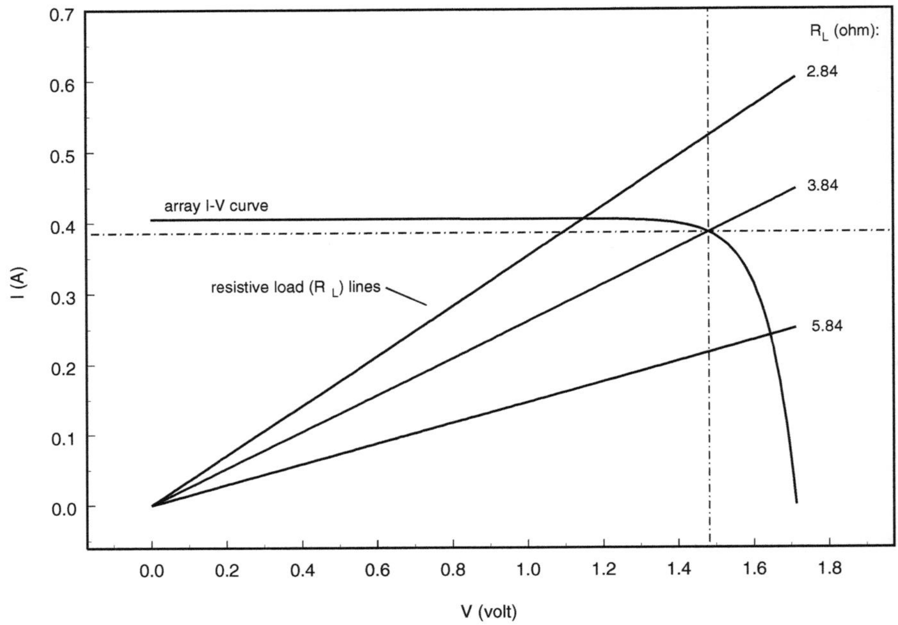

Fig. 3.21 Resistive load lines

Application In a typical solar car application, an MPPT would be connected to each string of a series-parallel array, as shown in Fig. 3.22. Each string, especially on a streamlined car, will have a different irradiance and a different temperature compared to the average irradiance and temperature of the array. Matching strings to the main bus yields a higher array efficiency than matching the entire array through one MPPT. Balancing this are cost and weight. Commercially built MPPT's are expensive, often several hundred dollars each. One per string will be a large investment. Also, some MPPT's are relatively heavy, weighing in the range of 2–4 lbf. One per string could increase the car's weight by 8–16 lbf, on a four-string array.

Operation DC–DC converters play the same role in DC circuits that transformers play in AC circuits: they transfer power between circuits that use different voltages. An MPPT is a specialized DC–DC converter that transfers power between an array string and the main electric bus, to which the battery and the motor are also connected. The MPPT keeps the string at its current maximum power point voltage while transferring power to the bus at the current bus voltage. As in AC transformers, some power is lost in the transfer. However, the efficiency of well-made MPPT's can lie in the 98–99 % range (Storey et al. 1994).

The maximum power point may be sensed by very rapidly moving the array voltage around the optimum value while observing the effect on the output power. A correction is then applied to the array voltage in the proper direction to increase the output power, if required. Another method is to assume that the optimal voltage

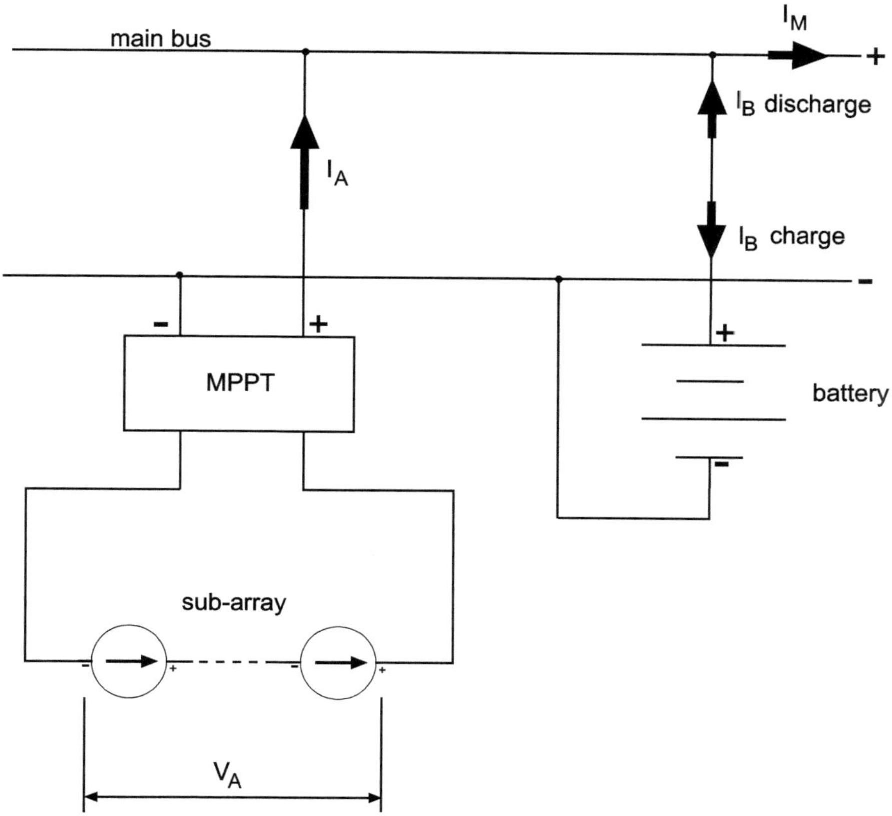

Fig. 3.22 Maximum power point tracker installation

is a fixed fraction of the open circuit voltage. The string is momentarily switched to open circuit to find V_{oc}. V_m is then obtained from a voltage divider. This method was proposed by Schoeman and van Wyk (1982). Storey et al. (1994) report that one 1993 World Solar Challenge team used 26 such units. This method perhaps lends itself to home-built, low-weight, low-cost units.

Types Two types of DC–DC converters are of interest: boost regulators and buck regulators. Boost regulators match a lower string voltage to a higher bus voltage and buck regulators match a higher string voltage to a lower bus voltage. Boost regulators may have a slightly lower efficiency than buck regulators. They can supply current as the array maximum power point voltage drops with increasing temperature or decreasing radiation; however, in the absence of a current load, the output voltage will rise to the point of destruction of the device, unless additional protective circuitry is included.

Reliability The reliability of the MPPT's must be high. If an MPPT fails, its string fails, cutting the power by a large fraction. This will require roadside repairs, such as replacing the MPPT or adding the orphaned string to the MPPT of another string. In either case, precious time will be lost.

Build or Buy Storey et al. (1994) reports that 10 of the 52 cars finishing the 1993 World Solar Challenge used homemade MPPT's. Building your own MPPT's is attractive because it is cheaper than buying commercial units and, if they can be built as an educational project, because of the experience it gives the student builders. However, commercial MPPT's have presumably been extensively tested and some brands are race proven. It may be difficult for students to design, build, and *adequately test* home-built MPPT's during the race cycle time, even in the 2-year span between Sunrayces.

3.29 Array Temperature Distribution

Flow Conditions When there is no relative motion between the car and the air, the cells in its array will be at their highest temperatures.[8] Most of the solar energy absorbed in a cell is not converted into electrical energy and consequently heats the cell. This heat is removed by radiation and *free convection* to the surroundings. The latter term refers to the buoyancy-driven motion of the air in contact with the cell. This air is hotter than the ambient air and rises from the car; cooler air from the surroundings takes its place. This circulation carries away heat.

When the car is in motion, heat is removed by radiation and *forced convection.* This is a continual flow of cooler air across the array, mostly from front to rear, forced by the relative motion between the car and the ambient air. Forced convection cools the array more effectively than free convection because the speed of the air is greater. At very low vehicle speeds, as in stop-and-go traffic, there will be a combination of free and forced convection. At higher speeds forced convection dominates.

Energy Balance At steady conditions, the cell's temperature rises until the rate of energy absorption is balanced by the combined rates of energy loss (radiation plus convection), and energy conversion. There will be a *temperature distribution* over the array because this balance between energy supply, loss, and conversion will be different for different positions in the array. For example, cells on the sunward side of the car will tend to be hotter than those on the opposite side, which see only diffuse irradiation. For a particular cell, denoted by the subscript "c," this balance between rates of energy transfer may be expressed as

$$G_T A_c(\tau\alpha) - U_c A_c(T_c - T_a) - P_c = 0. \tag{3.27}$$

[8] If the cells are not actively cooled, as by a water spray.

The first term denotes the solar radiation intercepted by the cell that has passed through the glazing and been absorbed by the cell. The second term represents the total thermal loss: conduction through the glazing then radiation and convection to the surroundings. Conduction from the back of the cell through the supporting surface is neglected. This loss is modeled as proportional to the difference between the cell's temperature, T_c, and the ambient temperature, T_a. The proportionality factor, the *overall heat transfer coefficient* U_c ($W/m^2 \cdot C$), incorporates all the loss modes. Because they are temperature-dependent, so is U_c. The final term, P_c, represents the electric power produced by the cell, which is also temperature dependent, as we have seen.

Temperature Relation The builders of a solar car apply the glazing to the cells while assembling the array. Hence, the *I–V* curves of the cells they use have been determined for the as-purchased cells with no glazing. In these circumstances, the first term of Eq. (3.27) replaces $G_T A_c$ in the denominator of Eq. (3.25), the definition of the cell efficiency. With this modification, we may use the definition of the efficiency to eliminate the cell power from Eq. (3.27):

$$G_T A_c(\tau\alpha)(1-\eta_c) - U_c A_c (T_c - T_a) = 0. \tag{3.28}$$

Note that this energy balance implicitly assumes that the cell is at a uniform temperature and that it stores no thermal energy (the energy stored would replace the zero). The cell is small, 10-cm on a side or smaller, and also quite thin, so that we ignore the temperature distribution over its face and across its thickness. The thermal energy it stores is proportional to the cell's mass. We neglect it because the mass is very low.

Solving Eq. (3.28) for T_c gives

$$T_c = T_a + \frac{G_T(\tau\alpha)(1-\eta_c)}{U_c}. \tag{3.29}$$

A cooler cell is more efficient. Reduction of the ratio on the right of T_a accomplishes this. U_c increases with the speed of the car, as mentioned previously. U_c could also be increased by arranging additional cooling under the mounting surface upon which the cell is mounted. However, schemes to do this usually cause additional drag and may also increase the weight of the car.

3.30 Example 3.9

Estimate the temperature of the solar cell of Example 3.6 if the solar car on which it is mounted is stopped in traffic in still air and the ambient temperature is 30 °C.

Solution Under these conditions, there is no forced flow over the array to keep the cell cool and the cell's temperature can become high. Suppose a conversion efficiency of 10% results and the loss coefficient is 15 or 20 W/m² K. The glazing is thin, so this value consists almost entirely of free convection and radiation from the glazing's surface. The back of the cell is assumed perfectly insulated by the array support structure. From Eq. (3.29),

$$T_c = 30^\circ\text{C} + \frac{612.3\dfrac{\text{W}}{\text{m}^2 \times K}(1-0.1)}{20\dfrac{\text{W}}{\text{m}^2 \times K}} = 52.6^\circ\text{C}.$$

This is 126.7 °F. For 15 W/m²AK the result is 66.7 °C, or 150 °F. Although representative values of the efficiency and the loss coefficient were assumed above, in reality they must be calculated by trial and error because both depend upon the temperature of the cell.

References

Angrist, S. W. (1982). *Direct energy conversion*. Boston: Allyn and Bacon, Inc.

Benford, F., & Bock, J. E. (1939). A time analysis of sunshine. *Transactions of the American Illuminating Engineering Society, 34,* 200.

Brandemuehl, M. J., & Beckman, W. A. (1980). Transmission of diffuse radiation through CPC and flat-plate collector covers. *Solar Energy, 24,* 511.

Collares-Pereira, M., & Rabl, A. (1979). The average distribution of solar radiation—correlations between diffuse and hemispherical and between daily and hourly insolation values. *Solar Energy, 22,* 155.

Cooper, P. I. (1969). The absorption of solar radiation in solar stills. *Solar Energy, 12*(3), 333.

Craparo, J. C., & Thacher, E. F. (1995). A solar-electric vehicle simulation code. *Solar Energy, 55*(3), 221.

Duffie, J. A., & Beckman, W. A. (1991). *Solar engineering of thermal processes* (2nd ed.). New York: Wiley.

Erbs, D. G., Klein, S. A., & Duffie, J. A. (1982). Estimation of the diffuse radiation fraction for hourly, daily, and monthly-average global radiation. *Solar Energy, 28,* 293.

Habte, A., Lopez, A., Sengupta, M., & Wilcox, S. (2014). *Temporal and spatial comparison of gridded TMY, TDY, and TGY Data Sets, NREL/TP-5000-60886*. Colorado: National Renewable Energy Laboratory, Golden.

Hodgman, C. D. (Ed.). (1957). *Handbook of Chemistry and Physics* (38th ed.). Cleveland: Chemical Rubber Company.

Hsieh, J. S. (1986). *Solar energy engineering*. Englewood Cliffs: Prentice-Hall Inc.

Hu, C., & White, R. M. (1983). *Solar cells*. New York: McGraw-Hill, Inc.

Klein, S. A. (1977) Calculation of monthly average radiation on tilted surfaces. *Solar Energy, 19,* p. 325.

Knapp, C. L., Stoffel, T. L., & Whitaker, S. D. (1980). *Insolation data manual, SERI/SP-755-789*. Washington, D.C.: Government Printing Office.

Liu, B. Y. H., & Jordan, R. C. (1960). The interrelationship and characteristic distribution of direct, diffuse, and total solar radiation. *Solar Energy, 4*(3), 1.

Liu, B. Y. H., & Jordan, R. C. (1963). The long-term average performance of flat-plate solar energy collectors. *Solar Energy, 7,* 53.

Liu, B. Y. H., & Jordan, R. C. (1977). Availability of solar energy for flat-plate solar heat collectors. In R. C. Jordan & B. Y. H. Liu (Eds.), *Applications of solar energy for heating and cooling of buildings*. Atlanta: ASHRAE.

McCarney, S., Olson, K., & Weiss, J. (1987). *Photovoltaics: A manual for design and installation of stand-alone photovoltaic systems*, appropriate technology associates, Carbondale, Colorado.

Rauschenback, H. S. (1980). *Solar cell array design handbook*. New York: Van Nostrand-Reinhold.

Schoeman, J. J., & van Wyk, J. D. (1982). A simplified maximal power controller for terrestrial photovoltaic panel arrays. IEEE PESC Record, p. 361.

Siegel, R. (1973). Net radiation method for enclosure systems involving partially transparent walls. NASA-TN-D-7384.

Smith, R. E., & West, G. S. (1983). Space and planetary environmental criteria for use in space vehicle development 1982 (Revision Vol. 1). NASA TM 82478.

Spencer, J. W. (1971). Fourier series representation of the position of the sun. *Search, 2*(5), 172.

TMY. (1981). *Typical meteorological year user's manual* (*TD-9734*). Asheville: National Climatic Center.

TMY3. (2008). National solar radiation data base 1991-2005 update: Typical meteorological year 3, http://rredc-nrel.gov/solar/old_data/nsrdb/1991-2005/tmy3/by_state_and_city.html.

Whillier, A. (1953).Solar energy collection and its utilization for house heating.ScD. Thesis, Massachusetts Institute of Technology.

Wilcox, S., & Marion, W. (2008). *Users manual for TMY3 data sets, technical report NREL/TP-581-43156*. Golden: National Renewable Energy Laboratory.

Chapter 4
Storing Electric Energy

4.1 Introduction

Storage Methods All autonomous cars must use propulsive energy stored in the car. Vehicles using *chemical* storage run on the energy stored in gasoline or other hydrocarbon fuels. Solar energy can be stored in several ways: by *sensible* storage (heating a mass), by *phase change* storage (melting a substance) by *electrochemical* storage or *capacitive* storage (conversion to electric energy and storage in a battery or capacitor, respectively), or by *flywheel inertial* storage (converting electric energy to rotational kinetic energy and storing it in a spinning flywheel). Note that inertial storage is intrinsic to all vehicles because each stores kinetic energy in its own mass and the rotating masses of its wheels and drive. To summarize compactly:

$$E_{\text{stored}} = E_{\text{battery}} + E_{\text{capacitor}} + E_{\text{sensible}} + E_{\text{phase change}} + E_{\text{translational}} + E_{\text{rotational}}. \tag{4.1}$$

Methods in Use The constraints imposed by and on solar racing[1] limit the electric-energy storage choice for solar racing cars to batteries, at present. The translational and rotational kinetic energy stored in the mass of the car may be recovered by *regeneration*; this will also be discussed.

4.2 Terms and Units

Energy *Energy* means the potential to do work. Energy, like water, can be accumulated. *Charging* means to accumulate energy in some storage mode. For example, suppose a car having some initial kinetic energy coasts up a hill, coming to a stop for an instant at its top, having thereby lifted itself a vertical distance, *h*. At this

[1] Chapter 16 contains sample battery regulations.

E. F. Thacher, *A Solar Car Primer*, DOI 10.1007/978-3-319-17494-5_4

moment, the car has stored in its position *gravitational potential energy* equal to the work that was done on the car in rising against the force of gravity.

Power If this work were done rapidly, more *power* would have been used because power is the rate at which work is done. Stored energy would have accumulated more rapidly at high power than at low but the amount of energy storage would be the same. Like the water flow rate, power cannot be stored; it is simply the work rate.

Efficiency Not all the original kinetic energy of the car can be charged. The energy expended in working against rolling resistance and air drag is lost, for example. The *charging efficiency* is the energy stored divided by the energy available, the initial kinetic energy at the bottom of the hill in this case.

Discharging means to remove energy from storage. As the car rolls down the other side of the hill, the stored energy is discharged: converted into the total kinetic energy of the car. However, after coasting down through a vertical distance h, the car will not have a kinetic energy equal to the stored energy because some of that was lost as drag and rolling resistance work. The *discharge efficiency* is the recovered kinetic energy divided by the stored energy at the hilltop. The whole charge-to-discharge transaction can be characterized by a *storage efficiency*, the ratio of the energy recovered to the energy available. It is also the product of the charge and discharge efficiencies.

Units The W·h (*Watt-hour*) or kW·h (*kilowatt-hour*) are convenient units of energy. Your home's electric bill is calculated according to how many kW·h you use. A current of 1.0 A flowing from a 1.0-V source represents a power of 1.0 W, or 1.0 J/s. (The Joule, equivalent to one N·m of work, is also a unit of energy.) If the current flows for 1.0 h, then 3600 J or 1.0 W·h have been released from the battery.

Electric charge stored in a battery is measured in Coulomb[2] or more conveniently (for our purposes) in Ampere-hours (A·h). One Ampere of electric current transports charge at the rate of 1.0 C/s. A current of 1.0 A flowing for an hour would transport a charge of 3600 C or 1.0 A·h. The charge-storing ability, or *capacity*, of a battery is usually stated in A·h. A battery's storage efficiency for electric charge is usually well above 90 %.

4.3 Battery Storage

We wish to store the excess electrical energy supplied by the solar array in a battery. But, as in the case of the hill-climbing car, work must be done to store this energy. In the hill-climbing case, the work was done to lift the mass M against the gravitational potential, gh (in N·m/kg). In the battery-storage case, the work must be done to "lift" the electric charge, q, entering the battery against the voltage of the battery

[2] The charge carried by an electron is $1.602(10^{-19})$ C.

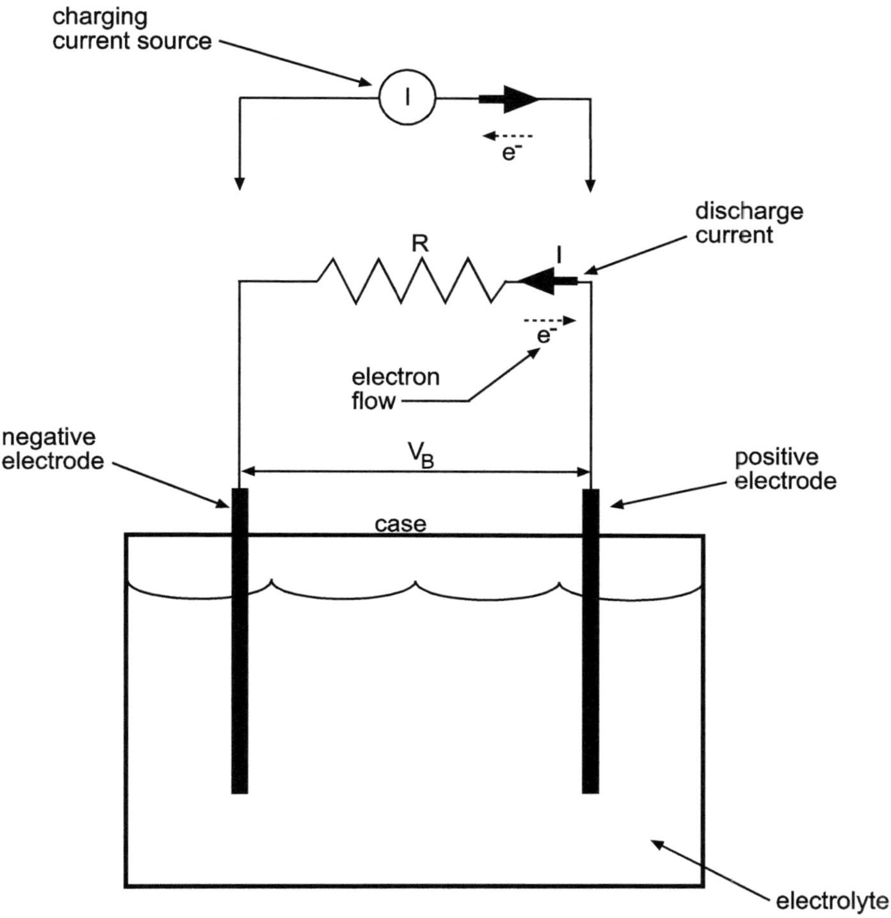

Fig. 4.1 Electrochemical cell

have been released from the battery, an electric potential. And, as in the former case, some of this work is lost in heating the internal electric resistance of the battery.

Cell Figure 4.1 shows the basic unit of a battery, called an *electrochemical cell*, or just a *cell*, for short.[3] The figure shows that a cell is made of a positive *electrode,* a negative electrode, and an *electrolyte.* The electrodes are dissimilar conductors. The electrolyte is an electrically conducting, dilute solution of an acid, base, or salt; it may be a liquid, as shown, or a paste. In a lead-acid cell, for example, the positive electrode is lead dioxide (PbO_2), the negative electrode is spongy lead (Pb), and the electrolyte is a solution of sulfuric acid (H_2SO_4) in water (H_2O).

[3] The conventional current, *I,* is conceived as positive charge flow. So it flows opposite to the electron flow.

Table 4.1 Some cell electrode reactions. (Crompton 1996)

Cell	+ Electrode	– Electrode
Nickel-cadmium	$2NiO(OH)+H_2O+2e^- \leftrightarrows 2Ni(OH)_2+2OH^-$	$Cd(OH)_2+2e^- \leftrightarrows Cd+2OH^-$
Nickel-metal hydride	$NiO(OH)+H_2O+e^- \leftrightarrows Ni(OH)_2+OH^-$	$H_2O+e^- \leftrightarrows 1/2H_2+OH^-$
Lead-acid	$PbO_2+3H^+ +HSO_4^-+2e^- \leftrightarrows PbSO_4+2H_2O$	$Pb+HSO_4^- \leftrightarrows PbSO_4+H^++2e^-$

"→" denotes discharging and "←" denotes charging
NiO(OH) is nickel oxyhydroxide

Table 4.2 Nominal voltages. (Crompton 1996)[a]

Couple	V (volt)
$LiFePo_4$	3.7
Pb-acid	2.0
Ni-MH	1.25
Ni-Cd	1.2

[a] $LiFePo_4$ is lithium iron phosphate or lithium ferrophosphate (LFP)

When the electrodes are immersed in the electrolyte, an open-circuit voltage difference is found to exist between them that is characteristic of the materials of the cell (see Table 4.2).

Connecting the cell to an electric load causes chemical changes in the *active material* of the electrodes. In the case of the lead-acid cell, the active materials are the lead and lead dioxide. During discharge of the lead-acid cell, lead sulfate ($PbSO_4$) is formed on each electrode. Table 4.1 gives the electrode reactions when charging and discharging for lead-acid and two other cells.

As the reactions in Table 4.1 show, when discharging, electrons are liberated from atoms of the negative electrode and bound to atoms of the positive electrode. However, these electrons cannot enter the electrolyte to flow to the positive plate but must travel through the external circuit and thus do work in the load.

Primary cells, which cannot be recharged, are used to power non-rechargeable devices such as flashlights. Often these are *dry cells* in which the electrolyte is in paste form. In *wet cells,* the electrolyte is a liquid solution. Primary cells can play a role in solar car racing. Consult Chap. 16 for examples.

Cells that can be both charged and discharged are called *secondary cells*. In these cells, the chemical changes in the active materials when the cell is discharged can be reversed, and the material restored to its original form, by supplying current to the battery. In the lead-acid cell, for example, this results in the lead sulfate being converted back to lead at the negative electrode and to lead dioxide at the positive electrode. The electrons are supplied to the negative electrode and removed from the positive electrode by an external current source, such as a solar cell array. Rechargeable cells are called *secondary* because the source of the charge stored or discharged is not the battery itself.

The cell's voltage is characteristic of the electrodes, the electrolyte, and the state of charge. There are many possible combinations. Table 4.2 shows the nominal terminal voltage when discharging for the cells of Table 4.1 and a lithium battery.

Battery The voltages of series-connected cells and the currents of parallel-connected cells add. There will usually be several series-connected cells in a case, with a single pair of external positive and negative terminals. The case with its contents is called a *module*. Solar racers use an array of several modules in series, or even in series-parallel. We will call this array the battery.

The higher discharge voltage of the lithium cell means that, other things being equal, lithium batteries can be lighter than the three other cells while still supplying the same energy.

4.4 Example 4.1

Small, AA-size, rechargeable, nickel-cadmium cells may be purchased locally. The cells are rated at 1.25 VDC and 0.045 A. At this rate, they can discharge for 14 h, and no depth of discharge (DOD) limit applies. What would the capacity be? How many of these would be necessary to power a small motor requiring 2 V at 0.08 A?

Solution The capacity would be (0.045 A)(14 h) = 0.63 A·h ("0.63 A·h at the 14 h rate"). Two batteries in series would be needed, for a 2.5 V nominal voltage. The number in parallel would be 0.08/0.045 = 1.78 or two branches. The direct current (DC) motor's effective resistance is 2 V/0.08 A = 25 Ω. The current would be 0.1 A. The resulting array of four cells is shown in Fig. 4.2.

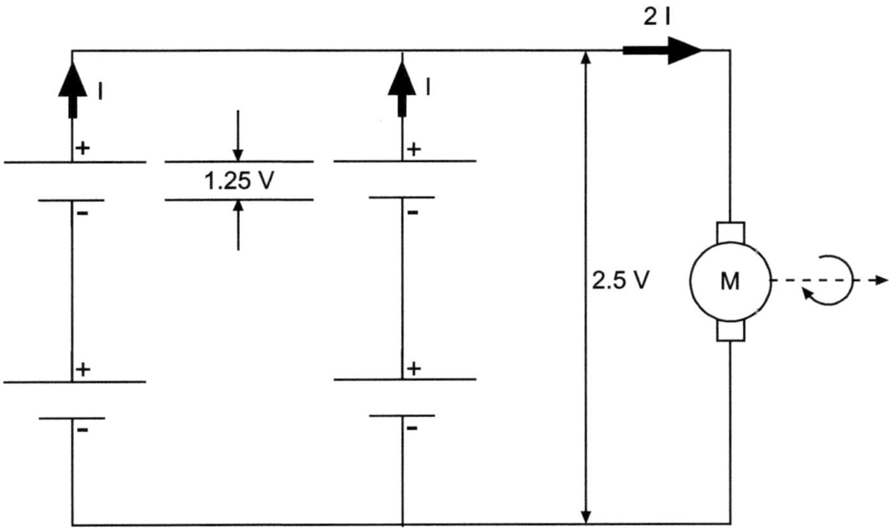

Fig. 4.2 Battery design for Example 4.1

Capacity The capacity of a series-connected string of battery cells is equal to the capacity of one cell in the string. If strings of identical cells are connected in parallel, the capacities of the strings add, as in Fig. 4.2.

Because it will reach its discharge cut-off voltage first, the cell in a series string having the lowest capacity controls the performance of the string. The capacities of the modules in the string should therefore be matched.

4.5 Example 4.2

Suppose a battery (assumed to be lossless) is charged at an average current of 5 A for 10 h at an average voltage of 2 V. How much charge and energy have been stored?

Solution The current flows at 5 C/s. Thus,

$$Q = \left(5\frac{\mathrm{C}}{\mathrm{s}}\right)(10\mathrm{h})\left(3600\frac{\mathrm{s}}{\mathrm{h}}\right) = 180,000\mathrm{C} = 50\mathrm{A{\cdot}h}.$$

Average rate of energy supply is

$$\dot{E} = (5\mathrm{A})(2\mathrm{V}) = 10\mathrm{W}.$$

And the amount stored is

$$E = (10\mathrm{W})(10\mathrm{h}) = (50\mathrm{A{\cdot}h})(2\mathrm{V}) = 100\mathrm{W{\cdot}h} = 360\mathrm{J}.$$

The *rated capacity* is given with reference to the discharge current that the battery can maintain before the terminal voltage falls below a value specified by the manufacturer. The DOD is the current fraction of full discharge.[4] The DOD of lead-acid batteries should be limited to prevent reduction of the life of the cell caused by repeated discharge below this limit. A typical limit for these batteries is a DOD of 0.8, or 80 %. However, battery life beyond the 15 or 20 deep discharge cycles typical of a solar car race is not as important to racing teams as it might be to owners of a fleet of electric-powered delivery vans. Discharging below a standard DOD limit may therefore be done occasionally.

Charge State Measurement The energy or charge in a battery cannot be measured directly. Measuring the terminal voltage when charging and discharging or measuring the electrolyte's specific gravity are in general means of ascertaining the state of charge of a battery. The method to use depends on the characteristics of the battery materials.

[4] The converse is the *state of charge*, F, the fractional amount (or percentage) of full charge in the battery; DOD + F = 1.

Either will work in the case of lead-acid cells (although measuring the specific gravity is not convenient on the road). The chemical reactions above show that when discharging and charging, lead-acid cells either create water in or remove it from the electrolyte, respectively. Thus, the specific gravity of the electrolyte is reduced on discharge and increased on charge. It is therefore a direct indicator of the state of charge. On the other hand, the specific gravity change of nickel-cadmium or nickel-metal hydride cell over the period of a race would be almost negligible because there is no net water formation in the electrolyte of these cells. Measuring the terminal voltage is the only practical option for the nickel-cadmium case. The cell voltage depends upon the discharge rate as well as the state of charge and varies nonlinearly with time at a particular discharge rate. However, the hydrogen pressure in a nickel-metal hydride cell is a function of its state of charge as well (Crompton 1996).

Most solar racers will measure the battery bus voltage (even if it is characteristically flat during discharge) and supplement this information with an Ampere-hour meter. Ampere-hour meters monitor the charge by sampling the current leaving or entering the battery over short time intervals and displaying the net charge held in the battery.

The curves of Fig. 4.3 show that the capacity is also lower at high discharge rates than at low discharge rates. So capacities are usually quoted at some standard rate of discharge. Rather than mentioning the actual rate, the time required to completely discharge the battery will be given instead. For example, "the capacity is 25 A·h at the 10 h rate." This rate would be 2.5 A.

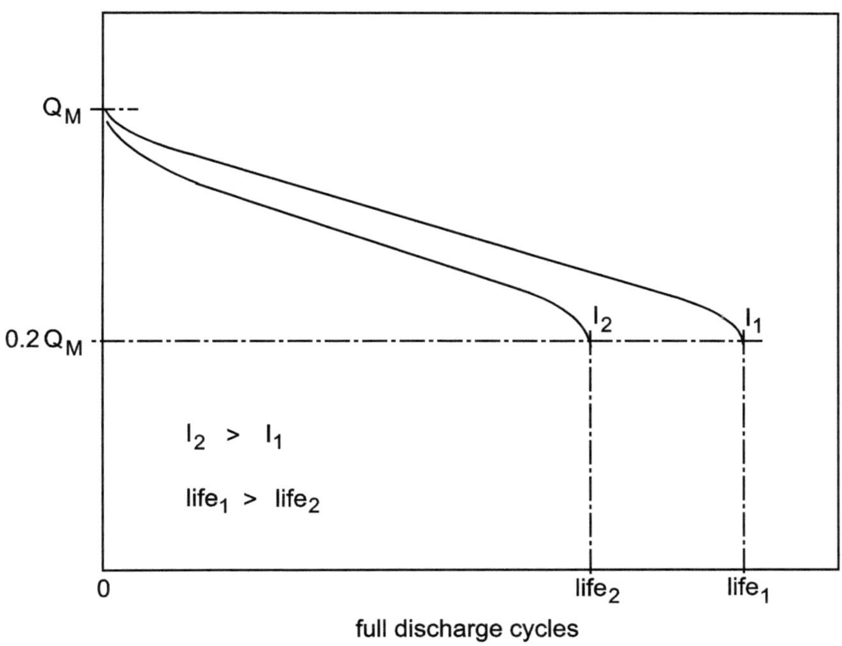

Fig. 4.3 Capacity and life

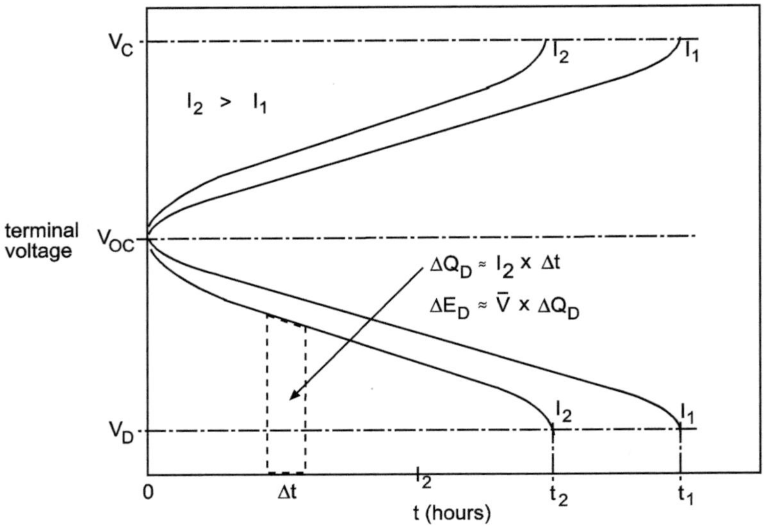

Fig. 4.4 Charge and discharge

Charge and Discharge Characteristics Figure 4.4 shows the terminal voltage of a typical battery being discharged at different constant rates. During discharge, the terminal voltage is

$$V(t) = V_{oc} - IR_B(t), \tag{4.2}$$

where V_{oc} is the open circuit voltage, I is the discharge current (A) and $R_B(t)$ is the internal resistance (Ohm). Note that the internal resistance is not a constant. When the current begins to flow, the terminal voltage drops suddenly to its initial value and then decreases nonlinearly for a short time. The terminal voltage then decreases linearly at a slower rate (in some cells, Ag-Zn for example, the voltage is nearly constant) for a longer period. The larger the discharge current, the shorter this period and the steeper the slope of the discharge characteristic. As the conversion of the active material nears completion, the chemical reactions at the electrodes change. In cells such as those in Table 4.1, the electrolysis of water increasingly dominates. Consequently, oxygen and hydrogen gas bubbles accumulate on the electrodes. These gases are electrical insulators and have the effect of increasing the internal resistance of the cell. This gas accumulation causes the terminal voltage to drop rapidly once again. The discharge is stopped at the voltage recommended for the cell by the manufacturer. This will be about 1.75 V for lead-acid cells.

Figure 4.4 also shows a typical voltage–time graph of the terminal voltage when charging. The terminal voltage is related to the open circuit voltage, current, and internal resistance by

$$V(t) = V_{oc} + IR_B(t). \tag{4.3}$$

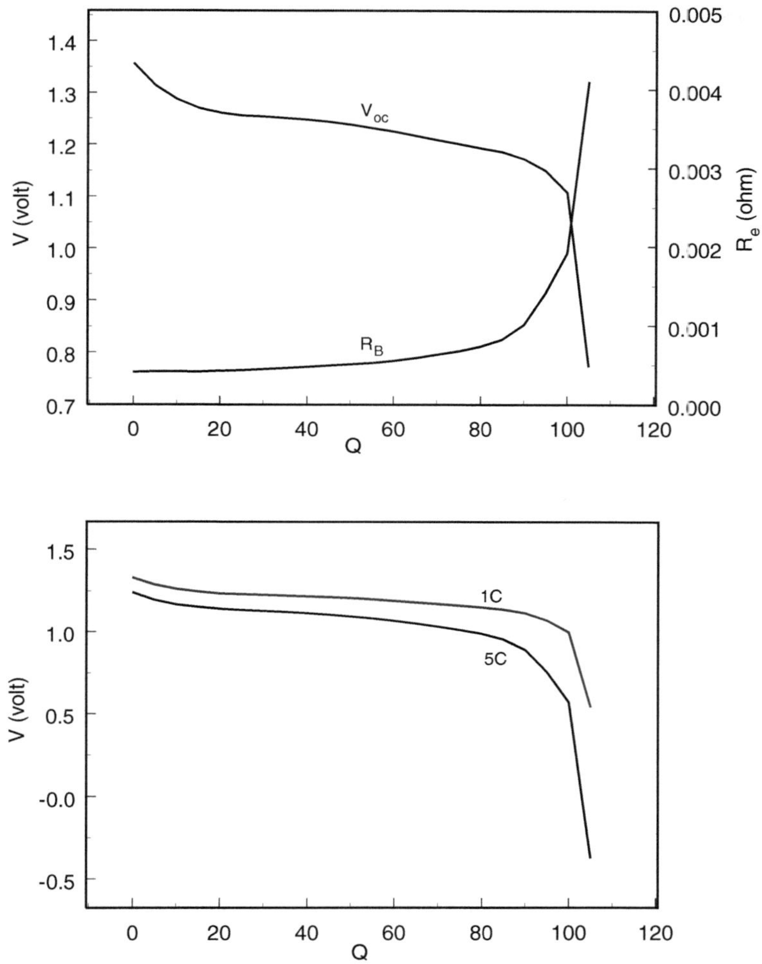

Fig. 4.5 Saft STX600 Ni-Cd battery characteristics. (Courtesy Saft Engineering)

These curves mirror the discharge curves, featuring a sudden rise in voltage to the initial value followed by a short, nonlinear voltage increase and a longer, linear voltage increase. Finally, a rapid, nonlinear voltage increase takes place. This last increase begins when most of the active material has been restored to its predischarge state. This rapidly increases the internal resistance, as in the case of discharge. Figure 4.5 shows V_{OC}, R_B, and two discharge curves for the Saft STX600 Ni-Cd cell ("1C": current, A, numerically equal to capacity, 56 A·h).

A current–voltage characteristic for the battery may be found by first plotting the charge and discharge curves as a function of capacity (Q). Then the voltage of each discharge and charge curve is sampled at the same capacity and plotted on

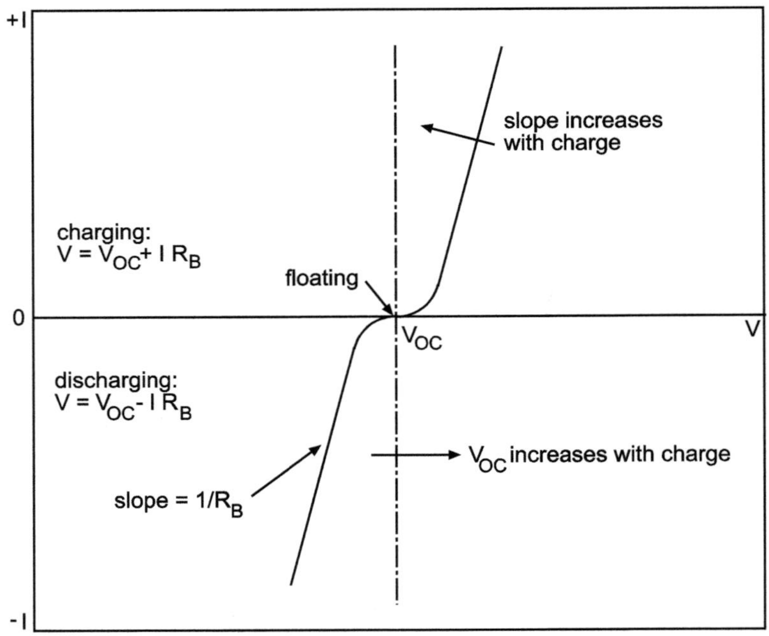

Fig. 4.6 *I–V* curve of a battery

current–voltage coordinates. The curves that result are the battery's *I–V* characteristics at different states of charge. Each of these characteristics has a discontinuity at zero current (open circuit) caused by the sudden increase or decrease in the terminal voltage when charging or discharging begins, respectively. The slope of a particular characteristic is $1/R_B$.

A terminal voltage to the left of the open circuit voltage will cause discharging of the battery; one to the right of it will cause charging of the battery. Figure 4.6 shows this, qualitatively.

Specific Energy Table 4.3 gives the *specific energy*, the energy stored per unit battery mass (e_B, W·h/kg), of some battery couples and also of gasoline, for comparison. The large gap remaining to be closed between the batteries and gasoline is evident.

A solar racer using a battery with a high specific energy can store the same energy as a car using a battery with lower specific energy, but weigh less. For example, Table 4.3 shows the specific energy of nickel-metal hydride batteries to be twice that of lead-acid batteries.

Specific Power The maximum instantaneous power per unit mass that can be delivered by a battery is called the *specific power*. Specific power relates to the rate at which energy may be delivered. Hence, it limits the speed when climbing hills, for example. Specific power and specific energy trade off against each other. A high specific power implies high current. This requires larger conductors inside the battery and therefore increases the mass, reducing the specific energy. Solar racers

Table 4.3 Specific energy. (Crompton 1996; Cook 1991)

Storage Type	e_B (W·h/kg)
Gasoline[a]	14,000
Zinc-air	160
Silver-zinc	100–400
Lithium-ferrophosphate	140–180
Nickel-metal hydride	70
Nickel-zinc	60
Nickel-cadmium	35–50
Lead-acid	35

[a] Lower heating value, a variable depending upon ratio of mass flow rate of combustion air to mass flow rate of fuel and other factors

Table 4.4 Specific power comparison. (Moore 1996)

Cell	Specific power (W/kg)
Ni-Cd	160
Ni-MH	230
Pb-acid	300

Table 4.5 Life comparison

Cell	Life in cycles
Li-$FePO_4$	>1000[a]
Ni-MH	1500 (Moore 1996)
Pb-acid	400 (Moore 1996)

[a] Conservative estimate from written statements by retailers

are slowly accelerated, whenever possible. However, in common situations, such as maneuvering in city traffic or when passing other solar racers on the highway, low acceleration is not prudent. Table 4.4 shows sample specific power ratings.

Age and Life A battery's capacity decreases with the number of full charge–discharge cycles it has experienced, referred to as the battery's *age*. The age at which the battery's capacity has been reduced to 80 % of its original is called its *life*, even though operation is still possible. Figure 4.4 shows curves of battery capacity as a function of its age. Battery life is not critical in solar racing because the number of full cycles accumulated in testing, qualification, and the race will probably be smaller than the shortest life shown in Table 4.5.

4.6 Operation and Control

Racing Figure 4.7 shows three modes of operation. If the car is operating at a speed just sustainable by the solar energy supply, the battery may simply *float*, neither charging nor discharging. However, when cruising, the usual condition is

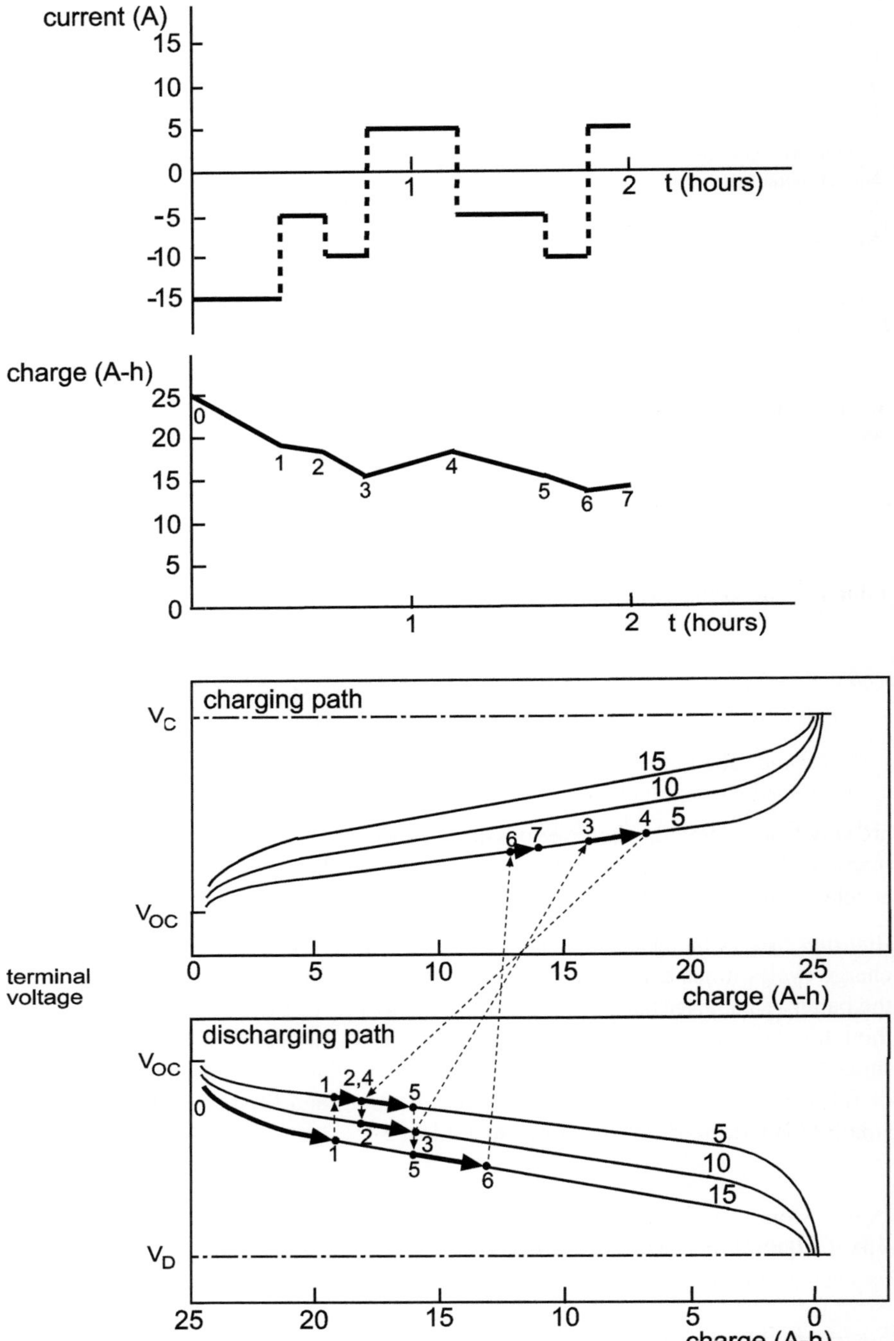

Fig. 4.7 Realistic operation

discharging. But this occurs at varying rates, with the highest rates occurring on uphill grades. The battery should be thought of as moving between the discharge curves of Fig. 4.4 so that its actual voltage–time characteristic consists of segments of these curves. Also, the battery alternates between charge and discharge in the stop-and-go traffic experienced when passing through towns. Figure 4.7 illustrates these shifts by a hypothetical, 2-h, battery current–time profile. In this figure, positive currents charge the battery and negative currents discharge it. Below the current profile is the resulting battery-charge profile as it would be kept by an Ampere-hour meter, and, below this, the path the terminal voltage follows is displayed. The dashed arrows indicate transitions to different currents at different states of charge.

The schedule of a racing day allows for static charging in the morning before racing begins, during the midday stop, and after arriving at the day's destination. During the morning and evening periods, the rules allow the car to be placed into a special charge configuration, with the array kept pointed at the sun. Here, the objective is to put as much charge into the battery as possible during the allotted time. The amount of current is controlled by the available sunlight and the solar cell array. No attention is paid to controlling the charge to extend the life of the batteries.[5] Given the typical solar conditions and the restrictions imposed by race rules on the size of the solar car's solar cell array, keeping the charging current below the maximum specified by the battery manufacturer will not be a worry.

The battery will usually be discharged well below 50 % DOD during a race day. If the weather on the current day is good, and if the weather for the next day is expected to be good, the car will run fast. Conversely, if the day is overcast, the speed will be low with a heavy reliance on battery energy.

Containment The battery may be the most massive object in the solar racer. Therefore, it must be mounted in an enclosure that can contain it during any of the emergency scenarios (front, side, and rear impact and rollover) contained in the design specifications (Chap. 8). Under these conditions, the objective is to protect the driver, not to prevent damage to the solar racer. During normal operation, the enclosure holds the battery securely in place, surrounds the battery with nonconducting materials, and must be continually supplied with fan-forced air taken directly from outside the car whenever the battery is electrically connected to the main bus. This ensures that no explosive gas accumulates in the enclosure.

4.7 Wiring and Testing

Wiring Connections to the module terminals should be tightened as required by the manufacturer's specifications. A good, tight connection with no air gap between the connector and the battery terminal, giving the maximum possible area of contact, is

[5] Extending the life of the battery beyond the race is not an important constraint. Often an entirely new car will be built for the next race, and the current car will be used for training and public relations.

very important. Coating the terminals with a conductive grease can further reduce the contact resistance. Loose connections have higher electrical resistance and can therefore heat up when current passes through them. At best, this wastes energy; at worst, it is a fire hazard. However, terminals should not ever be overtightened.

Avoid sharp bends in the wiring and plan the wire runs so that the module-to-module wire lengths are equal. If equal lengths are not possible, run multiple wires where necessary to equalize the wire resistances between modules throughout the battery. Install small sensing wires to monitor the voltage of each module. Fuses should be installed in these sensing wires.

Temperature It is important to keep batteries warm, especially lead-acid batteries. Other batteries, such as nickel-cadmium and nickel-metal hydride, are more tolerant of temperature extremes. Lay out the battery so that the modules are grouped as close together as possible to keep them warmer. The temperature should be kept below the maximum allowed by the manufacturer's instructions. For example, Delphi lead-acid batteries should be kept below 50 °C.

Test Bench test the modules, using the manufacturer's recommended procedure, to ascertain the discharge capacity of each module. Connect those modules in series which have discharge capacities that are most nearly alike. Then, test the battery using a dynamic load profile that closely approximates the racer's electrical load.

Some batteries need to be "worked" (cycled several times) when they are new before they can achieve their maximum capacity. For example, Delphi lead-acid modules require about 15 deep cycles before reaching peak capacity. Others may require as many as 50 deep cycles.

The battery should be given an equalizing charge the day before the race begins, again using the manufacturer's procedure. This reduces self-discharge (a problem especially with lead-acid modules) and brings all the modules to the same voltage (this means the modules must be charged in parallel).

Orientation and Venting The valve-regulated, lead-acid modules made for electric vehicles by Delphi may be installed with the active material plates horizontal (module case vertical). This increases their capacity by about 5 % by preventing acid concentration gradients from forming in the electrolyte at the bottom of the plates.

Any plastic tubing connected to the vents of sealed batteries should be impermeable to hydrogen gas.

4.8 Regeneration

The opening example was of car climbing a hill and then coasting down the other side. Suppose the desired speed were reached, and the grade were such that the racer could have continued to accelerate without discharging the battery. Using the brakes to maintain a steady speed wastes this available energy by creating friction heat. Some motor–controller combinations can be switched to allow the motor to run as

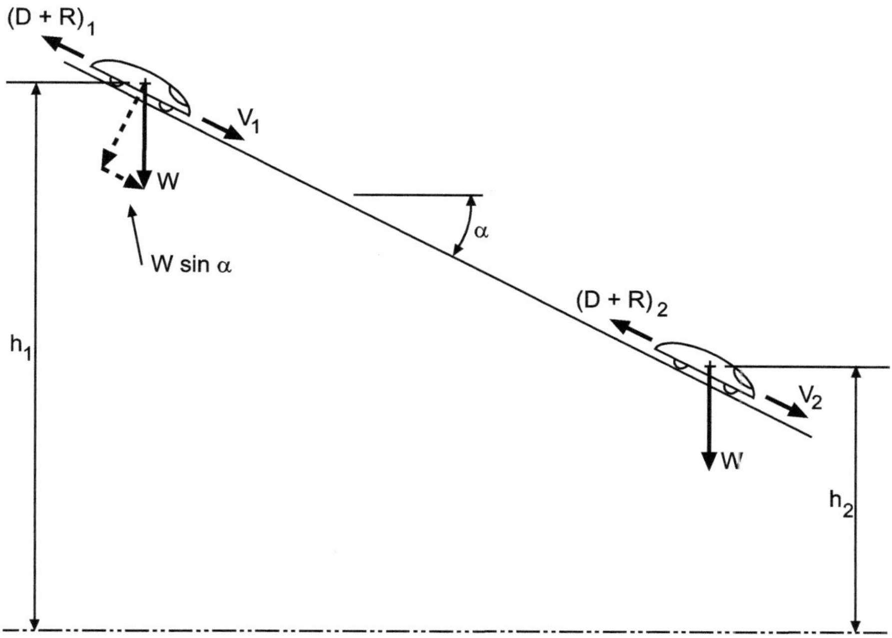

Fig. 4.8 A regenerating vehicle

a generator. This reversed current flow then charges the battery. Using the motor as a generator applies a braking torque to the driven wheel or wheels. The amount of this torque can be controlled to keep a steady speed. This mode of operation is called regeneration.

Regeneration can be used to advantage. If the team is approaching a long downgrade, regeneration can be set to maintain the desired coasting speed while simultaneously charging the batteries.[6] The solar array output would also charge the battery in this case because the speed would be maintained by gravity. Regeneration should be used to slow the car before applying the brakes. This recovers some of the kinetic energy stored in the car and reduces wear on the brakes.

Let us examine regeneration more analytically. Figure 4.8 shows a solar car coasting down a hill of constant grade, α. At the top of the hill the height of the car's center of gravity

(CG) is h_1 m and its speed is V_1 m/s. At some point down the hill, the height is h_2 and the speed V_2. Writing energy conservation between points 1 and 2 gives

$$\frac{1}{2}MV_1^2 + Mgh_1 = \Delta E_{CB} + \frac{1}{2}MV_2^2 + Mgh_2 + W_{12} + Q_{12}. \tag{4.4}$$

[6] With nearly exhausted batteries, Clarkson used this procedure to descend into the Ohio River valley during the 1990 Sunrayce. It allowed that team to charge enough energy to cross the river and make the midday stop.

The term ΔE_{CB} represents the energy stored in the battery, W_{12} is the work done against drag and rolling resistance, and Q_{12} is energy lost as heat in the drive and when charging the battery between positions 1 and 2. The net energy available for storage is

$$\Delta E_A = \frac{1}{2}M\left(V_1^2 - V_2^2\right) + Mg(h_1 - h_2) - W_{12}. \tag{4.5}$$

If the car in Fig. 4.8 were coasting at constant speed,

$$\Delta E_A = Mg(h_1 - h_2) - W_{12}. \tag{4.6}$$

Or, when braking to a stop on level ground, V_2 and $h_1 - h_2 = 0$, and

$$\Delta E_A = \frac{1}{2}MV_1^2 - W_{12}. \tag{4.7}$$

The energy charged may now be written more compactly as

$$\Delta E_{CB} = \Delta E_A - Q_{12} \tag{4.8}$$

or

$$\Delta E_{CB} = \eta_{RM}\eta_{CB}\Delta E_A. \tag{4.9}$$

Where η_{RM} is the average efficiency of the drive in regeneration mode, and η_{CB} is the average energy charging efficiency of the battery during the regeneration.

The average efficiency of the drive depends upon the torque and rotational speed of the motor during regeneration. When descending a hill at constant speed, the efficiency would be high, perhaps 90 %. During braking on a level surface from city traffic speed it would be lower, perhaps 50 %. The charging efficiency of the battery would depend on the charging current and the state of charge. If the regeneration current were not large, which is probably typical, the efficiency would be high, perhaps 95 % or more. Therefore, the energy charged would be roughly 45–80 % of the net available, depending on the maneuver.

The motor's braking torque may be found from Eq. 2.17. When coasting down a hill at steady speed, the magnitude would be

$$\tau = r_W(W sin|\alpha| - D - R) \tag{4.10}$$

Where D is the drag force, R is the rolling resistance force, and r_W is the wheel radius.

4.9 Example 4.3

Suppose that the downgrade, speed, weight, drag area, wheel radius, and rolling resistance coefficients were −5 %, 55 mph (24.59 m/s), 800 lbf (3558.1 N), 0.15 m², 0.25 m, 0.004, and 0.0001 s/m, respectively. If lift is negligible, find the magnitude of the regenerative braking torque applied to the driving wheel and shaft in still air at standard temperature and pressure. Then find the charging current.

Solution The air density would be 1.184 kg/m³ (Example 2.1) and the dynamic pressure would be

$$q = \frac{1}{2}\rho V^2 = \frac{1}{2}\left(1.184\frac{\text{kg}}{\text{m}^3}\right)\left(24.59\frac{\text{m}}{\text{s}}\right)^2 = 358.06\frac{\text{N}}{\text{m}^2}.$$

The drag force would be

$$D = c_D A_D q = (0.15\text{m}^2)\left(358.06\frac{\text{N}}{\text{m}^2}\right) = 53.7\text{N}.$$

The rolling resistance coefficient would be

$$\mu = \mu_1 + \mu_2 V = 0.004 + \left(0.0001\frac{\text{s}}{\text{m}}\right)\left(24.59\frac{\text{m}}{\text{s}}\right) = 0.00646.$$

And the rolling resistance force would be (−5 % grade = −2.28°)

$$R = \mu N_{\text{W}} = 0.00646 \times 3558.1\text{N} \times cos(-2.28^\circ) = 22.96\text{N}.$$

Therefore, the torque at the driving wheel would be

$$\tau_{\text{W}} = 0.25\text{m}[3558.1\text{N} \times sin(2.86) - 53.7\text{N} - 22.96\text{N}] = 25.22\text{N}\cdot\text{m}.$$

The torque delivered to the motor (now a generator) would depend upon the speed reduction and efficiency of the transmission (if any). Suppose these were 5 and 95 %, respectively. Then the torque at the motor would be

$$\tau_{\text{M}} = \tau_{\text{W}}\frac{\eta_{\text{G}}}{n_{\text{G}}} = 25.22\text{N}\cdot\text{m} \times \frac{0.95}{5} = 4.79\text{N}\cdot\text{m}.$$

Suppose that the efficiency of the motor–controller combination were 90 % at this torque, and the rotational speed corresponding to 55 mph (4696 rpm) and that the battery bus voltage were 120 V. The regenerative charging current to the battery would then be

$$I_{\text{B}} = \frac{\tau_{\text{M}}\omega_{\text{M}}\eta_{\text{M}}}{V_{\text{B}}} = \frac{4.79\text{W}\cdot\text{s} \times 491.8\frac{\text{rad}}{\text{s}} \times 0.90}{120} = 17.6\text{A}.$$

The car would travel a mile in 0.0182 h. Therefore, for every mile traversed down the hill a regenerative charge of 0.32 A·h would supplement the charge from the solar array to the battery.

References

Armstrong, L. (1996). Batteries not included. *Business Week,* 78 (September 23).

Cook, R. (1991). Electric car showdown in Phoenix. *Popular Science,* 64 (July).

Crompton, T. R. (1996). *Battery reference book* (2nd ed.). Pittsburgh/London: SAE/Butterworth-Heinemann.

Loew, E. A., & Bergseth, F. R. (1954). *Direct and alternating currents*. New York: McGraw-Hill.

McCarney, S., Olson, K., & Weiss, J. (1987). *Photovoltaics*. Carbondale: Appropriate Technology Associates.

Moore, T. (1996). The road ahead for EV batteries. *EPRI Journal, 21*(2), 6 (March/April).

Pickett, D. E. (1990). Nickel alkaline batteries for space electrical power systems. SAE paper 901055, SAE Aerospace Atlantic, Dayton, Ohio, April 23–26, 1990.

Prater, D. (1996). Batteries. Sunrayce Workshop, April, 1996.

Reisner, D. (1991) Batteries: It's a jungle out there! Proceeding S/EV 91 26–27 October 1991 Boxborough, MA, NESEA, Greefield, MA, p 52.

Storey, J. W. V. (1994). Batteries. In J. W. V. Storey, A. E. T. Schinckel, & C. R. Kyle (Eds.), *Solar racing cars* (p. 99). Canberra: Australian Government Publishing Service.

Wheeler, J. R., Cook, W. D., & Smith, R. (1990). Small capacity, low cost (NiH_2) design concept for commercial, military and higher-volume aerospace applications. Proceeding Fourth Annual AIAA/Utah State University Conference on Small Satellites, Logan, Utah, August 27–30, 1990. Center for Space Engineering, Utah State University, p. 4.

Chapter 5
Electric Motor Drives

5.1 Introduction

The *drive* is the electric motor, its controlling electronics, the speed reduction, and the driven wheel (solar racing cars usually have only one driven wheel). Figure 5.1 is a schematic of a typical drive connected to its solar-electric power source. This chapter discusses the operation of each of the drive's components.

Motors may be classified as alternating-current (AC) or direct-current (DC), according to the kind of current supplied to them. They are also classified by the means to sustain their magnetic fields: permanent magnets or electromagnets. The emphasis herein falls on drives using permanent-magnet, DC motors because these motors have found wide use in solar racing cars.

The chapter concludes by using the array, battery, and motor *I*–*V* curves to explain the electrical interaction of the drive with its two power sources.

5.2 Electric Motor

Motor Action An electric motor is a device that converts electric energy into mechanical energy. The interaction that causes this conversion to take place is as follows: When an electric current is flowing in a wire which is also in a magnetic field, the wire experiences a force perpendicular to the plane in which the magnetic field vector and the current vector lie. The mechanism of the motor is arranged in such a way that this force causes rotation of the shaft of the motor. This rotating shaft can then be used to perform mechanical work, such as moving a solar-electric car.

Figure 5.2 shows a two-pole, permanent-magnet, brushed, DC electric motor. Several loops of wire connected to a source of DC current are wound on a steel core and held inside the field set up by the north and south poles of a magnet (only the

E. F. Thacher, *A Solar Car Primer*, DOI 10.1007/978-3-319-17494-5_5

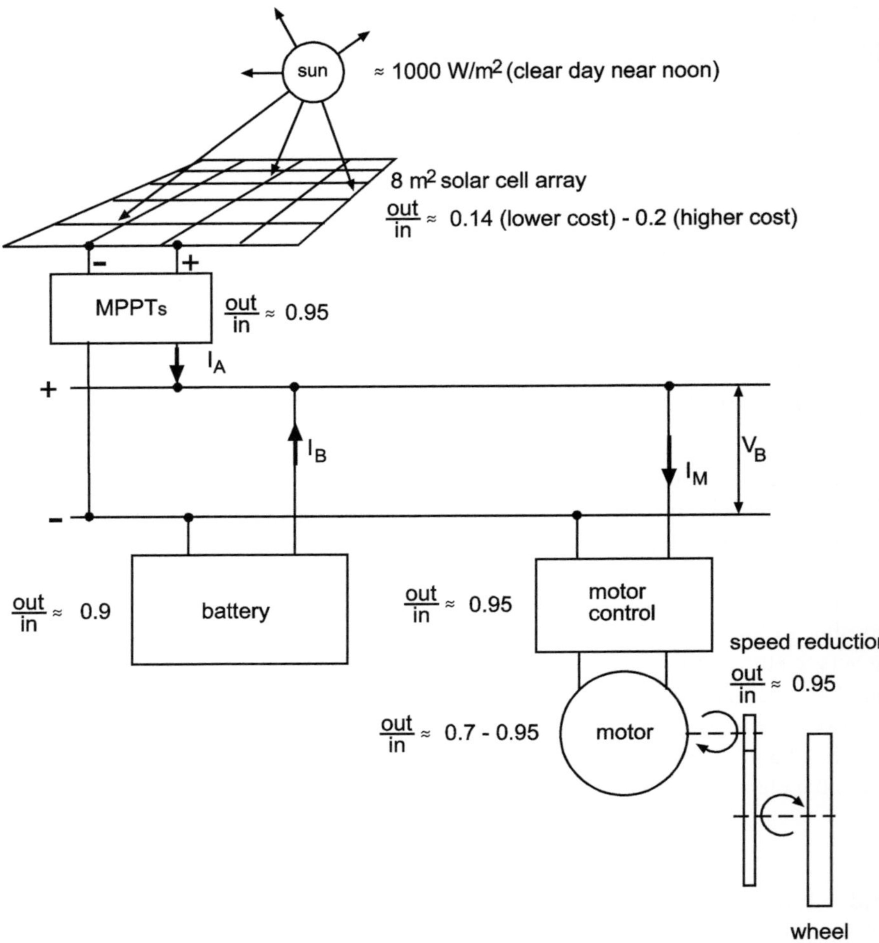

Fig. 5.1 Solar-electric drive

magnet's poles are shown).[1] When there is no armature current (arrow labeled "*I*"), there is no force on the conductors. When current is flowing, the force on each wire is exerted sideways—left or right—depending on the direction of the current relative to the direction of the field. The figure shows the current going away from the reader under the north pole ("+" symbol) and toward the reader under the south pole ("•" symbol). The field around the conductor reinforces the magnet's field on one side and opposes it on the other (armature reaction).[2] The force on the wire points toward the weakened air gap field.

[1] The picture is simplified for clarity. The windings are actually placed in slots so the force on them will not tear them free.

[2] Imagine the right hand grasping the wire with the thumb pointing in the direction of the current. The fingers curl in the direction of the wire's field. This is the "right-hand rule."

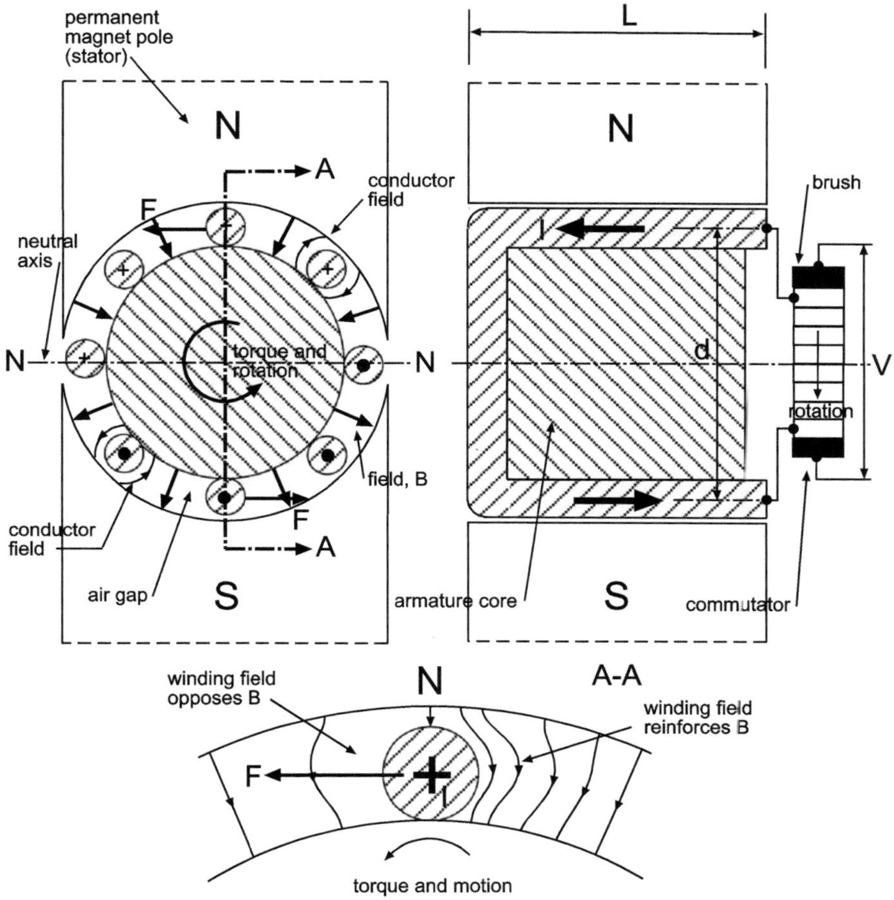

Fig. 5.2 Simple DC motor

Only the length, L, of the wire inside the field at each pole experiences a force. The force on each wire is directly proportional to the magnetic field intensity, B (Weber/m^2), the current, I (A), in the wire, and to the length:

$$F = BIL. \tag{5.1}$$

Torque The total torque on the armature at any instant is the sum of the moments of the forces on each wire:

$$T = n_P n_{WP} Fr, \tag{5.2}$$

where T is the torque, n_P is the number of poles, n_{WP} is the number of wires under a pole, and r is the radius to the centerline of a wire. The size of F, and therefore of T, depends on the strength of the magnetic field and the magnitude of the electric current. But for a given field and motor construction, the torque on the coil is directly proportional to the current.

$$T = k_T I, \tag{5.3}$$

where k_T is the *air gap torque constant.*

A distinction must be made between the torque on the armature, or air gap torque, and the torque delivered to the output shaft, the *shaft torque,* τ_S. The shaft torque will be less than the air gap torque because of mechanical losses such as friction in the motor bearings. These will be discussed in more detail later, but in general the losses depend on the rotational speed. Therefore, the air gap and shaft torques will be the same only when the torque is sufficient to prevent rotation; the motor is *stalled.*

The rotation moves the rotor through the field of one pole every half revolution, in the two-pole motor shown. The torque is constant under the poles and nearly zero when the loop has rotated to position A-A where the field is approximately zero. A-A is called the *neutral axis*. The torque on a winding drops to zero at the neutral axis, producing waviness in the shaft torque. However, the ripple is quite small because of the large number of windings employed.

Power The power, P_M, is the product of the average torque per revolution and the angular rotation rate in radians per second. If the motor is rotating at N rpm,

$$P_M = \frac{\pi}{30} \tau_S N, \tag{5.4}$$

where τ_S is the average shaft torque per revolution.

The shaft torque "constant," k_S, is the proportionality factor between the current and the shaft torque:

$$\tau_S = k_S I. \tag{5.5}$$

This quantity must be measured when power is being delivered. Therefore, it is dependent upon the current, and hence the torque load, because of the distortion of the air gap field caused by the field set up by the current in the armature. Also, it has speed dependence because it includes the rotational speed-dependent losses. This will be demonstrated when discussing the efficiency of a motor. Ibramim (1989) gives k_S as 0.2152 N·m/A ± 10% for a version of the Hathaway HSSS3810 permanent-magnet, DC motor.

Operating Limits The passage of current through the armature resistance dissipates energy in the armature at a rate proportional to the square of the current. Other energy losses also cause heating. The temperature of the armature will rise until the heat lost to the environment equals the generated heat. Excessive temperatures reduce the electrical resistance of the materials that electrically insulate the armature coils.

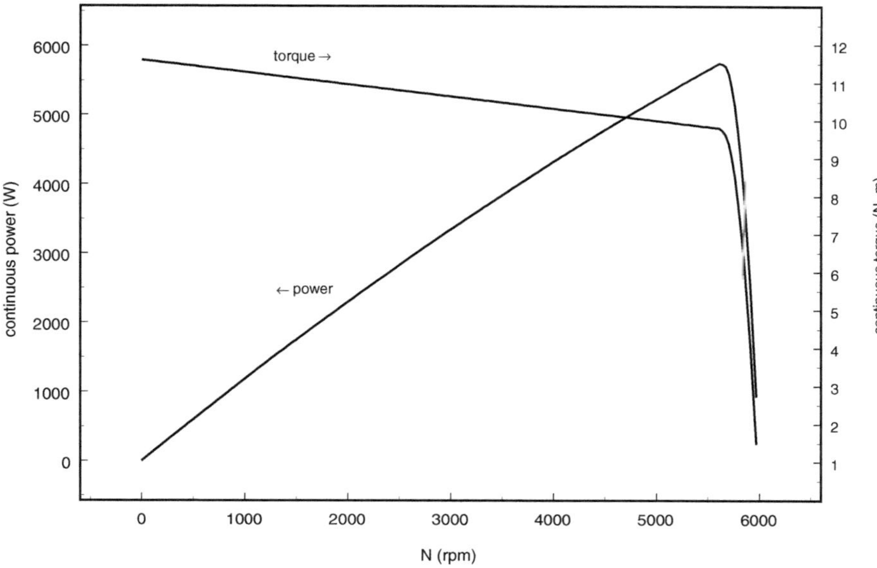

Fig. 5.3 Brushless, permanent-magnet motor performance

This reduction accumulates over time and may cause short circuits, thereby causing additional heating and possibly catastrophic failure. The torque–speed combinations at which the motor may be continuously operated without damage form the boundary within which the motor operating point must usually remain. The motor may be rated for operation outside these limits only for periods of a few minutes. Figure 5.3, adapted from manufacturer's data, shows the torque–speed operating limits for a brushless, permanent-magnet motor and the corresponding maximum continuous power envelope. The operating limit characteristic has also been plotted in Fig. 5.6 so the reader can see its relation to the motor's efficiency. (Because these data are for a brushless motor, the curves of Fig. 5.3 only qualitatively represent those of the brushed motor of Fig. 5.2.)

Limits on the rotational speed, winding temperature, and other quantities, such as the applied DC bus voltage, are imposed by the motor's manufacturer. Sensors report the condition of the motor to the controller. Exceeding a limit will usually cause the controller to shut down the motor.

Stall Torque Suppose the torque loading the shaft increases until the motor stalls. In this condition, the shaft power is zero and the current may be large. When stalled, all the electric power supplied to the motor is converted to heat. Consequently, the temperature is high, and damage may result.

The motor may stall at different torque loads depending upon the current available to it. The largest stall torque the motor can endure without damage is called the *maximum continuous stall torque*. For example, Ibrahim (1989) gives the maximum continuous stall torque of a version of the Hathaway HSSS3810 motor as 8.59 N·m. Under these conditions the temperature rise above ambient would be 42 °C.

Fig. 5.4 DC motor equivalent circuit

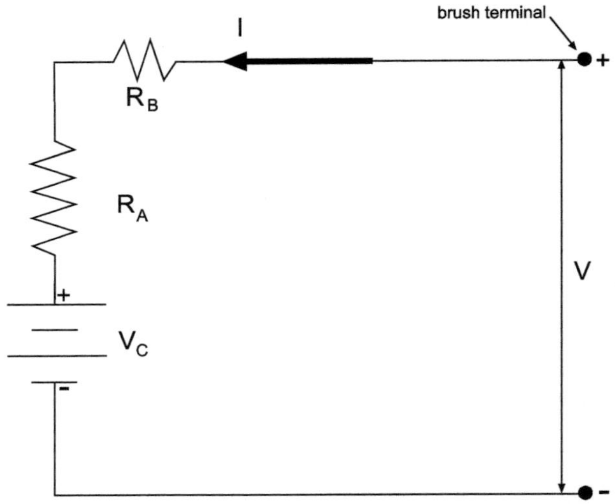

Commutation A means must be provided to reverse the current flow in a winding when it reaches the neutral axis. Otherwise, the forces will reverse their directions and stop the rotation. In *brushed* motors, a mechanical arrangement called a *commutator* switches the positive voltage to the lower side and the negative voltage to the upper side so the direction of current flow will not change under a pole whichever side is lower or upper. More discussion of commutation appears below in the discussion of losses. In *brushless* motors, the armature is the stator, and the field magnets rotate inside it. The switching is done by an electronic motor controller.

Circuit Figure 5.4 shows the equivalent electric circuit of the brushed motor of Fig. 5.2 operating at constant current. The current must pass through the armature resistance, R_A. This is the sum of the resistance of the windings, brushes, brush contact resistance, and the commutator segments. The battery or the solar array, or both, supplies the voltage, V, which drives the current, I, through the armature and makes it rotate.

$$V_C = k_C N \tag{5.6}$$

The little battery labeled "V_C" represents the *counter-voltage (*or *counter-electromotive force*). Because the wire rotates in the magnetic field, the field creates, or "induces," a voltage in the wire. The polarity of this voltage would cause current to flow in the wire counter to the direction of the current supplied to the wire. For a given motor, the counter-voltage is proportional to the rotational speed. The proportionality factor k_c is called the *counter-* or *back-emf constant.*[3] Thus, the faster the motor rotates, the higher V_C is. Ibrahim (1989) gives k_C as 0.02254 V/rpm ± 10 % for a version of the HSSS3810.

[3] The counter-emf constant is measured at no load, and therefore does not reflect the field distortion caused by the armature current.

The input voltage impressed on the equivalent circuit must be:

$$I = \frac{V - V_C}{R}, \tag{5.7}$$

where R is the sum of the armature and commutator resistances. Ibrahim (1989) gives 0.06 Ω ± 12 % for this total resistance for the HSSS3810 motor.

Note that

$$I = \frac{V - V_C}{R}. \tag{5.8}$$

When the motor starts, N is zero and therefore so is V_C. Consequently, the starting current can be very high at that instant. Ibrahim (1989) gives 66.51 V as the design voltage for the HSSS3810. Using R of 0.06 Ω, this means a starting current of 1109 A, or 7.39 times the peak current limit of Ibrahim's motor, 150 A. One of the functions of the motor controller is to keep the starting current below the motor's design limit. This could be done by introducing a series resistance of, say, an additional 0.4 Ω that would be gradually removed as N increased.

Multiplying Eq. (5.8) by the current gives the power that must be supplied to the motor as

$$P_{IN} = IV = IV_C + I^2 R. \tag{5.9}$$

The product IV_C does not represent a power loss but is equivalent to the shaft power plus a stray power loss (a function of rotational speed to be defined later).

$$P_{IN} = \frac{\pi}{30} \tau_S N + I^2 R + P_L \tag{5.10}$$

Equating Eqs. (5.9) and (5.10) and solving for IV_C gives

$$IV_C = \frac{\pi}{30} \tau_S N + P_L \tag{5.11}$$

as above.[4] A portion of IV_C supplies losses that occur "downstream" of the electric circuit, as it were, because they are associated with the motor's rotation. The remainder is the shaft power.

---

[4] Equation (5.11) leads to an interesting relation between k_S and k_C. Substituting from Eqs. (5.5) and (5.6) and rearranging gives

$$k_C = \frac{\pi}{30} k_S \left(1 + \frac{P_L}{P_M}\right).$$

The ratio P_L/P_M is usually small, so

$$k_C \approx \frac{\pi}{30} k_S.$$

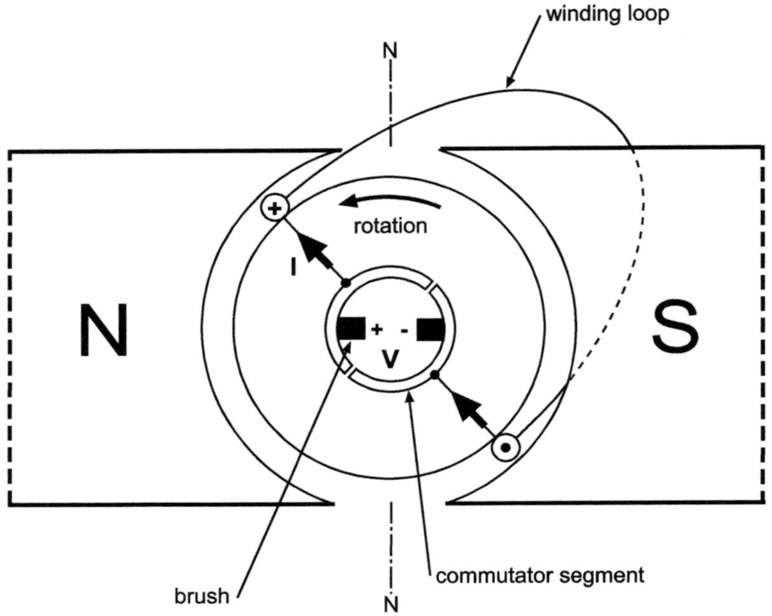

Fig. 5.5 Commutation principle

5.3 Losses

There are several loss mechanisms. They may be categorized as current-dependent or rotational speed-dependent. The current-dependent losses occur in the commutation device (mechanical or electronic) and in the resistance of the armature windings. A portion of the electronic commutation loss in brushless motors is speed-dependent. The remaining speed-dependent losses occur in the armature core, the rotor bearings, and in overcoming air rotational drag.

Mechanical Commutation Figure 5.5 shows a one-wire commutator. The wire current to the motor is connected to carbon brushes which are held against a segmented commutator ring that rotates with the armature. In actual motors, each armature wire loop begins and ends at a segment of this ring, as shown for one wire. These terminal segments are on opposite sides of the ring. The direction of current flow is shown by arrows. The electrical contact between the brushes and the segments of the ring is not perfect and therefore introduces resistance to current flow.

At the neutral axis, the commutator brush short-circuits the coil, the current drops toward zero, and the field it creates collapses. As the segment passes the brush, the current rises to its former value, but in the other direction. These changes, however, induce voltages in the coil tending to resist the current decrease or increase.

Multiplying k_S from Ibrahim (1989), 0.2152 N·m/A, by $\pi/30$ gives 0.02253 V/rpm, quite close to 0.02254 V/rpm, the value of k_C given by Ibrahim.

The voltages cause a current to circulate through the coil, the segments, and the brush. Arcing between the brush and the commutator segments may occur because of imperfect contact, such as at the trailing edge of the brush. The burning and melting actions of the arc cause damage to the brushes and the segments.

Motors may be fitted with *commutating poles* to combat arcing. These poles are wired in series with the armature. The arcing voltage induced by commutation is proportional to the load current. Hence, these small poles are not permanent magnets but derive their fields from the armature current. Thus, they may be designed to provide a compensating field which neutralizes the arc-producing field at any load. The resistance R_B in Fig. 5.4 represents the electrical resistance of the coils of the compensating poles, and of other materials in the current's path, such as the brushes, and the contact resistance of the brushes. The commutation power loss, P_C, is

$$P_C = I^2 R_B. \tag{5.12}$$

Electronic Commutation Solid-state devices are used for switching the armature coils of a brushless motor. Sensors detect the position of the field poles to initiate the switching action. A portion of the loss in the controller of a brushless motor arises from losses in the switching devices. It depends upon the switching rate, which is proportional to the rotational speed of the motor. The remainder of the loss is in the electrical resistance of the conductors. The commutation power loss model should therefore be the sum of a rotational speed-dependent term and a load current-dependent term.

$$P_C = k_{SW} N + I^2 R_B \tag{5.13}$$

The switching loss has been assumed directly proportional to N, with k_{SW} representing the constant of proportionality. The power consumption in electronic controllers may be made quite small. For example, AERL (1993) gives an efficiency of 96–99.5 % over the operating range of the controller it describes.

Electronic controllers can also provide various user-selectable control functions, such as cruising at constant speed or torque, regenerative braking, and current limiting when starting. They are also used with brushed motors to provide such features, as AERL (1993) describes.

Armature The power, P_A, lost in the armature resistance is proportional to the square of the current.

$$P_A = I^2 R_A \tag{5.14}$$

Note because the torque is proportional to the current, the armature resistance loss is also proportional to the square of the torque.

The magnetic field caused by the current in the armature windings interacts with the field of the poles, the armature reaction mentioned earlier. The resultant field in the air gap is distorted and weaker than that produced by the poles at no load, i.e., when the armature current is zero. Armature reaction thus reduces the torque available. The measured torque constant of the motor reflects the effects of armature reaction. Strictly speaking, therefore, it is not a constant but load-dependent.

Hysteresis The armature wire loops are wound over an iron core. The atoms of the core behave like small magnets. Each experiences a torque tending to rotate its magnetic poles into alignment with the magnetic field direction imposed upon it at any moment. Hence, the rotation of the armature forces realignment when the atoms in a region of the iron pass under a new pole. The atoms, however, cannot rotate freely because they are bound in the structure of the iron. The result is a kind of frictional heating of the iron core called *hysteresis*. The power lost in hysteresis, P_H, is approximately proportional to the rotational speed (Hanselman 1994).

$$P_H = k_H N \tag{5.15}$$

Eddy Current The changing magnetic field each conductor experiences because of the relative motion between the rotor and stator induces small circulating currents in the conductor. These are called *eddy currents*. The power P_E lost in eddy currents is approximately proportional to the square of the rotational speed (Hanselman 1994).

$$P_E = k_E N^2 \tag{5.16}$$

Bearing The mechanical friction torque in the bearings of the motor consumes power. The data of Kay (1988) suggest the friction torque speed dependence has a linear form with a constant term. The power, P_B, lost in the bearings would therefore be

$$P_B = (k_{B1} + k_{B2} N) N. \tag{5.17}$$

Drag Power, P_D, is also consumed in overcoming the air drag torque opposing the rotation of the armature. This loss depends upon the speed and upon characteristics of the rotor and stator surfaces bordering the air gap between them. These surfaces are rough compared to the air gap thickness. Because of the large relative roughness of the bounding surfaces and the low viscosity of air, we will assume the flow in the air gap may be categorized as turbulent at most rotating speeds, and that the rotational drag coefficient is independent of the rotational speed. The torque is therefore proportional to N^2, and the power consumed is

$$P_D = k_D N^3. \tag{5.18}$$

The proportionality factor, k_D, though independent of N, is dependent upon the pressure and temperature in the air gap.

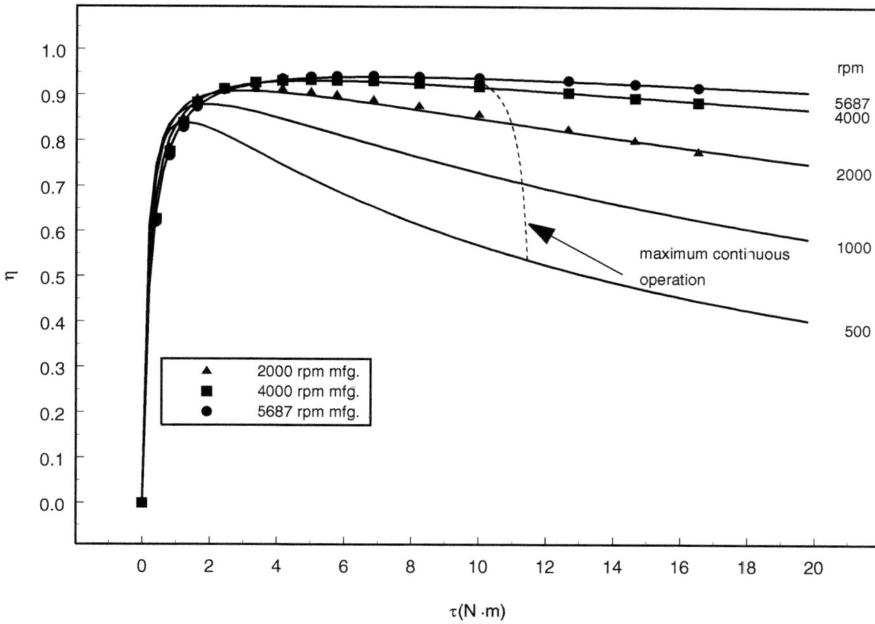

Fig. 5.6 Motor efficiency curves

5.4 Efficiency

The efficiency, η_M, of the motor is the ratio of the shaft power to the input power. Using the symbols defined above:

$$\eta_M = \frac{P_M}{P_A + P_B + P_C + P_D + P_E + P_H + P_M}. \tag{5.19}$$

Motors of interest to solar racers have efficiencies of 90 %, or greater, over portions of their operating range. Figure 5.6 shows the efficiency curves for the Solectria Corporation's BRLS8, a brushless, permanent-magnet motor. The curves were adapted from data furnished by the manufacturer. Each constant-rotational speed curve passes through a maximum, and the highest maximum efficiency occurs at the highest rotational speed allowed. We will use these curves to understand how the efficiency depends on rotational speed and torque. First, we substitute the loss definitions into the efficiency equation and transform current to torque using k_S. The result is

$$\eta_M = \frac{\frac{\pi}{30}\tau_S N}{\frac{R}{k_S^2}\tau_S + P_L + \frac{\pi}{30}\tau_S N}. \tag{5.20}$$

Table 5.1 Loss constants for BRLS8 efficiency model

k_0	4.0578 W/(N·m)^2
k_{01}	−3.4388(10^{-4}) W/(rpm·N^2·m^2)
k_{02}	1.8342(10^{-8}) W/(rpm·N·m)^2
k_1	0.01114 W/rpm
k_2	2.1575(10^{-6}) W/(rpm)^2
k_3	≈0

The coefficient R/k_S^2 has some speed dependence, represented by

$$\frac{R}{k_S^2} = k_0 + k_{01}N + k_{02}N^2. \tag{5.21}$$

The stray power loss, P_L, is represented by

$$P_L = k_1 + k_2 N^2 + k_D N^3. \tag{5.22}$$

The constants k_1 and k_2 are sums of the various constants defined in the stray power loss terms. Fitting Eq. (5.19) to the data (symbols) of Fig. 5.6 gives Table 5.1.

Note that the air drag constant was negligible. The solid lines in the figure were plotted from the curve fit.

The behavior of the efficiency can be understood from the behavior of the torque- and speed-dependent loss terms in Eq. (5.19). At a particular rotational speed, the losses that depend only on rotational speed are fixed. Therefore, at low torque, and therefore low armature resistance loss and low power, they form a proportionately greater part of the loss. Hence, the efficiency is low. The efficiency rises rapidly as the torque increases. However, the rapidly increasing armature resistance loss overcomes the shaft power increase, and the efficiency passes through a maximum and decreases.

Applying the condition that the slope of the efficiency curve must be zero at the maximum efficiency point at each rotational speed gives

$$\tau_{\text{optimal}} = \sqrt{\frac{P_L k_S^2}{R}} \tag{5.23}$$

for the torque at best efficiency at a particular speed. P_L increases, and R/k_S^2 decreases, with rotational speed. Hence, the optimal torque increases with rotational speed, as Fig. 5.6 shows.

5.5 Motor Types

Brushed, Permanent-Magnet Motors The motor of Fig. 5.2 is termed a brushed, permanent-magnet motor. Its magnetic field is supplied by permanent magnets in the stator, and the current to the rotor, inside the stator, is controlled by a commutator.

The use of permanent magnets avoids the power losses arising from supplying current to electromagnets in the stator to create the magnetic field. However, the magnets are very expensive compared to electromagnets.

The rotor is enclosed within the stator, which acts as a thermal insulator. This makes the rotor more difficult to cool. Brushed, permanent-magnet motors may come with a fan installed on the rotor shaft to provide forced cooling. The power consumed in cooling the motor in this fashion reduces the efficiency of the motor system.

Note how the maximum torque points of Fig. 5.6 cluster. The efficiency of brushed, permanent-magnet motors tends to have a narrow peak with respect to speed. This makes it more difficult to keep the drive operating near its best efficiency over a wide range of torques and speeds. Because of this, Bleck (1993) suggests this motor is best for cruising at a relatively steady speed. Hilly terrain, with its frequent requirement for low speed and high-torque operation, or urban driving, with its frequent speed changes, is not the best match to the efficiency characteristic. This conclusion applies to applications using a single-speed reduction. Employing a selectable-speed reduction, although adding weight and complexity, may give a net increase in energy efficiency during a race through such conditions because its use keeps the motor's efficiency higher, on average.

A controller is not required for commutation. One may be used, however, to provide the control functions mentioned earlier, such as low-current starting. The controller will be lighter and cheaper than those used with permanent-magnet, brushless motors.

Brushless, Permanent-Magnet Motors Placing the field magnets in the rotor, enclosing the rotor in the stator, and controlling the current to the stator electronically eliminate the brushes. Figure 5.7 shows this arrangement.

Commutation losses are reduced. Cooling of the stator is easier because it is on the outside of the motor, and a cooling fan may not be required. This also increases the efficiency.

The stator windings may be longer, for the same motor volume, because there is more volume available to them when on the outside. Therefore, the motor can produce two or three times the shaft torque than can a brushed, permanent-magnet motor of the same volume or weight (Bleck 1993). This characteristic is most desirable in solar racers, which must be light. But the addition of a large and complex controller counterbalances the smaller weight of the motor. Also, the controller is expensive, and its power losses must be charged to the drive. However, these losses are small because the controllers are typically very efficient.

The maximum efficiency of the motor–controller combination can reach 94% (Bleck 1993).

The efficiency characteristics at low rotational speed have relatively narrow peaks. Like the brushed motor, keeping the motor near its peak efficiency will be difficult in hilly terrain or urban driving.

Alternating-Current Induction Motors The stator windings of AC induction motors are placed around the rotor and supply the magnetic field in the air gap. The motor

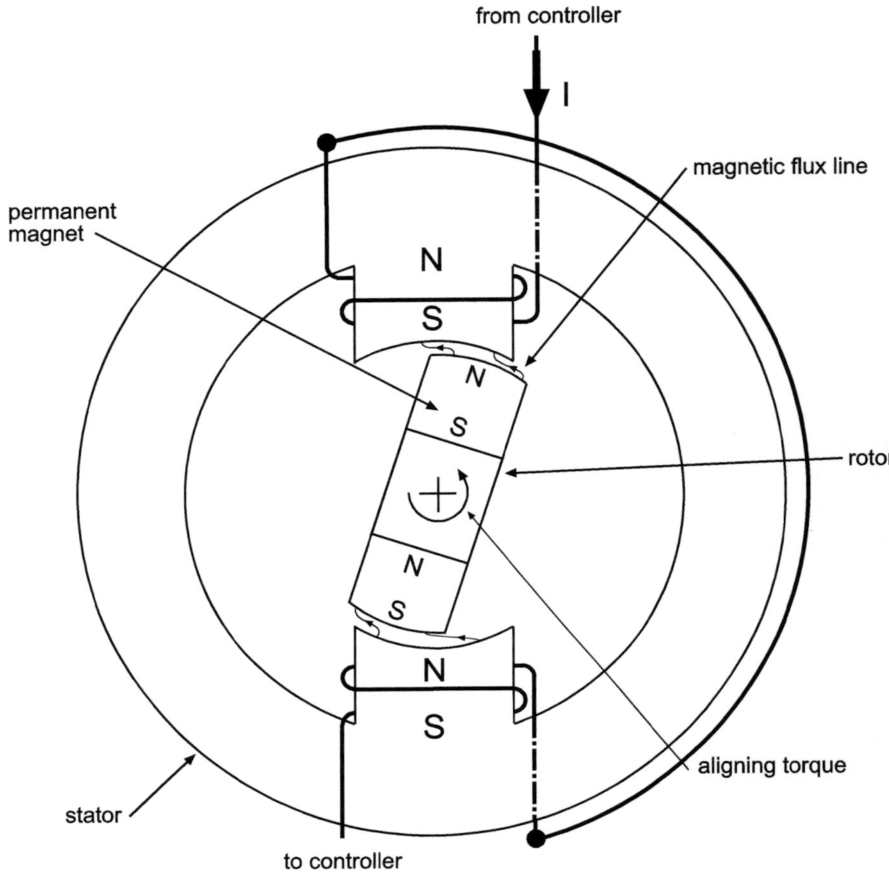

Fig. 5.7 A two-pole, brushless, permanent-magnet motor

controller transforms the DC current supply into two or three separate AC currents, called *phases*. Each of these phases is offset from the others in time by an equal fraction of the period corresponding to the frequency of the current. In the case of a two-phase motor, the offset is one-quarter period; in the three-phase motor, it is one-third period. Figure 5.8 shows the currents for the two-phase case. One phase is zero when the other is a maximum, because of the offset.

Each phase is connected to a winding of the stator. The windings are spaced about the stator such that the direction of the magnetic field in the air gap rotates as the current in each winding changes magnitude with time. This is illustrated in Fig. 5.9 for the two-phase case. Thus, instead of supplying current to the armature rotating in a stationary permanent-magnet field or a permanent magnet rotating inside an armature-stator, the field is electrically rotated. Figure 5.9 shows two poles rotating electrically; more than two poles are possible.

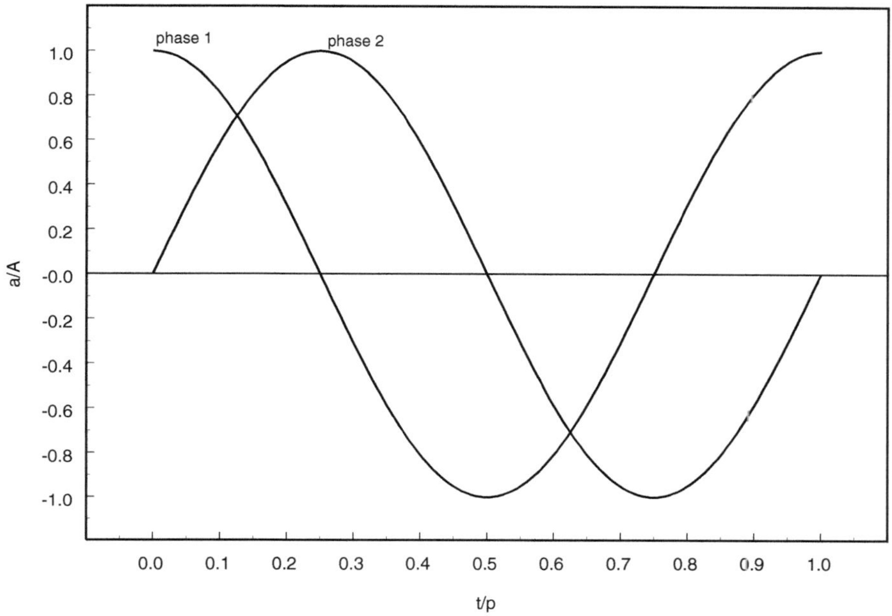

Fig. 5.8 Two-phase induction motor stator winding currents

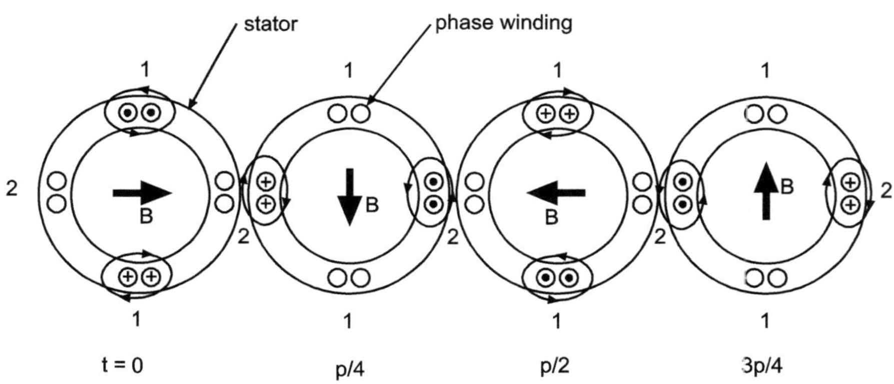

Fig. 5.9 AC induction motor field rotation. (Adapted from Loew 1954)

If f is the frequency of a phase and n is the number of poles, the rotational speed of the field in rpm is (Loew 1954)

$$N_{\mathrm{F}} = \frac{120 f}{n}. \tag{5.24}$$

Torque and shaft rotation are produced as follows. Current is not supplied to the rotor. Instead, the current *induced* in the rotor by the electrically rotated magnetic field creates a magnetic field which interacts with the field of the stator to produce torque and rotation. The voltage induced, and hence the current, is proportional to the rate at which the rotor conductors cut the air gap field. This rate is proportional to the difference between the field and rotor speeds. Let the *slip*, *s*, be the difference between the field and rotor rotational speeds, expressed as a fraction of the field speed.

$$s = 1 - \frac{N}{N_F} \tag{5.25}$$

At start-up, *s* is 1; as the rotor picks up speed, *s* is reduced. If there were no load nor bearing and air drag loss, the rotor speed would eventually equal the field rotational speed, the *synchronous* speed. However, in a real machine at constant load, the rotor speed eventually reaches that producing slip sufficient to generate a torque exactly meeting the load plus the losses. To do this, the field speed must always exceed the rotor speed.

The power lost in the resistance of the two windings (rotor and stator) reduces the efficiency compared to the two DC, permanent-magnet motors discussed because they each have just one set of windings. But the losses caused by hysteresis, which depend upon the slip, not the rotor speed as in the DC motor case, are diminished. The result is a motor with a flatter efficiency characteristic with respect to speed changes than either of the permanent-magnet motors. This suggests that AC induction motors may provide a higher average efficiency when a race route is characterized by urban driving or frequent hills.

Wheel Motors Electric motors may be integrated with the wheels they drive. King et al. (1995) describe such an AC induction motor-wheel integration. The need for a heavy rear axle and differential was thereby eliminated. Through careful design, permanent-magnet motors can also be built for high efficiency and torque at rotational speeds typical of the driven wheel, 1000 rpm and below. These characteristics allow the rotor to be connected directly to the driven wheel, thereby eliminating the transmission. Figure 5.10 illustrates the concept. Utilizing an axial field path allows adjustment of the thickness of the air gap between the rotor and stator. This gives the operator control over the torque constant of the motor, and the torque may be increased or decreased for a given current to the stator. NGM (1995) reports a 9-kW motor that can nearly triple its torque constant in this manner. Eddy current and hysteresis losses are lower because the motor must operate well below the rotational range of the radial-field designs already discussed. The motor, therefore, has a relatively flat efficiency characteristic. Figure 5.11, adapted from NGM (1995), shows the combined motor and controller efficiency characteristics at three rotational speeds. Note that the maximum efficiency is about 93 %. The design discussed by Lovatt et al. (1998) achieved a maximum efficiency of 97.5 %.

There are drawbacks. The rotor and stator are exposed to the environment and so must be carefully sealed against moisture and dust. From the viewpoint of the suspension, the motor becomes part of the driven wheel; its weight is unsprung. Therefore, its mass contributes to the inertial loading on the wheel when driving over potholes and bumps. And the structure of the motor must be designed to withstand the static and dynamic forces experienced by the wheel assembly.

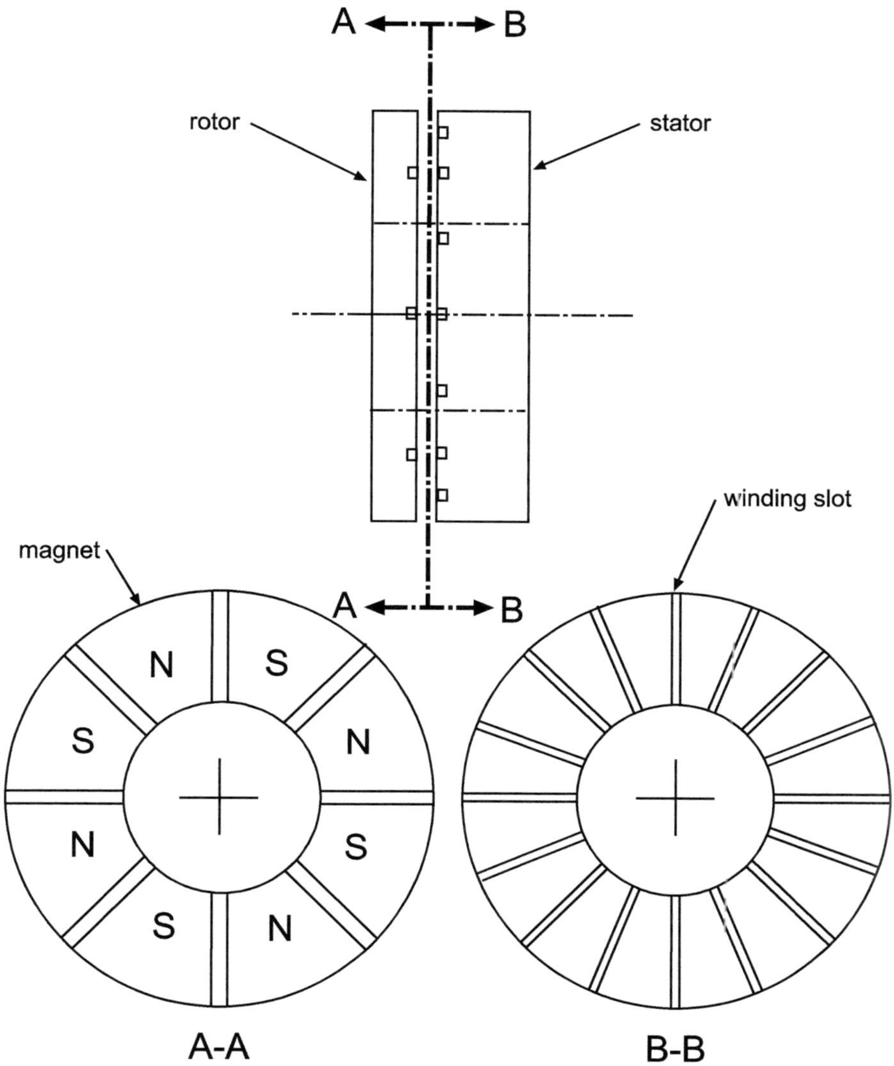

Fig. 5.10 Axial flux wheel motor. (Adapted from Hanselman 1994)

5.6 Speed Reduction

Suppose a conventional brushed or brushless motor is to be used, rather than a wheel motor, perhaps for economic reasons. The efficiency curves of Fig. 5.6, typical of conventional motors, show high efficiency at low torque and high rotational speed. This characteristic creates a mismatch with the higher torque–lower rotational speed requirement at the driven wheel of the solar racer. As Example 5.1 shows, a speed reduction between the motor and the driven wheel can be used to reduce this mismatch.

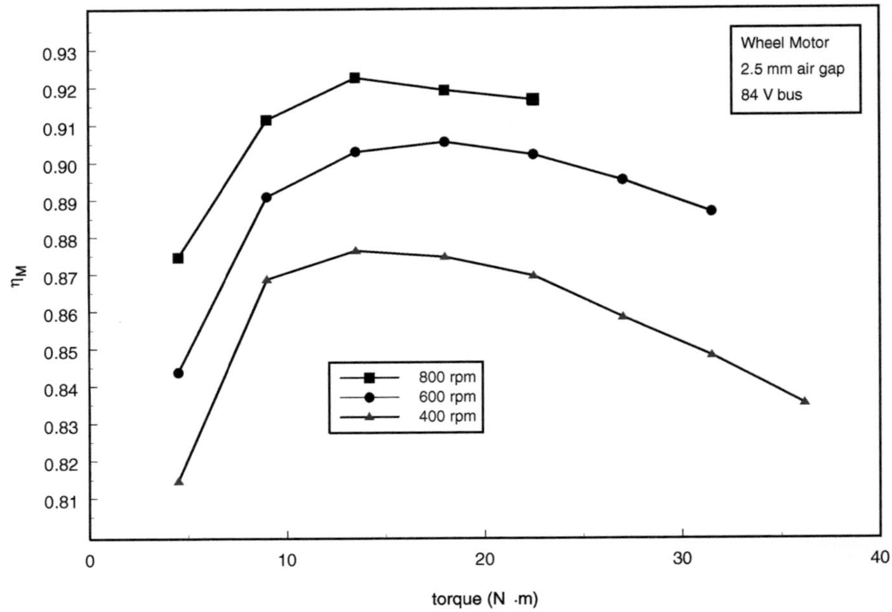

Fig. 5.11 Wheel motor performance. (Adapted from NGM 1995)

5.7 Example 5.1

Suppose the racer has the vehicle characteristics given in Example 4.3. It is to cruise at 55 mph on a horizontal road, climb a 5 % grade at 25 mph, and use the BRLS8 motor (Fig. 5.6). Select a speed reduction for this motor.

Solution The vehicle's wheels would turn at 940 rpm at 55 mph, and the driven wheel would require a torque of 19.18 N·m and a power of 1888 W (1.27 hp). A properly installed timing belt or chain drive reduction would have an efficiency of 95 % or greater. These are commonly used by solar racers because of their low weight and high efficiency. Using this value means the motor must produce 1987 W. The speed reduction must now be set to keep the motor at the highest possible efficiency. Multiplying Eq. (5.22) by $\pi N/30$ gives the power at maximum efficiency at any rpm. This power must be 1987 W.

$$P_{\text{opt}} = \frac{\pi}{30} N \tau_{\text{opt}} = \frac{\pi}{30} N \sqrt{\frac{P_L k_S^2}{R}} = 1987W \tag{5.26}$$

Solving this equation by trial and error gives 3824 rpm or a speed reduction of 4.068. The power and speed correspond to a torque of 4.96 N·m and an efficiency of 93.06 %. This point is plotted in Fig. 5.12. The power demanded from the battery and solar array would be 2135 W, a 13.4 % increase over the wheel power.

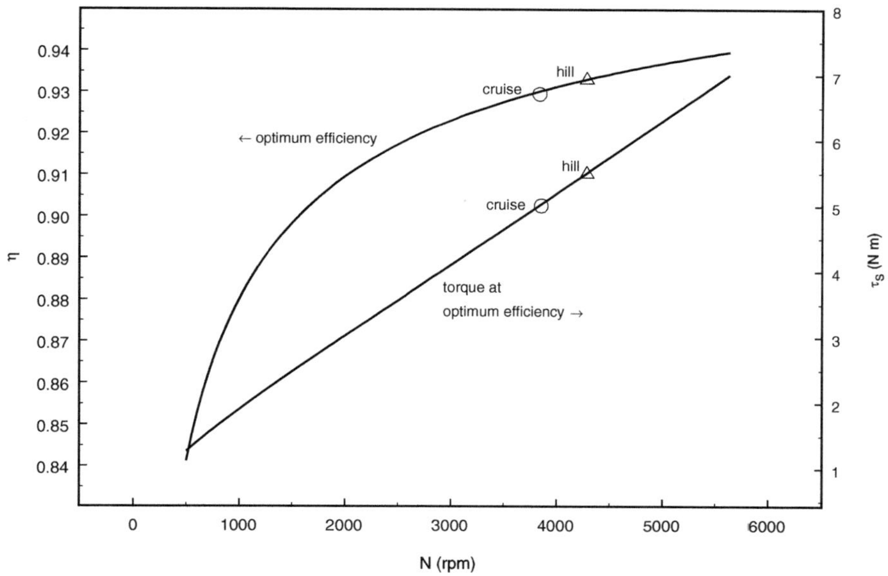

Fig. 5.12 Example 5.1 results

If, instead of cruising at 55 mph, the racer were climbing a 5 % grade at 25 mph, the torque and rotational speed required at the driven wheel would be 51.71 N·m and 427 rpm, respectively. The corresponding wheel power would be 2312 W. The corresponding motor power would be 2434 W, assuming the same speed reduction efficiency. Typically, solar racers run with a single reduction. If climbing the 5 % grade with the 4.068 reduction, the motor would operate at 1737 rpm, 13.38 N·m, and an efficiency of 78.83 %. Figure 5.6 shows this operating point to be outside the continuous operation region, far from optimal. The power demanded of the battery and solar array would be 3088 W, a 33.6 % increase over the wheel power. A compromise reduction for a hilly course would give more efficiency on hills but would limit the top speed when cruising on flats in order to keep the motor below its maximum allowable rotational speed.

If the racer were to be fitted with a two-speed reduction, with its design point chosen as climbing a 5 % grade at 25 mph, improvements would result. Assume that the two-speed reduction is 95 % efficient. Setting P_{opt} to 2434 W and solving Eq. (5.26) yields a motor rotational speed of 4266 rpm. Therefore, a speed reduction of 10 would be required. The torque would be 5.45 N·m and, from Fig. 5.12, the motor efficiency would be about 93.3 %. This operating point is comfortable within the continuous operating limit region of Fig. 5.6.

The power demanded of the battery and solar array would be 2608 W, a 12.8 % increase over the wheel power. This result, less than half that of the single-speed reduction, implies that, for the car under study, the greater weight and complexity of the two-speed reduction may be justified for a hilly course.

(The assumption that the one- and two-speed speed reductions have the same, constant, efficiency needs to be justified.)

5.8 I–V Curves

Drive An *I–V* characteristic for the drive, where *I* is the battery bus drive current I_M, and *V* is the battery bus voltage, can be combined with the solar array and battery *I–V* characteristics on a single plot. We can study this plot to acquire an understanding of the electrical dynamics of the solar array and battery, as their common load, the drive, changes.

For conceptual simplicity the motor will be a brushed, permanent-magnet unit, the *I–V* characteristic of which is described by Eq. (5.8). The controller feeding the motor in effect transforms[5] the current, I_M, and voltage, *V*, supplied by the battery bus to those values required by the motor according to

$$I_{MC}V_{MC} = \eta_C I_M V, \tag{5.27}$$

where I_{MC} and V_{MC} represent the current and voltage, respectively, supplied to the motor and η_C is the controller's efficiency. The bus voltage will be larger than the motor voltage. Consequently, the bus current will be smaller. The effect will be to shift the motor's *I–V* curve as seen by the battery bus toward higher voltages and lower currents. At a particular rotational speed, the curve is still a straight line with slope $1/R$. This load line has been drawn on Figs. 5.13–5.16. The curve shifts to the left or right as speed decreases or increases, respectively, because V_C is proportional to the speed.

Array The maximum power point trackers, symbolized by a single unit in Fig. 5.1, also act in effect as transformers, converting the array current and voltage according to

$$I_{MP}V_{MP}\eta_T = I_A V, \tag{5.28}$$

where I_{MP} and V_{MP} represent the optimal current and voltage, respectively, I_A is the array current to the bus, and η_T represents the efficiency of the trackers. The bus voltage is greater than the array voltage and the current less in the case of boost regulator trackers, and the reverse in the case of buck regulators.

Unlike that of the drive, the array's *I–V* curve viewed from the battery bus does not retain its characteristic shape. A particular solar irradiance and cell temperature define a unique maximum power characterized by unique values of I_{MP}, V_{MP}, and η_T. Thus, for these conditions, the array supplies a constant power to the bus. Hence,

[5] Zinger and Braunstein (1981) developed this concept for modeling maximum power point trackers. It was used later by Appelbaum (1989) and Appelbaum and Singer (1989).

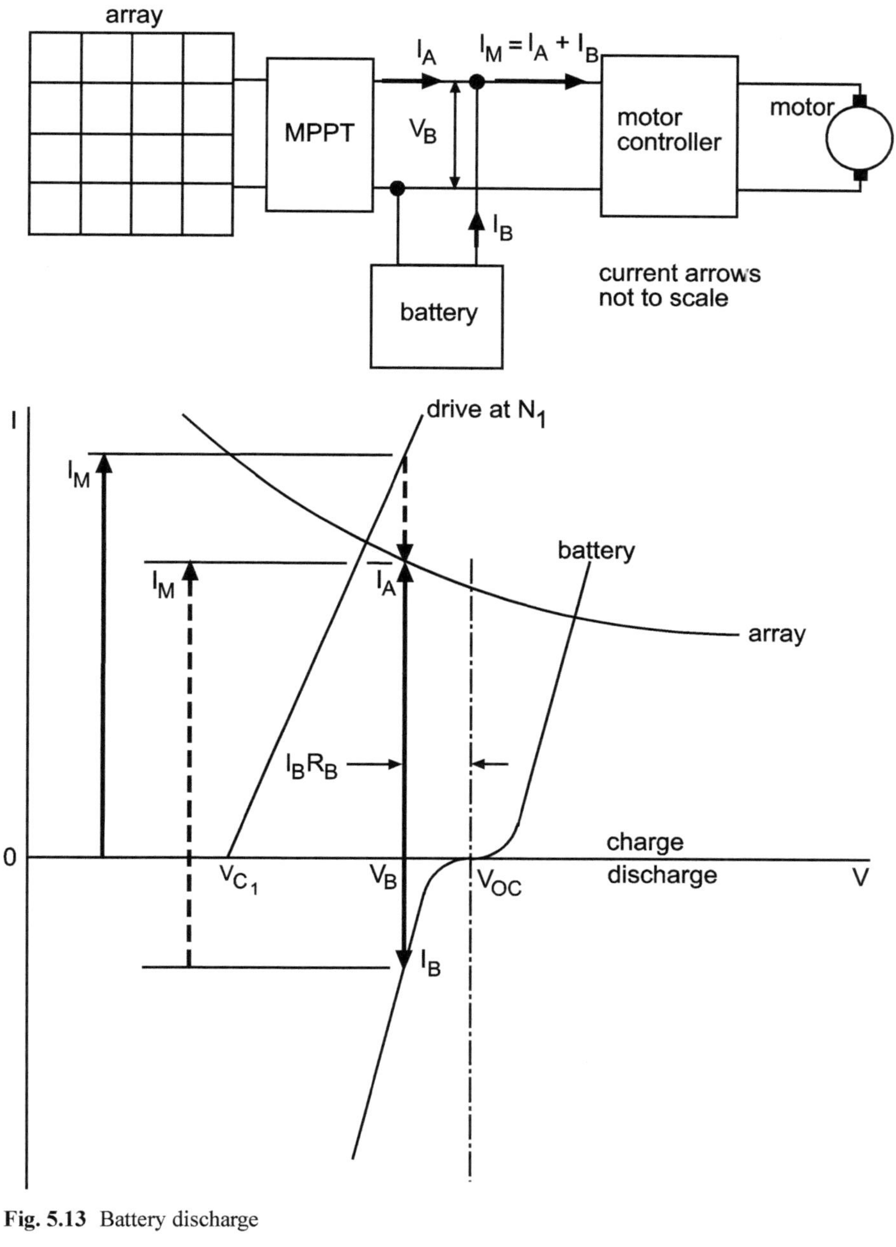

Fig. 5.13 Battery discharge

the array *I–V* characteristic appears to the bus as the hyperbola: *I–V* equals a constant. This curve is shown in Fig. 5.13. If the solar irradiation increases or the cell temperature decreases, the solar array maximum power increases. This results in a new hyperbola above the first; the reverse happens if the solar irradiance decreases or the cell temperature increases.

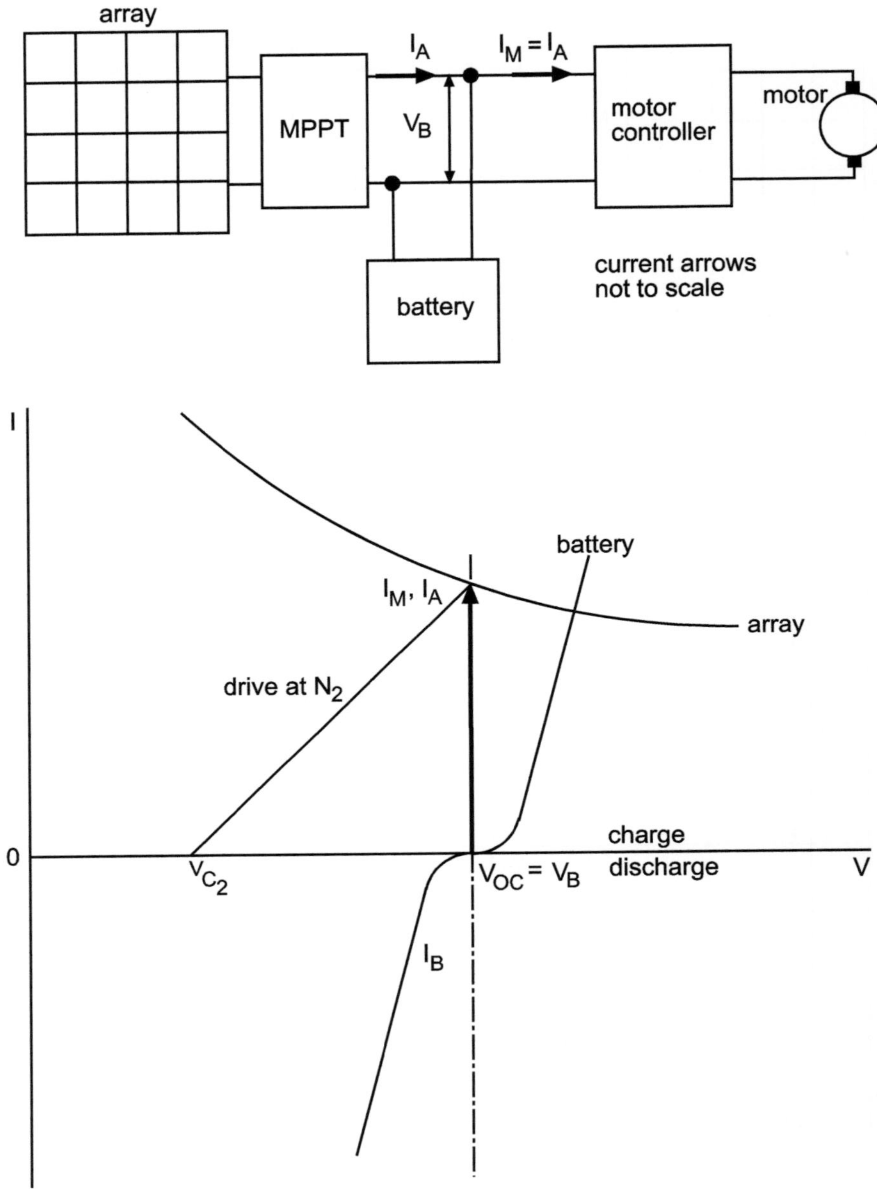

Fig. 5.14 Battery float

Battery The figures also show the battery's *I–V* curve. The curve intersects the voltage axis at the battery's open circuit voltage, V_{oc}. The slope of the discharge half of the *I–V* curve flattens toward the voltage axis as the charge in the battery decreases, so that the available battery current is less at a given voltage. The reverse happens

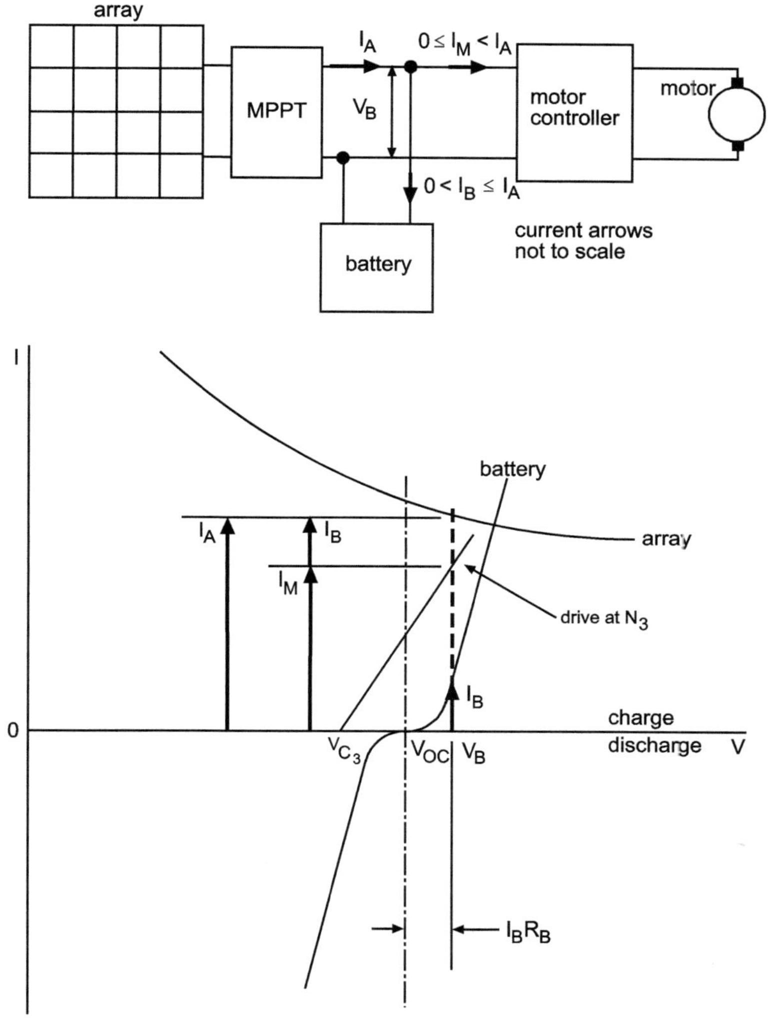

Fig. 5.15 Battery charge

as the battery gains charge. In the following discussion, we will at first assume that the slope is constant so we may focus the discussion on the main events. We will also ignore the effect of age (number of equivalent full charge–discharge cycles). In a 2-week-long race, age will not affect the *I–V* curves of the types of batteries presently allowed.

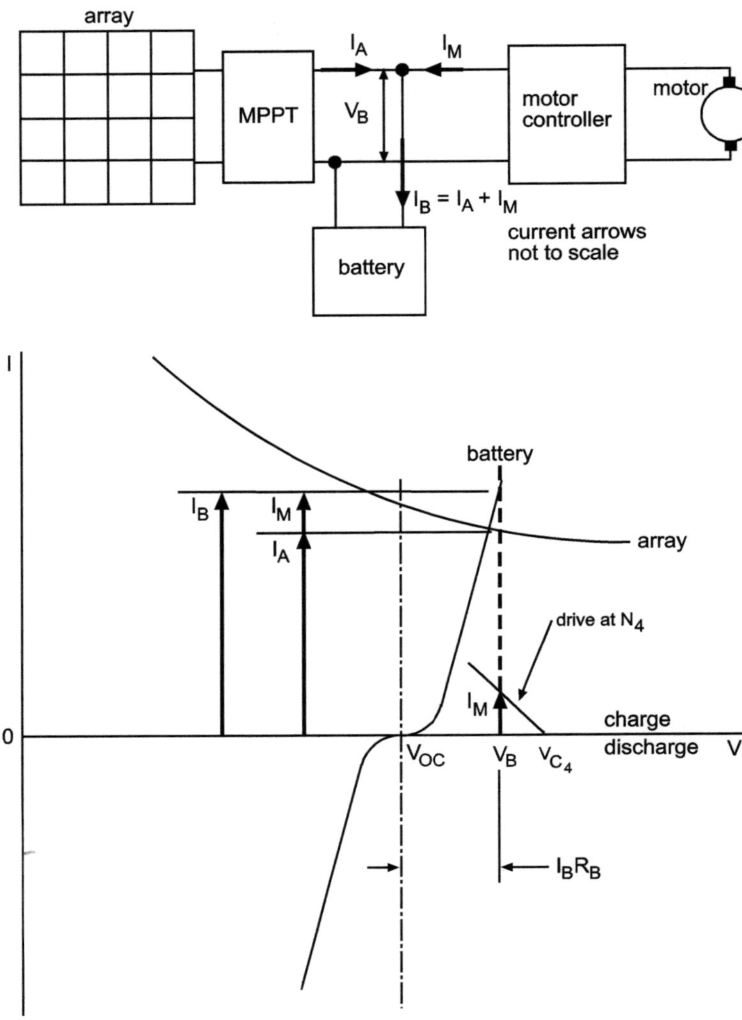

Fig. 5.16 Regeneration

5.9 Solar-Electric Drive Operation

A vertical line representing the bus voltage must intersect all three I–V curves in the figures. The intersection points imply the current supplied or used by the device represented by a particular curve or combination of curves. In this discussion, we will ignore the housekeeping loads on the battery bus, such as telemetry. These loads will be small compared to that of the drive.

Battery Discharge Suppose the car is cruising at a highway speed corresponding to a motor rotational speed of N_1. The cruise condition usually requires that current be drawn from the battery to supplement the array's supply. The sum of the array current, I_{A1}, and the battery current, I_B, supplies the current demanded by the drive, I_M. Figure 5.13 shows this situation schematically. In the *I–V* plane, the bus voltage, V_B, drops below the battery's open circuit voltage, causing the battery to discharge at a current of I_B.

If the charge on the battery were low, the battery's *I–V* curve would be flatter. The current the battery could supply at each voltage would then be less in any discharge scenario. The array *I–V* hyperbola is fairly flat, so the driver must reduce the demand, i.e., slow down. Otherwise, the bus voltage will drop further, rapidly increasing the battery current and rapidly decreasing the battery's charge. This will continue until the controller reaches its low-voltage cut off, or the motor stalls.

Battery Float Suppose that the racer enters a reduced speed zone. The river reduces speed until I_M is such that the solar array current, I_A, can supply it exactly: I_M equals I_A. Figure 5.14 shows this state. There is no excess array current and no load on the battery. Therefore, I_B is zero; the battery is in the open circuit condition, floating on the bus, so to speak. The drive's load line will then pass through the intersection of the battery's open circuit voltage with the array's load line, as shown.

Battery Charge When the car is moving slowly in traffic we suppose that the drive current demand can still be met by the array, but with some excess. This excess is charged into the battery such that I_M is I_{A2} less I_B. Figure 5.15 shows this schematically. Suppose the car waits at a stop light. There is no current demand from the drive. Therefore, the bus voltage V_B is greater than the battery's open circuit voltage and passes through the intersection of the battery's *I–V* curve with the array's *I–V* curve. The battery is now charged at the rate I_B equal to the array current, I_A.

Regeneration If the car begins to coast down a hill, as in Example 4.3, the motor requires no current but is instead driven by the torque from the net force component pointing down the hill. The motor continues to rotate in the same direction and picks up speed as the car accelerates. V_C, therefore, rises and overcomes the applied voltage. The armature current then reverses direction and is supplied to the bus; the motor becomes a generator. The torque created by the armature current reverses, resisting the shaft torque driving the generator. This brakes the car. The generated current is supplied to the bus, charging the battery. The sign of the slope of the drive's *I–V* curve is reversed. This situation is illustrated in Fig. 5.16.

The controller must allow regeneration to be applied gradually, perhaps through a regeneration pedal. If it is applied rapidly, the armature current rises rapidly, and the battery may be charged too quickly, causing gassing, and the large braking torque may cause the driven wheel to lock. The braking effect tends to be stronger at high rotational speeds and weaker at low.

It is possible in a day of stop-and-go and hilly driving to recover from less than 10 % to as much as 25 % of the battery's capacity, according to Korff (1980). Korff also observes that frequent short charging reduces stratification in the battery's electrolyte, tending to increase the battery's life and capacity.

References

AERL. (1993). *New AERL MOTORMAX*. Queensland: Australian Energy Research Laboratories P/L.

Appelbaum, J. (1989). The operation of loads powered by separate sources or by a common source of solar cells. *IEEE Transactions on Energy Conversion, 4*(3), 351.

Appelbaum, J., & Sarma, M. S. (1989). The operation of permanent magnet DC motors powered by a common source of solar cells. *IEEE Transactions on Energy Conversion, 4*(4), 635.

Bleck, O. (1993). *Electric drive systems: Understanding and selecting the proper drive system for an electric vehicle*. Wilmington: APN2001 (preliminary), Solectria Corporation.

Hanselman, D. C. (1994). *Brushless permanent magnet motor design*. New York: McGraw-Hill Inc.

Ibramim, F. (1989). *Polyphase brushless DC motor (HSSS3810)*. Tulsa: Hathaway Corporation.

Kay, R. (1988). The new ball bearings. *Bike Tech, 10*.

King, R. D., Haefner, K. B., Salasoo, L., & Koegl, R. A. (1995). Hybrid electric transit bus pollutes less, conserves fuel. *IEEE Spectrum, 32*(7), 26.

Korff, W. H. (1980). *Motor vehicle design*. Burbank: M-C Publications.

Loew, E. A. (1954). *Direct and alternating currents*. New York: McGraw-Hill Book Company, Inc.

Lovatt, H. C., Ramsden, V. S., & Mecrow, B. C. (1998). Design of an in-wheel motor for a solar-powered electric vehicle. *IEE Proceedings-Electric Power Applcations, 145*(7), 402.

NGM. (1995). *NGM wheel-motor and controller series*. Vienna: New Generation Motors Corporation.

Singer, S., & Appelbaum, J. (1993). Starting characteristics of direct current motors powered by solar cells. *IEEE Transactions on Energy Conversion, 8*(1), 47.

Zinger, Z., & Braunstein, A. (1981). Dynamic matching of a solar-electrical (Photovoltaic) system, an estimation of the minimum requirements of the matching system. *IEEE Trans Power Apparatus and Systems, PAS-100*(3), 1189.

Chapter 6
Electric Power Conversion and Distribution

6.1 Introduction

The electrical system connects the electric power sources to the high- and low-voltage loads. It is shaped by the current and voltage demands of the various loads, the requirements for grounding, switching, and electrical overload protection imposed by good practice and race rules, the need for low weight, low power loss, reliability, rapid maintenance and repair, and low cost.

The rules of solar-electric vehicle races differ. Therefore, to avoid having to consider many cases, the 2001 American Solar Challenge (ASC) rules (Chap. 16) will be applied herein. However, as previously mentioned, be sure to consult the current rules for the race you are considering.

6.2 Power Supply

Under ASC rules, once the race begins, the solar cell array and the battery are the only sources of power allowed, with three exceptions. Power for the radio, electronic panel meters, and telemetry equipment, etc., may come from "supplemental, replaceable" batteries.[1]

6.3 Loads

Typical Loads Table 6.1 lists the typical electric loads of a solar car electric system. The ventilation could be supplied by a single fan, as was done for the Sunraycer (MacCready et al. 1990). Also given are representative quantities for each kind of load and whether the load operates continuously (C) or intermittently (I).

[1] See Chap. 16, page 320, sub-section Supplemental Batteries.

E. F. Thacher, *A Solar Car Primer*, DOI 10.1007/978-3-319-17494-5_6

Table 6.1 Electric loads

Load	Quantity	Duty
Motor controller	1	C
MPPT	4	C
Battery fan	1	C
Cockpit fan	1	I
Brake lights	2	I
Turn signals	4	I
Horn	1	I

No headlights are shown in Table 6.1 because only daylight operation is required during a race.[2] All the loads must operate from the direct current (DC), otherwise an *inverter*, which converts DC to AC, must be employed. Inverters consume power and add mass.

Supplemental Loads Powering the radio, some panel meters, and the telemetry equipment from the main storage batteries will reduce the energy available for propulsion and increase the mass of the car by the mass of the connecting wiring. Alternatively, powering these devices from supplemental batteries will increase the mass of the car by the mass of the supplemental batteries. The mass involved is not large (except possibly that of the telemetry equipment battery). Nevertheless, every kilogram increase should be thought through to see if it gives a net advantage. The trade-off question is: Will the reduction in range from powering the equipment from the main battery be greater than that from carrying the extra mass of supplemental batteries?

6.4 Basic Interconnections

Block Diagram Figure 6.1 shows a simplified block diagram of a typical electrical system. The battery and solar array are connected in parallel with the motor controller so that either or both of these power sources can supply power to the motor controller, and excess power can be sent to the battery. The voltage of this *main bus* or *battery bus* is therefore applied equally to the solar array, the battery, and the motor controller.

Low-Voltage Loads Fig. 6.1 shows that all the loads in the electrical system do not require the same voltage. Generally, there is a low-voltage subsystem (the "low-voltage taps," typically 12 V, of Fig. 6.1) that powers loads such as, the turn signals, backup lights, horn, and the cockpit-ventilation fan. The DC–DC converters shown in Fig. 6.2 reduce the main bus voltage to the lower voltage required. They feed two *branch circuits*.

[2] However, the state where the car is built and tested may require headlights.

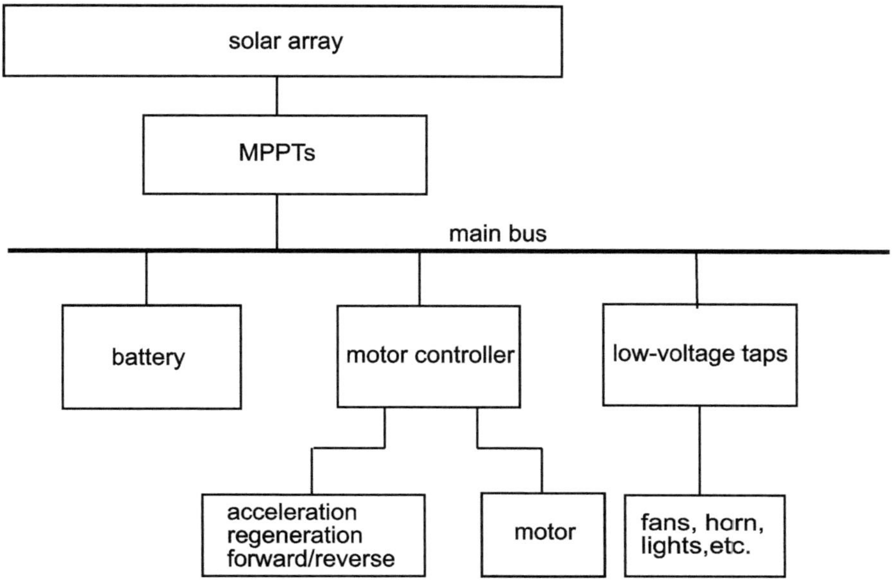

Fig. 6.1 Electric system block diagram

6.5 Efficiency and Voltage

Efficiency The low power available to the solar racer makes it essential to pay close attention to minimizing electric power losses. Power losses in wiring are proportional to the product of the square of the current and the electric resistance (or to the product of the voltage drop over the wire and the current). For a fixed resistance, halving the current reduces this loss by a factor of four. Motor operation and battery charge and discharge are also more efficient at low current and high voltage. Therefore, the rule is to operate the electrical system at high voltage and low current.

Design Main Bus Voltage According to the thumb rule discussed above, the *design* main bus voltage should be as high as practical.[3] ("Design value" means the value of the bus voltage under specified design conditions, such as those of Chap. 16). Beside meeting this general goal, the design value must be an integral multiple of the battery module voltage, be within the allowable input-voltage range of the motor controller, and be within the allowable input-voltage range of the DC–DC converters supplying the low-voltage subsystem.

If the MPPTs are boost regulators, then the goal should be to have the array voltage always[4] below the main bus voltage. For buck regulators, it should always be

[3] The voltage at any instant will depend on the current draw at that instant and the I-V characteristics of the two power sources and the connected loads; this is developed in more detail in Chap. 5.

[4] Read et al. (1990) reports a boost regulator design in which an overvoltage can occur when the array is in full sun and the main bus voltage is low because the batteries are nearly drained and the motor current is large.

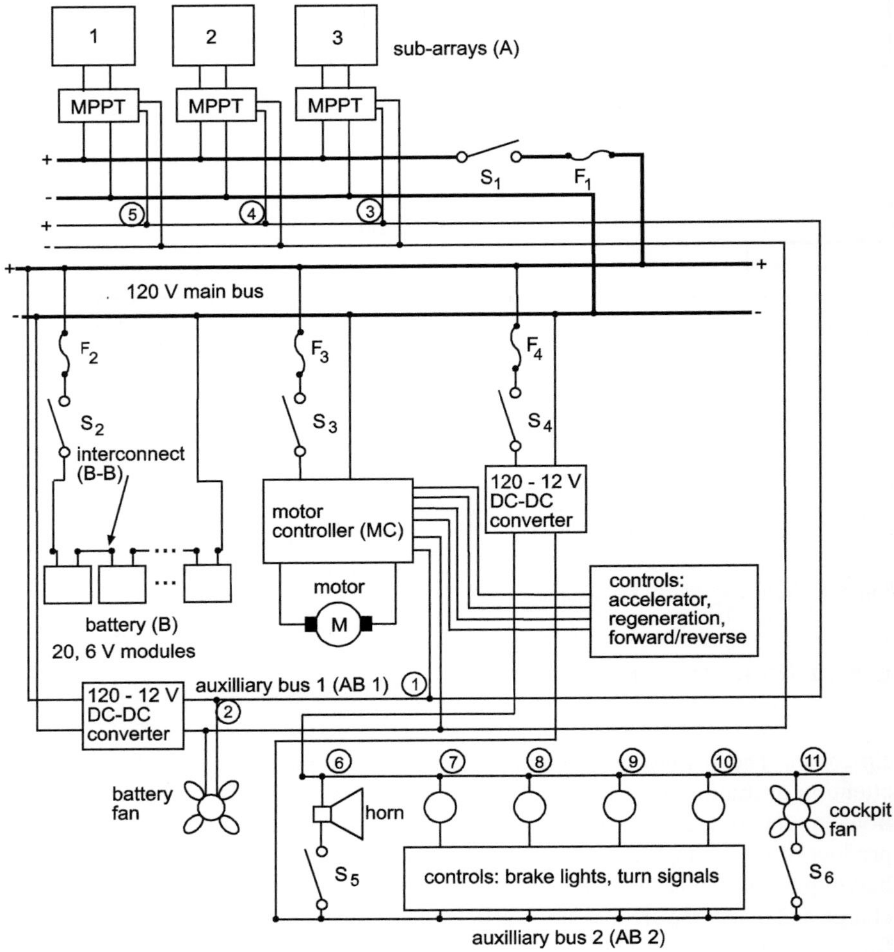

Fig. 6.2 Power system schematic

above the main bus voltage. The main bus voltage thus also influences the number of solar cells in the series-connected strings of the solar array.

6.6 Mass

Target Chapter 11, *Solar Racer: Construction,* advises that the mass specified for the car be apportioned among the car's systems. Thus the electrical system should have a mass target, most of which will be consumed by the batteries. Nevertheless, every opportunity to reduce the mass should be exploited; many small reductions add up to a big reduction. And a solar car has many small parts.

Table 6.2 Bare copper wire characteristics

AWG	Diameter (in)	Ft/lb	Ohm/1000 ft
22	0.0	514.2	16.46
20	0.0320	323.4	10.35
18	0.0403	203.4	6.510
16	0.0508	127.9	4.094
14	0.0641	80.44	2.575
12	0.0808	50.59	1.619
10	0.1019	31.82	1.018
8	0.1285	20.01	0.6405
6	0.1620	12.58	0.4028
4	0.2043	7.914	0.2533
2	0.2576	4.977	0.1593
1	0.2893	3.947	0.1264

Current Low current demand, in addition to improving efficiency, allows smaller wires, smaller fuses, and switches, which reduce the mass of the electric system. Wire size is selected by the amount of current to be carried, larger currents requiring larger wire, and mass is directly proportional to wire size. Size is indicated by the wire's American wire gauge (AWG) size, with large AWG meaning small wire. The data in Table 6.2 for solid, bare copper wire show how weight and length are related (ARRL (1973)).

Note that for a given length, decreasing the resistance by reducing the AWG increases the mass. However, the layout of the electrical system controls the length of the wire, and both the mass and the resistance are directly proportional to the length. Careful attention to the layout can reduce both mass and power loss by keeping wire runs short.

6.7 Wiring

Sizing Wire The size of the wire affects not only the efficiency, but also safety. Undersized wire can overheat, resulting in damaged insulation, short circuits, or fires. The maximum possible current allowed in a particular wire (the *ampacity*) depends upon the wire size, the kind of insulation on the wire, and whether the wire is a single conductor in free air (better cooling), or bundled in a conduit or cable (poorer cooling). For example, Table 6.3, adapted from the National Electric Code (NEC) and presented in McCarney et al. (1987), shows that 14 AWG wire has an ampacity of 15 A when in a conduit or cable and covered with thermoplastic insulation, but has an ampacity of 20 A when in free air with the same insulation.

Sizing Rule A thumb rule recommended by McCarney et al. (1987) to size the wire is to allow a maximum of 2% voltage drop in branch circuits (fed from load centers) and an overall maximum of 5% voltage drop from the power source to the

Table 6.3 NEC copper wire ampacity

	Conduit and cable		A wire; free air	
AWG	T and TW	THW	T and TW	THW
14	15	15	20	20
12	20	20	25	25
10	30	30	40	40
8	40	50	60	70
6	55	65	80	95
4	70	85	105	125
2	95	115	140	170
Code	*Tmax (C)*	*Env*	*Insulation*	
T	60	Dry	Flame-retardant, thermoplastic	
TW	60	Dry or wet	Flame-retardant, moisture-resistant thermoplastic	
THW	75–90	Dry or wet	Flame-retardant, moisture- and heat-resistant thermoplastic	

load (applied to runs between the array, the battery, or the motor and the main bus). Smaller limits may be required by manufacturer's instructions. Wider limits may be allowed, too. For example, the allowable input-voltage range to a commercial DC–DC converter producing 12 V might be 100–200 V, or the allowable input-voltage range to a motor controller might be 65–130 V.

For branch circuits, using the wire length to the farthest load and the peak current may result in too much mass. If so, divide the branch into segments and size the wire in each segment according to its length and peak current.

Wire Types Copper is the preferred metal for use in wires because of its low electrical resistance. Wires may be solid or braided. Braided wires are more flexible and therefore, are preferred for large sizes such as those used for battery module interconnections. Wiring in solar electric cars must always be insulated. The insulation should be color coded to show the wire's service assignment. As Table 6.3 implies, the type of insulation must be chosen for the temperature range and whether dry or wet conditions prevail. Summer operation dictates a high specified design temperature, at least 50 °C. Moisture inside the car should be expected.

Connections To promote rapid maintenance and repair, modular construction and quick-release electrical connectors, and tie-downs should be employed, when possible. This allows rapid replacement of malfunctioning modules.

6.8 Switches and Fuses

Switches Switches are used to connect power sources to the main bus, and to turn loads on and off. They are rated for a particular voltage, current type, and amount of current. Current type matters because DC current tends to arc (jump) across the contacts of a switch as the switch opens.[5] Switches intended to interrupt DC current are designed for this. If the switch is not properly rated, it may burn out from repeated arcing, or in heavy current applications the contacts may weld together (McCarney et al. 1987).

Fuses Fuses are the means of overcurrent protection required by race rules; circuit breakers are not allowed. Fuses should always be placed in the positive wire; the grounded side of a circuit should never be switched or fused. If a fuse was placed in the negative wire, an overcurrent condition causing the fuse to open the circuit would disconnect the load from ground, not from the voltage supply. Thus, a person touching the load could be electrically shocked.

6.9 Grounding

Motivation and Definition Grounding helps to prevent electric shock to people working on or touching the body or components of the solar car. In the case of vehicles, it does not refer to an actual connection to the earth. Instead, it means that a common point is provided in the electric system where the negative (return circuit) wires of the power and instrumentation circuits are electrically tied together. This point will be the common voltage reference for the car, the zero voltage point.

Grounded Frame Sometimes solar cars are constructed using a "space frame" of metal tubing inside a streamlined, composite shell. The frame should be connected to the electrical ground point. Otherwise there will be an electric potential between the frame and any conductor in the car, even between the frame and the ground point. Also, an ungrounded frame can accumulate an electric charge, which when discharged can damage sensitive electrical devices. The frame should be connected to ground at only one location to prevent current flow between multiple ground points (*ground loops*).

When the frame is connected to ground, electric shock hazard exists only at high-voltage points, such as the battery and solar array terminals or the main bus. Race rules require high voltage warning signs at such dangerous locations.

Electrical Noise Electrical noise interferes with radio voice communications because it is picked up by radio receivers along with the transmitted voice signal, causing the voice to be less intelligible. Sources of electrical noise are sparking from propulsion motor and ventilation fan commutator brushes (if present), operating turn signals, and wheel and tire static. ARRL (1973) discusses solutions for these problems. The "receiver" for the electrical noise is the wire on board a solar car, which may be of considerable length. Consider using electrically-shielded wire

[5] For example, when power to an inductance (i.e., a motor) is turned off, the current decreases quickly, inducing a large voltage in the windings, which appears across the switch contacts creating a momentary arc.

with the shielding grounded. As usual, this must be balanced against the additional mass added to the car.

6.10 Wiring Diagram

Figure. 6.2[6] shows one way of connecting the power sources to the loads. The figure shows no instrumentation to measure the voltage and current in different parts of the system. Instrumentation will be added in Chap. 7, *Instrumentation.* The two power sources, the motor controller, and Auxiliary Bus 2 incorporate a fuse and a disconnect switch in the positive wire. Auxiliary Bus 1, which powers the battery fan, has no disconnect switch nor fuse but is energized whenever the battery or the solar array is connected to the main bus, as required by race rules.

6.11 Example 6.1

Size the wire, using the sizing thumb rule given above, fuses, DC–DC convertors, and circuit breakers for the solar car electric system of Fig. 6.2. Estimate the weight of the wiring. Add allowances for insulation, connectors, and uncertainty in wire run length.

Loads (main bus at 120 V):

1. A 2-hp DC motor with an overload capability of 4 hp for 15 min (30-A overload).
2. 12 V, 1 A for the electronics, motor controller, and battery fan.
3. 12 V, 9.25 A for the turn signals, running lights, cockpit fun, and horn.

Power sources:

1. Solar cell array giving 9 A at 120 V, (under standard testing conditions).
2. 20, series-connected, 6-V, Ag–Zn batteries.

Solution Figure 6.3 shows the approximate distances for the wire runs and the locations of the components. The main bus is a terminal strip housed in a waterproof, plastic junction box.

The motor controller is located near the motor in the rear of the car close to the driven wheel. The two DC–DC converter connectors are near the main bus box. The battery box is mounted just forward of the main bus box.

The data in Table 6.1 were used to size the wires. The codes in the table of results, below, are those next to the component names in Fig. 6.2. Somewhat larger wires were chosen to allow for uncertainties in wire run lengths.

[6] A simplified version of the electrical system of Kalakwaneha, Clarkson University's first solar racer, as presented in Read et al. (1990).

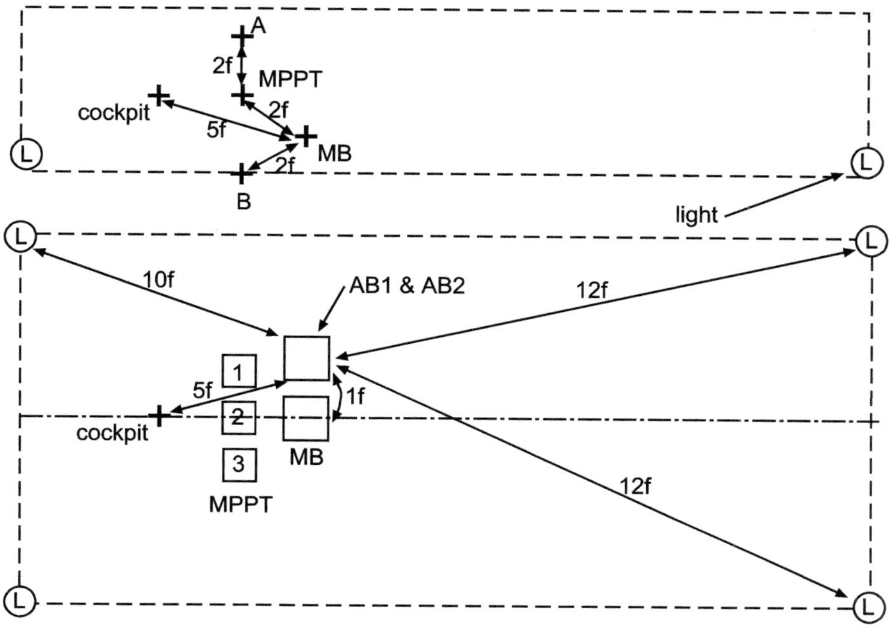

Fig. 6.3 Wire runs for example 6.1

Wire run	Max. volts	Max. amps	One-way length (f)	Loss (%)	AWG	Type	Wte. (lb)
A-MPPT	120	3	1	2	18	Solid	0.01
MPPT-MB	120	3	2.5	2	18	Solid	0.05
B-MB	120	35	2	2	14	Braid	0.05
MB-MC	120	30	12	5	12	Braid	0.48
MB-AB1	120	1	1	2	18	Solid	0.01
AB1-1	12	0.25	12	2	22	Solid	0.03
AB1-2	12	0.15	2	2	22	Solid	0.005
AB1-3	12	0.2	2	2	22	Solid	0.005
AB1-4	12	0.2	2	2	22	Solid	0.005
AB1-5	12	0.2	2	2	22	Solid	0.005
MB-AB2	120	2	1	2	18	Solid	0.01
AB2-6	12	5	5	2	16	Solid	0.08
AB2-7	12	1.0	10	2	18	Solid	0.1
AB2-8	12	1.0	10	2	18	Solid	0.1
AB2-9	12	1.0	12	2	18	Solid	0.12
AB2-10	12	1.0	12	2	18	Solid	0.12
AB2-11	12	0.25	5	2	22	Solid	0.012
B-B	120	35	15 × 0.33 ft	Small	2	Braid	1.005
						Total	2.20

The calculation for branch AB2-9 was as follows. The maximum two-way length for a particular AWG and peak current should yield less than the specified percentage voltage drop

$$\frac{\Delta V}{V} > \frac{I_{peak} R' 2L}{V} \tag{6.1}$$

where R′ is the resistance per foot from Table 6.1, L is the one-way wire run length, and ΔV/V is the fractional voltage drop allowed over 2 L. Substituting the numbers for the example branch,

$$0.02 > \frac{1A \times 6.510(10^{-3})\frac{\Omega}{ft} \times 2L}{12V}$$

or, L < 18.4 ft. Because the run is 12 ft, this gives a 6.4 ft, one-way allowance for run length uncertainty, using #18 wire.

The battery interconnects (B–B) were made much larger than necessary, following the advice of McCarney et al. (1987).

Allowing 20 % for connectors and insulation brings the total from 2.20 to 2.64 lbf.

All switches and fuses are sized for 150 % of the peak current and the relevant DC voltage. This will satisfy the sample race rules which restrict the battery fuse to not more than 200 % of the maximum expected current draw. Thus, in our example, the battery fuse must be rated for 53 amps. S_2, the battery disconnect, must be rated for at least this current.

The DC–DC converter for Auxiliary Bus 1 must supply 12 V at 1 A, or 12 W. The converter for Auxiliary Bus 2 must supply 12 V at 9.25 A, or 111 W. The manufacturer's data sheets show that the converters will weigh about 0.5 lbf each, bringing the system weight to 3.64 lbf.

6.12 Final Thought

When planning the wiring of the car, viewing the actual interior space may inspire ideas for improving the layout, that is, reducing the wiring weight and power loss. A vehicle mock-up, in addition to the uses suggested in Chap. 11, is *very* handy for planning wiring layouts.

References

ARRL. (1973). *The radio amateur's handbook, Amateur radio relay league*. Newington: Connecticut.

Chryssis, G. (1984). *High-frequency switching power supplies: Theory and design*. New York: McGraw-Hill Book Company.

Dewan, S. B., & Straughen, A. (1975). *Power semiconductor circuits*. New York: Wiley.

MacCready, P., et al. (1990). *Sunraycer case history*. Warrendale: Society of Automotive Engineers.

McCarney, S., Olson, K., & Weiss, J. (1987). *Photovoltaics: A manual for design and installation of stand-alone photovoltaic systems*. Carbondale: Appropriate Technology Associates.

Read, R. K., Raponi, D., Kemp, B., Thacher, E., Rutkaukas, P., & Sherwin, T. (1990). *The electric and electronic systems of clarkson's solar car*. Proceedings future transportation technology conference and exposition, San Diego, California, August 13–16, 1990. Paper 901511, Society of Automotive Engineers, Warrendale, Pennsylvania.

Chapter 7
Instrumentation

7.1 Introduction

A solar racing car cannot be managed during a race without knowledge of its speed and the state-of-charge of its battery. Other parameters are also important, such as the array current, the motor current, the main bus voltage, the motor temperature, and the cockpit temperature. The present chapter covers the means for making the foregoing measurements.

In addition to display in the cockpit, measurements may be sent to a computer in the chase vehicle by a telemetry system. The information then becomes part of the database used to adjust the speed of the car to maximize its average speed over the day's route. This chapter will also discuss telemetry.

7.2 Voltage

Main Bus The main bus voltage must be measured. It is a pointer to the state of charge of the battery, although it usually will not be an accurate measure of it, and is one indicator of the general electrical health of the car.

Figure 7.1, shows the electrical system of Fig. 6.2 with instruments added. It shows a voltmeter connected to the main bus; this may be physically done at any convenient point and the meter itself installed in the cockpit display. The current flowing through the voltmeter will be almost zero because the internal resistance of the voltmeter is very high. Therefore, the length of the wire running to the meter is not critical.

Auxiliary Bus Voltages Figure 7.1, shows voltmeters connected across the two auxiliary buses. Although not essential, this arrangement gives fault cause information should some of the equipment supplied by these buses malfunction. Instead of sepa-

E. F. Thacher, *A Solar Car Primer*, DOI 10.1007/978-3-319-17494-5_7

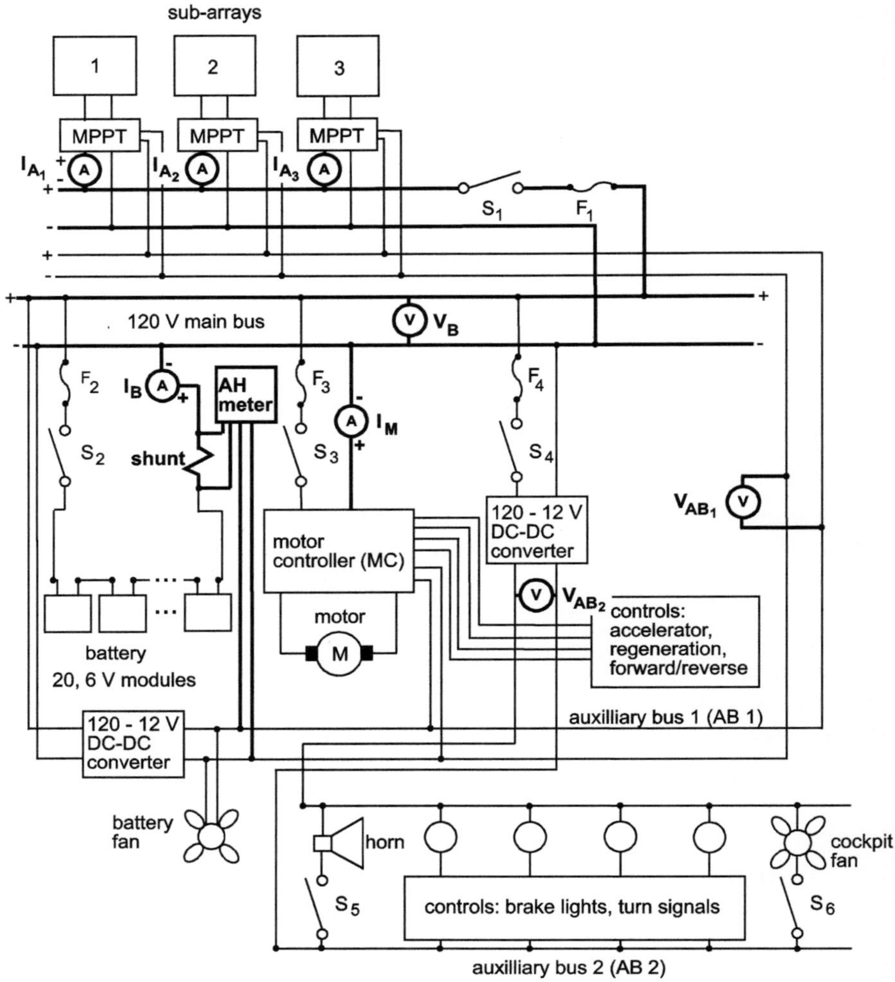

Fig. 7.1 Instrumentation example

rate meters for each bus, a single meter with a selector switch would save weight, money, and simplify the instrument panel.

7.3 Current

Array Current Figure 7.1 shows ammeters measuring the currents from each string of the array. The current (or lack of it) from each string is an indicator of possible faults, such as an open circuit. The total array current multiplied by the main bus voltage is the power supplied by the array, an important energy management parameter.

Battery Current The battery current should be displayed by an ammeter that indicates both positive (charging) and negative (discharging) currents. When the motor is not regenerating, the battery ammeter is a sensitive indicator of the car's energy demand relative to the solar energy available from the array. Consider the following example.

Suppose the road is level and the sky is clear. At slow speed the battery ammeter will indicate charging, showing that the rate of supply of solar energy (the solar power) exceeds than what is required to maintain the current speed. This excess is being stored in the battery. Let the speed increase to a new steady value. The charging current decreases. If this continues, a speed will be found at which the battery current is zero. This is the *solar speed*, the speed sustainable under the prevailing conditions by solar power alone. If the incremental speed increases are continued, the battery current will show a discharge. This implies that the solar power supplied by the solar array is less than that required for the current steady speed, and the drive must draw on the stored energy to maintain the speed.

The forgoing example is also discussed in Chap. 5, *Electric Motor Drives*, with the aid of the I–V curves of the solar array, drive, and battery.

Motor Current Motor current instrument, like the battery ammeter, must display both positive (in to controller) and negative (out of controller during regeneration) currents.

The product of the motor current and the main bus voltage is the power consumed by the motor and controller at the current speed. The current multiplied by the motor's torque constant gives the torque of the motor (Chap. 5). Constant motor current, that is, constant torque, operation is a way of managing the drain on the battery directly. The power delivered to the transmission may be found by multiplying the torque times the motor's angular speed (radians per second). This information is useful in adjusting the solar car's speed for the most efficient drive operation.

7.4 Temperature

Thermocouples A *thermocouple* is formed by a cable composed of two dissimilar-metal wires, such as copper and constantan (copper-nickel), joined at each end. If one junction is heated and the other is cooled, a voltage between the junctions is produced that is proportional to the temperature difference between the junctions. The constant of proportionality is called the *Seebeck coefficient* and the voltage produced is on the order of a millivolt.

If the cold junction is maintained at a known temperature, as by an ice bath, for instance, then the temperature of the hot junction can be computed by adding the quotient of the thermocouple voltage and the wire pair's Seebeck coefficient to the cold junction temperature. Ice baths are not convenient for solar cars, being heavy, bulky, and difficult to keep at 0°C under summer race conditions. Small, battery-operated, electronic temperature references can be used, instead.

Thermistor The electrical resistance of a substance is a function of the substance's temperature, often increasing as the temperature increases. The resistance of a *thermistor* is especially sensitive to temperature and varies linearly with it. Therefore, supplying a thermistor with a small current will yield a voltage drop proportional to the thermistor's temperature. Thermistors are not usable at temperatures much above the 100–150°C range. However, they do not require a temperature reference.

Motor It is prudent to measure the motor's temperature because motors can be easily destroyed by overheating. This has happened in past Sunrayces when motor cooling had failed.[1] Manufacturer's instructions should be followed for the placement of the temperature sensor.

Cockpit The ambient temperature in the cockpit should be monitored. In stop-and-go traffic in June, when the outside temperature is high, cockpit temperatures of 120°F have been measured in solar cars with high cockpit solar gain. Clearly, the ventilation system and the cockpit glazing should be designed to prevent such dangerous extremes. Nevertheless, it is prudent to monitor the cockpit temperature.

Array It is not generally possible, especially when the car is in motion, to speak of an "array temperature" because as Chap. 3 points out, there will be a temperature distribution over the array. So, measurement of the temperature of a single cell cannot characterize the entire array. The temperatures of cells at key locations, such as the leading edge and the trailing edge of the array, might give the lower and upper bounds of the temperature distribution of the array. Thermocouples, because of their small size, are best adapted for this measurement.

7.5 Speed

Speedometer A convenient method of measuring the solar car's speed is to use a digital bicycle speedometer. These small devices can display speed, elapsed distance, average speed, maximum speed, and time. A sensor detects the passage of a magnet mounted on the rim of a wheel.

Motor Tachometer If the motor controller supplies a voltage proportional to the motor rotational speed, this signal is directly proportional to the speed of the car.[2] The voltmeter sensing the signal must be calibrated to account for the transmission rotational speed reduction and the diameter of the driven wheel. In the absence of a manufacturer-supplied rotational speed signal, a bicycle speedometer magnet could

[1] Colloquially known as "frying" your motor.

[2] Read et al. (1990) reports such an output accurate to within 5 %.

be mounted on the rim of a small, lightweight, nonmagnetic disk fixed to the motor shaft. Then a bicycle speedometer pickup could sense the rotation. Fraser (1991) suggests using a ferrous, 60-tooth gear and a magnetic pickup. The frequency of the pulses from the pickup will be numerically equal to the motor's rotational speed in revolutions per minute. In either case, the pulsing signal must be converted to a DC voltage.

An advantage of the tachometer method over the bicycle speedometer is that a voltage proportional to speed is available not only for cockpit display as the speed, but also to the telemetry system.

7.6 Battery Charge

The following remarks emphasize and supplement those in Chap. 4.

Methods The state-of-charge of the battery is the most important information to know, but it is also the most difficult to measure. A hydrometer can be used to measure the specific gravity of the electrolyte (if the batteries are not sealed), which Chap. 4 showed was a direct indicator of the state-of-charge for some types of batteries. There may be time for this during the morning or evening charging periods, but stopping the car during the day to make this measurement is out of the question. Then, reliance must be placed on indirect methods of which there are two: measurement of the battery open circuit voltage and measurement of the net charge flow into the battery.

Bus Voltage Figure 7.2, shows charging characteristics for AgZn, and Pb-Acid batteries. For the Pb-Acid battery a unique correlation between the fractional state of charge and the bus voltage is evident. However, the long, flat region of the Ag-Zn curve shows that this unique relationship is not guaranteed. Additionally, the state-of-charge indicated by the bus voltage depends upon the age of the battery, being lower at greater ages. This potential ambiguity introduces only a small inaccuracy during a race because the battery will usually not have aged significantly during the race, with at least one exception; the life of Ag-Zn batteries is quite short, perhaps 15 deep cycles. Thus, at the end of a 10-day race, an Ag-Zn battery will be nearly at the end of its useful life.

Ampere–Hour Meter An ampere–hour meter sums the charge flow in and out of the battery, with the charge flow in being positive, and displays the net charge sent to the battery. Therefore, when the battery is charging, the number displayed increases. This number usually changes slowly,[3] hence, a meter may also have a light display that shows whether the battery is charging or discharging. The unit of measure is

[3] Although, when a short circuit, or a semi-short circuit, happens (caused by rain water entering the openings you have not properly sealed) the ampere–hour meter reading unreels like a slot machine display after the handle is pulled.

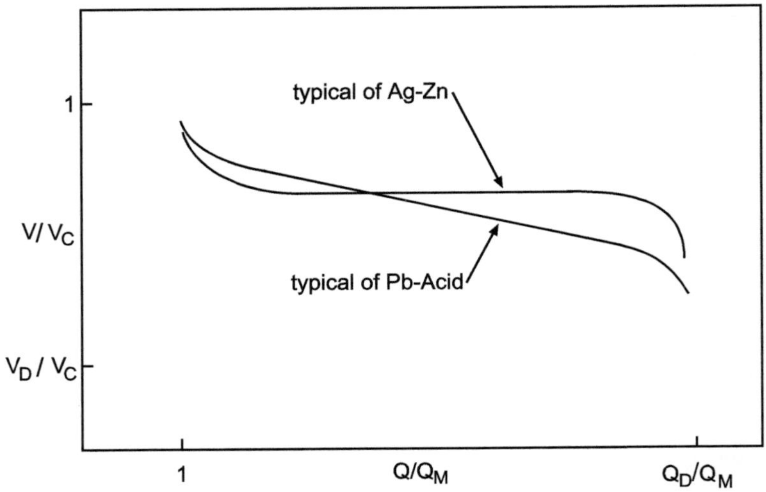

Fig. 7.2 Voltage and charge

the ampere–hour rather than coulombs.[4] The charge storage efficiency of batteries is high, perhaps 97 %, so the actual charge is close to that registered on the ampere–hour meter.

Stored Energy The product of the ampere–hour reading and the bus voltage (W-h), if computed when the battery is charging or discharging, is an inaccurate measure of the energy stored. When charging, the bus voltage will be higher than the internal voltage at which the charge is stored; when discharging, it will be lower. The product instead measures the rate of energy delivery to the battery or to the bus, respectively. The net energy delivered to the battery or to the bus over a time interval may be computed from this product. If the time interval is short enough, the charging and discharging components of the net energy can be separated and the transaction efficiency, energy discharged divided by energy charged, of the battery may be estimated. The accuracy of this estimate will depend upon the time interval, which will be controlled by the sampling rate of the telemetry system.

7.7 Instrument Panel

Forethought must be given to the instrument displays' appearance, and to the cockpit layout and lighting, especially, if driver-reported instrument readings are to be used for energy management. All instruments and controls should be clearly marked. The instruments should be readable at a glance, not shaded nor in glare, and should not divert the driver's eyes from the road excessively. Switches should

[4] 3600°C is equal to 1 A-h.

be within easy reach. The race rules give requirements for the location and marking of the battery and motor switches. Storey et al. (1994), show several photographs of cockpit layouts.

7.8 Telemetry

Function The telemetry system gathers data about the condition of the solar car and transmits the data to the chase vehicle, where it is received and entered into a computer. A computer program processes the data to assist the race team in managing the solar car's energy supply.

Manual System The simplest, lightest, least power-consuming, and cheapest system is to have the solar car driver radio and the readings of the cockpit that displays at regular intervals. The recorder notes the time, and then enters the readings manually into a spreadsheet program. The time interval between readouts should be 10 or 15 min (and sometimes longer in difficult traffic), and the number of instruments read should be minimized, to avoid excessively distracting the driver. The time resolution and precision of the readings will be coarse, and the number of parameters monitored will be small, compared to automatic systems. But, with *practice*, a solar car can be adequately managed in this manner.

Automatic System An automatic system can obtain, transmit, and record many measurements every few seconds. Figure 7.3, shows a block diagram of such a system based on a *wireless data logger*. The currents are measured using shunts and the voltages using taps. These could be the same shunts and taps that are used for the cockpit meters.

A data logger is essentially a voltmeter that can very rapidly scan (read and record) many channels of data and send the data to a computer. Wireless units use a low-power radio signal, instead of a cable, to send the data to a remote receiver. The receiver sends the data to a computer. Commercial wireless loggers can be battery-operated and accept inputs from many AC or DC voltage sources (ranging in amplitude from millivolts to 200 or 300 V), resistance temperature detectors (RTDs), and thermocouples. They may be expensive and weigh around 3–4 lbs. The more input channels you desire, the more expensive the logger becomes.

Lower-cost wireless systems require more construction by the users, or donations from sponsors (see Chap. 14, *Fundraising and Public Relations*). Read et al. (1990), outline a partly-home-built-and programmed wireless system that used a donated laptop computer fitted with an analog-to-digital (A/D) conversion board. The analog (continuous) voltage signals from the shunts, and voltage taps were connected to the A/D board which converted them to digital (sampled) signals for the computer. Going even further, Fraser (1991) and Storey et al. (1994), present outlines of designs for racing team-constructed computers for wireless telemetry systems.

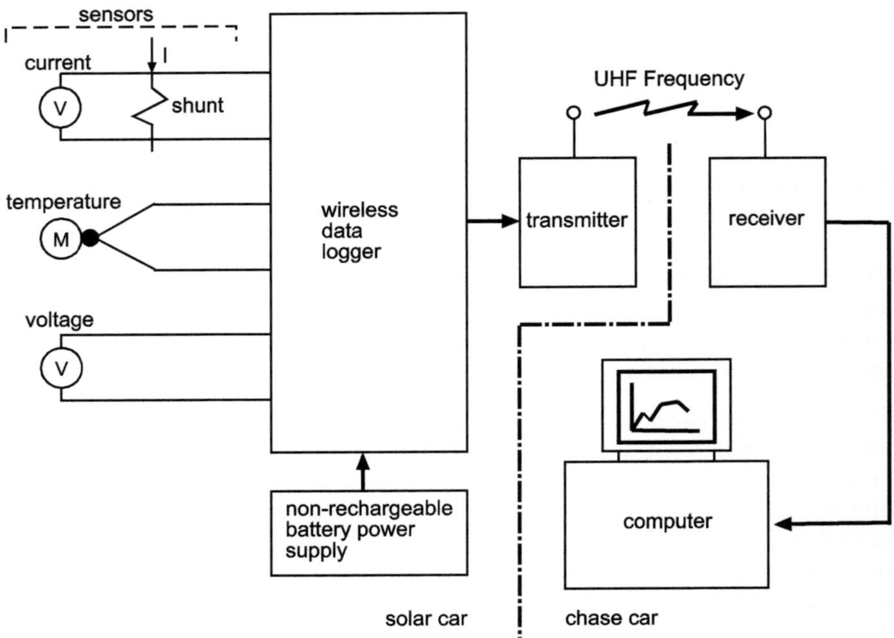

Fig. 7.3 Telemetry system

References

Fraser, D. A. (1991). Basic electric vehicle instrumentation. Proc. S/EV 91 Solar and Electric Vehicle Symposium, Car and Trade Show, 26–27 October 1991, Boxborough Host Hotel, Boxborough, Massachusetts, Northeast Sustainable Energy Association, Greenfield, Massachusetts.

Read, R. K., Raponi, D., Kemp, B., Thacher, E., Rutkaukas, P., & Sherwin, T. (1990). The electric and electronic systems of Clarkson's solar car. Proceedings of Future Transportation Technology Conference and Exposition, San Diego, California, August 13–16, 1990, paper 901511, Society of Automotive Engineers, Warrendale, Pennsylvania.

Storey, J. W. V., Schinckel, A. E. T., & Kyle, C. R. (1994). *Solar racing cars*. Canberra: Australian Government Printing Office.

Chapter 8
Solar Racer—Specification

8.1 Introduction

This chapter and the succeeding chapters draw on experience with the five solar racing cars designed and built by students at Clarkson University for the 1990, 1993, 1995, 1997, and 1999 Sunrayces.

8.2 Preliminary Design

The term "preliminary design" refers to decisions about the shape, weight, drive, etc., made early in the design process. Some will be based on estimated information and so may be revised later; design is a recursive process. Preliminary design may be thought of as having three phases: planning and specification writing, concept generation, and concept selection. When the concept has been selected, detailed design begins. Of course, design, construction, and test are really a continuous process. No one stands up at a meeting and formally announces the beginning of the next phase. Each phase blends into the next, and they often go on in parallel. The process will bifurcate, at least, with the electrical and mechanical parts of the car following parallel time lines and therefore entering different phases at different times.

8.3 Project Planning

Putting the car on the road early is usually of transcendent importance because it allows time to test the car, to improve it, and to train the race team. Also, the project is complex, involving much more than just building a car; the designers must view their job as constructing a racing system. To accomplish this task early requires a plan.

E. F. Thacher, *A Solar Car Primer*, DOI 10.1007/978-3-319-17494-5_8

Documents A project plan might consist of a bar chart, weight-control sheet, engineering drawings, procurement sheets, and a design notebook containing supporting calculations, manufacturer's data, and the like.

The bar chart shows the tasks to be done, when they start and stop, how they relate to the other tasks, and who will do them. It is a way of taking a bird's eye view of the whole enterprise. The first plan will be made with a number of assumptions about resources, the time required to accomplish the tasks, etc. Some of these assumptions may turn out to be wrong, so the plan must be changed. Hence, it is a dynamic document, not a sacred, chiseled-in-stone set of commandments.

Drawings Fabrication planning requires engineering drawings because they describe what is to be built. The production of engineering drawings takes time up front. Korff (1980) observed that it is possible to build an experimental vehicle "… with very few engineering drawings." This may be true of seasoned professionals and to some extent of student teams, in proportion to their experience. Nevertheless, without engineering drawings it is very difficult to work in parallel because there is no set of instructions to give to each sub-team. Working in parallel saves time. Also, with no drawings, designs tend to be seat-of-the-pants affairs and to be built without proper review. There is also an increased tendency to lose control of the weight, and those managing tend to lose the overall picture. The design and fabrication burden tends to fall on the few who have the work in their heads; nothing can go forward without their personal presence and direction. Drawings do not have to be formally produced using CAD software. They can be done with drawing tools, or even be freehand, but *must* be careful, neat, clear, and scaled. Chapter 10 will show examples of engineering drawings.

Procurement and Weight Procurement sheets are used to keep track of items to be acquired. They have columns for part number, name, source, price, procurement document identity and date, and the date of receipt. Weight-control sheets facilitate the control of the design's weight.

Decisions The decisions in the design process can be made by trial-and-error guesswork or by careful analysis. In practice, both are used. When cars are built by amateurs, such as college students[1] who have an academic agenda to pursue beyond the car project, there may not always be sufficient time (especially as the project deadline approaches) for careful analysis and the production of engineering drawings (particularly the latter), or even testing. Trial and error will be employed. If the guess is right and the part does not break, at least not immediately, time has been saved. But nothing has been learned nor documented: The part is a mystery. If the guess is wrong, even though failure did not occur immediately, it may occur at a critical moment, such as during the race, and more than time may be lost.

[1] However, always bear in mind that persons attracted to solar car projects may be highly skilled in some areas; these skills must be identified.

Meetings and Communication Ideally a team should have an office[2] where the time line and the major assembly drawings can be posted on a wall. Everybody should have a personal copy of the bar chart and specification. There should be an all-team meeting at some convenient interval, but at least monthly. Subunits of the team should meet more frequently. This is difficult in a school setting, but it is necessary to keep communication lines intact.

8.4 General Objective

The general objective in solar racing vehicle design is to minimize the energy required by the car to traverse the race route at an average speed high enough to win. This energy is consumed by the opposing forces arising from the vehicle's interactions with its environment: gravity, the atmosphere, and the road surface. The designer minimizes these opposing forces by adjusting vehicle characteristics such as shape and weight.

Racing begins the moment design begins. Even though your competitors are not visible, they are there, working hard. So the organization, staffing, and operations of the project, but particularly its pace, should reflect this racing attitude. This is a difficult lesson to learn because the start of the race is many months away. But every hour lost pushes back the start of construction and the time available to test the car.

Designing, building, and testing the car are the most important parts of the competition. Every mistake or omission made during the period before the race will cause a problem during qualification or on the road. There is no such thing as a minor mass addition; *every* kilogram added to the car reduces its range. If your shape is well-designed, then the drag is in the details; pay attention to the surface finish, to smoothing every crack and ridge. If the cockpit noise is high, the driver will have difficulty hearing and being understood on the radio. If the car is not sealed, rain (and it *will* rain) will shut it down—even through the wheel wells. If the battery airflow is not sufficient, you will be red-tagged during inspection of the car by the race management team, called *scrutineering*. You may fail to take advantage of rules that allow more performance, but your competitors *will not*.

8.5 Specification

The car cannot be designed without specifying what performance and features it must have and without knowing the conditions under which it must perform. The specification should be developed by research in three areas: the salient features of

[2] For a high-end view of the office of a solar car team, see that of the University of Michigan's team at http://solarcar.engin.umich.edu. The objectives of an office can be achieved with less space. But the ample space indicates the support the team receives from the university.

the top cars from the last race ("benchmarking"), the race rules, and the characteristics of the race route, including its weather.

The specification should be as detailed as possible from the outset, even though it may have to be revised later. Design begins with assumptions replacing the unknowns. This may require iteration should facts learned later contradict the early assumptions. The designers' first idea about how fast or far the car should go may have to be revised as they learn more about how much solar energy is usually available along the race route. Or, it may not turn out to be feasible to require the racer to climb 40 % grades at 30 km/h. Race rules are sometimes changed or reinterpreted. It is likely that the specification will be revised.

Individual specifications may conflict, such as low weight and stability. Thus, not all portions of the specification document carry equal priority in the design process. Some features must be emphasized at the expense of others, and flexibility must therefore be present in the specification so that these trade-offs can be made. Other specifications, such as those associated with safety, are demands. So the specification is divided into demands (D), which must be met, and wishes (W), which may be met.

The mission of the vehicle helps to define the priority and content of each characteristic specified. A solar-electric commuter car might be intended for a 40-km round trip, incorporating low–average speed, stop-and-go driving, with a long period of solar charging during the day while sitting in the company parking lot. A cross-country racer might be intended for a 400-km, one-way trip at high average speed with charging periods in the morning before racing and in the afternoon and before sunset after arrival and race rules to obey. The specifications of these vehicles will give different weights and values to such characteristics as the vehicle's drag coefficient and battery capacity.

The characteristics in the specification should be couched as limits, not design elements. Limiting metrics should be included. (These may not be known at first.) To write, "The drag area in straight-ahead flow shall be 0.11 m^2 at most" is a limit. To write, "The car shall have [a particular shape]" is a design element. But, and this is the disadvantage of specifying design elements, it may not be the *best* way of meeting the requirement of low drag. Specifying design elements closes off possible solutions.

Rather than simply writing "the drag shall be low," the sample drag specification above gives a numerical limit below which the vehicle's drag must fall: a metric. Each characteristic should be so written at first, if possible. Unknown metrics should be denoted TBD (to be determined) and research conducted to supply the metric needed. Vague statements destroy the utility of the specification as a ruler to measure the success of the design.

The specification should not be secret, but published. All persons working on the car should know it—even those involved in less technical areas such as fund raising. For some projects, there may be budget metrics. Shared information makes unity of effort possible. A policy of sharing information creates a climate of ownership; people feel the project is theirs. Then they are inclined to work more effectively. Finally, one never knows where a good idea will originate. Including people increases the number of minds working in parallel and encourages them to share their ideas.

Benchmarking Learn the important features and the performance of the top three finishers in a previous running of the race.[3] Assume that the performance of your car must exceed that of these cars, particularly the average speed. The car with the highest *average* speed will win, but this is not necessarily the car with the highest top speed. The average speed is affected by the characteristics of the car and by how the car is managed.

The facts you should obtain are:

1. The average speed
2. The empty weight
3. The shape and drag area in straight-ahead flow, at least
4. The tire size, type, and pressure
5. The rolling resistance coefficients
6. The number of wheels
7. The efficiency and components of the drive
8. The features of the solar array
9. The battery type and mass
10. Special features that influenced performance

Race Rules The organizers of the race will publish criteria that the car must meet, such as visibility from the cockpit and braking performance. These will probably be similar to the requirements for the previous race.

Route Characteristics For the American Solar Challenge, only the race period and the general direction and location of the race route are known at first; details appear later.

The best way to gather information is to survey the route, but this may not be possible. Much can be learned by consulting geographical and topographical maps (once the route is specified) and weather records, questioning local chambers of commerce, the American Automobile Association, persons you may know who have lived along the race route, and of course the race organizers.

You will need the following information:

1. Road surface type and condition
2. Direction of road from true north
3. Slope, length, and number of hills
4. Atmospheric pressure and temperature ranges
5. Wind speed and direction ranges
6. When and how long it is likely to rain
7. Typical cloud-cover range and solar energy data
8. Longitude and latitude ranges

Ideally, items 1, 2, and 3 should be known as functions of distance along the road. Item 3 could also be elevation above a datum (such as mean sea level) as a function of distance along the road.

[3] Team Nuon of the Delft University of Technology, Delft, Netherlands, has won the World Solar Challenge (WSC) four times. Its web site, http://www.nuonsolarcarteam.nf, is a good source of benchmark information. For earlier WSC cars, see Storey et al. (1994).

Table 8.1 Objective

D	1. The solar car must achieve an average speed of at least 66 km/h over the design race route on the design race day after starting with a fractional state-of-charge (FOC) of 1.0 and finishing with at least an FOC of 0.2

8.6 Sample Specification

The sample document that follows lists specifications to be met in addition to those in the race rules (Chap. 16). Section 16.5 of Chap. 16 governs the electrical aspects of the car and Sect. 16.6 the mechanical. These requirements are not therefore restated, with a few exceptions.

Objective Specification begins with an objective, clearly stated and understood by all. Otherwise, teamwork will be difficult, and the actions and level of effort required to reach the objective will not be known.

The objective should be to produce a car that is capable of winning the race. Such an objective is not unrealistic and has nothing to do with ego; it is the necessarily unambiguous beginning of the design process. The team may not win, but unless you attempt to win, you will not learn the level of effort required. You take a risk in setting a high goal. If your race performance is considerably below your expectation, the disappointment will be keen (Table 8.1).

Design Race Route The information gathered about the race route is the basis for this part of the specification. The route should be a composite of features culled from the entire actual race route. This composite may consist of typical features encountered at an average, or at a conservative, frequency. This frequency could be ascertained by compiling statistics of the actual route, such as the distribution of hills in certain grade ranges. However, it is likely that the route will not be known in sufficient detail at this stage to do this. Therefore, the design route may have to be invented and in this case should be conservative.[4]

The route shown in Table 8.2 is not intended to be representative of any particular race, but to suggest how to specify the information. The total distance traveled in the sample is 261.4 km, and the net maximum elapsed time is 3.91 h, neglecting acceleration time. (Note that the time at the midday stop is not counted in computing the average speed.) The minimum average speed is therefore 66.83 kph, or 41.77 mph. The speed specified for a leg is in most cases a minimum speed and sometimes a speed limit, such as 88.5 kph (55 mph). Therefore, 3.91 h, though neglecting acceleration time, is close to the maximum elapsed time.

The following nomenclature is used in Table 8.2: *a*, acceleration, g-units; α, grade, degrees; C_V, course angle, degrees true; *F*, fractional state of battery charge,

[4] Once known, the entire route may be used in a computer simulation as a succession of "design days."

Table 8.2 Design race route

D	1. The following route shall be used to design the vehicle. Only *changes* are shown at each distance, *S*	
Leg	*Description*	*Time period*
0–1	*Battery charging with manual tracking* $S=0$; $V=0$; $Z=0$; $\alpha= 0°$ latitude = 38° N; longitude = 86° W	06:00–09:45
1–2	*Battery charging in racing configuration* Parameters as for leg 0–1	09:45 – 10:00
	City stop-and-go driving (2–20)	
2–3	$S=0$: $V=40$; $C_V=270°$ T, $a=0.1$ g; $\rho_G=0.2$	
3–4	$S=0.1$: $V=0$ [stop signal]	
4–5	$S=0.1$: $V=40$; $a=0.1$ g	
5–6	$S=0.2$: $V=0$	0.5 min
…	[continue, stopping every 0.1 km]	
19–20	$S=0.8$: $V=40$	0.5 min
20–21	*Climb* $S=0.9$: $\alpha = 10\%$; $V=40$; $C_V=315°$ T	
21–22	*Descend* $S=1.0$: $\alpha = -10\%$; $V=88.5$; $C_V=225°$ T	
22–23	Cruise $S=1.1$: $V=72$; $C_V=270°$ T; $\alpha= 0°$	
23–24	Pass $S=10.4$: $V=88.5$	
	Cruise-climb-descend (24–56)	
24–25	$S=11.1$: $V=72$	
25–26	$S=21.1$: $V=40$; $\alpha = 5°$	
26–27	$S=22.1$: $V=88.5$; $\alpha = -5°$	
27–28	$S=23.1$: $\alpha = 0°$	
…	[repeat nine more times]	
55–56	$S=131.1$: $V=72$; $\alpha = 0°$	
	City stop-and-go driving (56–60)	
56–57	$S=131.2$: $V=0$	0.5 min
57–58	$S=131.2$: $V=40$	
58–59	$S=131.25$: $V=0$	0.5 min
59–60	$S=131.25$: $V=40$	
60–61	*Midday stop* $S=131.3$; $V=0$	0.25 h
61–65	*City stop-and-go driving* $S=131.3$; repeat 56–60	
65–66	*Cruise* $S=131.5$; repeat 22–23	
66–67	*Pass* $S=140.5$; repeat 23–24	

Table 8.2 (continued)

D	1. The following route shall be used to design the vehicle. Only *changes* are shown at each distance, *S*	
67–76	*Cruise-climb-descend* $S=141.2$; repeat 24–27 ten times	
76–80	*City stop-and-go driving* $S=261.2$; repeat 56–60	
80–81	*Battery charging with manual tracking* $S=261.4$; $V=0$; charge until $F=1$, or charging time ends	Arrival until ≤8:30 p.m.
D	2. The design road surface shall be smooth asphalt	

1 meaning full charge, nondimensional; *V,* vehicle speed, kph; *S,* distance traveled from starting line, km; *Z,* altitude above mean sea level (msl), m; μ_1, static rolling resistance coefficient; and r_G, ground reflectivity. The distance at which variables change is specified.

Design Day The design race day was chosen for about the midpoint of the 1995 Sunrayce. The cloud cover (CC, denotes the fractional amount of cloud cover) is conservative but not unusual for the time of year and location (Table 8.3).

Payload and Mass Race terrain is often hilly and sometimes mountainous, consequently, low mass is very important. The largest force on the vehicle when going uphill is the weight component opposing the motion. The mass limit, 338 kg, was derived from the gross mass of the second place car in the 1993 World Solar Challenge, the *Spirit of Biel/Bienne III* (Storey et al. 1994; Table 8.4).

Aerodynamics The car must be able to cruise economically at or near the speed limit when the terrain is relatively flat. Under these conditions, the dominant force on the car is aerodynamic drag. The drag area limit is the drag area of the Spirit of Biel/Bienne III (Storey et al. 1994), to nearest 0.01 m^2 (Table 8.5).

Rolling Resistance Frictional rolling resistance is directly proportional to the weight and to the coefficients representing the road–tire interaction and opposing moments generated by the rotation of the wheels. The value of the wheel-road coefficient was that of the Honda Dream, first in the 1993 World Solar Challenge (Storey, et al. 1994). The coefficient associated with rotation-induced moments was based on the value at the low end of the range given by Steeds (1960) (Table 8.6).

Table 8.3 Design race day

D	1. The date shall be June 23
D	2. The ambient conditions shall be: temperature, 25 °C; pressure, 101.3 kPa; wind speed and direction, 8 kph from 270°T; cloud cover (CC), 0.5
D	3. The hourly clearness index shall be computed from: $k=0.832-(0.05)(\mathrm{CC})$
D	4. The ground reflectivity shall be 0.2

Table 8.4 Payload and mass

W	1. The vehicle's gross mass shall not exceed 338 kg
W	2. The driver's mass shall not exceed 80 kg
W	3. The battery mass shall follow ASC rules 5.8.1 or 5.8.2n

Table 8.5 Aerodynamics

W	1. The drag area shall not exceed 0.11 m^2
D	2. The pavement force normal to the windward wheels' contact patches shall be greater than zero when the car is driven at 88 km/h through a wind gust of 50 km/h blowing perpendicular to the direction of travel

Table 8.6 Rolling resistance

W	1. The static rolling resistance coefficient shall not exceed 0.004 on a smooth asphalt road surface
W	2. The dynamic rolling resistance coefficient shall not exceed 0.0001s/m

Table 8.7 Load cases and safety factors[a]

Notes: 1. One "g" is 9.807 m/s^2 2. Normal weight (1.0 g) present in each case	
W	1. No structural damage shall occur from traversing a 3.0-g bump while steering straight
D	2. No structural damage shall occur from 1.0-g braking while steering straight
D	3. No structural damage shall occur from the combined effect of a 1.0-g turn, a 2.0-g bump, and 1.0-g braking
D	4. Welded joints shall be designed with a minimum safety factor of two applied to the reduced yield strength in the weld zone
W	5. No structural damage shall occur from operating over the race distance times a factor greater than one that accounts for prerace and postrace driving and trailering
D	6. Regions of the shell subjected to flexing during operation or trailering must be designed with stiffness adequate to prevent cracking of composite structures or welds

[a] Chapter 22 suggests a straight-ahead bump-load calculation method

Loads The loads in Table 8.7 were based on a bump-induced vertical acceleration of 3 g, a 1-g centrifugal acceleration when turning, a 1-g braking deceleration, and a combined 1-g braking, 1-g turn, and 2-g bump.

Transport by trailer ("trailering") gives the car its worst beating, particularly in the suspension and in parts of the body that can flex. Trailering miles should be included with driving miles when estimating loads and fatigue during design (Table 8.7).

Cockpit Racing experience has shown that cockpit temperatures approaching 49 °C (120 °F) are possible in stop-and-go traffic when relying primarily on ventilation flow caused by the car's motion. This cannot be allowed. Experience has

Table 8.8 Cockpit

W	1. The cockpit temperature shall not exceed the ambient temperature
D	2. The interior of the car shall be sealed against rain and against water thrown up by the wheels
W	3. Components shall be individually sealed against moisture
D	4. The signal-to-noise (S/N) ratio,[a] when transmitting or receiving on the radio at any speed, shall be at least 24 dB
D	5. Instruments shall be easily readable by the driver without significant head motion and when the sun is directly overhead or shining into the driver's eyes

[a] S/N=20 $\log_{10}(V_s/V_n)$, (dB). If the signal voltage, V_s, were equal to the noise voltage, V_n, then S/N=0. If V_s/V_n=15.85, then S/N=24

Table 8.9 Electrical

D	1. The minimum array power-to-vehicle mass ratio at 1.0 sun (1000 W/m^2) and 25 °C ambient air temperature shall be 3.5 W/kg measured at the battery bus
D	2. The minimum life of the battery shall be 15 equivalent full-discharge cycles
D	3. The battery bus voltage and interconnecting wire size and lengths shall be chosen to minimize transmission losses and wire mass
D	4. Each sub-array shall be operated at its maximum power point
D	5. Components must withstand constant vibration at ambient interior car temperatures of 50 °C for the driving distance of specification 7.5
W	6. The array structure must prevent cracking of the cells during setup for charging, transport in the trailer, or design bump loads
D	7. The array structural surface must be such that all cells laid upon it will be supported over the cell's entire undersurface
D	8. The array materials and construction must be such that electrical short circuits do not occur

also shown that great care must be taken to protect the electrical system from water-caused short circuits. Teams must expect rain and design against it (Table 8.8).

Electrical The array power-to-mass ratio was derived from data on the 1993 World Solar Challenge Honda and Biel cars (Storey et al. 1994).

The battery specifications were based on the Eagle-Pitcher cells employed by Clarkson in the 1995 Sunrayce and so represent a low-budget minimum. Note that the battery life is only 15 full-discharge cycles. It is not necessary to specify more cycles than this (although the life of most commercial batteries exceeds 15 cycles). The batteries used in the race will not undergo more full-discharge cycles than this in a 9- or 10-day race. "Full-discharge" means about 80% for lead-acid cells, although Clarkson's experience shows that occasional discharges beyond this limit may be made with recovery to full charge capacity later (Table 8.9).

Transport The transport of the car when not racing is part of the racing system. The interior dimensions of the trailer used for this purpose will impose constraints on the size of the car. Also, properly securing the car in the trailer is of utmost importance. Finally, thought should be given to a system for loading the car into

Table 8.10 Mechanical and transport

D	1. The vehicle shall fit within the existing transport trailer and have no contact with it in the transport position
D	2. The vehicle shall be secured within the trailer such that it cannot be damaged during normal transport
D	3. A means of loading the vehicle into the transport trailer that minimizes lifting and other handling by humans, or disassembly of the car, shall be provided
D	4. The vehicle shall comply with the Sect. 16.6 (Chap. 16)

Table 8.11 Stability

D	1. The c.g. for a 2F-1B car shall be in the first third of the wheel base, as measured from the front axle
D	2. The c.g. of a 1F-2B car shall be in the last third of the wheel base as measured from the front axle
D	3. The c.g. of a 2F-2B car shall be in the first half of the wheel base as measured from the front axle
W	4. All points on the boundary of the car shall remain within its traffic lane for at least a period of 0.8 s when acted upon by the wind gust of 5.2
D	5. In a turn in still air, the car will skid before it rolls over, whether it is accelerating or braking

the trailer smoothly and with relative ease. It should not require ten people and a partial disassembly of the car to load it aboard its trailer (especially when it is raining) (Table 8.10).

Stability The specifications in Table 11 deal with the stability of the car when turning at constant speed and when subjected to side gusts. The specifications apply to a car operating on the design day and on a dry design road surface. The terms 2F-1B, 1F-2B, and 2F-2B refer to two wheels in front and one in back, one wheel in front and two in back, and the standard four-wheeled car, respectively. "Center of gravity" is abbreviated as "c.g."

The car must remain stable in side gusts, which frequently occur when crossing open reaches such as those in states like Kansas and Colorado. Little benchmarking information on side-gust stability is available. The side-gust specification was based on a test administered to Clarkson University's 1990 Sunrayce car. The specified center-of-gravity locations have been shown to be necessary for stability (see Huston and Graves 1982, for example).

The sensitivity to side gusts in Table 8.11 specification 4 is based on an assumed driver reaction time of 0.8 s. With regard to Table 8.11 specification 5, a skid is the lesser of two evils in that there is some chance of control and subsequent recovery. A rollover is a disaster (Table 8.11).

Steering For safe, predictable operation, the vehicle should understeer (see Chap. 21, pages 411–414) over the possible range of the lateral acceleration. Failing this, drivers should at least understand the steering characteristics of the car, so they do

Table 8.12 Steering

D	1. The steering angle on each steered wheel shall be such as to minimize tire wear, at least at low speeds
D	2. The car must understeer when turning

Table 8.13 Energy consumption rate

W	1. The tractive energy consumption rate shall not exceed 14.5 W. h/km at 55 mph, no wind, zero grade, and design atmospheric temperature and pressure

not unknowingly blunder into the oversteer lateral acceleration range and roll the car before they can recover (Table 8.12).

Energy Consumption The drag area and mass limits are both aimed at one target: low energy consumption per kilometer. To reach a winning energy consumption rate, the effect of shape and mass of the car must be optimized. Thus, each cannot be rigidly proscribed; again, the best racing system wins, not necessarily the one with the lowest drag.[5] A design team may focus excessively on drag and lose control of weight or produce a poor surface for the solar array (Table 8.13).

The tractive energy is that delivered to the contact patch of the driven wheel. The "design conditions" include the design race route and the design day. It could be argued that the data in Table 8.13 is superfluous: The overall objective, (Table 8.1), essentially includes energy consumption. But it is useful to have a target consumption rate during design.

References

Huston, J. C., & Graves, B. J., 1982, "Three wheeled vehicle dynamics," paper 820139, Society of Automotive Engineers, Warrendale, Pennsylvania.

Korff, W. H., 1980, Motor vehicle design. Burbank: M-C Publications.

Storey, J. W. V., Schinckel, A. E. T., & Kyle, C.R., 1994, Solar racing cars. Canberra: Australian Government Printing Office.

[5] The Honda *Dream,* winner of the 1993 World Solar Challenge, had a drag area of 0.114 m^2, whereas the second-place car, the *Spirit of Biel/Bienne III,* had a drag area of 0.105 m^2 (Storey et al. 1994).

Chapter 9
Solar Racer—Concept Generation and Selection

9.1 Introduction

In this book, "conceptual design" describes the second stage of the design process in which a vehicle's physical envelope, its boundary, is defined. Our goals are to:

1. Select the car's shape
2. Set its major dimensions
3. Estimate the gross mass and the location of the center of gravity (CG)
4. Specify the main requirements for the drive
5. Set the main features of the driver–car interface
6. Set the main requirements for the electrical power subsystem
7. Set the number of wheels
8. Examine the stability

The design selected for continued development will satisfy, or will at least be judged to have the potential to satisfy, all of the required and more of the wished-for characteristics in Chap. 8 than its competitors.

9.2 Concept Sketch

The requirements that the solar racer must meet, in the designer's judgment, to win a cross-country solar car race have been written in the specification. Now, take a blank sheet of crosshatched paper, set the scale of the smallest square, and begin to design to meet these requirements. First, lay out the specified bounding dimensions; this gives the box in which the car must fit. Then, make a to-scale, freehand concept sketch in three views that fits within the bounding box and incorporates the main features of the car. But what car shape should be drawn?

E. F. Thacher, *A Solar Car Primer*, DOI 10.1007/978-3-319-17494-5_9

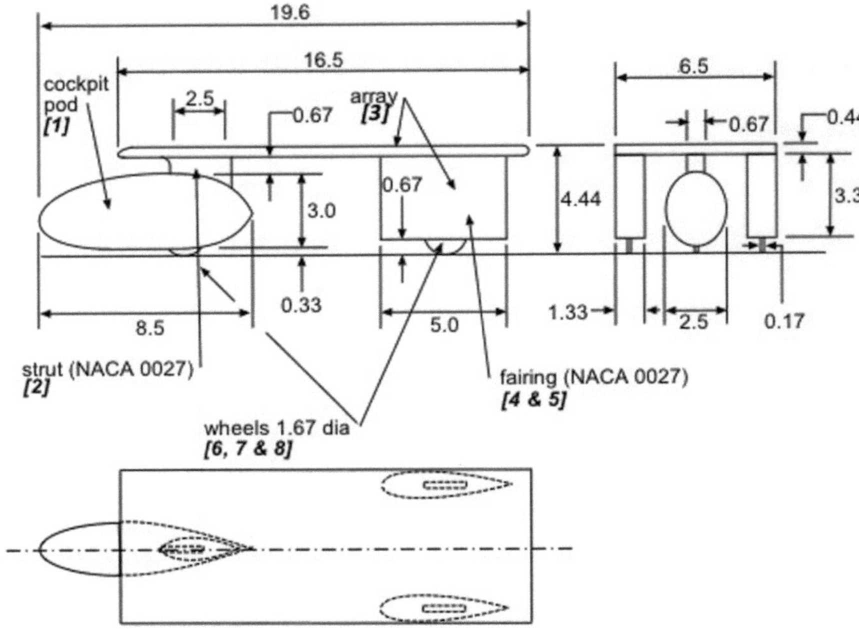

Fig. 9.1 Proposed solar racer (dimensions in feet)

9.3 The Table Top

The study of other solar racers that underlies the specification showed the great variety of solar racer shapes and provided starting ideas. Some cars emphasize solar collection and are essentially rolling tabletops. Others emphasize low drag and so have the solar array integrated into their streamlined body. However, even though the top cars are often streamliners, suppose the ideas uppermost in thought are that solar collection must be maximized and that the extra drag that this may cause could be compensated for by large wheel fairings which could supply thrust in cross-winds. Suppose research has shown that the prevailing winds blow across the race route; hence, the design's "secret weapon" will be the ability of the car to sail.

Figure 9.1 shows a drawing done from a freehand sketch of a three-wheeled[1] solar racer with a flat solar array on top, large rear sailing fairings, and a streamlined cockpit pod located in front. We will call it the "Table Top." Using only three wheels reduces the weight. The tricycle wheel arrangement allows the front structure of the car to be simplified with the front wheel housed in the cockpit pod. To provide sufficient room for the rear wheels and low drag, the cross-section profiles of the rear fairings and the pylon supporting the front of the solar array are symmetric, 27%-thickness-ratio airfoils with the maximum thickness at 30% of the chord from the leading edge. The car complies with Chap. 16, Sect. 16.1.1.1.

[1] Three-wheeled cars were not allowed in the 1999 Sunrayce.

9.4 Table Top Drag Estimate

The concept sketch will now be used to estimate the as-built drag area. If the drag area is satisfactory, the weight and other characteristics will be estimated.

We will make the estimate at 40 mph (17.88 m/s). A quarter-scale model of this vehicle was tested in a wind tunnel[2] at the Reynolds number corresponding to 40 mph (160 mph tunnel wind speed) and an atmospheric pressure and temperature of 99.6 kPa and 23.3 °C, respectively. We will estimate the drag coefficient for these conditions so we can compare it with the tunnel measurements.

Experience has shown that the drag area of the actual vehicle will usually be larger than that found by calculation, even those done using sophisticated computer programs.[3] This is caused by the many features, small and large, that computer models may overlook: seams between parts of the car, surface finish, the edges of solar cells, finite-radius trailing edges, the scooping of air into wheel fairings, etc. Therefore, the actual, as-built, drag area of the car must be measured. This may be done by full-scale wind tunnel, coast-down, or constant-speed testing.

Full-scale wind tunnel services are very expensive, and large wind tunnels are not usually nearby, nor accessible to solar car teams. Smaller wind tunnels for testing scale models are more accessible, and testing time in them is less expensive than that of full-scale tunnels. Scale-model testing can come closer to the as-built drag area because some of the features missed by simulations can, if properly scaled, be included.

Method As described in Chap. 17, the *drag build-up* method estimates the drag area and drag coefficient by modeling the car as a composite of shape elements which have known drag coefficients. The drag areas of these shape elements are corrected (as appropriate) for interference, ground effect, and roughness, and added to give the drag area of the car at a particular speed.

We study the Table Top and divide it up into eight component shapes. The profile areas of these shapes are designated A_1–A_8 in Fig. 9.1 (bracketed numbers in italics). The corresponding drag forces are D_1–D_8. Equation (17.2) is then applied to each shape.

At the wind tunnel ambient conditions Eq. (2.3) gives

$$\rho = \frac{p}{RT} = \frac{99.6\text{kPa}}{0.287\dfrac{\text{kJ}}{\text{kg}\cdot\text{K}} \times 296.45\text{K}} = 1.17\frac{\text{kg}}{\text{m}^3}$$

as the air density. The Reynolds number per unit characteristic length will be handy. The letter "*L*" symbolizes any characteristic length.

$$\frac{R_e}{L} = \frac{\rho V}{\mu} = \frac{1.17\dfrac{\text{kg}}{\text{m}^3} \times 17.89\dfrac{\text{m}}{\text{s}}}{18.39(10^{-6})\dfrac{\text{kg}}{\text{m}\cdot\text{s}}} = 1.14(10^6)\text{m}^{-1}$$

---

[2] See Chap. 12.

[3] Kurtz (1980) suggests on the order of 20 % higher.

The free-air dynamic pressure is

$$q = \frac{1}{2}\rho V^2 = \frac{1}{2} \times 1.17\frac{\text{kg}}{\text{m}^3} \times \left(17.89\frac{\text{m}}{\text{s}}\right)^2 = 189.23\text{Pa}.$$

Pod, A_1 The pod has an elliptical cross-section and is somewhat more elongated at the downstream end. We select an ellipsoid as the shape most closely approximating the pod, and for which data are available.

1. ***Free Air:*** The profile area of the pod, A_1, is

$$A_1 = \frac{\pi ab}{4} = \frac{\pi \times 3\text{ft} \times 2.5\text{ft}}{4} = 5.89\text{f}^2 = 0.547\text{m}^2.$$

The drawing shows the length of the pod as 8.5 ft, or 2.59 m. This is the characteristic length. The Reynolds number is

$$R_{\text{eL}} = 1.14(10^6)\text{m}^{-1} \times 2.59\text{m} = 2.96(10^6)$$

This implies turbulent flow, as Fig. 2.6 shows. The ellipsoid data of Table 17.1 are for a circular cross-section. Therefore, we calculate the pod's fineness ratio (length over diameter) as 3.1, using the diameter of the circle having an area equal to A_1, 2.74 ft. This technique was used by Kurtz (1980) when characterizing the fineness ratios of passenger cars. The turbulent flow column of Table 17.1 gives an estimate for $c_{\text{D}\infty}$ of 0.11.

2. ***Ground Effect:*** From Fig. 9.1, the minimum clearance ratio is about 0.13. The pitch angle is zero. The zero-pitch-, zero-camber-ratio teardrop in Chap. 17 is the closest to the ellipsoid shape. Using Eq. (17.17) and the zero-camber-ratio constants from Table 17.6 gives

$$\frac{\Delta c_{\text{DG}}}{c_{\text{D}\infty}} = -0.1256 + 0.1541\left(\frac{0.33\text{m}}{2.5\text{m}}\right)^{-1} - 0.0041\left(\frac{0.33\text{m}}{2.5\text{m}}\right)^{-2} = 0.807.$$

Figure 17.10 shows $c_{\text{D}\infty} = 0.078$. Hence,

$$\Delta c_{\text{DG}} = 0.078 \times 0.807 = 0.063.$$

3. ***Interference:*** There is interference between the pod and the strut and the front wheel. The extra drag is apportioned between the pod, the strut, and the wheel. We cannot estimate how much is apportioned to each. So, we will place all the interference drag on the strut or the wheel. Hence, for the pod: $\Delta c_{\text{DI}} = 0$.
4. ***Roughness:*** .For the smooth-skinned model, there is no extra roughness, so $\Delta c_{\text{DR}} = 0$.
5. ***Protuberances:*** There are no protuberances, such as rear view mirrors or antennae: $\Delta c_{\text{DP}} = 0$.

Strut, A_2 The strut has a NACA 0027 cross-sectional shape.

1. ***Free Air:*** Figure 9.1 shows the chord length as 2.5 ft and the thickness and average span as 0.67 ft. This chord gives a Reynolds number of $8.7(10^5)$. Tables 17.2 and 17.3 give drag coefficients, based on the planform area (chord × span), of several NACA airfoils at zero angle of attack. These data apply to wings of infinite span having the listed profile over the entire span. However, we take these to be an adequate approximation and note that the drag coefficient increases approximately linearly with the thickness ratio, but at the higher thickness ratios it decreases with Reynolds number. Using these observations to extrapolate gives a drag coefficient of 0.0092 at $Re_C = 1.75(10^6)$ for the NACA 0027. Finally, using the NACA 0025 Reynolds number dependence, because its thickness ratio is closest to 27 %, we estimate a drag coefficient of 0.0096 at the operating free-air Reynolds number for the NACA 0027. The drag on the airfoil must be the same irrespective of the reference area. Therefore, we correct this planform-referenced drag coefficient to a profile-area-referenced value by dividing by the fractional thickness ratio.

$$c_{D\infty} = (c_{D\infty})_{\text{planform}} \frac{A_{\text{planform}}}{A_{\text{profile}}} = (c_{D\infty})_{\text{planform}} \frac{c \times \text{span}}{t \times \text{span}} = \frac{0.0096\text{m}}{0.27\text{m}} = 0.036$$

2. ***Ground Effect:*** The strut is shielded from the ground by the pod: $\Delta c_{DG} = 0$.
3. ***Interference:*** The strut causes interference drag at each surface to which it is attached. Equation (17.12) gives the interference drag for a strut with a wall at one end for t/c a little higher than 27 %. An adequate estimate may be obtained by multiplying this equation by two. We use Eq. (17.14) to reference Δc_{DI} to the profile area, A_2.

$$\Delta c_{DI} = \frac{\Delta D_I}{qA_2} \approx 2[0.8(.27)^3 - 0.0003]\frac{(0.762\text{m})^2}{(0.204\text{m})^2} = 0.43$$

The interference drag is large. It can be reduced by using fillets at each end of the strut and reducing t/c considerably. This raises a question of strength. However, let us complete the estimate for the entire car first.

4. ***Roughness and Protuberances:*** We neglect these two effects as we did for the pod.

Solar Array, A_3 The drag of this component will be estimated for a smooth surface (no solar cells) so a comparison with wind tunnel data on the model car can be made. There are no shape element data for the array as a whole. Total drag could be estimated as the sum of friction drag and pressure drag from trailing edge separation: $D_3 = D_{3f} + D_{3P}$. However, the nose of the array is rounded, its trailing edge is tapered, and it has a large length-to-thickness ratio (greater than 37). Therefore, since the friction is proportional to the surface area, and the pressure drag depends approximately only on the shape of the leading and trailing edge, we assume $D_{3P}/D_{3F} << 1$, that is, the pressure drag force will be negligible compared to the friction drag.

1. ***Free Air:*** Figure 9.1 shows the length as 16.5 ft (5.03 m), width as 6.5 ft (1.98 m), and the thickness as 0.44 ft (0.134 m). Using the length gives a Reynolds number at the trailing edge of 5.73(10^6). This means that the flow is turbulent when it leaves the trailing edge.[4] We assume that the laminar–turbulent transition takes place at a Reynolds number of 3(10^6), that is, the plate is very smooth. Using this "critical" Reynolds number gives the distance from the leading edge to the transition as 2.63 m. Equation (17.20) gives the friction drag coefficient.

$$c_{DF} = \frac{1.328}{(3\times10^6)^{1/2}}\frac{2.63\text{m}}{5.03\text{m}} + \frac{0.031}{(5.73\times10^6)^{1/7}}\left[1-\left(\frac{2.63\text{m}}{5.03\text{m}}\right)^{7/8}\right] = 0.00185$$

This must be turned into a drag coefficient referenced to the profile area of the array. We assume that c_{DF} is uniform over the entire surface. The area of the array, including the sides but less the area of its underside taken up by the strut and the two fairings, is 20.74 m^2. The drag force is then

$$D_{F3} = c_{F3}A_{S3}q = 0.00185\times 20.74\text{m}^2 \times 187.23\frac{\text{N}}{\text{m}^2} = 7.18\text{N}.$$

The profile area of the array is 0.266 m^2. The friction drag coefficient, referenced to the profile area of the array, is therefore

$$c_{D3} = \frac{7.18\text{N}}{0.266\text{m}^2 \times 187.23\frac{\text{N}}{\text{m}^2}} = 0.144.$$

This will be $c_{D\infty}$ for the arra.

2. ***Ground Effect:*** None (reasoning as for the strut).
3. ***Interference:*** Interference is present, but we will continue to add it to the other shape elements that attach to the array.
4. ***Roughness and Protuberances:*** None (as explained).

9.5 Fairings A_4 and A_5

1. ***Free Air:*** Proceeding as for the strut, we get a free-air-drag coefficient of 0.036 for each fairing.
2. ***Ground Effect:*** Assume that the effect of proximity to the ground is not large since most of the area of the fairing is at a large clearance ratio. This correction will then be zero.
3. ***Interference:*** The junction of each fairing with the solar array causes interference drag, as does the proximity of the fairings to each other.

[4] Critical Reynolds number 5(10^5)–3(10^6), with upper limit for a very smooth plate.

In the first case, notice that Fig. 9.1 shows that the outer surface of each fairing is approximately even with the outer edge of the array. We judge that this arrangement reduces the total interference drag from the two fairings to that for one alone. Using the "one end" wall interference correction as $\Delta c_{DI} = 0.087$, or 0.0435 per fairing.

Figure 17.5 shows the mutual interference between two airfoils in parallel. The ratio of the minimum spacing to the airfoil thickness is 2.887. From the free-air-drag coefficient in Fig. 17.4 and Eq. (17.8) we get for each of the pair of foils:

$$\Delta c_{DI} = 0.199 \times 0.0662 = 0.0132.$$

The total interference on each fairing is $\Delta c_{DI} = 0.0562$.

4. ***Roughness and Protuberances:*** None.

9.6 Wheels, A_6, A_7, and A_8

The wheel wells of the model were filled, and the wheels themselves were simply fixed, partial discs with small profile areas. Also, when tested in the wind tunnel, these disks were suspended a small distance above the tunnel's ground plane. We nevertheless treat them as real, stationary wheels in contact with the ground. Figure 9.1 gives the wheel diameter of 1.67 ft (0.508 m), the wheel width as 0.17 ft(0.052 m), and the projected un-faired height, 0.33 ft(0.1 m) and 0.67 ft(0.2 m) for the front and rear wheels, respectively. We take the c_D of an isolated, stationary wheel in contact with the ground to be 0.526 (Fig. 17.3). Following the wheel drag discussion in Chap. 17, we take the drag coefficient of the exposed portion of the shielded wheel (a wheel in a fairing or wheel well) to be approximately the same as the unshielded wheel. The drag area of a shielded wheel is therefore equal to the profile area of the unshielded part of the wheel times 0.526.

We ignore interference between the pod and the front wheel because the wheel is small compared to the pod.

9.7 Results

Table 9.1 summarizes the results. Note that the projected areas, $A_{D\#}$, may be summed to give A_D because all of each $A_{D\#}$ appears in the front view.

The estimated c_D of the car is therefore 0.141.

Discussion The Table Top's drag coefficient was measured at 0.139 in a wind tunnel at the National Research Council of Canada's Advanced Aeronautical Laboratory under the conditions mentioned at the beginning of the preceding calculation.[5] The calculation overestimated the drag coefficient by 1.4 %.

[5] Chapter 12 "Testing," describes the construction of the quarter-scale model of the Table Top and presents some of the data from the tests.

Table 9.1 Table top drag area

#	$A_{D\#}$	$c_{D\infty}$	Δc_{DG}	Δc_{DI}	c_D	$c_D A_D$
1	0.547	0.110	0.064	0	0.174	0.0952
2	0.042	0.036	0	0.43	0.466	0.0200
3	0.266	0.141	0	0	0.141	0.0383
4	0.412	0.036	0	0.056	0.092	0.0380
5	0.412	0.036	0	0.056	0.092	0.0380
6	0.0063	–	0	0	0.526	0.0033
7	0.0063	–	0	0	0.526	0.0033
8	0.0052	–	0	0	0.526	0.0029
$A_D = 1.697\text{m}^2$				$c_D A_D = 0.2390\text{m}^2$		

The results are within an expected error of about 10%. Changes in some of the assumptions could certainly alter the results by several percent. Recall the earlier remark that because construction adds flaws and unavoidable features not included in the calculation, such as seams between body panels, the drag coefficient of the as-built, full-scale car will be higher still, and therefore also its drag area. And note that the drag area is greater than the specified limit of 0.11 m^2 (specification 8.5.1). The Table Top concept is rejected because its predicted drag area will be irredeemably greater than the limit specified.

9.8 The Shark

Another shape concept must now be considered. Morelli (1983) reported extensive tests of basic body shapes designed to have low drag near the ground. Some configurations of these shapes had extraordinarily low drag coefficients, around 0.05. And these figures were from tests, not calculations, albeit for smoothed, basic shapes. But beginning with such a low drag shape would provide a large margin between the specified drag area and the basic shape drag area to allow for the drag increase caused by transforming a smoothed model into a real car. However, the sloping, compound-curved Morelli shape would reduce the power of the solar cell array compared to one installed on the Table Top.

Figure 9.2 is a hand-drawn sketch of the Morelli shape concept, which we will call the "Shark," showing its principal features. A cross marks the estimated CG for each mass. No internal structural members are shown, other than the roll frame, a substitute for the rule-described roll bar, against which the driver reclines. The crosshatched paper enables projected areas to be estimated by counting squares.

The wheels are enclosed in fairings for streamlining and sailing and are arranged in a two in front–one in rear (2F-1R) wheel configuration. A three-wheel configuration will be lighter than a four-wheel one. The 2F-1R configuration is more stable

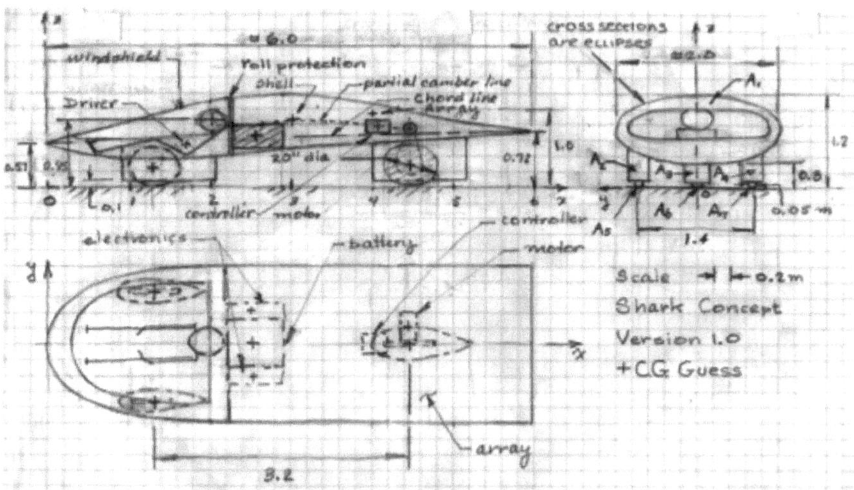

Fig. 9.2 Shark concept (dimensions in meters)

when turning and braking, which is the most common emergency maneuver.[6] A means for the fairings to accommodate large steering angles must be developed. The driver is in a near-reclining position. Achieving the view angles from the cockpit required by the race rules looks problematical. However, the low-drag potential of the shape is enormously attractive, and so we will continue to develop the concept. The shape has been drawn with a slight pitch-down attitude. This may slightly increase the drag above that for zero pitch. However, it may give a more negative lift coefficient, which may be a safer condition. Also, the drag of the ventilation system, array, or rear view mirrors will not be estimated at this stage.

9.9 Shark Drag Estimate

Let $p_\infty = 101.3$ kPa and $T_\infty = 25^\circ$C, then $\rho_\infty = 1.18$ kg/m^3. The drag will be evaluated for straight-ahead flow, 55 mph, or 24.58 m/s, and no wind. The Reynolds number per unit characteristic length and the dynamic pressure are

$$\frac{R_e}{\lambda} = \frac{1.18\dfrac{\text{kg}}{\text{m}^3}\times 24.58\dfrac{\text{m}}{\text{s}}}{18.46(10^{-6})\dfrac{\text{N}\cdot\text{s}}{\text{m}}} = 1.57(10^6)\text{m}^{-1},$$

$$q = \frac{1}{2}\times 1.18\times(24.58)^2 = 356.5\frac{\text{N}}{\text{m}^2}.$$

[6] See Chap. 21.

Body, A_1 Counting squares in Fig. 9.2 gives A_1, the profile area of the basic shape, as about 1.45 m^2. The pitch angle, α, the angle between the chord line and the horizontal is about −1.43°. Since the body's length is 6.0 m, the chord, *c,* is slightly longer, but still about 6.0 m. A low camber ratio[7] is best because it provides a flatter surface for the array while still giving low drag. The maximum camber is about 0.22 m. Thus, the camber ratio is 3.67 %. The lowest camber ratio reported by Morelli (1983) was 3.7 %.

The clearance ratio is about 0.15. The design Reynolds number, chord times Re/λ, is 9.4(10^6), twice the value at which the Morelli data were taken. Therefore, the c_D will be somewhat underestimated because it trends upward as *Re* increases past the transition to turbulence. (See the ellipse curve in Fig. 2.6, for example.)

1. ***Free Air:*** Figure 17.8 gives $c_{D\infty}=0.045$ for 3.7 % camber ratio and $\alpha=0°$. Since the pitch is outside the ±1° range, this will be rounded up to 0.05.
2. ***Ground Effect:*** Equation (17.16) with the 3.7 % camber ratio data of Table 17.5 gives $\Delta c_{DG}/c_{D\infty}=0.1$; hence,

$$\Delta c_{DG} = 0.1 \times 0.05 = 0.005.$$

9.10 Front Fairings, A$_2$ and A$_4$

1. ***Free Air:*** The chord is 1.15 m with a maximum thickness of 0.24 m. The thickness ratio is therefore 0.21. The average span is about 0.35 m. The profile shapes approximate trapezoids. Thus, the product of the thickness and the average span gives the profile area, A_2 or A_4, as 0.084 m^2. Using Re/λ and the chord as the characteristics length, the Reynolds number at the fairing is 1.81(10^6). Extrapolating in Table 17.2 using NACA 2421 gives $c_{D\infty}=0.0081$ based on the planform area, and dividing by t/c, 0.21, gives $c_{D\infty}=0.039$ referenced to the profile area.
2. ***Interference:*** The interference drag between each fairing and the underbody of the basic shape and the mutual interference between the front fairings will be found in the same way as for the rear fairings of the Table Top.

From the sketch of the Shark, the minimum spacing between A_2 and A_4, y_{MIN}, is about 1.15 m. Hence, $y_{MIN}/t=1.15/0.24=4.79$. For this dimensionless spacing, Figure 17.4 shows a nearly zero mutual interference drag between A_2 and A_4. For the interference between A_2 or A_4 and the underbody, Eq. (17.12) gives

$$\Delta c_{DI} = [0.8(0.21)^3 - 0.0003]\frac{(1.15\text{m})^2}{0.24\text{m} \times 0.35\text{m}} = 0.11.$$

3. ***Roughness, etc.*** None, as before.

[7] Camber ratio (%) is (maximum camber/chord) × 100; see Chap. 17.

Table 9.2 Initial Shark drag area

Piece #	$A_{D\#}$	A_P	$c_{D\infty}$	Δc_{DCG}	Δc_{DI}	c_D	$c_D A_{D\infty}$
1	1.500	1.500	0.05	0.005	0	0.055	0.0828
2	0.085	0.085	0.039	0	0.110	0.149	0.0126
3	0.080	0.154	0.033	0	0.146	0.179	0.0275
4	0.085	0.085	0.039	0	0.110	0.149	0.0126
5	0.005	0.005	0.526	0	0	0.526	0.0026
6	0.005	0.005	0.526	0	0	0.526	0.0026
7	0.005	0.005	0.526	0	0	0.526	0.0026
A_D	1.763					$c_D A_D$	0.1430

Rear Fairing, A_3 The rear fairing also has a chord of 1.15 m. It was given a thickness of 0.32 m because it may house the rear brake disk and caliper and portions of the drive (and possibly a wheel motor). It has a thickness ratio of 0.28 and a profile area of 0.154 m². We assume that the front fairings do not significantly affect the flow around A_3. Thus, the same chord Reynolds number calculated for the front fairings will be used for A_3. Extrapolating for NACA 0028 at a chord Reynolds number of 2.94(10⁶) in Table 17.3 gives $c_{D\infty}=0.033$. Equation (17.12) gives $\Delta c_{DI}=0.146$. We make neither roughness nor protuberance corrections.

Wheels, A_5–A_7 We treat these in the same fashion as those of the Table Top. This gives a c_D of 0.526 per wheel.

Drag Area Table 9.2 summarizes the Shark results. The free-air profile areas, A_P, are in the third column. These do not necessarily sum to the front-projected area of the car.

The total drag area of 0.143 m² (rounded up) exceeds the specified drag area of 0.11 m², with the ventilation drag, array drag, and incidental drags from roughness introduced during manufacture still to be added.

Revision to Concept We will reduce the t/c ratio of the rear fairing, bearing in mind the drive elements it may contain besides the wheel. Fortunately, a relatively small reduction in thickness pays big dividends because the interference drag is a function of the thickness ratio cubed. Reducing t to 0.25 m gives t/c of 0.217 and a front-view profile area of 0.0625 m². The free-air profile area drops from 0.154 to 0.120 m².

The reduced thickness increases $c_{D\infty}$ to 0.037 but decreases the interference drag coefficient of the fairing from 0.146 to 0.087. The free-air profile area of the fairing shrinks to 0.120 m² and the front-view profile area from 0.080 to 0.0625 m². The latter reduction shrinks the profile area of the car to 1.746 m². The total drag area drops from 0.143 to 0.128 m².

Fillets extending downstream from the maximum thickness location of each fairing cause additional reduction in drag area. The fillet radius should be 4–8 % of the chord of each fairing (0.046–0.092 m in this case) and the length of the fillet downstream of the fairing trailing edge equal to the chord (Hoerner 1965). This trailing length and radius choice can reduce the interference drag to 1/10 (or less), of its no-fillet value, at a t/c of 0.2–0.25.

Recalculating the drag area for the car with the fairing interference drag at 1/10 of its no-fillet value gives 0.104 m^2. This range leaves about a 5.5 % margin for the other sources of drag not included in the current estimate. Also, after considering the wheel and suspension design, we may find that the profile areas of the fairings can be reduced further, thus increasing this margin. Figures 9.10 and 9.11 show the car with the fairing fillets added.

The aerodynamics are marginally satisfactory, so far. We pass to consideration of the weight and CG.

9.11 Shark Weight and CG

Weight Control "Weight control" means not just how much weight but the distribution of weight. The weight and its distribution is much easier to control if the car is finished early in an orderly way, say by the summer or fall preceding the race (American Solar Challenge schedule). It is more difficult to control if the car is built at the last minute, that is, during the spring before the race. Control requires a procedure to track the weight of the car and to allot weights to different systems; and a firm resolve[8] to stick to the procedure, specification, and allotment.

Results The mass of the vehicle must satisfy specification 8.4 and the CG location must satisfy specification 8.11.1.

The method of Chap. 21 was employed to estimate these parameters. The shell was assumed to be made of a Kevlar®–Nomex® composite sandwich having a density of 185.3 kg/m^3 (see Table 9.5). The masses of some of the components in Fig. 9.2 can be estimated from previous experience or manufacturer's data. The driver's mass is known to be 80 kg because the driver will be ballasted to that mass.

To make the estimate more realistic, set the construction of the internal stiffening structure of the car as shown in Fig. 9.3. The stiffener masses will be estimated based on carbon–foam composite construction. The plate-like shapes were assumed to have a uniform composition through their thickness and therefore that the CG is on the center plane of the plate. Hence, we calculate only the x, z or y, z coordinates of the CG, as appropriate. The left and right cage walls have a nonuniform construction, so they must be broken down into pieces. Table 9.3 shows the results for the stiffeners and the shell. Coordinate x is measured from the car's nose. Once the stiffeners' CG locations and masses are known, the structures may be treated as isolated masses, like the driver or the battery. Table 9.4 summarizes the guesses and estimates for the other components.

The calculation gives the vehicle's gross mass as 320.2 kg, with the CG located at an x-coordinate of 2.618 m, a y-coordinate of 0.008 m, and a z-coordinate of 0.641 m. The off-centerline CG location is caused by the off-centerline location of

[8] That is, management must be really hard-nosed about it.

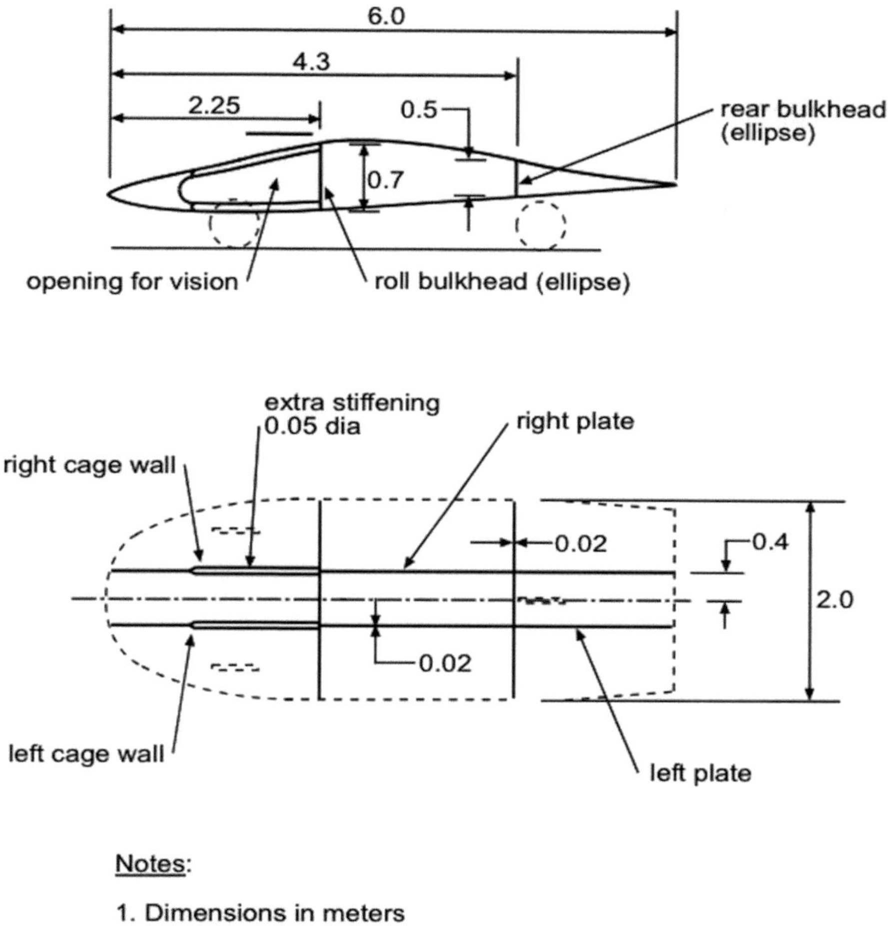

Fig. 9.3 Tentative internal structure (dimensions in meters)

Table 9.3 Stiffener and shell CG coordinates

Stiffener	x(m)	y(m)	z(m)	M(kg)
Left	3.5	0.4	0.85	2.0
Right	3.5	−0.4	0.85	2.0
Rear	4.3	0	0.8	2.0
Roll	2.25	0	0.8	2.5
Left cage	1.0	0.4	0.6	1.0
Right cage	1.0	−0.4	0.6	1.0
Shell	3.05	0	0.85	38.6

Table 9.4 Mass summary

Category	Mass (kg)
Wheels	5.19
Driver	80.00
Battery	140.00
Motor and Controller	24.60
Array and MPPTs	15.30
Shell	38.60
Fairings	6.00
Frame	10.50
SUBTOTAL (kg)	320.20
ALLOWANCE (kg)	17.80
Specification 4.3 (kg)	338.00

MPPT *maximum power point tracking*

Table 9.5 Shell moment of inertia

Seg.	$I_{CG}(\text{kg}\cdot\text{m}^2)$	$x_j(\text{m})$	$x_jM_j(\text{m}\cdot\text{kg})$	$M_j(\text{kg})$	$h^2(\text{m}^2)$	$I_{Zj}(\text{kg}\cdot\text{m}^2)$
1	1.146	0.50	2.330	4.660	4.554	22.370
2	1.146	1.50	10.525	7.017	1.286	12.090
3	3.293	2.50	18.234	7.294	0.018	3.427
4	3.064	3.50	24.558	7.017	0.750	8.329
5	2.606	4.50	29.079	6.462	3.482	25.109
6	2.125	5.50	32.339	5.880	8.214	50.425
Totals:			117.066	38.329		121.750

the CG of the motor; the other off-centerline CGs are balanced. Note that the x- and z-coordinates are close to the corresponding coordinates of the battery's CG.

The wheelbase is 3.2 m and begins 1.3 m from the nose, so the CG is spaced 41.1 % of the wheelbase from the front wheel contact patches, or 7.9 % outside of the specified zone of 33.3 % (specification 8.11.1). The steering, suspension, instruments, cabling, and lights were left out of the CG calculation. However, the battery and the driver are the strongest influences on the CG location.

The mass is 17.8 kg below the target, but, as noted above, a number of components were not included. The 17.8 kg is then an estimate of the mass allowance for these components (Table 9.3).

9.12 Rollover and Skid

Specification 8.11.5 governs here. Once the weight and CG calculations have been completed, sufficient information is known to estimate the rollover and skid stability. Use the traction limits of Chap. 21 ($a_x/g=0.85$, $a_y/g=0.75$), absent knowledge of the actual tires to be used by the Shark. Using the CG location, L, and T with

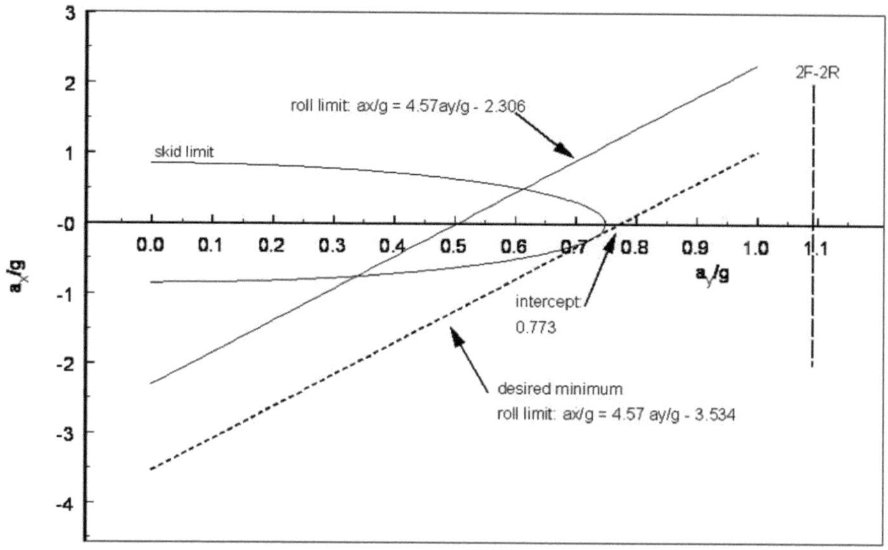

Fig. 9.4 Initial roll and skid limits

Eq. (21.32), the limiting rollover line has been plotted in Fig. 9.4. The prohibited operating regions are points to the right of the line.

The positive ordinate is the braking inertial force in *g*'s. The negative ordinate is the acceleration inertial force in *g*'s. Both are tangent to the path of the car's CG. The abscissa is the centrifugal inertial acceleration perpendicular to the path of the CG.

The car fails the specification in that the roll limit line is well inside the skid ellipse for $0.34 \lesssim a_y/g \lesssim 0.61$. For transverse accelerations to the right of this line, the car will roll before skidding as long as the braking or longitudinal acceleration is less than the skid limit.

A more subtle point is that even when turning at steady speed at a *g*-load less than the roll limit, the car could initiate a roll if the driver accelerates. For example, at a transverse acceleration of 0.45 *g*, the car will roll if the driver accelerates at about 0.25 *g*. This can only happen, however, between 0.34 and 0.5 transverse *g*: Below 0.34 *g*, the 0.25 *g* acceleration has no effect; above 0.5 *g* the car has already begun to roll. The destabilizing effect occurs because the moment of the rearward inertial force component adds to the tipping moment of the transverse acceleration in the 2F-1R wheel configuration.

The opposite effect happens if rolling has begun and the brakes are applied. If rolling at slightly more than 0.5 *g*, a little braking, say 0.1 *g*, would stop the roll, assuming the braking can be at least this effective with the outside front tire not in contact with the road. The moment of the forward inertial force opposes the tipping moment of the transverse acceleration. This effect makes the 2F-1R car safer in emergency obstruction avoidance, if the brakes are applied first. Suppose a deer enters the road and the driver first brakes at 0.6 *g* and then turns right at 0.5 *g*, the Shark will not roll.

Using Eqs. (21.34) and (21.35), the line labeled "minimum desired" rollover limit has also been plotted in Fig. 9.4. This line has a 0.773 *g* intercept with the a_y/g axis. The wheelbase and the track set the slope (2 *L*/*T*) of the limit lines. If these are fixed, then this intercept means that

$$\frac{b}{h} \geq 3.534$$

to meet specification 8.11.5. The distance $b = L - a$, or 1.887 m.

Hence, *b*/*h* is actually 2.944. This constraint will require some changes in the Shark concept, if three wheels are retained, just to get the smallest acceptable *b*/*h*, 3.534. For instance, if *b*/*L* were increased to 0.67 as required for directional stability, say by moving the batteries forward beyond the roll bulkhead, then *h* should be no more than 0.607 m. This is a reduction of 3.4 cm. This could be done. The car could be lowered, for example, although this would increase the drag somewhat.

Increasing *T*/*L* would steepen the roll limit line and move it to the right. Race rules limit the width of the car to 2 m, so there is not much room to increase *T*. The car could of course be shortened. Although making the car smaller (which should reduce the weight, an important advantage) and lower is certainly an option, completely exploring all the implications of this is a lengthy process. Let us examine a direct way of meeting the stability requirements.

Two in Front–Two in Rear (2F-2R) The slope of the rollover line can be made vertical, independent of the CG location, and moved to the right by shifting to four wheels, as Eq. (21.31) shows. For the present design:

$$\left[\frac{a_y}{g}\right]_{a_x=0} = \frac{T}{2h} = \frac{1.4\text{m}}{2 \times 0.641\text{m}} = 1.09g.$$

This rollover limit line is the vertical broken line in Fig. 9.4.

If we were to move to a four-wheeler, and assuming the CG height were about the same, the above intercept would be adequate. It would provide an allowance for the compliance of the suspension and the tires (neglected in Chap. 21), and for a transverse skid limit greater than 0.75 *g*. Notice also that *b*/*L* satisfies now-governing specification 11.8.3 because (if adding another wheel does not shift the CG much farther rearward) it is already less than 0.5. We would be out of the woods.

The downside of four wheels is somewhat increased weight, rolling resistance, drag, complexity, and cost. However, we may be able to compensate for the first three of these negative effects by reducing weight (perhaps by moving to a shell-and-space frame design) and drag elsewhere and in the tire design. The extra complexity and cost we will have to accept. The gain in safety overrides these considerations. Let us evaluate the 2F-2R design.

Drag Area Keeping the wheelbase the same, we place the two rear tires directly behind the front tires. The rear track will then be 1.4 m. The current rear wheel fairing design, NACA0028 will be retained.

There will be, as in front, negligible interference between the rear wheels. However, interference between the front and rear fairings on each side is now possible. Using a dimensionless spacing (see Fig. 17.5) of 2.70 in Eqs. (17.10) and (17.11) for a front and rear wheel pair on each side gives decrease of −0.042 for each one of the front wheels but a drag increase of 0.012 for each of the rear wheels. (Here, we have assumed that the fillets have no effect on the interference. This may not be conservative.)

Using the 1/10 rule to reduce the interference drag with the shell and then superimposing the interference corrections gives the total profile drag area of

$$c_D A_D = 0.106\text{m}^2.$$

The drag area margin that existed in the 2F-1R configuration has been reduced to 3.6 %. The drag area will probably exceed 0.11 m^2, as-built. Using the 2F-2R wheel arrangement reduces the profile area to 1.684 m^2 so that the car's drag coefficient is 0.062.

Mass and CG Adding another wheel and fairing increases the total mass to 323.6 kg, rounded up. This is still acceptable. We place the motor over on the left rear wheel. Accounting for the extra mass of the new wheel and fairing then gives $x_{CG}=2.624$ m, $y_{CG}=-0.020$ m, and $z_{CG}=0.639$ m. And $a/L=0.414$, within the 2F-2R stability limit, as expected.

9.13 Side Gust

The rollover limit is specification 8.5.2, which sets $V_W=50$ kph, and $V_X=88$ kph. We will use the ambient temperature of 25°C and pressure of 101.3 kPa required by specification 8.3.2 and set the ratio of the gust slope width to the vehicle's length at 1. Figure 21.5 shows the gust scenario.

Method Equations (21.3) will be solved for the car's trajectory using the "little tap" method of Chap. 2. The force and moment used in these equations will be approximated for each new time by the method in Chap. 21. Finally, the accumulated distance moved in the global *y*-direction (transverse to the traffic lane) and the change in the heading of the car will be calculated for each time. All the information needed for the calculation is known except the cornering stiffness of the front and rear wheels and the moment of inertia of the car about its CG.

Cornering Stiffness Roland (1973) gives an empirical equation for the cornering force developed on a wheel contact patch at zero camber angle as

$$\frac{F_Y}{F_Z} = (1 - 4.88 \times 10^{-4} F_Z)(12.95\alpha + 436.37\alpha^3).$$

We will assume the constants, derived for a bicycle tested by Roland, are typical of the bicycle-type tires used on solar racing cars.

Multiplying by the vertical force, F_Z, and dividing by the slip angle, α, gives an equation for the cornering stiffness as a function of the slip angle. Taking the average of this expression up to a slip angle of 3° and using the current weight distribution (929.8 N/wheel, front, 656.5 N/wheel, rear) gives the average front (F) and rear (R) cornering force coefficients (see Eq. (21.1))

$$\overline{C}_{\alpha F} = 114.5 \frac{\text{N}}{\text{dgr}} \text{ and } \overline{C}_{\alpha R} = 101.1 \frac{\text{N}}{\text{dgr}}.$$

Therefore, the understeer gradient (Eq. (21.5)) is

$$K = \frac{W_F}{\overline{C}_{\alpha F}} - \frac{W_R}{\overline{C}_{\alpha R}} = \frac{929.8\text{N}}{114.5 \frac{\text{N}}{\text{dgr}}} - \frac{656.5\text{N}}{101.1 \frac{\text{N}}{\text{dgr}}} = 1.630 \frac{\text{dgr}}{\text{g}}.$$

There is no critical speed because the understeer gradient is positive.

Moment of Inertia The moment of inertia, I_Z, about the CG, will be estimated using the method of Chap. 21. I_Z must be known to estimate the effect of a side gust. The estimate for the shell is shown below. The shell was assumed to be made of a Kevlar®-faced Nomex® composite with a density of 93 kg/m^3 and a thickness of 3/8 in. (0.00953 m).

The canopy, which is integral with the shell, was assumed to have about the same density. The shell was divided into six 1-m-long segments. The cross-sectional shape of each segment, an ellipse, was given the major and minor axes dimensions read from Fig. 9.2 at the center of the segment. Thus, the approximated shape would have a stepped appearance. Table 9.5 shows the resulting I_Z of the shell: 121.75 kg·m^2.

The second column gives the moment of inertia of the segment with respect to its own CG, I_{CG}. This is equal to the difference between the I_{CG} of the elliptic cylinder having the major and minor axes of the segment and that of the elliptic cylinder concentric with it but having its major and minor axes reduced by the segment's thickness. For segment 1, the mass of the outer cylinder is

$$M_{O1} = (\rho\pi abL)_1 = \pi \times 185.24 \frac{\text{kg}}{\text{m}^3} \times 0.7\text{m} \times 0.15\text{m} \times 1\text{m} = 61.11\text{kg},$$

where "*a*" and "*b*" represent half the major and minor axes, respectively. Subtracting the thickness from *a* and *b* and recalculating gives

$$M_{I1} = \mathbf{56.45kg and} M_1 = (M_O - M_I)_1 = 4.66\text{kg}.$$

I_{CG1} is then (using a standard formula)

$$I_{CG_1} = \left(M \frac{L^2}{12}\right)_1 + \frac{1}{4}(M_O a_O^2 - M_I a_O^2)_1 = 1.146\text{kg} \cdot \text{m}^2.$$

The squared distance from the CG of the segment to the CG of the car is

$$h_1^2 = (x_1 - x_{CG})^2 + (y_1 - y_{CG})^2 = (0.5\text{m} - 2.634\text{m})^2 + [0 - (-0.02\text{m})]^2 = 4.554\text{m}^2.$$

The parallel axis theorem then gives I_{Z1} as

$$I_{Z1} = I_{CG} + h_1^2 M_1 = 1.146\text{kg}\cdot\text{m}^2 + 4.554\text{m}^2 \times 4.66\text{kg} = 22.37\text{kg}\cdot\text{m}^2.$$

Completing the calculation of I_Z by adding in the other parts of the car (j=2, 3,…6) gives $I_Z = 389.88\text{kg}\cdot\text{m}^2$. The radius of gyration of the vehicle's mass is

$$r_g = \sqrt{\frac{I_Z}{M}} = \sqrt{\frac{389.88\ \text{kg}\cdot\text{m}^2}{323.6\ \text{kg}}} = 1.097\ \text{m}.$$

Shell CG At the end of the fourth column of Table 9.5 is the sum of the products of the segments and the distances of their CGs from the nose of the car. Dividing this sum by the total mass of the shell gives the coordinate of the shell's CG.

$$x_{CG} = \frac{117.066\ \text{m}\cdot\text{kg}}{38.33\ \text{kg}} = 3.045\text{m}$$

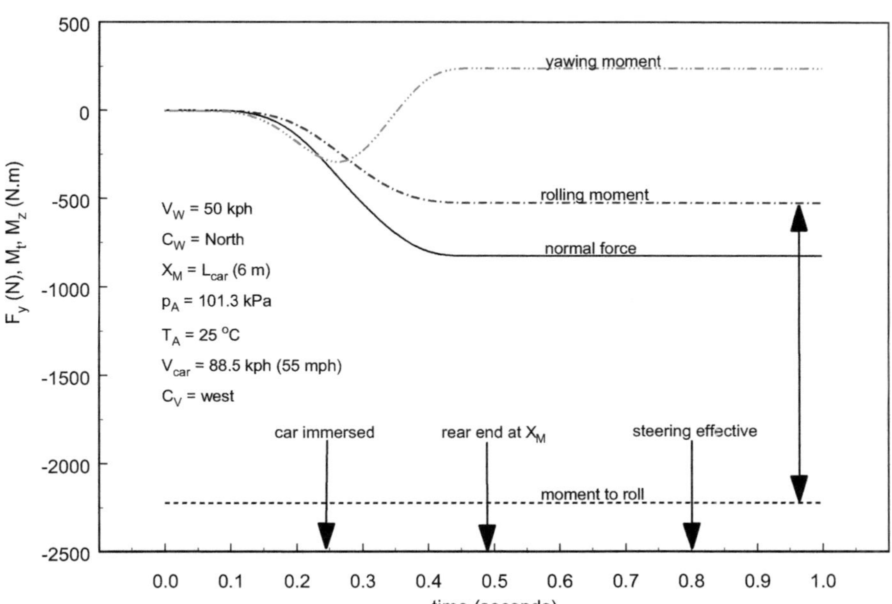

Fig. 9.5 Side force and moments on the Shark

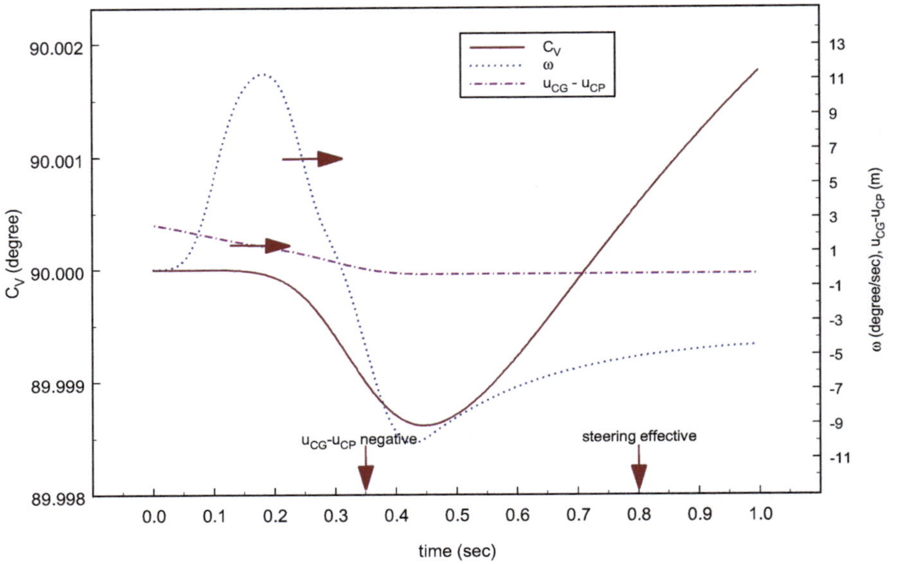

Fig. 9.6 Heading and heading velocity of the Shark

Results Figure 9.5 shows the side force, F_y, the yawing moment, M_z, and the rolling moment M_t. Also shown is the moment causing rollover, calculated as −2222.6 N m by Eq. (21.20). Note that, as the double-headed arrow indicates, there is a large clearance between the moment causing rollover and the maximum rolling moment.

The yawing moment is at first counterclockwise (less than zero) and then shifts to clockwise. This behavior is caused by the yawing moment arm,[9] a function of the car's shape, which is positive until about 0.35 s, then passes through zero and is negative thereafter. Hence, as Figs. 9.6 and 9.7 show, this causes the car to turn slightly north at first and then slightly south. The displacement of the car's CG in the traffic lane was minuscule, effectively zero, and therefore was not plotted.

The simulation results show that the car will be stable and will deviate only a negligible amount from a straight path under the design side gust conditions. However, the designer should bear in mind the limitations of the calculation. For example, Figure 9.7 shows that the car will not roll in a 100 kph gust. This is clearly not credible. Under such heavy aerodynamic loads, effects ignored in the model, used in the calculation will come into play. Lift is one such effect. A positive angle of attack to the relative wind would be produced by flexing of the suspension (also neglected in the model). This would cause upward lift and generate an additional aerodynamic rolling moment that would cause roll. The calculations presented here are intended as aids to design, not as revelators of absolute truth. They are not equivalents to wind tunnel testing of scale models. Nor should their results justify operating a solar car unwisely, such as in 100 kph wind gusts.

[9] The yaw moment arm, $u_{CG}-u_{CP}$, where u is the distance from the nose of the car and CP denotes the longitudinal coordinate of the center of pressure, is plotted in Fig. 9.6.

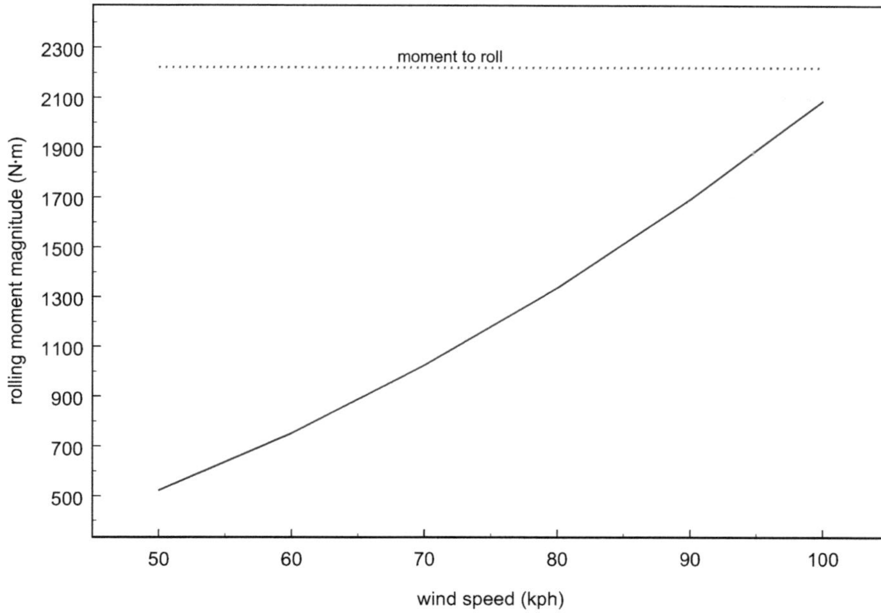

Fig. 9.7 Aerodynamic rolling moment study

9.14 Drive, Battery, Energy Rate and Range

Drive The design maximum grade and the design speed up this grade establish the maximum tractive torque and power required to climb hills. The design maximum grade should therefore not be less than the steepest grade in the design race route. The maximum cruising speed on a zero grade sets the maximum rotational speed.

The maximum grade in the design race route is 10 % (specification 8.2.0). The speed up this grade is 40 kph. The maximum speed on a zero grade is 88.5 kph (55 mph). To allow for mass gain we will generate the requirements using the mass limit of 338 kg in specification 8.4.1. Multiplying the current maximum drag area estimate by an "as-built" factor of 1.2 gives a drag area 0.127 m^2. See Chap. 10, Table 10.2 for the drive procurement performance points.

Battery The details of the battery will be left to Chap. 10. However, in computing the weight of the vehicle, we have assumed that the battery's mass is 140 kg. Good-quality lead-acid batteries have a specific energy of about 35 W·h/kg. Using these batteries gives a total capacity of about 4900 W·h, of which about 80 % (about 4000 W·h) is useable. Lead-acid batteries are the cheapest of the battery types allowed by the sample rules of Chap. 16.

Energy Rate and Range As an example, we will estimate the tractive energy consumption rate of the Shark using the drag areas estimated above and the conditions

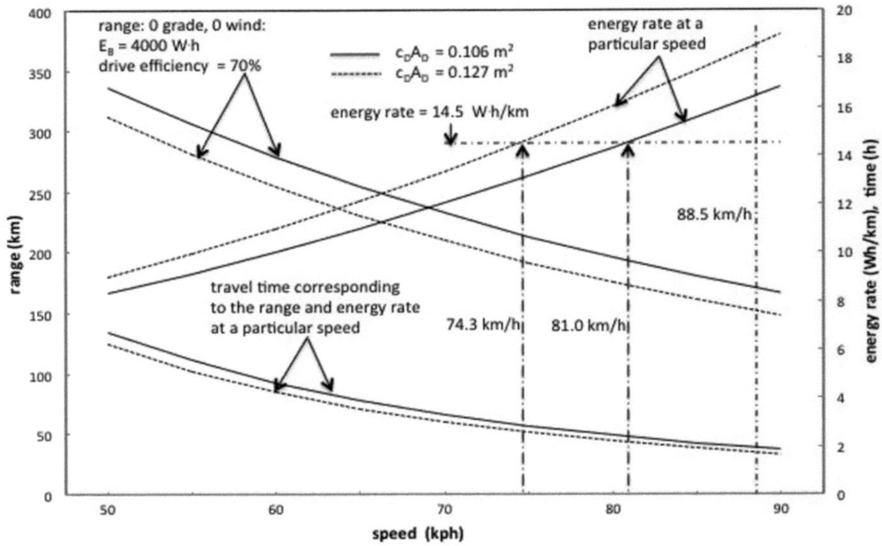

Fig. 9.8 Range and energy rate (battery only)

required by specification 8.13.1. This estimate will of course be revised several times during the design process as the information about the car becomes more complete and accurate.

Suppose the battery contains 4000 W·h of useable energy,[10] the efficiency, $\eta_{OVERALL}$, between the battery and the contact patch of the driven wheel is 70 %, and the rolling resistance coefficients are at the limits set in specifications 6.1 and 6.2. Let the mass be the maximum, 338 kg. Using these numbers, we will calculate the energy rate, range, and running time of the Shark on a horizontal road as a strictly battery-powered car. The energy rate, e_R, will be found from Eq. (2.17) and the range, S, from

$$S = \eta_{OVERALL}\frac{E_B}{e_R}.$$

Where E_B is the usable battery energy in W·h, and e_R is the energy rate in W·h/km.

The results are shown in Fig. 9.8. Notice that at the design speed the energy rate exceeds the specified value at both drag areas.[11] The car must cruise at 74.3 kph (46.2 mph) at the higher drag area and at 81 kph (50.1 mph) at the lower drag area to not exceed the design rate.

[10] This is assumed constant for convenience. The capacity of the battery is actually a function of the current discharge rate; see Fig. 4.5.

[11] Recall that specification 8.13.1 is a "wish."

Reducing the speed to about 50 kph (31 mph) will give an energy rate of about 9 W·h/km and a range of about 320 km (200 miles). Travel time would be 6.4 h.

9.15 Array Concept

In Fig. 9.2 we have already established the underlying shape and, by implication, the maximum area of the array. And, for the weight calculation, we assumed four parallel strings, each with its own maximum power point tracking (MPPT). Further array details will be left to the next chapter.

9.16 Driver Interface

Visibility The next step in the development will be to check the driver position for minimum height-of-eye and the side and forward view angles from that position (see Chap. 16, **Sect. 16.6.6**). The rear view system will be postponed to the next chapter. Figure 9.9 shows, except for the rear vision requirements, that the Shark concept is within the rules if the canopy encompasses the angles shown.

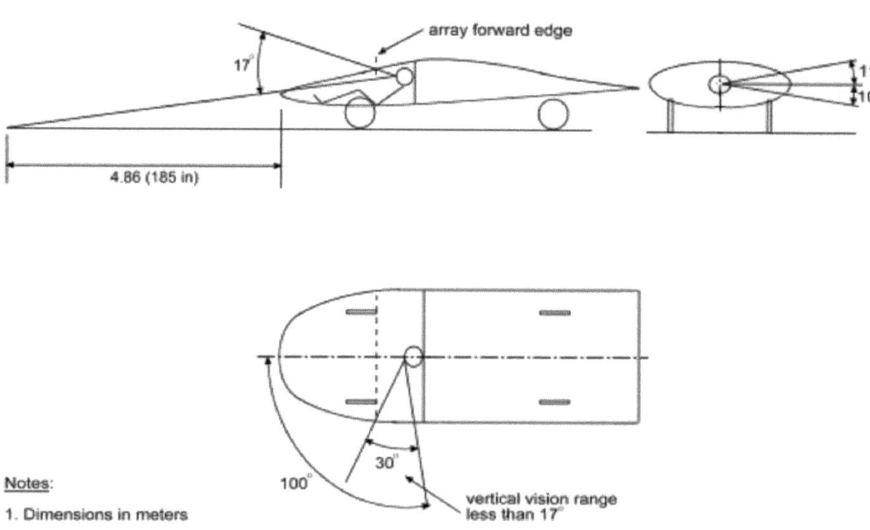

Fig. 9.9 Driver visibility study

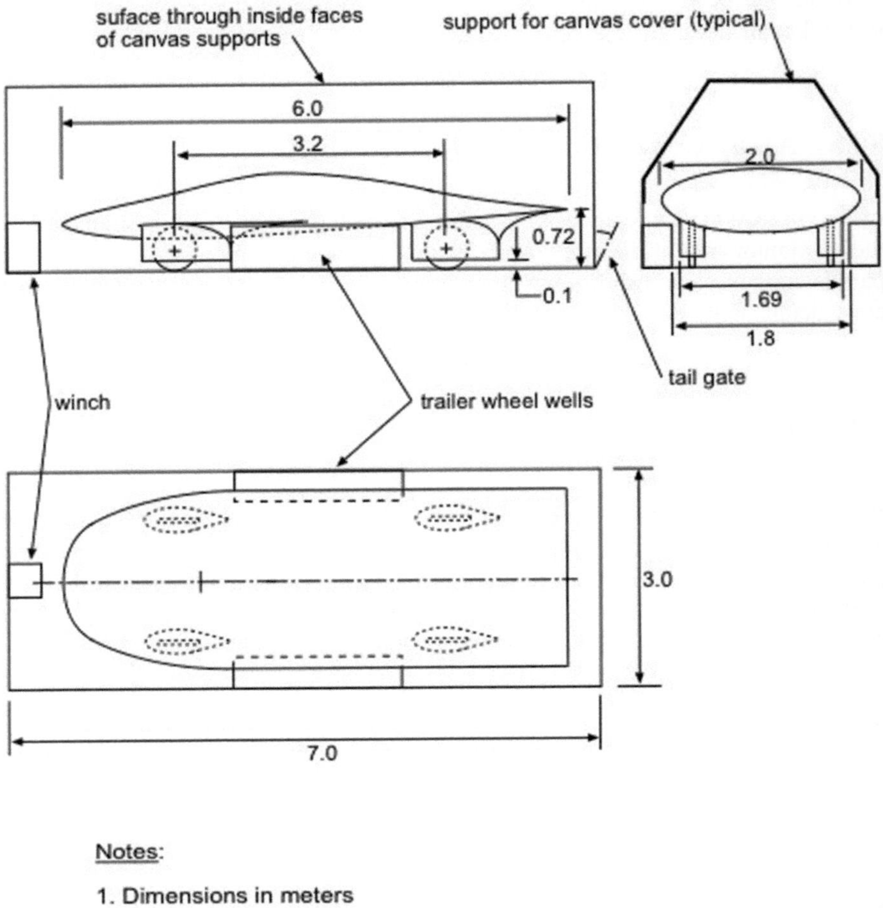

Fig. 9.10 Transport clearances

9.17 Transport Compatibility

Figure 9.10 shows the spacing between and the height of the wheel wells of an actual transport trailer. Comparison of these dimensions shows that the Shark will fit inside the trailer, although it is a close fit in some places. The vehicle will have to be carefully restrained and possibly padded in some places to prevent damage.

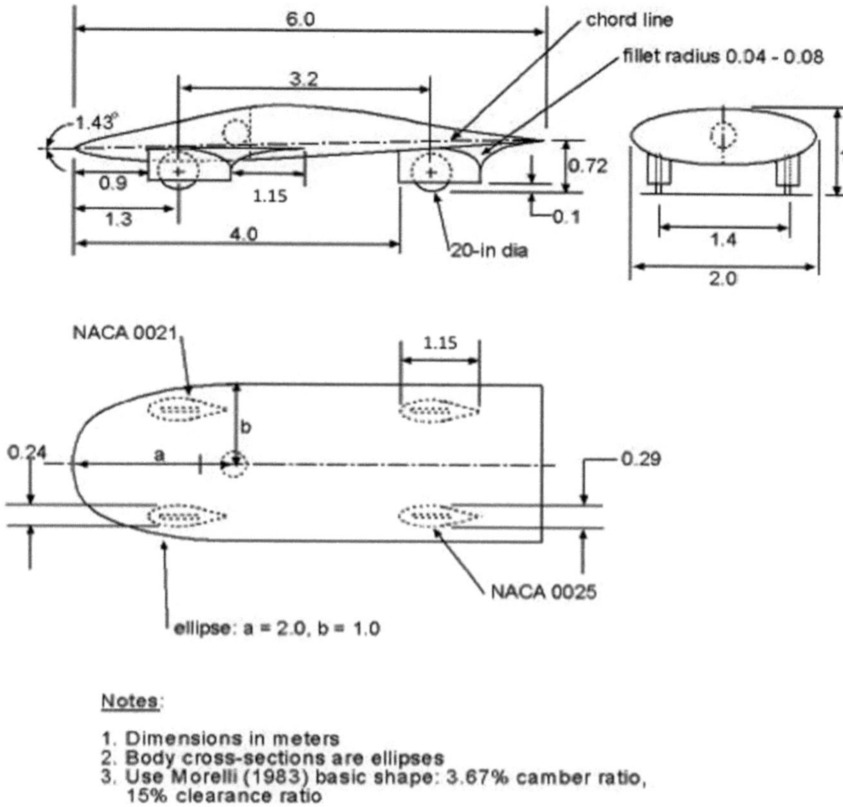

Fig. 9.11 Shark Layout

9.18 Concept Summary

Layout A three-view layout drawing is the end product of the conceptual design phase. Information for the drawing comes from the concept sketch and the calculations completed above. The drawing for the Shark is shown in Fig. 9.11. As Raymer (1989) points out, this drawing is the basis for further studies of the design and may be revised several times. Some way to accommodate the fairings to large steering angles must be devised.

Summary Table 9.6 summarizes the principal findings for the Shark, so far.

Table 9.6 Concept summary

Shape	Morelli “Shark”: 3.7 % camber ratio, 15 % clearance ratio, −1.43° pitch angle.
Dimensions	Length = 6 m, max. height = 1.2 m, wheelbase = 3.2 m, track = 1.4 m, max. width = 2.0 m
Inertial	Mass = 323.6 kg, CG at $x = 2.624$ m, $y = -0.02$ m, $z = 0.639$ m
Aerodynamic	$c_D A_D = 0.106\ \text{m}^2 - 0.127\ \text{m}^2$, $A_D = 1.68\ \text{m}^2$
Array	Integral to Shark shape; four parallel strings each with an MPPT
Battery	Capacity ≈ 4000 W·h, efficiency ≈ 80 %
Drive	See study displayed in Fig. 9.8
Tractive energy rate	About 16.6–18.6 W·h/km at 88.5 km/h
Wheels	2F-2R, 20 in. outside diameter
Problems	1. Fairings at large steering angles 2. Rear vision from cockpit 3. Energy rate too high

MPPT *maximum power point tracking,* 2F-2R two in front–two in rear

References

Hoerner, S. F. (1965). *Fluid dynamic drag.* Bakersfield: Hoerner Fluid Dynamics.

Kurtz, D. W. (1980). Aerodynamic design of electric and hybrid vehicles: A guidebook. JPL Publication 80–69 (NASA-CR-163744 and NTIS N81–12943), Jet Propulsion Laboratory, Pasadena, California.

Raymer, D. P. (1989). *Aircraft design: A conceptual approach.* Washington, DC: American Institute of Aeronautics and Astronautics.

Roland, R. D. (1973). “Computer simulation of bicycle dynamics,”. Mechanics and Sport, ASME Winter Annual Meeting, Detroit, Michigan, November 11–15, 1973, ASME, New York, pp. 35–83.

Chapter 10
Solar Racer—Detailed Design

10.1 Introduction

This chapter continues the Shark design by developing some details or examples of procurement specifications for certain components. The components considered are: the wheels, drive, array, battery, ventilation system, and brakes. Some examples of student-produced assembly drawings for a solar car are shown to illustrate the final product.[1]

As the design process continues, greater and greater detail is generated. Each item moves through the process on its own schedule. Each of these schedules is coordinated by a master manufacturing plan. The outcomes of this process are procurement specifications for items to be purchased and (ideally) engineering drawings of items to be constructed.

10.2 Procurement

A number of components, including the motor and controller,[2] the solar cells, the battery, the brake system master cylinders and calipers, the maximum power point trackers, the direct-current-to-direct-current (DC-DC) converters, the rack and pinion for the steering system, and the signal lights will usually be obtained from commercial sources. Most of these components are too specialized or require too much development time for typical teams to design and build them in-house. Others, such as the signal lights (and other such small parts), are available from local auto or electronic parts stores.

In general, the literature furnished by the manufacturer of each purchased item will contain dimensioned drawings, operating characteristics, and instructions for

[1] The drawings are not for the Shark, but for a later car.

[2] However, Storey (1994) reports that some teams in the 1993 World Solar Challenge built their own wheel motors. A necessity: At the time there were no commercial sources for such motors.

E. F. Thacher, *A Solar Car Primer*, DOI 10.1007/978-3-319-17494-5_10

its installation, operation, and maintenance. Installation information should be transferred to, or referenced by, the engineering drawings of the car. Some manufactured parts can be designed only after this information becomes available. All the original information should be organized in notebooks or files and thus preserved for later reference.

10.3 Wheels

The work of Chap. 9 assumed 20-in. wheels. However, suppose that three wheel diameters are being considered for the Shark. The actual diameters (as opposed to the advertised nominal diameters) of the tires are 17, 20, and 26 in.

Suppose further that each has a smooth tread, a tire pressure of 80 psig, a solid rim, the tire K-factor of Chap. 20, and is operating at the maximum racing speed. The theory of Chap 20 will be used to study the influence of wheel diameter and inflation pressure on rolling resistance.

Here is a sample calculation for 20-in. diameter wheels. From Table 9.6, the mass is 323.6 kg, or 713.5 lb, 178.4 lb/wheel on average. Equation (20.4) gives the sinking rate as:

$$h = 0.5\frac{26\,\text{in}}{20\,\text{in}}\left(\frac{80\,\text{psig}}{220\,\text{psig}}\right)^{0.3072}\left[20\,\text{in} - \sqrt{20^2\,\text{in} - \frac{4\times 178.4\,\text{lbf}}{\pi}\frac{2.456+0.251\times 20\,\text{in}}{19.58+0.5975\times 80\,\text{psig}}}\right] = 0.3048\,\text{in}.$$

The rolling resistance of the tire is therefore:

$$R_{\text{T}} = kh = 2.47\frac{\text{lbf}}{\text{in}}\times 0.3048\,\text{in} = 0.753\,\text{lb}.$$

We will use the Suntour® bearing data reported by Kay (1988). At 55 mph, the wheels are turning at 923.9 rpm. Kay (1988) gives the test load as 104.5 lb. A linear fit to the data and a quadratic load correction (see discussion of Eq. (20.12)) gives

$$\tau_{\text{B}} = \left(0.0256\,\text{in}\cdot\text{lbf} + 0.000292\frac{\text{in}\cdot\text{lbf}}{\text{rpm}}\times 923.9\,\text{rpm}\right)\left(\frac{178.4\,\text{lbf}}{104.5\,\text{lbf}}\right)^2 = 0.861\,\text{in}\cdot\text{lbf}.$$

The rolling resistance force equivalent to the bearing torque therefore is:

$$R_{\text{B}} = \frac{0.861\,\text{in}\cdot\text{lbf}}{10\,\text{in}} = 0.0861\,\text{lbf}.$$

We will use Eqs. (20.6) and (20.8)–(20.10) to find the rolling resistance equivalent to the air drag resisting rotation of the wheel. The Reynolds number defined by Eq. (20.10) is (using data in m-k-s units):

$$\mathrm{Re}_{rW} = \frac{Vr_\mathrm{W}}{\nu} = \frac{24.58\dfrac{\mathrm{m}}{\mathrm{s}} \times 0.254\mathrm{m}}{16.19 \times 10^{-6}\dfrac{\mathrm{kg \cdot m}}{\mathrm{s}}} = 385{,}660.$$

The moment coefficient from Eq. (20.9) is therefore:

$$C_\mathrm{M} = \frac{0.146}{\mathrm{Re}_{rW}^{1/5}} = \frac{0.146}{385{,}660^{1/5}} = 0.0111.$$

Using "q" from the Sect. 9.8 of Chap. 9 and Eqs. (20.6) and (20.8) yields

$$R_\mathrm{D} = qr_\mathrm{W}^2 C_\mathrm{M} = 357.8\frac{\mathrm{N}}{\mathrm{m}^2} \times 0.254^2\,\mathrm{m}^2 \times 0.0111 = 0.257\mathrm{N} = 0.0578\,\mathrm{lbf}.$$

The rolling resistance coefficient therefore is:

$$\mu = \frac{R_\mathrm{T} + R_\mathrm{B} + R_\mathrm{D}}{W_\mathrm{W}} = \frac{0.753\,\mathrm{lbf} + 0.0861\,\mathrm{lbf} + 0.0578\,\mathrm{lbf}}{178.4\,\mathrm{lbf}} = 0.00503.$$

Figure 10 .1 shows the rest of the results. The reduction in μ between wheel diameters of 17 and 26 in. is about 38.7%. The 26-in. wheel should be chosen, considering rolling resistance alone. But other factors must be taken into account. Smaller wheels are easier to streamline. Therefore the designer must consider whether the reduction in the energy consumed in overcoming rolling resistance exceeds the possible increase in the energy to overcome drag caused by larger wheels. Larger wheels may be weaker under cornering loads, when efforts are made to keep the wheels light, but may have fewer flats because the contact patch pressure is lower. Larger wheels tend to increase the height of the center of gravity, thus reducing the resistance to roll over. We decide to continue with a 20-in. wheel.

Increasing the inflation pressure will lower the rolling resistance. Figure 10.2 shows the effect of inflation pressure on rolling resistance for the 20-in. wheel at 55 mph. An inflation pressure of 110 psig would lower the rolling resistance coefficient to 0.0045. This is about 12.5% above the specified value, 0.004. In keeping with a general philosophy of optimizing the whole system, we judge this to be close enough at this point.

The side and vertical loads the wheel must bear will be estimated.

The center of gravity is 2.634 m from the front axel. Hence, a is 1.324 m, and b is 1.876 m (refer to Figs. 9.11 and 21.1). Thus, b/L is 0.586 and the 1-g vertical load on each of the front tires is (using a 338 kg mass):

$$W_\mathrm{F} = \frac{Wb}{Ln_F} = \frac{3314.8\,\mathrm{N} \times 1.876\,\mathrm{m}}{3.2\,\mathrm{m} \times 2} = 971.2\,\mathrm{N} = 218.3\ \mathrm{lb}.$$

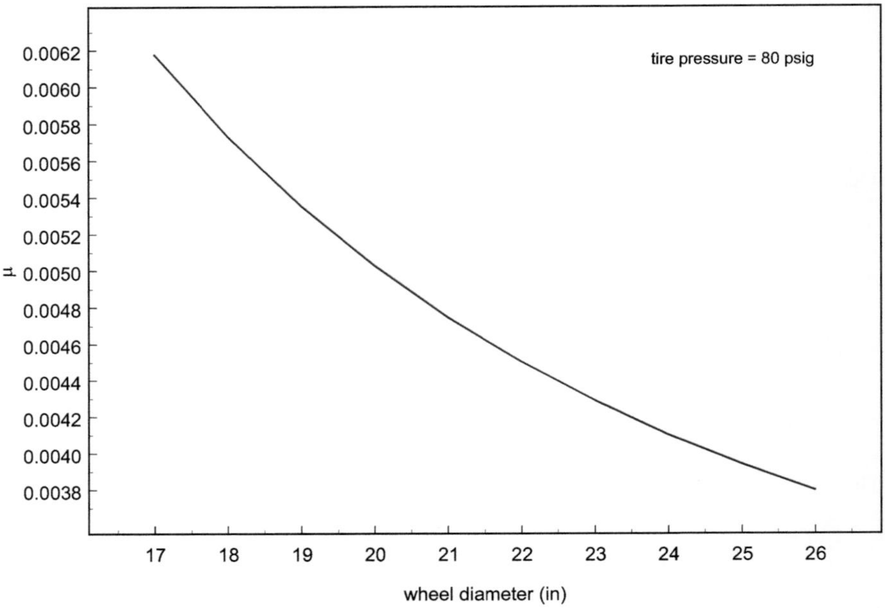

Fig. 10.1 Rolling resistance vs. wheel diameter

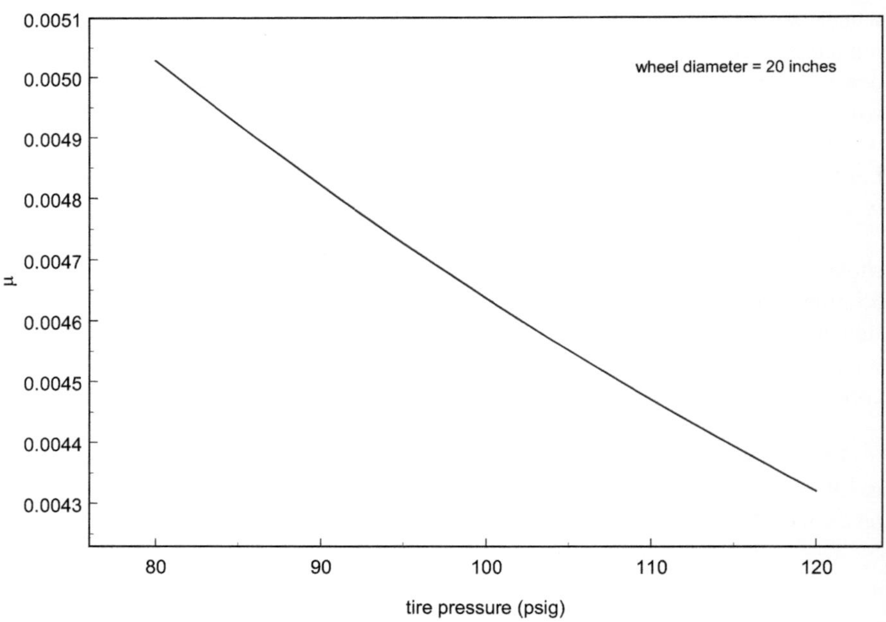

Fig. 10.2 Rolling resistance vs. tire pressure

Table 10.1 Wheel procurement requirements

Rim	Solid aluminum
Inflation pressure	≥110 psig
Actual diameter	20 in., bicycle-like width
Tire type	Tubeless
Tread	Smooth
Bearing torque	At or below Suntour™ bearings
Weight of wheel	≤16.95 N (3.81 lb)
Design side load	971.2 N (218.4 lb), certified
Design vertical load	≥1942.5 N (436.2 lb), certified 2913.7 N (655.0 lb) if no trade-off

Specification 8.7.1 asks for a 3-g design vertical load, or 2913.6 N (655.0 lb), on either of the front wheels. This, however, is not a "demand" but a "wish." The demand is the 2-g bump of specification 8.7.3, which implies a 1942.5 N (436.7 lb) vertical load. The same paragraph requires a 1-g side load, 972.2.2 N (218.4 lb; Table 10.1).

10.4 Large Steering Angles

The problem of accommodating large steering angles with the fairings (Table 9.6) will be solved by incorporating doors on both sides of each fairing. The doors will be spring loaded to shut and will be pushed open by aluminum bars attached to each wheel. This was the actual solution adopted for *Helios,* alias the Shark, for the front wheels.

10.5 Drive

Figure 5.6 implies that the optimum operating conditions for the motor may not match the operating conditions of the wheel. A transmission between the motor and the wheel allows the motor to be operated nearer its optimum torque and speed while driving the wheel over its torque and speed range. On the other hand, the transmission consumes energy, adds weight, and increases the number of ways the drive can fail.

The simplest transmission is a single-reduction chain or a toothed-belt drive.[3] This is a light-weight, reliable, and efficient solution, if the motor can be kept near its optimal efficiency by a fixed gear ratio. Clean, lubricated, and properly tensioned chains may have efficiencies of 98.5 %, as a source cited by Whitt and Wilson (1974) asserts. Toothed belts are slightly less efficient, but require no tension adjustment, are much quieter than chains, and need no lubrication.

[3] Or, as in the case of Clarkson's 1990 Sunrayce car, a hockey puck directly driving the left-rear wheel.

The race course will strongly influence the design of the transmission. Do steep grades predominate, as in the Swiss Tour de Sol? Then a motor driving a single-reduction transmission will spend most of its time at high torque and low speed, will run hot, and be less efficient. Or is the course mostly gently rolling terrain, as in the 1995 and 1997 Sunrayces? Then choosing a cruising condition allows the car to spend more of its time near its cruising speed and the motor will be cooler and more efficient. In the former case, employing a transmission with a steep-hill gear and a cruising gear may give a net energy gain, compared to a single-speed reduction chain drive. In the latter case, the extra weight and power loss of a multiple-speed transmission may result in a net loss of energy over the race course, compared to the single-reduction unit.

Clearly, there are a number of ways the motor, controller, and transmission can be combined to produce a drive that will, through some combination of weight and efficiency, cause the smallest net loss in energy over the race course, meet the torque and rotational speed requirements of the race, and remain within the budget of the builders.

We recalculate the torque and power requirements listed in Table 9.6 at a mass of 338 kg using the new rolling resistance coefficient. Recall that from the Specification Table 8.2, Design Race Route, the Shark must climb a 10 % grade at 40 kph and cruise at 88.5 kph on a level road. Take the drag area to be 0.127 m^2 at either speed. The rolling resistance coefficient has been found to be 0.0045 at 88.5 kph. We find it to be 0.004 at 40 kph, using the same method. As in Chap. 5, we assume a transmission efficiency of 95 %. Table 10.2 shows the steady-speed results for each operating condition.

The available motors must be surveyed. Chapter 5 describes advantages and disadvantages of generic motor types. In solar racing, high efficiency over a wide range of torque and rotational speed (the efficiency should include the power supplied to the motor's cooling system), reliability, and low weight are very important. The weight to consider is the weight of the entire drive: controller, motor, transmission (if any), and cooling system (if separate). The cost of the motor and controller may also be a factor. Select likely candidates and obtain the manufacturer's data for each. Be sure the data contains characteristic curves equivalent to those in Fig. 5.6.

Suppose that we are choosing between a radial-flux motor and an axial-flux, wheel-mounted motor. Both are brushless, permanent-magnet machines. The price difference could be US$ 3000–5000 or more in favor of the radial flux motor. Cost constrains design choices just as rigorously as physics. Let us assume that we select the radial flux machine because it is cheaper. For illustrative purposes we choose the Solectria BRLS8, already examined in Chap. 5.

Table 10.2 Drive design points

V (kph)	Grade (%)	τ_W (N·m)	N_W (rpm)	P_W (W)	P_M (W)
88.5	0	15.3	924	1480.0	1557.8
40	10	89.5	418	3913.8	4119.8

The speed reduction will be set to keep the motor at the highest possible efficiency. For the zero-grade condition, the power delivered to the transmission, $P_{M_{opt}}$, at maximum motor efficiency must be 1557.8 W (Table 10.2). Use Eq. (5.23) as in Example 5.1:

$$P_{M_{opt}} = \tau_{M_{opt}} \frac{\pi N}{30} = \frac{\pi N}{30} \sqrt{\frac{P_L k_S^2}{R}} = 1557.8\text{W}.$$

Table 10.3 summarizes the optimal results for both 0 and 10 % grade conditions. P_B symbolizes the power from the battery bus and n_G the rotational speed reduction of the transmission.

Solar racers may run with a fixed reduction. If climbing the 10 % grade with the 3.62 reduction, the motor would operate at 1513 rpm, 26.0 N·m, and an efficiency of about 62.7 %, extrapolating Fig. 5.6. This operating point is outside the continuous operation region and far from optimal. The power demanded of the battery and solar array would be 6571 W, a 67.9 % increase over the wheel power. A compromise reduction for a hilly course would give more efficiency on hills but would limit the top speed when cruising on flats in order to keep the motor below its maximum allowable rotational speed. The BRLS8 is limited to 5687 rpm, so the maximum possible reduction would be

$$n_{G_{max}} = \frac{N_{M\,MAX}}{N_{W_{MAX}}} = \frac{5687\,\text{rpm}}{924.2\,\text{rpm}} = 6.15.$$

Suppose we set the reduction at 5.64. This gives the driver an 8 kph (5 mph) cushion above the maximum speed allowed by race rules, which we have taken to be 88.5 kph, before the motor overspeeds. Table 10.4 shows the motor efficiencies resulting from this choice.

Note that the design race route contains 24 km of 5 % grades but only one 0.1-km, 10 % grade. It would seem prudent to examine the effect of a two-ratio transmission using the optimal gear ratios for the zero and 5 % grades. We also ask: Because the grade is smaller, could we travel at, say, 50 kph up this grade?

The two-ratio reduction improves the efficiency for each case. At speeds less than 88.5 kph on horizontal roads, the 3.62 reduction yields efficiencies in the

Table 10.3 Motor design points

V (kph)	Grade (%)	τ_M (N·m)	N_M (rpm)	n_G	η_M (%)	P_B (W)
88.5	0	4.44	3348	3.62	91.76	1680.7
40	10	6.98	5640	13.5	94.0	4282.7

Table 10.4 Single and two-speed reduction results

V (kph)	Grade (%)	n_G	η_M (%)	n_G	η_M (%)
88.5	0	5.64	91.62	3.62	92.70
50	5	5.64	89.89	8.84	93.54
40	10	5.64	87.95	8.84	90.84

Table 10.5 Drive requirements

Weight (controller, motor, fan)	≤24.6 kg
Reduction to wheel	3.62 and 8.84, cockpit-selectable
Efficiency: transmission to wheel	≥95 %, over operating range
Efficiency vs. torque map, including cooling	Like Solectria BRLS8 or better
Environment	50 °C continuous, water spray

89–92 % range. These results imply that, for the car under study, the greater weight and complexity of the two-speed reduction may be justified for a hilly course.[4]

Multi-gear transmissions based on existing bicycle transmissions have been used by solar racing teams. Storey et al. (1993) reported that three teams in the 1993 World Solar Challenge used chain drives in combination with multispeed transmissions. One team used bicycle-hub gears, and two used a derailleur. The efficiency of the derailleur is "not greatly different" from a conventional chain reduction, and the estimated efficiency of the hub system ranged from 93 % (lowest ratio) to 85 % (fifth gear), according to Storey et al.

For the highest efficiency, chain-driven systems require the chain tension to be checked and adjusted and the chain to be well lubricated. Toothed-belt ("timing belt") reductions do not require tension adjustment or lubrication. They are also quieter than chain systems. In either system, however, the driving and driven sprockets must be in alignment to achieve the highest efficiency (Table 10.5).

The motor controller must function over the range of battery voltages expected. The controller cut-off voltage for Clarkson's 1999 car was 47 V, for example. When this voltage is reached, the controller, and therefore the motor, will stop operating. However, the battery will usually be the limiting device because its recommended minimum discharge voltage may be reached first.

10.6 Battery

As in Chap. 4, *module* will refer to a number of cells grouped in a single case, and *battery* will mean the entire assembly of modules.

Race Rule Effect Consider the rules of the race in which the car is to compete when selecting the size of the battery. The distinguishing feature is whether the race day is terminated by a time of day, that is, the car runs during a fixed interval, or by a destination, that is, the car travels a fixed distance.

Suppose, as in the World Solar Challenge, racing may continue until a time of day. Teams should consider limiting the mass of the battery to that which can be recharged during the time interval beginning after racing stops and lasting until it begins the following morning. Extra, presumably unused, battery mass would cause increased energy consumption but contribute no additional range.

[4] The efficiency of the actual two-speed transmission over its operating range should be determined to insure transmission losses do not cancel the gains in motor efficiency.

Now suppose a race day is ended when a team reaches a particular destination. This arrangement is typical of the original Sunrayce rules. Therefore, if a team reaches the destination before the final stopping time there will be more daylight available for recharging the battery than in the stop-at-time-of-day case. Carrying more battery mass would then allow higher average speeds.

However, nine of the first ten finishers, and fourteen out of the thirty-four teams in the 1993 World Solar Challenge carried close to the maximum allowed battery capacity of 5 kW·h (Storey et al. 1993). Other considerations besides those in the preceding two paragraphs play a role in choosing the battery capacity. Clouds can reduce the solar irradiance to 300 W/m^2, or even less. Carrying extra energy allows a higher speed in cloudy conditions, or at least to maintain speed and avoid trailering. In sunny weather, high speeds rapidly increase the power demand beyond that supplied by the solar array and thus favor increased energy storage.

Using high specific-energy batteries reduces the battery mass required for a given energy capacity, reducing the energy penalty for carrying that capacity. However, it increases the cost of the battery.

Battery Study Using the design race day and route, we will estimate the battery capacity to run the route, and some other important battery parameters, to illustrate how procurement specifications for the battery may be set. We follow the suggestions of Chap. 19, with the following simplifications introduced to make the modeling easier:

1. The bus voltage is fixed at a mean value of 120 V, which we assume is in the middle of the allowable range for the motor controller. We do not yet know the actual dependence of the battery voltage on discharge rate.
2. The battery energy efficiency is 90 % for charge or discharge.
3. The solar irradiance is a constant, average value. This avoids the complication of modeling the weather–sun interaction. On a clear day, this model would overestimate the irradiance in the morning and afternoon and underestimate it around solar noon. But on a cloudy day, it would be closer to reality. A value of 500 W/m^2 will be taken as typical of the average solar irradiance on a clear day and 200 W/m^2 as typical of an overcast and rainy day.
4. The effective tilt correction factor of the solar array is 1.0. The Shark's array, while not flat, is not greatly curved. The route course is generally west. Therefore, the tilt correction factor on the south (left) side of the array will be somewhat greater than one and that on the north side of the array will be somewhat less than one.
5. The array efficiency is constant at 14 %. This assumption avoids the complications of calculating the actual efficiency of the array (a function of geometry, speed, air temperature, and irradiance). This is a typical conversion efficiency at the maximum power point under standard conditions. This will overestimate the efficiency in the morning and afternoon, but will be closer to the actual value around solar noon. The area of the array will be set at 8 m^2, somewhat less than the actual area. The maximum power point tracker efficiency will be 98 %.
6. The Shark's mass and drag area are 338 kg and 0.127 m^2, respectively. The former is the specified limit, and the latter is the current maximum design estimate.

The rolling resistance will be calculated as a function of speed using the results for the 20-in., 110-psig, bald tire specified earlier.

7. The dependence of the motor resistance and stray power losses on rotational speed and torque are the same when acting as a generator as when acting as a motor. The efficiency of the motor acting as a generator is:

$$\eta_G = \frac{P_S - P_{TL}}{P_S}. \tag{10.1}$$

The symbol "P_{TL}" represents the total power loss and "P_S" the power delivered by the shaft in generator operation.

Positive torque at the driven wheel means that torque must be supplied by the motor. However, when coasting downhill at a constant speed, the wheel torque may be negative. This means that some of the excess gravitational potential energy must be dissipated as heat in the brakes to keep the speed constant. Or if regeneration is on, it means that the motor will act as a generator and supply power to the battery bus.

Figure 10.3 shows a flow chart of the program incorporating the foregoing assumptions and rules.

Figure 10.4 is a graph of the battery energy used with regeneration both on and off. The results show the energy recoupment value of regeneration over the design route: a reduction in required energy capacity of 0.48 kW·h. The solar irradiance is also shown to display its effect.

Table 10.6 gives the specification for the battery. To allow for overcast weather, the charge and energy capacity at an irradiance of 200 W/m^2 were chosen. The transit time for the entire design route was 3.7 h. This sets the average current, 8 A, at which the capacity is specified. The maximum discharge rate of 36 A occurs when climbing the 10% grade at 50 kph. The duration of this current is short; however, at that rate the battery would be discharged in less than 47 min (the capacity will be less than 28 A·h at 36 A).

The simulation did not account for acceleration. Moving quickly with traffic from a full stop, or passing other vehicles, may require larger discharge currents than the 36 A calculated for climbing the 10% grade at a steady speed of 40 kph. Suppose the limiting acceleration from a full stop is set at 0.1 g. Immediately after leaving the stop, the car must climb a 10% grade at 40 kph under overcast conditions (taken as 200 W/m^2), using the climbing gear ratio, 8.84. This maneuver will be used to specify the maximum discharge rate. The time required to accelerate would be 11.5 s. The maximum battery discharge current for this maneuver, found by including acceleration (via the "little tap" method) in the program outlined in Fig. 10.3, would be 76.6 A, and 0.0186 kW·h would be removed from the storage.

If the cruising gear ratio, 3.62, were used to accelerate and climb, the maximum discharge current would be 95.3 A, and 0.03 kW·h would have been removed. These results emphasize the usefulness of a climbing and a cruising gear. When traversing several urban areas during a day the energy consumed by accelerating

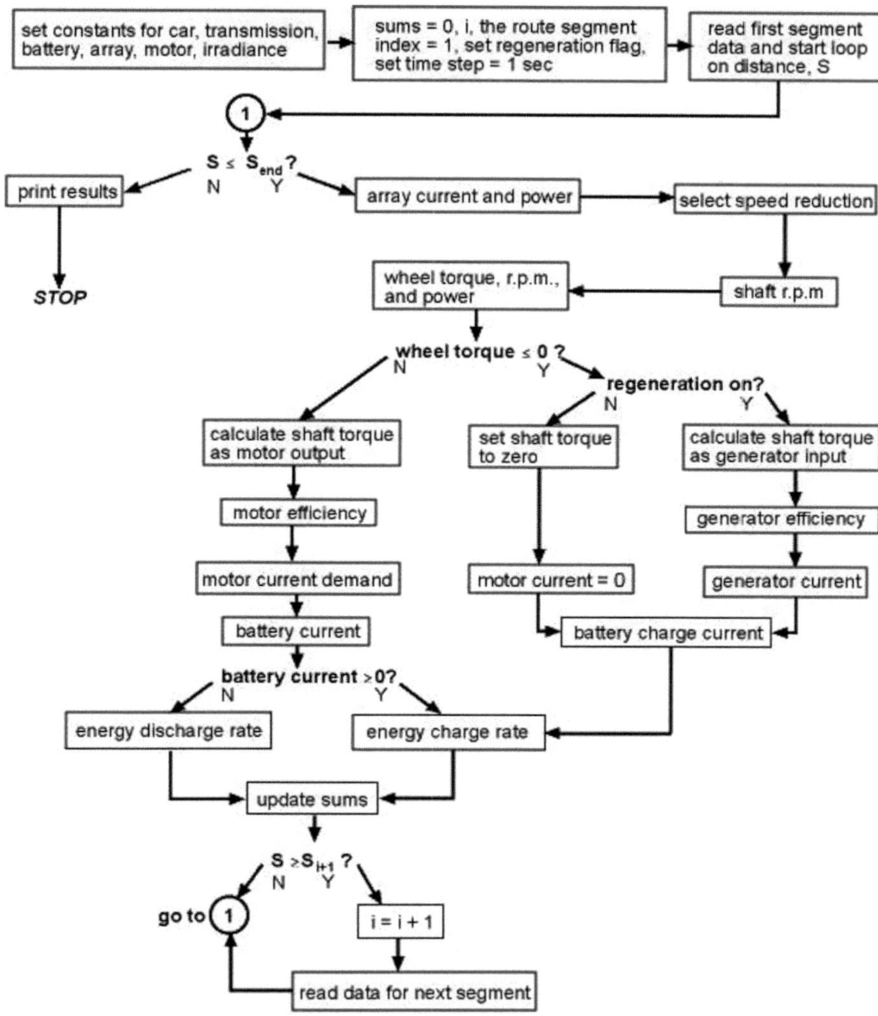

Fig. 10.3 Simulation flow chart

can be considerable. Even though it would be partially offset by regenerative braking, we specify an extra 0.2 kW·h to allow for it.

Many of the requirements in Table 10.6 have been written as limits. For example, the minimum specific energy appears as "≥27 W·h/kg," where "≥" means "greater than or equal to," that is, "at least." The last five rows of the table contain specimen requirements that were not derived from the simulation, such as the operating temperature, relative humidity, and the minimum acceptable life (set to include prerace testing). These must be developed from analysis, measurement, or by querying other solar car builders.

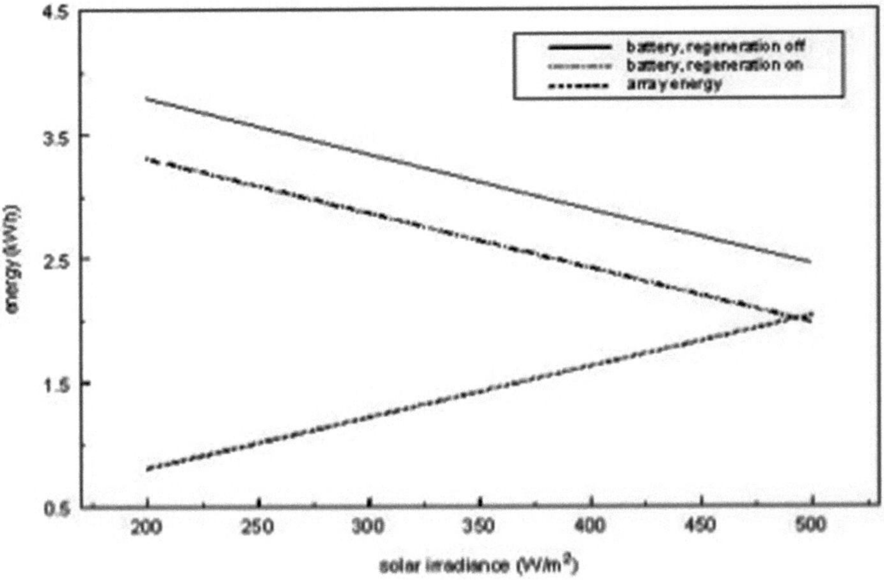

Fig. 10.4 Simulation results

Table 10.6 Battery and module requirements

Battery	
Bus voltage	120± 15 V DC
Energy capacity	≥3.9 kW·h, 3.7 h rate
Total mass	Per Chap 16
Module	
Energy efficiency	≥90%, charge or discharge
Coulomb efficiency	≥95%, charge or discharge
Maximum discharge rate	≥77 A for ≥11.5 s
Average discharge rate	8 A, the 3.7-h rate
Charge capacity	≥28 A·h at the 3.7-h rate
Minimum usable capacity	≤20%
Specific energy	≥27 W·h/kg at the 3.7-h rate
Specific power	≥34 W/kg
Charge method	Standard and opportunity
Life	≥75 deep cycles
Operating environment	40°C and 70% RH
Storage environment	20°C and 50% RH

DC direct current

Table 10.6 sets the minimum acceptable technical requirements. Apparently a lead-acid battery would fulfill these requirements. This is fortunate because advanced battery technologies with high specific energy storage capacity are expensive. However, using them does not guarantee winning. The winner of the 1999 Sunrayce carried lead-acid batteries.

Let us assume, then, that the budget will only allow purchase of lead-acid technology. The nominal voltage of such a couple (Table 4.7) is 2.0 V. Chapter 4 gives the lower voltage limit (at the knee of the discharge curve drop-off) as 1.75 V. This is a decrease of 12.5 % from nominal. Then the range of bus voltages we will take as 120 ± 12.5 % V, or 105–135 V.

10.7 Battery Ventilation Details

Procedure The minimum flow required by the race rules in Chap. 16 is 280 L/min. Figure 10.5 is a drawing of a possible battery ventilation system based on Fig. 2.7. Most of the dimensions in this drawing were estimated from the scaled sketches in Chap. 9. Some were arbitrary choices. We will apply the material in Chap. 18 to estimate the total pressure drop from inlet to outlet, the flow, and the ventilation drag. First, we must find the system characteristic, the head loss per unit flow of the system outside of the fan. We assume that p_{10} is approximately p_∞.

First Design We put the inlet in the front stagnation region to provide a high inlet pressure and give it rounded edges to reduce pressure loss as the flow enters. The air flows through a straight run of flexible duct to the battery box plenum. The flow passes into the box through a sudden expansion, leaves through a sudden contraction, and enters the fan. Downstream of the fan, another section of flexible duct directs the air to the diffuser which discharges the air into the external flow through the underbody of the car (not shown in Fig. 9.2).

At 55 mph, the maximum allowed speed, the ventilation system will probably supply plenty of flow to the battery. But it will also contribute its largest drag. So the design should be adjusted to minimize drag at 55 mph. Then the fan must be sized to give at least the minimum required flow (280 L/min) when the car is stopped.

We will neglect the air density variation in the system.

Rounded Inlet Figure 18.6 gives the loss coefficient of a rounded inlet. Assume that V_1 is 55 mph, or 24.6 m/s. This is high and must be checked after the first sweep through the design. Taking $r/d_2 = 0.1$ in Eq. (18.19) gives $K_{12} = 0.12$. The flexible duct which fits within the available interior passages of the Shark has a nominal inside diameter of 10 cm, or 0.1 m. The radius at point 1 will therefore be 0.06 m and the diameter at the discharge end 0.1 m to match the duct. A_1 is therefore 0.0113 m^2 and Q_1, the volumetric flow rate, is A_1V_1, or 0.278 m^3/s.

Flexible Duct There is a total of 4.5 m of the 10-cm nominal inside diameter flexible duct. We can lump it all together because there are no branches. The volumetric flow rate and the flow speed are therefore the same in all of the pieces of the duct. The duct is built like an accordion. We model this by using a roughness of about 2 cm, or $e/d = 0.2$. A_2, based on the nominal diameter, is 7.854(10^{-3}) m^2. Using the design temperature and pressure, we find the airspeed and Reynolds number in the duct, V_2 and R_{d2}, as 35.4 m/s and 2.26(10^6), respectively. The flow is well into

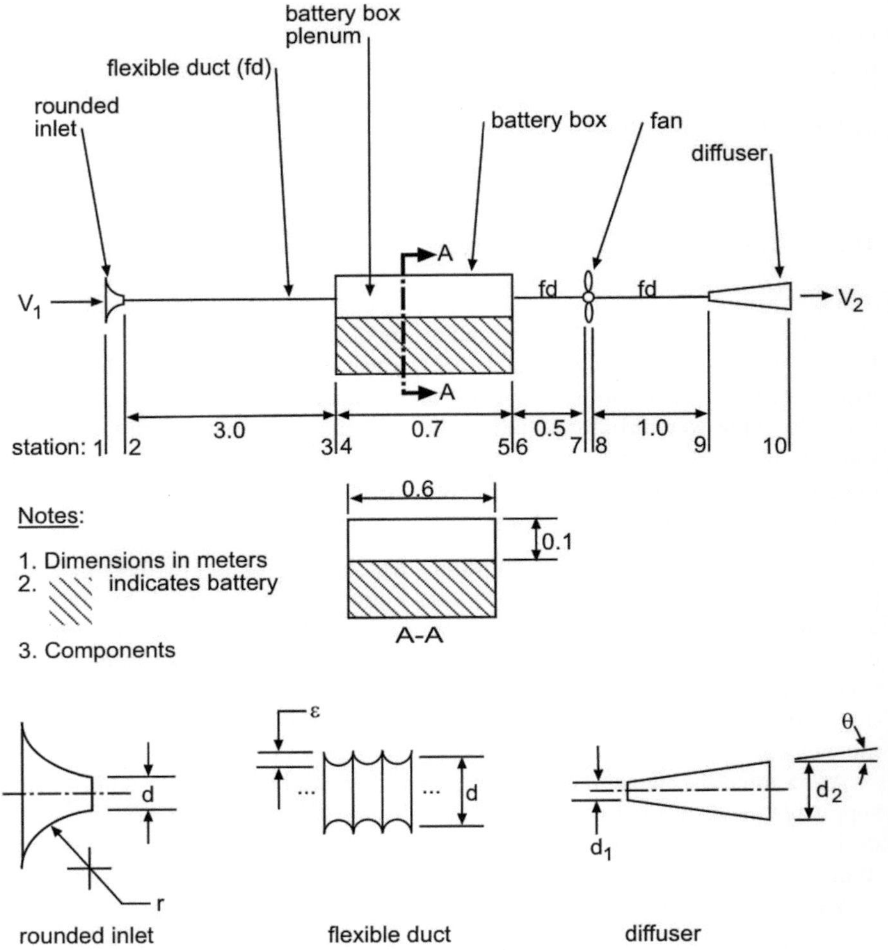

Fig. 10.5 Battery ventilation system

the turbulent regime (not surprisingly). From Eq. (18.28), $f \approx 0.156$. Entering the total length of duct into Eq. (18.26), we find K_{DUCT} to be 7.02. Note that increasing d_2 would cause a large reduction in the duct pressure loss, which varies inversely with d_2^5 in turbulent flow.

Sudden Expansion The sudden expansion as the air enters the battery box plenum causes non-recoverable pressure losses. Equation (18.27) is employed to find the hydraulic diameter, d_h, of the box plenum, a noncircular duct, to be 0.171 m. Then Eq. (18.21) gives K_{34} as 0.658.

Friction in Battery Plenum The relative roughness in the plenum is dominated by the battery terminals, their connecting wiring, and the spaces between the batteries.

We guess that this is equivalent to an average roughness of 1.0 in, or 0.0254 m, on the battery side and zero on the remainder of the surfaces parallel to the flow. This gives an area-averaged roughness of 0.0078 m. Using d_h from above, e/d becomes 0.046. Eq. (18.28) gives f_{45} as 0.069. K_{45} is then 0.282.

Sudden Contraction There are losses at the exit of the battery-box plenum caused by the sudden contraction of the flow back to the flexible duct. Equation (18.22) gives K_{56} as 0.276.

Fan We bypass the fan for the moment.

Diffuser If the half-cone angle of the diffuser becomes too large, the flow separates inside the diffuser, and the loss coefficient increases dramatically, as Fig. 18.8 illustrates. On the other hand, if the half-cone angle is small, the diffuser becomes rather long, and friction loss increases causing the loss coefficient to rise. There is an optimum half-cone angle of about 2.5° at which the loss coefficient is a minimum with a value of 0.3. Because of space constraints, we will specify a length of 0.5 m and choose a half-cone angle of 5°, for which we estimate a loss coefficient, $K_{9,10}$, of about 0.4. The cone angle and length imply an exit diameter and area of 0.1875 m and 0.0276 m^2, respectively. We neglect the biased truncation of the diffuser exit by the underbody of the car and assume the length and exit diameter to be where the diffuser's centerline penetrates the underbody.

System Characteristic The system characteristic is the sum of the head losses from each component. Note that $V_2 = V_9$.

$$h_L = (K_{12} + K_{DUCT} + K_{56} + K_{9,10})\frac{V_2^2}{2g} + K_{45}\frac{V_4^2}{2g}$$

It is convenient to reference the system loss coefficient to the volumetric flow, Q_1, at the inlet, station 1. With constant density, Q_1 is the same at every station. This is done by applying Eq. (18.16), giving

$$V_2 = \frac{Q_1}{A_2}, \quad V_4 = \frac{Q_1}{A_4}.$$

Adding the loss coefficients of the components of the system gives the system characteristic as[5]

$$\Delta p_{1,10} = 81.06 Q_1^2, \quad \text{kPa}.$$

As Chap. 18 explains, the pressure difference $\Delta p_{1,10}$ is dependent upon the shape of the car and the relative airspeed, not on the components of the ventilation system. The system characteristic may therefore be thought of as defining the flow at different

[5] This cannot be as conveniently done with a branching system like that in Fig. 18.11 because the exit pressures are different.

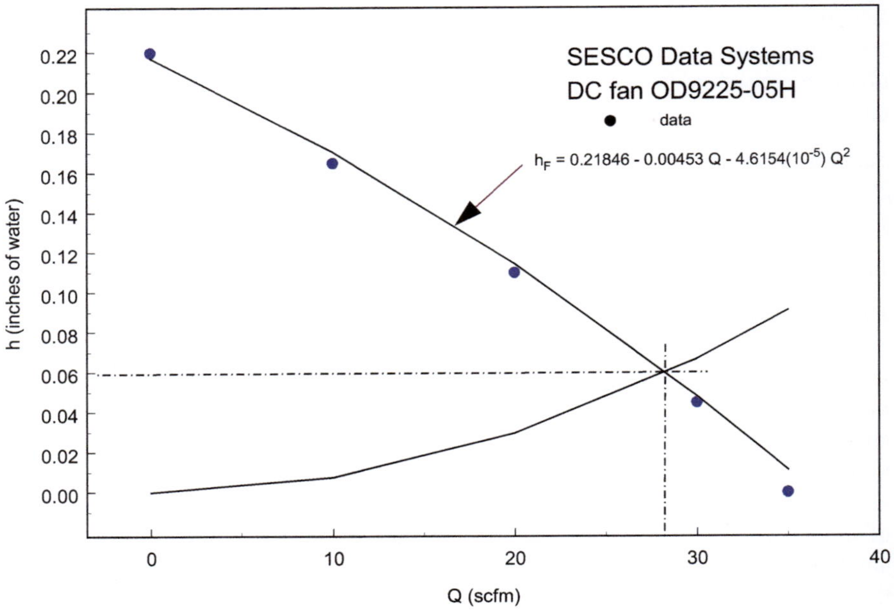

Fig. 10.6 Fan and system characteristic

car speeds. As long as the total loss coefficient is constant,[6] we can find the pressure difference causing a Q_1 different from Q_{55}, the volumetric flow at 55 mph, as

$$\frac{\Delta p}{\Delta p_{55}} = \left(\frac{Q_1}{Q_{55}}\right)^2.$$

This system characteristic is the lower curve in Fig. 10.6.

Actual Flow Rate and Velocity The stagnation pressure can be calculated without making a flow plot at described in Chap. 18. This pressure, p_1, is $p_\infty + \Delta V_\infty^{\ 2}/2$, which is 101.652 kPa at 55 mph and standard atmospheric pressure ($p_\infty = 101.325$ kPa). If p_{10} is equal to p_∞, then the gauge pressure at the exit, $p_{G10,}$ is zero and $\Delta p_{1,10}$ is 0.352 kPa. Solving the system characteristic for Q gives 0.0659 m^3/s, or 3954 L/m. The inlet velocity is Q_1/A_1, or 5.83 m/s. The exit velocity, Q_1/A_{10}, is 2.39 m/s.

The initially assumed value of 24.6 m/s was considerably in error. Another pass through the calculation should be made. However, recalculating the minimum Reynolds number (station 4, just inside the battery box) shows that the flow would still be fully turbulent. Hence, the total loss coefficient will not change much (see Footnote 5). Let us therefore use the velocities given in the preceding paragraph to estimate the drag of the system.

[6] If turbulent, separated flow dominates in the system, and heating effects are not large, the total loss coefficient will be nearly constant.

Drag Let us suppose that the diffuser discharges approximately antiparallel to the direction of motion. The acute angles between the outward normals at the inlet and exit are therefore both zero. Substituting in Eq. (2.8)

$$D_{\mathrm{V}} = 352\frac{\mathrm{N}}{\mathrm{m}^2}\times 0.0113\mathrm{m}^2 + 1.18\frac{\mathrm{kg}}{\mathrm{m}^3}\times 0.0659\frac{\mathrm{m}^3}{\mathrm{s}}(5.83-2.39)\frac{\mathrm{m}}{\mathrm{s}} = 4.24\mathrm{N}.$$

Applying Eq. (2.9), the definition of the ventilation drag coefficient, and using the revised A_D of 1.684 m^2, gives

$$c_{\mathrm{V}} = \frac{D_{\mathrm{V}}}{q_\infty A_D} = \frac{4.24\mathrm{N}}{352\frac{\mathrm{N}}{\mathrm{m}^2}\times 1.684\mathrm{m}^2} = 0.0072.$$

Therefore $c_{\mathrm{D}}A_D$ would be increased by 0.012 m^2 at 55 mph, about 11 %. About 6 % of the drag is contributed by the change in momentum between the inlet and outlet. The rest is from the applied pressure difference.

Now consider the fan.

Fan As previously observed, when the car is stopped, at least the minimum allowable flow must be present. This flow, 280 L/m (9.89 cfm), will reduce the component velocities considerably. The smallest, V_4, would be 0.078 m/s. At this low velocity, the Reynolds number in the battery plenum would be 869, indicative of laminar flow. A new, low-speed system characteristic should be found for this condition. Nevertheless, we estimate the pressure change required from the characteristic just obtained, getting 1.77 Pa. The fan thus chosen should have more than enough capacity to provide the minimum flow under laminar conditions.

The pressure under a 1-ft square, 33-ft-tall column of water is equivalent to one standard atmosphere. Therefore, 1.77 Pa is equivalent to about 0.01 in. of water, or "inches water gage," as it is usually expressed by fan manufactures' data sheets. So, we must look for a fan that will supply at least 280 L/min at 0.01-in. water gauge.

Figure 10.6 shows the characteristic of an actual fan, one similar to those used to ventilate computers. As Chap. 18 explains, the operating point is defined by the intersection of the fan and system characteristics. Clearly this fan has more than enough head. The operating point is at about 0.08 in water gauge and a flow of 28 scfm, or 793 L/min.

The system must be redesigned to reduce its drag. This reduction would allow a drag margin for the cockpit ventilation flow, which has not been considered. Moving the inlet underneath the car and closer to the battery would reduce the pressure at the inlet and also reduce the length of the flexible duct required.[7] This would reduce the weight of the system and the total loss coefficient, thus at least partially offsetting the reduced applied pressure. The streamline plotting method of Chap. 18

[7] Hoerner (1965) contains a section on inlet and outlet designs. See also Mossman and Randall (1948) for information about low-drag National Advisory Committee for Aeronautics (NACA) submerged inlets, or "NACA ducts."

would be used to estimate the inlet and exit pressures. The fan flow at zero vehicle speed should again be found, this time with a revised low-flow system characteristic. Figure 10.6 implies that a smaller-capacity, lower-power fan could be used. This will reduce the ventilation system current load.

10.8 Braking

Total Braking Torque The rules in Chap. 16 require the brakes to provide a deceleration of at least 17 kph/s from an initial speed of at least 50 kph. Equation (22.12) gives the total braking torque required to stop the car from some initial speed at a constant deceleration, ignoring drag and rolling resistance. It also incorporates a delay between the time the signal to stop is given and the time the brakes are applied.

The mass of the car to be used is the effective mass, estimated as follows: We estimate[8] the tire moment of inertia, I_W, from the wheel parameters already established to be the sum of the moment of inertia of a hollow rubber torus of 1/8-in. thickness, inner radius 7.5 in. and outer radius 10 in. and an aluminum disk of 1/8-in. thickness and 7.5-in. radius. The result is 0.0523 kg·m². We estimate the motor moment of inertia, I_M, by approximately scaling up from the known rotor moment of inertia of a Hathaway HSSS3810 brushless DC motor, $7(10^{-4})$ kg·m². The rotors were modeled as uniform cylinders. The moment about a cylinder's axis is $Mr^2/2$, where M is the mass and r the radius. The moment then scales as the product of the ratio of the masses and the ratio of the radii squared. The result was $1.25(10^{-3})$ kg·m².

Equation (22.15) was used to find M_E with the gear ratio set to 8.84, which gives the largest effective mass:

$$M_E = M + \frac{4n_W I_W}{d_W^2} + \frac{4n_G^2 I_M}{d_W^2} = 338\text{kg} + \frac{4\times 4\times .0523\text{kg}\cdot\text{m}^2}{(0.508\text{m})^2}$$
$$+\frac{4\times(8.84)^2\times 1.25(10^{-3})}{(0.508\text{m})^2} = 342.7\text{kg}.$$

Note that $M_E/M = 1.014$. Now, using Eq. (22.12), a delay of 0.5 s, and the minimum acceptable deceleration, 17 kph/s, and initial speed, 50 kph, given in Chap. 16, paragraph 6.12.1, gives

$$\tau_B = \frac{0.278\frac{\text{m}}{\text{kph s}}\times 342.7\text{kg}\times 17\frac{\text{kph}}{\text{s}}\times 0.508\text{m}}{1-\frac{0.5\text{s}\times 17\frac{\text{kph}}{\text{s}}}{50\text{kph}}} = 991.3\text{N}\cdot\text{m}$$

[8] By standard methods; see Hibbeler (1998) or similar texts.

or a total braking force, F_{XB}, of 1951 N.

The race rules require the stop to be made on a wetted surface. From Fig. 22.3, the peak braking coefficient, μ_{BP}, is 0.59 for this surface. The maximum gross vehicle weight is 3315 N. Therefore, if the peak braking coefficient is approximately the same for the front and rear wheels, Eq. (22.20) gives

$$F_{XB,\text{wet}} = \mu_{\text{BP}} W = 0.59 \times 3315 N = 1956\text{N}.$$

Under the conditions assumed, particularly the delay time, the car may not pass the braking test.

The aerodynamic drag and rolling resistance will add to the deceleration. Let us examine their effect on the 0.5-s delay case. During either the delay or the braking period the deceleration would be:

$$a = -\frac{F_{XB} + \mu_1 W + c_{\text{D}} A_D \dfrac{\rho V^2}{2}}{M_{\text{E}}},$$

where we have neglected the speed-dependent portion of the rolling resistance coefficient because it is about an order of magnitude less than the constant part and assumed a constant drag coefficient. The rolling resistance force is now a constant and the same during both the delay and braking periods.

Using the "little tap" method, we take small steps in time, Δt, of 0.01 s, say. Over this small interval, we take the deceleration to be constant. Then, using Eq. (2.20):

$$V_{n+1} \approx V_n + a_n \Delta t,$$

where "*n*" is the number of steps taken up to the present time. The force F_{XB} in the deceleration must be set to zero during the delay and to $F_{XB}+\mu_1 W$ thereafter. The rolling resistance is:

$$R = \mu_1 W = 0.004 \times 3315\text{N} = 13.26\text{N}.$$

F_{XB} was found from Eq. (22.12) for each case. The results are shown in Table 10.7, where V_1 is the initial speed, t_{S0} and t_S are the time to decelerate without and with drag and rolling resistance, respectively. Dry pavement was assumed.

Including drag and rolling resistance yields only small reductions in stopping time and corresponding increases in deceleration. But note that beginning the stop at a higher speed also reduces the total braking force because it lessens the effect of

Table 10.7 Braking effect of drag and rolling resistance

V_1 (kph)	t_{S0} (s)	t_S (s)	a (khp/s)	F_{XB} (N)
50	2.941	2.911	17.175	1951.3
60	3.529	3.488	17.199	1886.9
70	4.118	4.064	17.226	1843.4

the delay. Reducing the delay would also lessen the braking force, but keeping the delay less than 0.5 s requires practice.

Lockup and Proportioning Lockup of the front brakes initiates a skid and some loss of steering control. Lockup of the rear brakes makes the car susceptible to spinning if yawing forces are present (Gillespie 1992). Because of this, it is better to design to lock the front brakes first. The term "proportioning" refers to adjusting the ratio of the front and rear brake forces such that the deceleration goal can be met and the front brakes will lock first.

These issues can be made clear by plotting the maximum front brake force against the maximum rear brake force and adding a line of the minimum-acceptable, constant deceleration. We will employ the method outlined by Gillespie, again neglecting drag and rolling resistance. Two cases will be plotted: a dry surface represented by a μ_{BP} of 0.81 and a wet surface represented by a μ_{BP} of 0.59.

Equations (22.16) and (22.17) give the total normal force on the front and rear wheels, respectively. The total braking force, F_X, appearing in each equation, is the sum of the front force, F_{XF}, and the rear force, F_{XR}. Lockup boundary lines for the front and rear brakes can be found by multiplying each equation by μ_{BP} and solving for $(F_{XF})_{MAX}$ and $(F_{XR})_{MAX}$, respectively.

$$\begin{aligned} \left(F_{XF}\right)_{MAX} &= \frac{\mu_{BP}}{1-\mu_{BP}\dfrac{h}{L}}\left(W_F + F_{XR}\frac{h}{L}\right) \\ \left(F_{XF}\right)_{MAX} &= \frac{\mu_{BP}}{1-\mu_{BP}\dfrac{h}{L}}\left(W_R + F_{XF}\frac{h}{L}\right) \end{aligned} \tag{10.1}$$

As before, h is the CG height and L is the wheel base. W_F and W_R are the total static loads on the front and rear wheels, equal to (see Fig. 21.1 and Eq. (21.7))

$$\begin{aligned} W_F &= W\frac{b}{L} = 3315\text{N} \times \frac{1.876\text{m}}{3.2\text{m}} = 1943.4\text{N} \\ W_R &= W - W_F = 3315\text{N} - 1943.4\text{N} = 1371.6\text{N}. \end{aligned}$$

The ratio h/L is:

$$\frac{h}{L} = \frac{0.639\text{m}}{3.2\text{m}} = 0.1997.$$

Figure 10.7 shows the resulting braking diagram for the Shark.

The dry surface design requirements are represented by the triangle 4-5-6. It is bordered by the front and rear lockup lines for a dry surface and by the 17 kph/s, 0.5-s delay line. (Along this line, the sum of the front and rear braking forces is equal to 1951 N.) The wet surface requirements are represented, essentially, by the small triangle with its vertex at point 2.

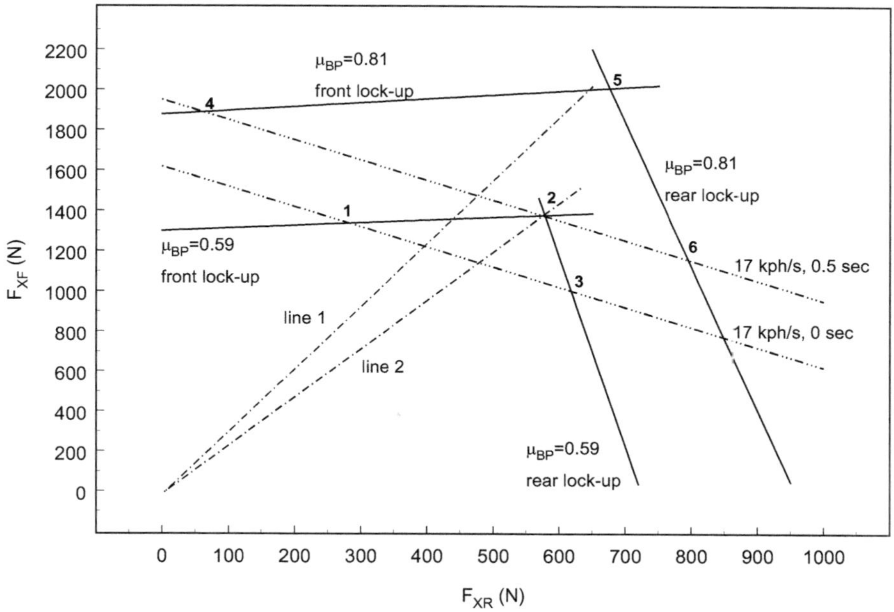

Fig. 10.7 Braking diagram

Lines 1 and 2 are brake proportioning lines. Along these lines the ratio of the front to the rear brake forces is constant. To satisfy the requirements for both wet and dry surfaces, a brake proportioning line must pass through triangle 4-5-6 and the small triangle at point 2 and cut the front brake limit line of each triangle. Line 1 can satisfy only the dry surface requirements and line 2 only those of the wet surface. Satisfaction of both the dry and wet surface requirements dictates a change in slope of line 2 at point 2 such that it steepens and passes through triangle 4-5-6, cutting the front brake limit line to the left of point 5. Let us call this steeper section "line 3." A two-stage *proportioning valve* will be needed. A proportioning valve adjusts the relation between the front and rear brake pressures.

Suppose caliper brakes are selected. The *brake gain, G,* is defined such that

$$F_B = \frac{(\mu r A)_B}{\overset{*}{r}_W} p_A = \frac{G}{\overset{*}{r}_W} p_A, \tag{10.2}$$

where μ is the coefficient of kinetic friction between the brake pad and the brake disk, $\overset{*}{r}_W$ is the radius from the wheel axle to the centroid of the brake pad contact area, A is the area over which the hydraulic pressure acts, and p_A is the actuation pressure.

The slope, S_1, of line 1 in Fig. 10.7 is 2.4. Therefore, assuming that the same pressure is supplied to both sets of brakes while on line 1 below point 2:

$$S_1 = \frac{F_{XF}}{F_{XR}} = \frac{G_F}{G_R} = 2.4. \tag{10.3}$$

After point 2, the hydraulic pressure must be adjusted by the two-stage proportioning valve. The slope of line 3 is:

$$S_3 = \frac{F_{XF} - F_{XF_2}}{F_{XR} - F_{XR_2}} = 7.1, \tag{10.4}$$

where F_{XF_2} and F_{XR_2} are the coordinates of point 2. By applying the definition of brake gain to Eq. (10.4), we find that along line 3 the actuation pressure to the rear brake must be adjusted according to

$$p_{AR} = p_{A_2} + \frac{S_1}{S_3}(p_A - p_{A_2}) = p_{A_2} + 0.34\,(p_A - p_{A_2}). \tag{10.5}$$

The proportioning valve should be able to reduce the pressure increase allowed at the rear brakes by no more than 0.34 of the gain in p_A after it senses that the actuation pressure has reached p_{A_2}.

On the wet surface, the margin between lockup and satisfying the stopping requirement with a 0.5-s delay is too small; lockup of all wheels is virtually assured. Changing *a*/*L* and *h*/*L* has no practical effect on the size of the small triangle. Practicing braking so that the delay time is consistently less than 0.5 s will enlarge the small triangle the most, as triangle 1-2-3, which incorporates a zero delay, implies.

10.9 Engineering Drawings

The engineering drawings presented in this section describe portions of Clarkson University's 1999 solar racer. They are reproduced here to illustrate some end products of the design process. The assembly drawings were further supported by drawings for each part. The drawings were taken from Evans et al. (1999), the structural report required for the 1999 Sunrayce.

To keep this book to a sensible size, the structural design of the car has not been treated in detail. The structural analysis of the assemblies and parts was done by hand calculations and by modeling the parts using a commercial computer code. Simulated crash testing to meet the impact g-loads imposed by Chap. 16 was done with a code specially adapted for impact analysis. Load sets were derived from calculations (as in Chap. 22), engineering judgment (which was not always correct—see Chap. 22 again), and Chap. 16.

Brake Assembly This assembly is shown in Fig. 10.8. The caliper subassemblies were purchased; the other parts were manufactured in Clarkson's machine shop by the team. Aluminum was used where possible.

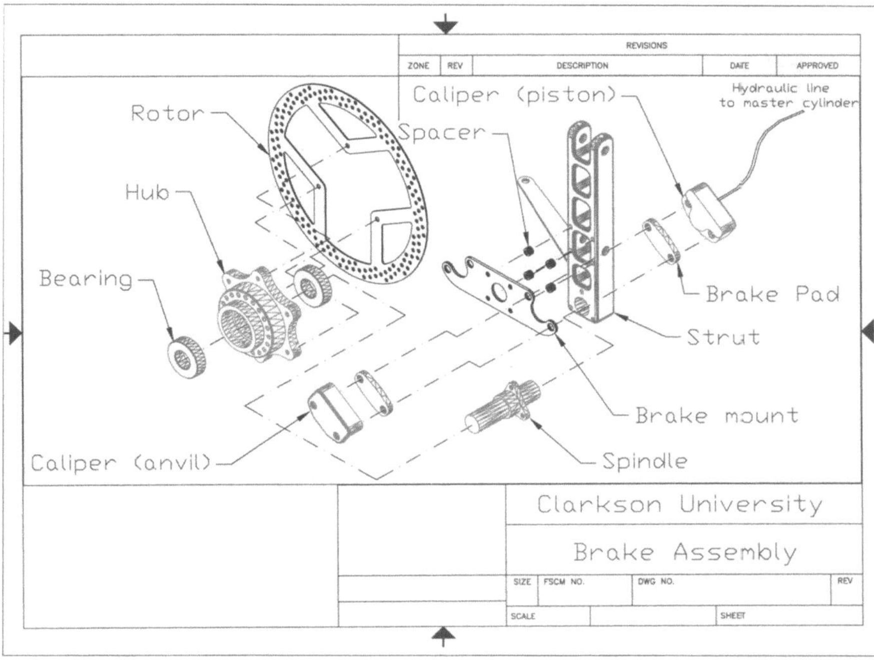

Fig. 10.8 Brake assembly

Suspension Figure 10.9 shows the suspension design. Each wheel used an independent, double A-arm suspension incorporating a mountain bike shock absorber (shown in lighter lines behind the A-arms). The A-arms were made of chrome-molybdenum steel tubing.

Chassis Preliminary design of the chassis was greatly aided by building a mock-up made of plastic tubing and duct tape. It was easy to add and remove stiffeners to qualitatively check the relative effect and to put a driver inside the cockpit portion of the frame for ergonomic checks. The welds were performed by Rome Laboratories, Rome, New York.

Overall View Figures 10.10, 10.11, and 10.12 present front, side, and top views of the car. Critical dimensions are shown (Figs. 10.10, 10.13).

Crash Simulation Figure 10.14 is a side view of a solar car after a simulated collision at 30 mph with a rigid wall. The blue material is the Kevlar®-Nomex® body shell, the red is the chassis, and the green is the battery box. The simulation code was LSDYNA3D®, an automobile-industry standard crash simulation code. The average deceleration was 5.01 g with a peak of 8.24 g lasting about 0.1 s. Note that the chassis member above the driver's legs has deflected the shell away from the driver. The driver's compartment is essentially intact.

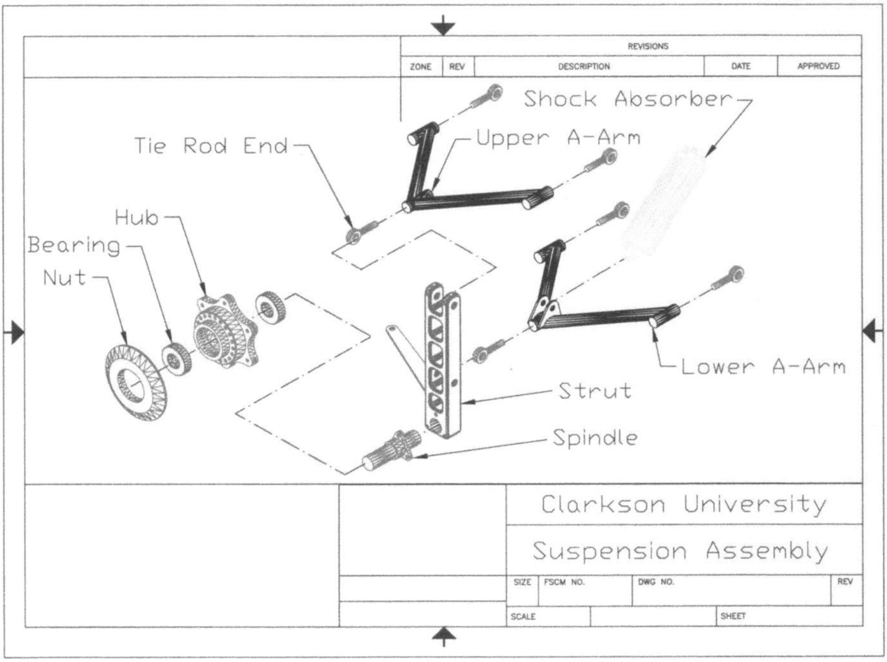

Fig.10.9 Front suspension

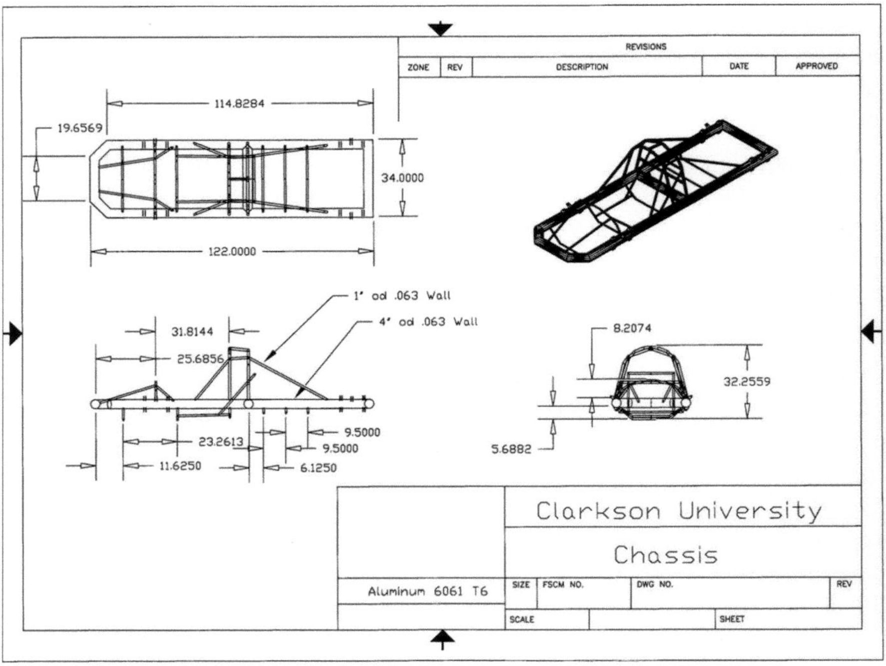

Fig. 10.10 Chassis

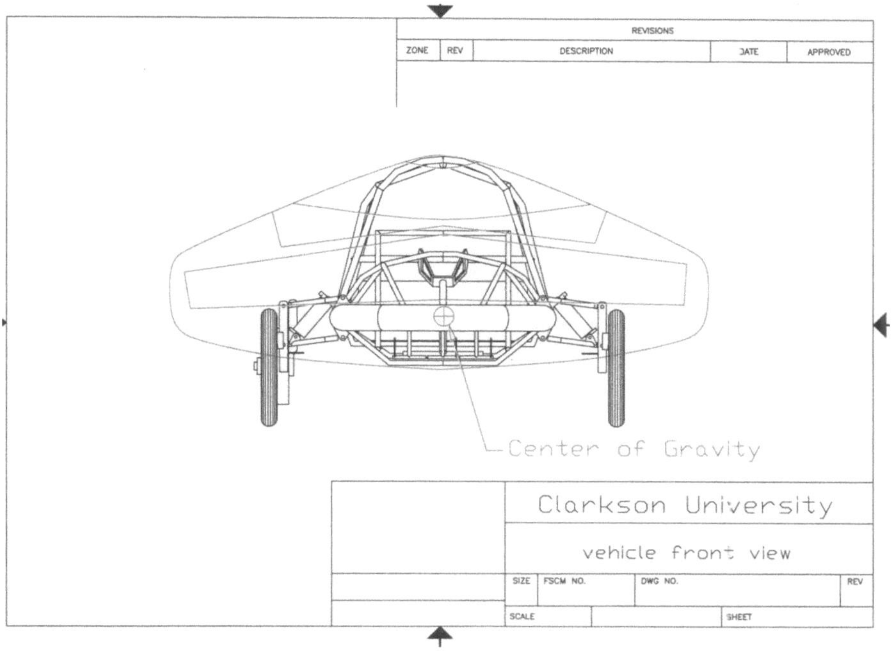

Fig. 10.11 Front view

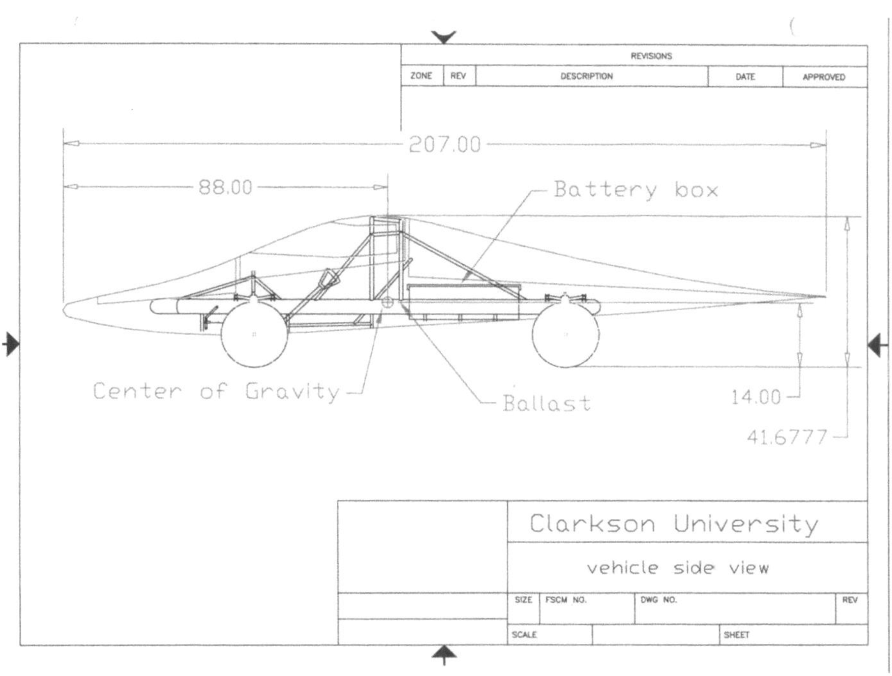

Fig. 10.12 Side view

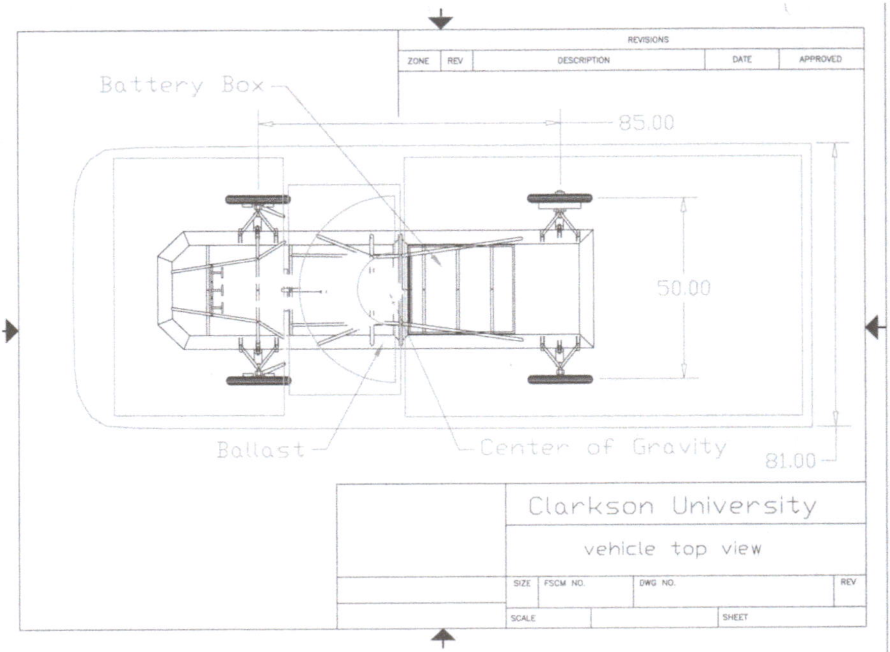

Fig. 10.13 Top view

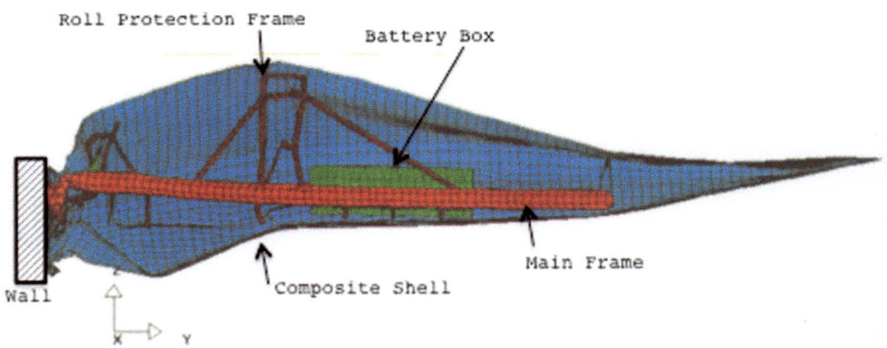

Fig. 10.14 Sample crash simulation result. (Evans et al. 1999)

10.10 Solar Array

Dimensions Table 9.6 requires a body-integral array consisting of four parallel sub-arrays each connected to the battery bus through a maximum power point tracker (MPPT). Chap. 16 requires that the array fit within a box 5 m long by 2 m wide by 1.6 m high. The product of the length and width of the array, less "any single rectangular region not occupied by array components," may not exceed 8 m^2. That is, the area of the array projected on a horizontal surface may not exceed this limit. The actual area available for the array is greater than this because the portion of the body of the Shark within the above box is convex.

Power Specification 8.9.1, Chap. 8, requires the ratio for the minimum array power supplied to the bus to the mass of the car when irradiated at 1000 W/m^2 to be at least 3.5 W/kg. At the current mass of 323.6 kg, this is a power of 1132.6 W. At the mass limit of 338 kg, it is 1183 W, which we round up to 1200 W. As a precaution, let us use this higher power to size the array.

Cells The rules in Chap. 16 do not directly restrict the cell technology.[9] But they limit the cost of bare cells to US$ 10 per Watt and require that the cells be available to all teams. This rules out high-performance cells, such as high-grade silicon cells made for the space program, and in effect restricts the array to commercial, silicon-based technology. Although more expensive, crystalline silicon cells are more efficient than amorphous cells. We select the former for our array.

A typical commercial silicon cell is 10 cm on a side, giving an area of 0.01 m^2. Under the specified irradiance, at a temperature of 25 °C and at its maximum power point, a commercial-grade, silicon-based solar cell of this size produces about 3 A at 0.5 V, or 1.5 W. Therefore, the array will have 800 cells.[10] There will be 200 cells in a series-connected sub-array, and each sub-array will operate at 100 V. The total current from the array will be 12 A. We keep in mind that the actual number of cells and the number in each sub-array will vary depending upon how the cells must be placed on the car to accommodate its curved surface. Also, we must leave about 1 mm between the cells on all sides to allow for thermal expansion. Chapter 11 shows a solar car team working through this mechanical aspect of array design. Herein, we will focus on the electrical aspect.

Trackers The current design will use four MPPTs boost regulators. Buck regulators require more cells per sub-array and therefore fewer sub-arrays. Three sub-arrays, two with 266 cells and one with 268 cells, would give 133 V and 134 V, respectively. Fewer strings means that it is more difficult to have each cell in a string; "see" the same irradiation on a curved surface like the Shark's. But the weight contributed by the tracker's will be less. We will keep the design at four sub-arrays because we believe this is a good compromise between cost (MPPTs are expensive), weight, and array performance. (Note, however, that Fig. 11.32 shows an installation of seven MPPTs. This implies fewer cells per sub-array and consequently a lower sub-array voltage. A boost regulator's efficiency may be less when the difference between the sub-array voltage and the battery bus voltage is greater. For the same number total of cells, the loss of one sub-array reduces the array power by 1/7th, not 1/4th.) The MPPTs must be able to accommodate the range of bus voltages expected during the

[9] Much more efficient cells can be used, if the money is available, under American Solar Challenge rules and giving array powers of 1600 W, or more.

[10] The number purchased should allow about 25 % for breakage.

race. The budget for the Shark will only allow purchase of lead-acid battery modules for which this range was previously estimated to be 105–135 V.

The sub-array voltage may be increased, and the cells better accommodated to a curved surface, by cutting them in half. The current from such a sub-array will be half of that using full-size cells. An array might include both full-sized and half-sized cells. Cutting is best done by the cell manufacturer and included in the purchase contract. A sub-array should consist of cut cells or whole cells, but not of both.

Diodes Placing bypass diodes in parallel with every cell would minimize the effects of cracked or shadowed cells. Placing the diodes underneath the array substrate would require a hole between each pair of cells and one at the beginning and end of each sub-array. Placing the diodes on top of the array next to the cells they bypass would reduce the area available for cells. Eight hundred diodes would be needed in either case. Instead, we will place diodes underneath the array and in parallel with ten-cell segments of each sub-array. This requires 80 diodes in all. The number of cells bypassed was arbitrarily chosen. Each diode must be rated for the short circuit current of the sub-array and the open circuit voltage of the string of cells it is connected to.

Mass Table 9.4 allocates 15.3 kg to the array, which includes the cells, interconnecting ribbon, diodes, and MPPTs. Cells of about 0.5-mm thickness and about 1.0-mm thickness are available. The former have a mass of about 7 g and latter a mass of about 14 g, giving a total cell mass of 5.6 kg and 11.2 kg, respectively. Thicker cells are less susceptible to cracking. But we resolve to make the substrate rigid enough to prevent this without a 5.6-kg structural mass increase. Taking the mass of an MPPT[11] as 0.9 kg gives a total tracker mass of 3.6 kg. Typical (0.004-in. × 0.055-in. tinned copper) interconnect ribbon has a mass of 0.00155 kg/m. Allowing 40 cm per cell (see Fig. 11.26) gives an interconnect ribbon mass of 0.5 kg. Taking the mass of a diode and its interconnecting wires as 2 g gives a diode mass of 0.16 kg. The estimate of the total mass is then 9.86 kg.

Summary Table 10.8 summarizes the array design.

Table 10.8 Array electrical design

Cells	1000, 10-cm x 10-cm, commercial silicon, four parallel sub-arrays, 200 cells per sub-array (200-cell breakage allowance)
MPPT	Boost regulator, match to 105–135 V bus, rated a t≥3.2 A, four required
Diodes	80, one per 10 cells, rated at≥3.2 A and 6 volts
Interconnect	0.004-in. x 0.055-in. tinned copper ribbon, 40 cm/cell (320 m required)
Total mass	9.86 kg

[11] The actual mass of a Solectria MPPT.

References

Bleck, O. (1993). "*Understanding and selecting the proper drive system for an electric Vehicle,*" *APN2001*. Arlington: Solectria Corporation.

Evans, C., Leonard, S., & Redick, M. (1999). *Sunrayce 1999 structural design report*. Potsdam: Clarkson University Solar Knights.

Gillespie, T. D. (1992). *Fundamentals of vehicle dynamics*. Warrendale: Society of automotive engineers.

Hibbeler, R. C., (1998). *Engineering mechanics dynamics* (8th ed.). Upper Sadle River: Prentice-Hall.

Mossman, E. A., & Randall, L. M. (January 1948). "An Experimental Investigation of the Design Variables for NACA Submerged Duct Entrances," NACA RM-A7130.

Rauschenback, H. S. (1980). *Solar cell array design handbook*. New York: Van Nostrand-Reinhold.

Storey, J.W.V., Schinckel, A. E. T., & Kyle, C. R. (1993). *Solar racing cars*. Canberra: Australian Government Publishing Service.

Whitt, F. R., & Wilson, D. G. (1974). *Bicycling science*. Cambridge: The MIT Press.

Chapter 11
Solar Racer—Construction

11.1 Introduction

Emphasis has been placed herein on the main steps in the construction process. Within this domain, most of the material focuses on construction of the composite body shell and the solar cell array. These processes are probably the most unfamiliar to the reader. Remarks on chassis and moldless shell construction, the management of construction and the skills, facilities, and equipment needed to support it have also been included.

For details of composite fabrication, the reader should consult specialized publications. Constructors of homebuilt light aircraft and solar cars share some goals: the need to fabricate a streamlined, lightweight, but strong and stiff structure. Hence, the composite materials and the construction techniques used for homebuilt aircraft may also be used for solar cars. The solar car builder may therefore tap the wealth of technology and information accumulated by the homebuilt aircraft industry.

Books on steering, braking, and suspension construction have been published. Most of the key components of these systems, such as steering wheels, rack-and-pinion assemblies, rod-end bearings, hydraulic brake caliper units, master cylinders, and shock absorbers, can be purchased ready-made. Construction of these has therefore not been included.

11.2 Body Subassemblies

A solar racing car body provides aerodynamic streamlining, supports the solar array, and keeps the weather out of the car. The array is often detachable so it can be mounted on a moveable stand that points it at the sun when charging batteries. Figure 11.1 shows a three-subassembly car: array, canopy, and chassis during battery charging.

A chassis is a structure made of metal tubes, composite panels, or a combination of composite panels and molded shapes, to which are attached the suspension ele-

E. F. Thacher, *A Solar Car Primer*, DOI 10.1007/978-3-319-17494-5_11

Fig. 11.1 Body subassemblies. (Jules LeDuc)

ments, the steering system, the battery, motor, the aerodynamic shell, etc. Thus, the chassis carries the main structural loads.

11.3 Space Frame

If the chassis is a frame made of tubes, it is called a "space frame." The tubes of a "pure" space frame are only in compression or tension. Space frames are relatively easily and quickly constructed or modified (provided the team has the necessary skills) because they are assembled by cutting tubing elements to length, and welding or brazing the elements together. When a space frame is used, the body usually consists of a shell with the space frame inside it.

The frame materials are usually aluminum or steel tubes (6061-T6 aluminum and 4130 chrome–molybdenum aircraft tubing are frequently used) of well-known geometry and physical properties. Whereas composite components, especially those made with a wet layup process by relatively unskilled workers, can have quite variable structural properties, including built-in flaws. Thus, a prediction of where and how a composite will fail carries far less assurance than does a similar prediction for a space frame. Marshall (1998) strongly recommends load testing of sample composite pieces and finished articles.

The strength of aluminum is reduced about 50–80 % in the weld-affected zone. The designer must compensate for the weakening at the welds by using more, possibly thicker-walled, tubes or other reinforcements, such as gussets. These additions make the space frame heavier. Chrome–molybdenum tubing is more dense than

Table 11.1 Steel and aluminum properties

Material Property	4130 Chrome–molybdenum steel	6061-T6 aluminum
Property		
Density (pci)	0.284	0.094
Specific weight	7.9	2.6
Young's modulus (psi)	29.0 (10^6)	10.0 (10^6)
Ultimate tensile strength (psi)	207,000	42,000
Yield strength (psi)	197,000	35,000
Strength/specific wt. (ksi)	26	16

aluminum but far stronger, so that the space frame can have fewer members made with smaller tubing (see Fig. 11.3). Furthermore, the frame can be brazed together, thus avoiding the effects of welding on the joint's strength. Table 11.1 gives room-temperature properties for 6061-T6 aluminum and 4130 chrome–molybdenum steel.

It is easy to predict the space frame's weight during design, and to control it during construction, because it is made of elements of a known weight and geometry. Wet layup composites tend to be heavier than intended. (See the remarks on "pre-preg" cloth in the Sect. 11.4 below.)

A space frame can be built independently of the shell. This means that a battery-powered, operable, but shell-less car can be finished before the shell and the solar array. Testing of the vehicle can begin. This is a very important advantage. Figures 11.2 and 11.3 show cars with aluminum and steel space frames, respectively. Contrast these with the chassis shown in Fig. 11.4 made of aluminum honeycomb composite sheets.

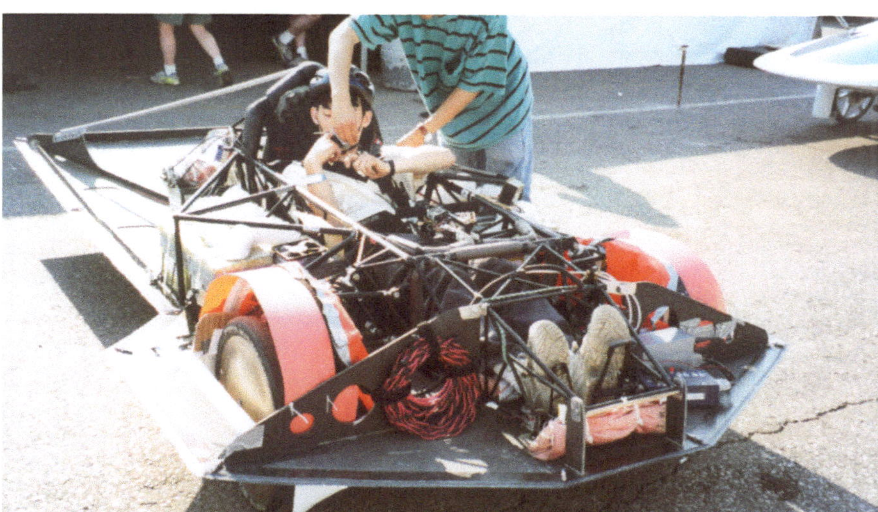

Fig. 11.2 Steel space frame—M.I.T. 1995. (Jules LeDuc)

Fig. 11.3 Aluminum space frame—Missouri-Rolla 1999

Mock-Up Mock-ups are temporary prototypes quickly constructed of cheap materials. Plastic tubing and duct tape were used to construct the mock-up shown in Fig. 11.5 made for checking the stiffness of the frame and its interface with other parts of the car. The figure shows the interface with the lower shell being assessed.

Fig. 11.4 Aluminum honeycomb chassis—Minnesota 1995. (Jules LeDuc)

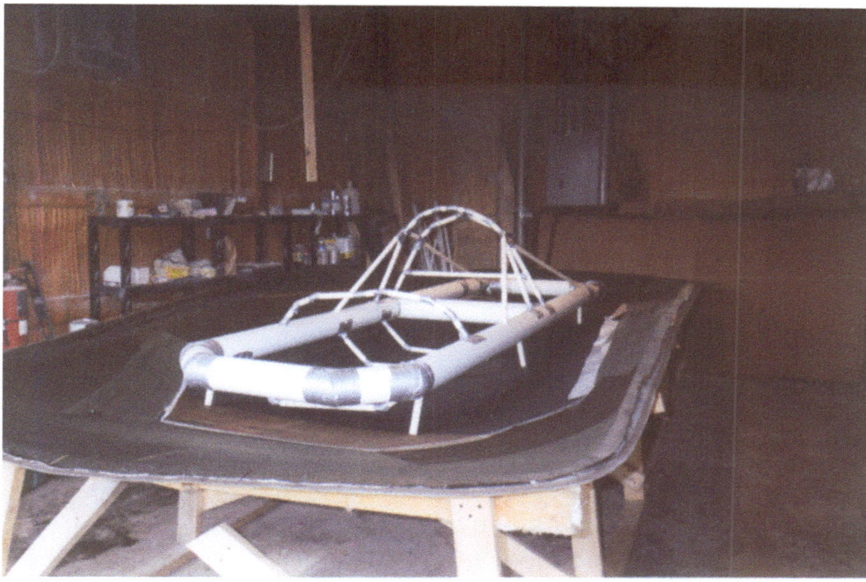

Fig. 11.5 Space frame mock-up. (Brian Lisiescki)

Mock-ups can be particularly helpful in the cockpit area. Helmeted drivers can sit in the mock-up's cockpit and evaluate the utility of the layout and clearance to the roll bar. Mock-ups are very helpful in planning the wiring and instrumentation layout.

Alignment Jig As the welded or brazed joints cool, the frame may twist or bend out of its intended alignment. To minimize this, a stiff fixture should be constructed to hold the members in alignment during joining. The frame will become rigid enough to hold its shape without the jig after a number of its members have been connected.

11.4 Shell Materials

Cloth Marshall (1998) defines a composite as an assembly of dissimilar materials combined to do a job that individually they cannot do. The composites used in solar cars are made from cloths woven from special yarns and wetted with a resin, sandwiching foam or honeycomb reinforcement. Table 11.2 shows some of the properties of three yarns commonly used in homebuilt aircraft and solar car composites. (Compare some of the properties given in Table 11.1). The cloths have been numerically ranked highest (1) to lowest (3) in the "impact resistance" and "toughness" characteristics. S-glass (the "S" denotes a higher silica content) is frequently used in homebuilt aircraft applications.

Table 11.2 Yarn comparison

Material Property	S-glass	Kevlar® 49	Carbon
Specific weight	2.49	1.48	1.78
Young's modulus (psi)	12.4 (10^6)	11.0 (10^6)	35.0 (10^6)
Ultimate tensile strength (psi)	665,000	525,000	500,000
Strength/sp. weight (ksi)	267	355	281
Cost, plain weave ($/yd)	5	19	29
Electrically conductive	No	No	Yes
Impact resistance	2	1	3
Toughness	2	1	3

The function of the cured resin is to bind the composite structure together so that the load is distributed over the reinforcing yarns. Thus, the strength of the structure also depends on the uniformity of the resin content, how well it has bonded to the cloth, and the strength of the resin. The resin also bonds the composite to surrounding or embedded structures.

Cloth already wetted (impregnated) with just the proper amount of resin, "pre-preg," may also be purchased. Pre-preg must be kept cool to prevent its resin from curing before the cloth is used. The wet layup process, described in great detail by Rutan (1983) and Marshall (1998), in which the builder adds resin to the cloth, tends to produce a heavier-than-intended car. The amount of resin added is difficult to control precisely, especially when the builders are inexperienced. Pre-preg cloth solves this problem. But pre-preg is much more expensive than the dry cloth.

Of the three fibers in Table 11.2, S-glass has the lowest strength-to-specific-weight ratio and Kevlar® the highest. Therefore, solar car builders, in their quest for light-but-strong cars, tend to pass over glass in favor of Kevlar®. But this choice incurs about four times the cost of S-glass.

Flexing of the shell may crack the delicate solar cells on its surface, so the shell must be stiff. But it must also be light. Hence, carbon cloth-based composites are attractive because they offer the possibility of low weight, high strength, and greater stiffness. The stiffness of the carbon fibers is about twice that of Kevlar®, the next largest. However, the carbon fiber's cost is more than five times that of S-glass, and it is electrically conductive. If it is used in the construction of the shell, special care must be taken to make sure the final structure does not cause short circuits in the solar cell array. The electric conductivity of the cured structure will be lower than the cloth by itself. But the distribution of the resin may be nonuniform, especially if a wet layup is used, so that local areas of higher electrical conductivity may be present. Also, the fibers of the finished composite are quite stiff and can easily cut through the insulation on wires (and through flesh). Holes where wires from the array pass through the shell should be smooth and lined to prevent shorts. Having tried both Kevlar® and carbon, the author advocates making the shell of the former, nonconductive, material.

Table 11.3 Resin systems comparison

	Polyester	Vinylester	Epoxy
Strength	1	2	3
Cost ($/gal)	8	20	50
Shrinkage amount	1	2	1
Ease of use	2	2	1

11.5 Resins

A comparison of frequently used resin systems appears in Table 11.3. Except for the cost per gallon, the resins are rated 1, 2, or 3, indicating lowest to highest. For example, epoxy resin systems have the highest strength and cost, the least shrinkage during cure, and are the most difficult to use.

The entries are called "resin systems" because chemicals are available to mix with the basic resin to change its properties. For example, catalysts may be added to reduce the interval between mixing the resin with the catalyst and the onset of thickening of the mixture, the "gel" time. For any given mixture, the gel time varies inversely with temperature and directly with humidity. Marshall (1998) recommends cessation of resin mixing or fabric wetting below 60 °F and 60 % relative humidity.

The epoxy systems were rated lowest in "ease of use" because of the care that must be taken when mixing them. The cured strength of epoxy systems will be reduced considerably by small errors in the proportions of the mixture. Errors in mixing the other two systems will affect the cure time, but the cured composite will have the expected strength.

All the resin systems are toxic and subject to strength loss from exposure to ultraviolet radiation.

11.6 Core Materials

The salient characteristics of the core materials more frequently used in solar racing cars are discussed below. The material below was adapted from Marshall (1998). More details may be found therein.

Polystyrene Foam Two-lb/f^3 (pcf), large-cell type (not insulation-grade) Styrofoam was specified by Rutan (1983) for moldless construction of the wings and some other portions of a homebuilt light aircraft. It has adequate strength properties and can be smoothly cut by the hot-wire process into almost any required shape. It does not emit any toxic gases during hot-wire cutting as do urethane foams. However, it has the disadvantage of being soluble in Vinylester resins and polyester-based body filler materials and resins.

Do not use beaded styrene foam material. The bonds between the beads may be weak, making the strength of the foam unreliable.

Polyvinyl Chloride Foam Two common core materials based on polyvinyl chloride (PVC) are Klegecell® and Divinycell®. These materials are sold in densities as low as 2.5 pcf with good strength properties. They may also be purchased as "diced" sheets: sheets of small rectangular pieces of foam bonded to a fabric backing. These are intended for layups on curved surfaces.

Honeycomb Honeycomb cores are an array of small, open cells. Other cell shapes exist, but the hexagonal shape generally has the best strength. Honeycomb cores yield a composite structure of higher strength-to-weight ratio than do foam cores. Honeycomb cores made of aluminum and of Nomex® have been used in solar racing cars (see Fig. 11.4). Aluminum is cheaper, but Nomex® has better corrosion resistance and toughness.

11.7 Molded Shell Construction

The top shell section will support the solar array and be removable from the car so that it can be placed on a special stand for charging the batteries, as shown in Fig. 11.1. The two sections will each be made as a composite "sandwich" with outside and inside layers of Kevlar® cloth glued to a middle layer of structural foam or of a honeycomb structure such as Nomex®. To give this composite structure the proper shape, its two halves will be cured in female molds so that the outside layer of Kevlar® will have the shape and dimensions, and will approximate the smoothness desired, of the outside surface of the shell.

Plug In the process illustrated in Fig. 11.6, the interior surface of the female mold is first formed on a male mold called the *plug*. The plug must be made, as nearly as possible, to the exact exterior dimensions and shape of the shell. The surface of the plug must therefore be smooth; any defects in it will be reproduced in the mold and therefore in the shell. Figures 11.7–11.13 show a plug being constructed. The engineering drawings of the shell were used to produce templates of the cross-sectional shape of the shell at many stations along its length. The templates were used to cut these shapes from insulating foam. The shapes were mounted on rails, glued together, and smoothed. The surface thus formed was sealed,[1] coated with an automobile body filler material, and sanded smooth again. The smoothed surface was cleaned, spray-painted with a primer, wet-sanded, and then coated with a mold release compound.

Mold The top mold was created by spraying glass fiber onto the top half of the plug. The top of the assembly was then fitted with a stand and the assembly inverted. The original stand was removed and the bottom of the plug completed. Then the top part of the mold was constructed. Figure 11.14 shows the plug being loaded for shipment to make the top half of the mold. The mold was made as an in-kind donation by Empire Fibreglass Products, Inc., of Little Falls, New York.

[1] To protect it from the polyester-based body filler.

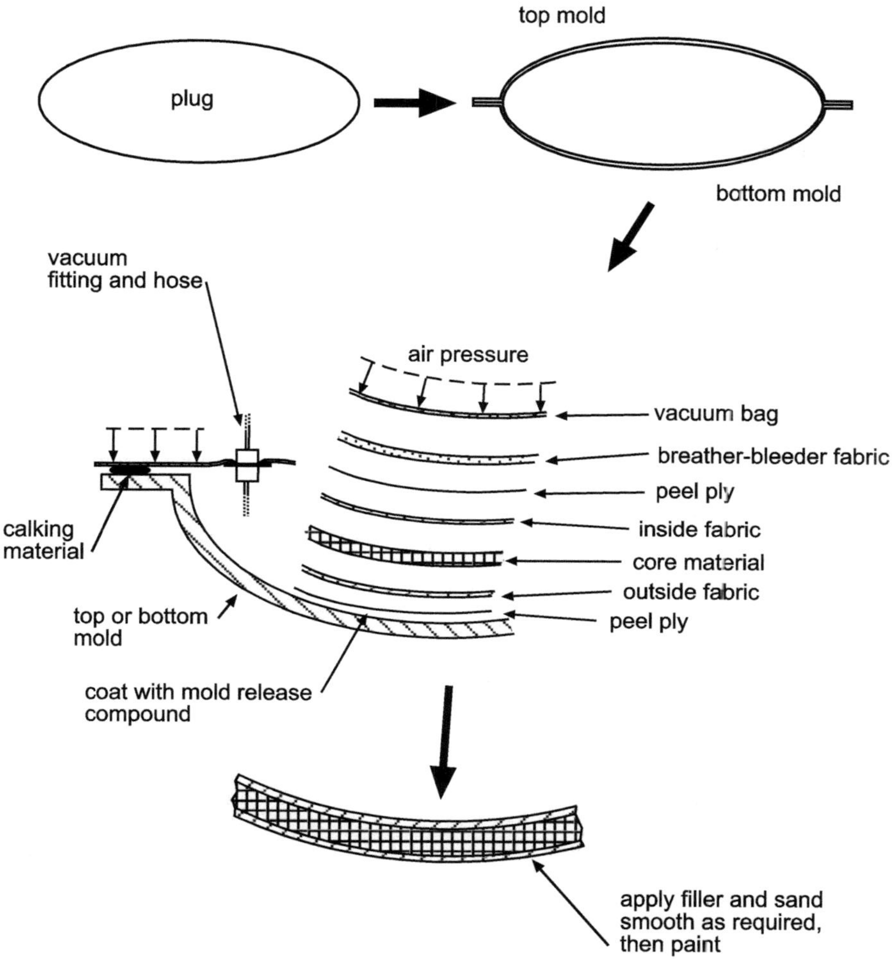

Fig. 11.6 Molded shell fabrication process

Figure 11.15 shows the finished molds prior to separation. The final steps in preparing the molds to make the shell were to thoroughly clean them and to coat them with three or four layers of a mold release compound. Figure 11.16 shows a mold being cleaned. After cleaning, the molds should be covered with a clean plastic sheet until used.

Layup Marshall (1998) and other practical books contain many layup details and suggestions. Before attempting to build a shell, obtain one of these books and perform some practice layups.

In Fig. 11.6, the first layer next to the mold is a release film that can pass through the curing process but not bond to the composite; it may be torn off without damag-

Fig. 11.7 Cutting plug foam segment. (Brian Lisiescki)

Fig. 11.8 Foam segments before sanding. (Brian Lisiecki)

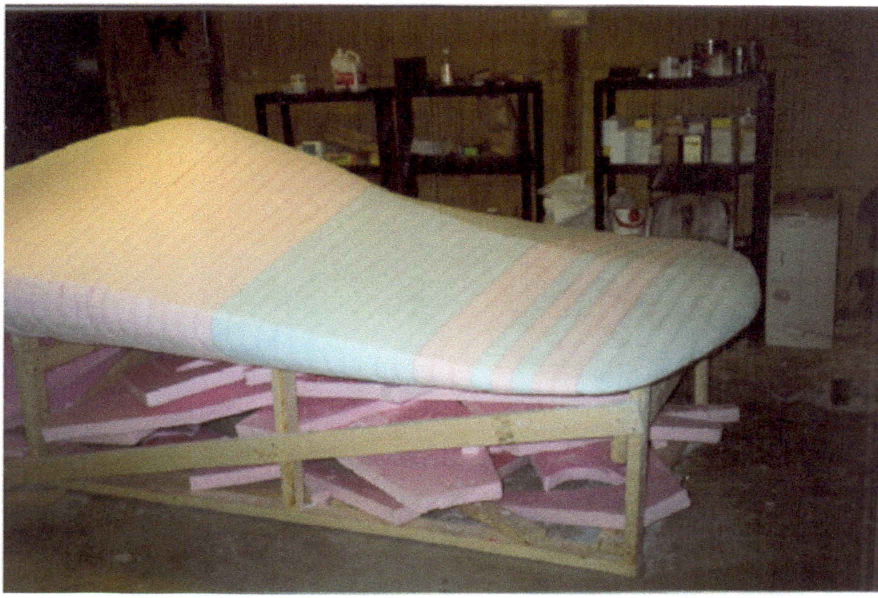

Fig. 11.9 Shaped foam core. (Brian Lisiecki)

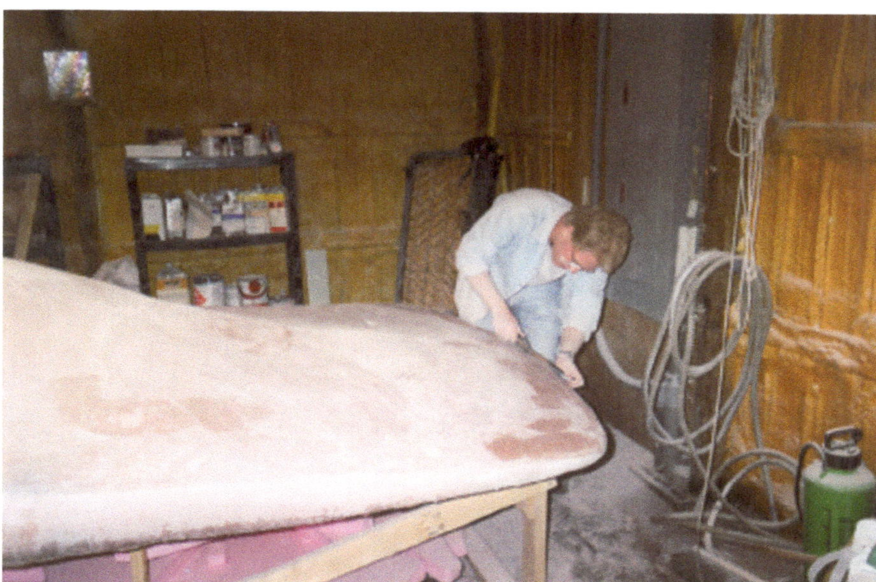

Fig. 11.10 Smoothing body filler. (Brian Lisiecki)

Fig. 11.11 Check flatness. (Brian Lisiecki)

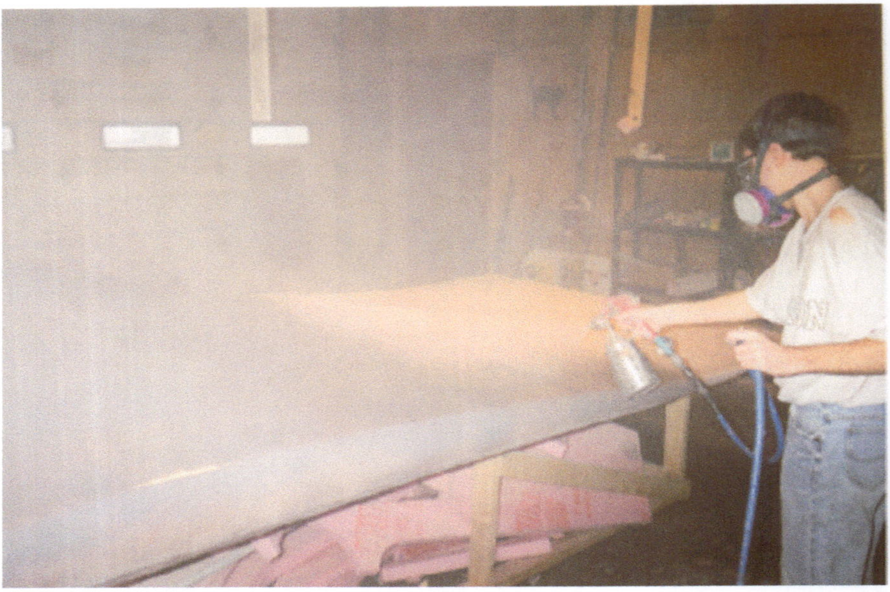

Fig. 11.12 Spraying primer. (Brian Lisiecki)

Fig. 11.13 Wet-sanding plug. (Brian Lisiecki)

Fig. 11.14 Shipping plug. (Brian Lisiecki)

Fig. 11.15 Finished molds. (Brian Lisiecki)

Fig. 11.16 Smoothing a mold. (Brian Lisiecki)

Fig. 11.17 Laying down outer peel ply. (Brian Lisiecki)

ing the cured parts.[2] The outside layer of fabric may be positioned in the mold and wetted with resin; however, the mold side of the fabric will be hidden and consequently also any dry spots on that side. Marshall recommends placing the fabric on a plastic film on a table next to the mold. The fabric should then be wetted with the proper amount of resin and positioned in the mold using the plastic backing. This backing protects the fabric while it is worked into close contact with the mold. Then it is removed. The core material is then put in place, followed by the outer layer of fabric, then the outer peel ply and bleeder layers, and finally the vacuum bag itself.

If the mold is sufficiently smooth, the cured shell also will be. But if not, application and sanding of a body filler material may be necessary. The density of most conventional body filler materials is between 25 and 50 lb/f^3 (Marshall 1998). This will add some weight and increase the fabrication duration.

Figures 11.17–11.20 show the layup process. The fabric in this case was prepreg. It was cured at an elevated temperature in a large oven at Stafford Machining/Cyclotherm Inc. of Watertown, New York.

Stiffeners Add stiffness by installing longitudinal ribs and transverse bulkheads inside the shell. These members are made by constructing flat composite panels, drawing the stiffeners' shapes on the panels, cutting the shapes out, and tabbing them in place with a resin-impregnated structural cloth. These reinforcements will of course add weight. Figure 11.21 shows stiffeners in the underside of a solar car shell.

[2] Marshall (1998) recommends testing the release agent you have chosen with your resin system. He points out that the ability of each resin system to penetrate the release agent is different.

Fig. 11.18 Laying down bleeder cloth. (Brian Lisiecki)

Fig. 11.19 Vacuum-bag on layup. (Brian Lisiescki)

Fig. 11.20 Bottom half of shell. (Brian Lisiescki)

Fig. 11.21 Shell stiffening—Missouri-Columbia 1999

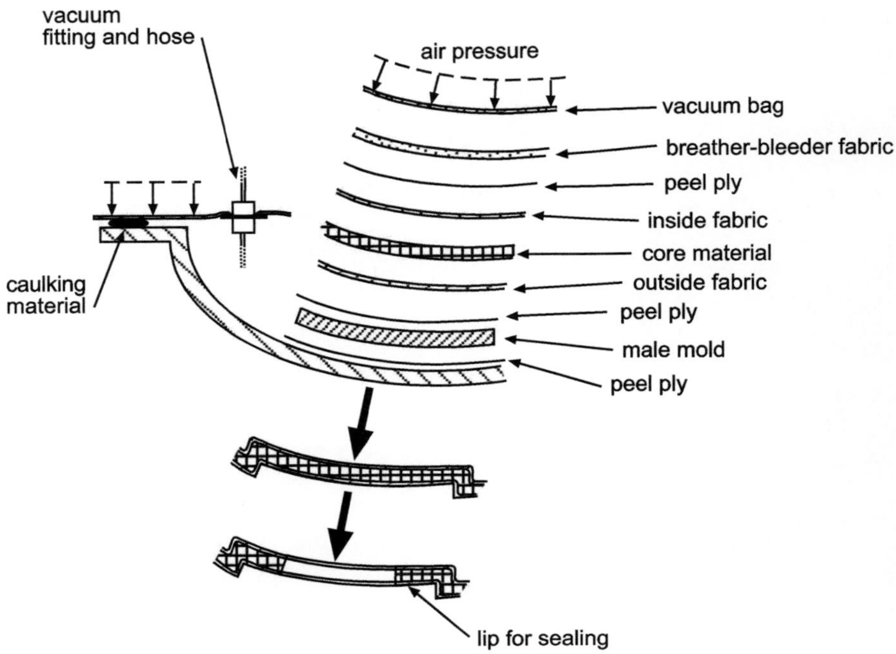

Fig. 11.22 Molding a flush-fit door

11.8 Doors and Windows

Figure 11.22 shows a method for making an opening in the shell for a flush-fit panel. The technique is the same as before except that a male mold slightly larger than the size (to provide clearance) and of the thickness of the door is placed as shown. Curing the shell in this manner produces a recessed area of the size and shape of the door. The center of this area is cut out to make the opening. A sealing lip around the opening remains.

11.9 Moldless Shell Construction

As the foregoing demonstrates, making a compound-curved shape using a female mold is an involved and therefore lengthy process. It may require 3 or 4 months, in an academic environment. Rutan (1983) describes the "moldless" construction process for homebuilt aircraft. The process has been successfully applied to solar racing cars. The essential idea is to form a surface of suitable lightweight foam that has been given the desired shape, but is smaller than the desired final dimensions. A "skin" of one of the cloths of Table 11.2 is resin-bonded to this core using a wet

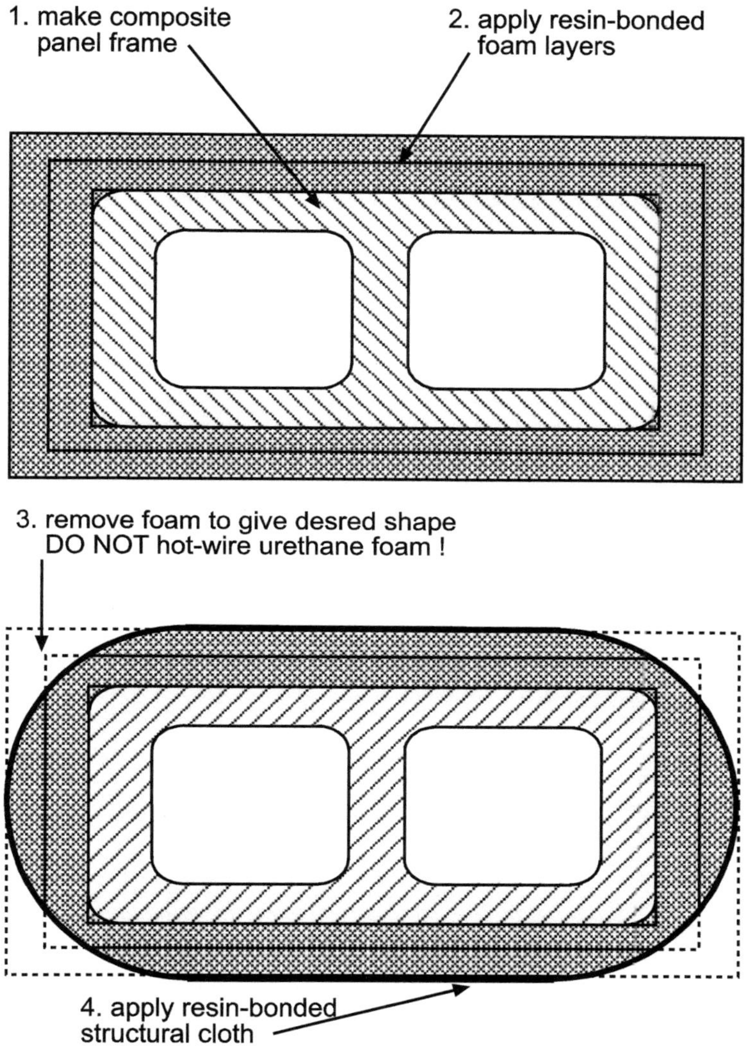

Fig. 11.23 Moldless construction process

layup process. This amounts to making a plug that becomes part of the car's structure. Finally, the outside surface of the final part is smoothed and finished. This may require much effort if insufficient attention has been paid to smoothing the plug's surface. The dimensions of solar cars must lie within certain limits and this must be allowed for when planning the dimensions of the core materials.

Figure 11.23 shows a simplified body frame constructed of composite panels. The panels would be made from one of the cloth–foam systems already described.

Foam is resin-bonded to the frame and sanded to the shape and dimensions required, less an allowance for the external cloth layers and other surface treatment. Then structural cloth is bonded to the foam and the surface finish applied.

11.10 Ultraviolet Protection

Resins decompose when exposed to ultraviolet (UV) light, causing a measurable loss in strength of unprotected composites over periods of a few weeks or months. UV inhibitors in the resin will slow the deterioration, not prevent it (Marshall 1998). Painting the car will block the UV light. Of course, painting will normally be done to improve the appearance. Until this is done, the body should be kept out of sunlight.

11.11 Solar Cell Array

Design and Layout The output of the design phase should be documents showing the number of cells required, the number of strings, their preliminary layout on the car, the wiring design (including the by-pass diodes), the encapsulant and underlayment materials, the cell-to-cell spacing, and the design area efficiency.

The final layout of the array, drawing the outlines of the parts of the array and the locations of wire holes directly on the shell's surface, should be done after the shell has been built and painted. Then account can be taken of the as-built dimensions, smoothness, and curvature.

Number of Cells An 8-m^2, rectangular, flat array may require 800, 10 × 10-cm cells. However, because solar cells are delicate, it is quite difficult to avoid breaking a number of them during array construction. Therefore, the quantity purchased should include extra cells to allow for this and also to provide replacements for cells damaged during operation of the car. The team may wish to practice tabbing, test sample modules, or try out methods of underlayment and encapsulation. Some cell suppliers furnish free sample cells that have defects and so could be used for experimenting with fabrication techniques. The need to replace broken cells in the array (see Fig. 11.1) or cells which were to be used in performance tests increases the quantity purchased by about 25 % for a total of 1000 cells.

Sub-arrays An array consists of a number of parallel-connected strings of series-connected cells. These will be called *sub-arrays*. The sub-arrays themselves are formed from subsets of series-connected cells, or *modules,* placed together on a particular part, or *facet,* of the shell's surface and connected in series with the other

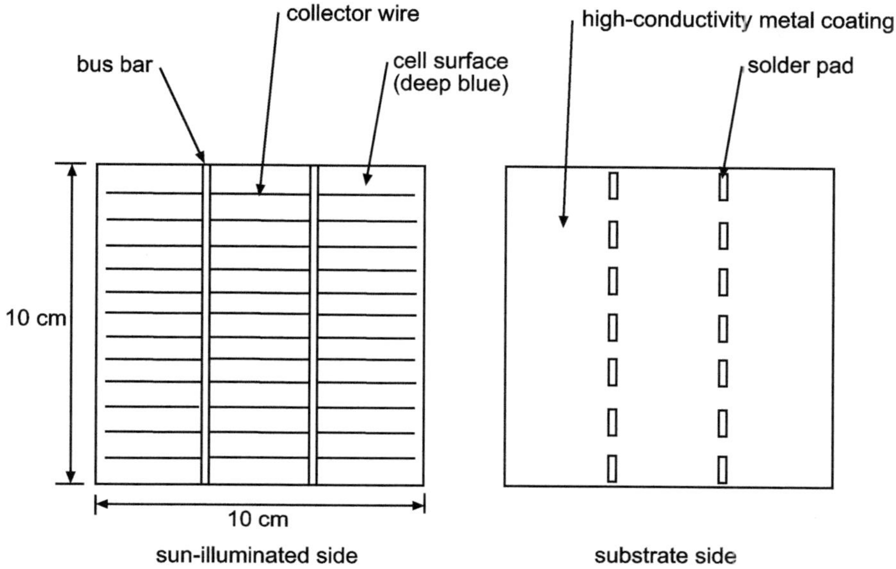

Fig. 11.24 Typical solar cell

modules of the sub-array. A facet is an approximately flat region molded into the shell's surface and designed to hold a particular module.

The cells in the modules should be connected together first. The team may stipulate in the purchase contract that this be done by the cell supplier.[3] This may save time and yield a better product; however, it will approximately double the cost of the array.

The first step in module construction will be *tabbing* the cells: soldering interconnect ribbon to the negative-side bus bars. Figure 11.24 shows the top, or illuminated negative side, and the bottom, or positive side, of a typical solar cell. Figure 11.25 shows a solar cell being tabbed. The worker is applying solder paste to the top bus bars. To save time, the tabbing operation should be done by several workers at once. The interconnect wires are flat, tinned (solder-coated), copper ribbons. Photon (1995) gives practical details about tabbing.

After tabbing, the cells are connected into strings. Figure 11.26 shows a cross section through part of a string. Consult Rauschenback (1980) for some variations on this standard "z-step" connection which were designed to reduce the thermal expansion between the cells in a string should be 1–1.5 mm to allow for thermal expansion. A fixture should be constructed to hold the cells at this spacing and to keep

[3] The terms of this contract should describe the manufacturer's responsibility in case he damages the cells.

Fig. 11.25 Tabbing a cell. (Brian Lisiecki)

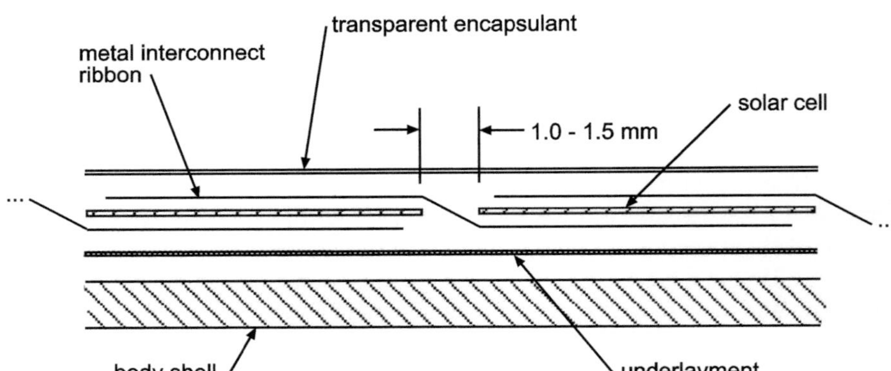

Fig. 11.26 Two cells in series

the edges of the cells in alignment while they are being connected. Figure 11.27 shows such a fixture; it was milled from a sheet of plastic.

Encapsulation The solar array must be protected from water, dust, and small stones that may be thrown against it in traffic. This may be done by applying an optically transparent encapsulant with a thermal expansion characteristic simi-

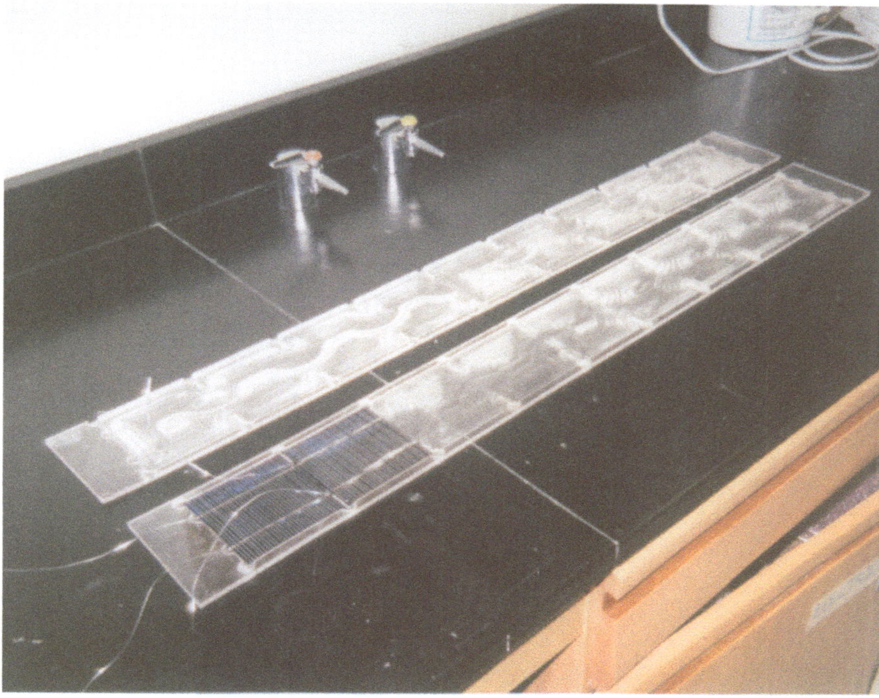

Fig. 11.27 Cell connection fixture

lar to that of the cells. Figure 11.28 shows workers brushing on a two-part silicone encapsulant to the back (foreground) and top (background) of modules. The two-part silicone materials will hold dust and are not washable. A one-part silicone that produces a washable surface is Dow-Corning 1-2577 Conformal Coating. Because the product cures by evaporating toluene solvent, it should be applied only while wearing solvent-rated breathing masks in well-ventilated spaces. Consult the materials safety data sheet and other instructions furnished by the manufacturer.

Each completed module should be tagged with a code indicating its place in the array, then stored and covered pending its installation on the solar car. The utmost care must be taken to prevent cracking of the cells when handling and transporting the modules.

Installation With the final layout of the array marked on the car, wiring holes drilled and lined, and the completed modules stored near the car, installation may begin. If access to the underside of the array substructure is not convenient when it is mounted on the car, place it in a fixture so that it is so.

Fig. 11.28 Applying encapsulant. (Brian Lisiescki)

Before applying its underlayment, give each module to be installed a final inspection. Then prepare and apply its underlayment, as shown in Fig. 11.29. The underlayment should be thick enough to fill in around the interconnect ribbon and solder joints on the backs of the cells. Also, if the depth is not elsewhere uniform, voids will form underneath the cells, leaving portions of the cells unsupported. These defects may result in cracked cells later. The underlayment depicted in Fig. 11.29 is "Tra-Bond 2151," a two-part silicon adhesive manufactured by Tra-Con of Bedford, Massachusetts.

After application of the underlayment, remove the module from storage, pass the interconnecting ribbons through the wiring holes, lower the module onto the underlayment, and *gently* settle it into place, as shown in Fig. 11.30. This process must be done with care to preserve the intended alignment, avoid cracking, and ensure the backs of the cells are uniformly wetted by the underlayment, as mentioned above. Underlayment that may be squeezed onto the cells out of the gaps between them should be removed using the solvent recommended by the manufacturer.

Fig. 11.29 Applying underlayment. (Brian Lisiescki)

After the underlayment has cured, wiring may begin. The design should incorporate proper wire sizing to minimize power losses and voltage drops, color-coded insulation, and the construction of wire harnesses. McCarney et al. (1987) offer practical advice in these matters. The harnesses should be supported at frequent points from the underside of the shell so the electric connections do not bear mechanical loads and breaks do not occur from cold working.

Fig. 11.30 Installing module

Fig. 11.31 A drive assembly

11.12 Electric System

Usually, the components of the electric system, from connecting wire to the propulsion motor and its controller, will have been purchased rather than fabricated from primary materials. The system will be built by mounting these components to the chassis or shell and connecting them according to the circuit design. During the design phase, thought should be given to the location and design of the mounts and the wiring layout.

It may be necessary to build mounts for certain components, such as the external lights, into the shell or the chassis during its manufacture, rather than add them afterward. Figure 11.31 shows a motor driving the rear axle through a chain drive. Kyle (1990) and Storey et al. (1993) show many examples of drive installations. Metal plates may be bonded into a composite structure to strengthen it for motor (or suspension) mounts. Marshall (1998) shows how to build such "hard points" from composite materials alone.

Fig. 11.32 Maximum power point trackers. (Jules LeDuc)

Short wire runs, low weight components, and ease of access should be guiding principles. Figure 11.32 shows maximum power point trackers conveniently mounted together in a Kevlar®-composite box.[4] A mock-up of the chassis will be very helpful in planning the electric system layout, as pointed out earlier.

Figure 11.33 shows a direct-current-to-direct-current (DC-DC) converter being connected to the circuit. Some of the circuit layout drawing can be seen.

11.13 Facilities

Central Office There should be a location near the car where the weight chart, a time line, and the engineering drawings can be displayed. Corrections to the time line, the chart, and to the engineering drawings to reflect any design changes found necessary combined with meetings of the construction crew serve to keep everyone informed, on schedule, and working together.

Array Construction, Composite Layup Figure 11.34 shows a fabrication facility. Ideally, the rooms in which these activities occur should be isolated from casual foot traffic, kept clean, and be well-ventilated. The depicted facility was shared with other student projects. But it was located next to a well-equipped machine shop, a great advantage.

[4] However, the box was mounted on top of the battery compartment, making access to the batteries difficult.

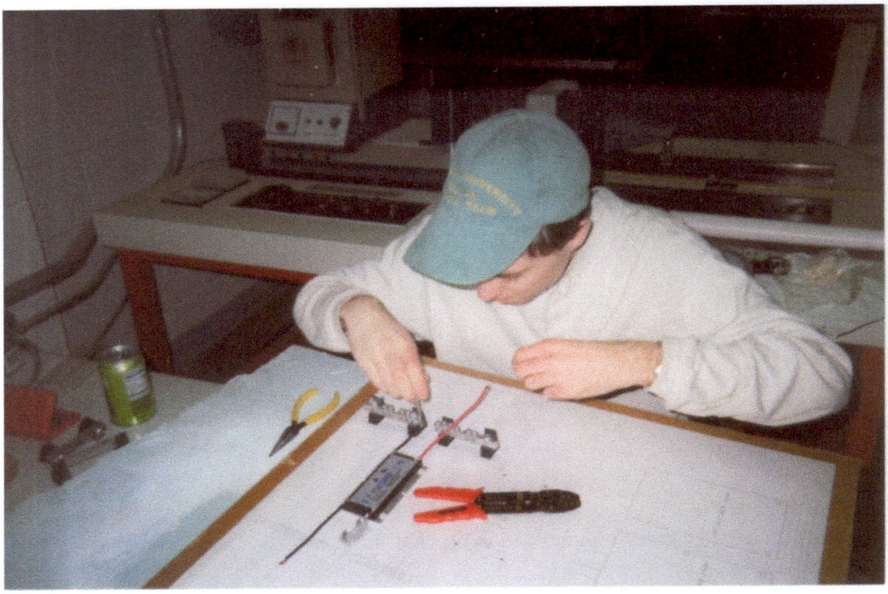

Fig. 11.33 Wiring begins. (Brian Lisiescki)

Fig. 11.34 Assembly facility. (M. Griffin)

Encapsulates, underlayments, resins, and other chemicals associated with array and composite construction are hazardous chemicals. They are fire hazards, toxic, and give off noxious, and sometimes injurious, fumes during cure. Learn and follow the special precautions necessary when working with these substances. The manufacturers of hazardous substances send material safety data sheets with their products. Read these sheets and keep them on file.

Workers must be provided with protective clothing, such as rubber gloves with thin cotton gloves underneath (to absorb sweat) and breathing masks. (Notice the mask worn by the worker spraying primer on the plug shown in Fig. 11.12.) Wear a barrier cream. Do not clean skin with a solvent. Solvents absorb natural skin oils that prevent the resins from penetrating. Cleaning should be done instead with a waterless hand cleaner. Finish cleaning with soap and water.

The school must have a system in place for proper storage and disposal of hazardous wastes. Contact the individual designated as the school's health and radiation safety officer to learn the proper procedures and to obtain training in the use of these materials, and to have your workspaces inspected for compliance with regulations.

Electrical Electrical fabrication facilities are much easier to acquire. Circuit wiring does not require lots of space nor expensive machine tools. Soldering equipment, crimping tools, wire strippers, and the like are required instead. The solar cell array is the single most difficult and time-consuming component to make. It requires a clean and dust-free environment and table space to lay out cell-string soldering jigs and trial panels.

The major electrical safety issues are those associated with electric storage batteries. The project team must be trained in how to work around and with them, how to install them, how to charge and discharge them, and how to store them. Crompton (1996) (see page 98) is a handbook of battery technology. Manufacturers provide information on the proper installation, maintenance, and operation of their batteries. These instructions should be acquired, disseminated, and followed.

At some point during construction, the batteries must be placed in the car and wired. This must be carefully planned for because all subsequent work must account for the presence of the batteries. For example, no welding should take place once the batteries are present. Wiring the batteries should be done with insulated tools, and two persons should always be present. Once the batteries are installed, they must be covered. This will prevent metal objects from dropping into the battery box and causing a short. Figure 11.35 shows batteries being installed in a solar car.

People continually working around the batteries may become accustomed to them and consequently act carelessly. Supervisors must insure that the workers remain aware of the presence of the batteries and the potential hazard they represent.

11.14 Construction Management

Plan The construction plan should incorporate parallel effort. This implies that all the fabrication skill not be concentrated in a small number of persons. It also implies that the information on how the car is to be built be recorded on drawings, rather

Fig. 11.35 Installing batteries. (Brian Lisiecki)

than in the minds of one or two persons. The need to share information cannot be overstressed.

The construction crew will mostly be students. That means that the plan must account for their other commitments, such as job-interview trips, exams, and weekend trips home. However onerous it may seem, work must continue during holidays and the summer before the race. These times are precious: During them, the workers will be relatively free of other commitments. This means that arrangements must be made to house and feed them on-campus.

Construction generally seems to require more time than one anticipates. Therefore, the plan should incorporate allowances for this typical underestimate. A rule of thumb is to multiply one's original estimate by three.

The skills and equipment that must be available will become evident during design. If these are not present either within the team or within the school, then plans must be made for "outsourcing," that is, sending work to agencies outside the school. Businesses thus employed must of course be capable of doing the work and be able to meet the construction schedule. The construction plan must include ways to pay for outsourced work in cash or for donation of its services by the business.

Controlling Weight The design empty weight of the vehicle and a table giving the portion of this weight allocated to each system of the car should be available before construction begins. The engineering drawings will show the location of the various components. This sets the weight distribution, which is crucial to realizing the vehicle's intended stability characteristics.

Unless a careful watch is kept, the weight will tend to increase beyond the design value. As a means of controlling this, frequent weighing should take place

Fig. 11.36 A tire jack. (Jules LeDuc)

and be recorded on the table mentioned above. A cause of overweight is purchased components exceeding their manufacturer's advertised weight. Components should be weighed before they are installed to verify that their weight is as expected. Another cause is the accumulation of many small weights. Screws, nuts, washers, wire ties—a host of small parts must be installed. Often these small items are not planned for but installed at the whim of the workers. Their accumulated weight can add several pounds.

The weight chart should show the present weight and center-of-gravity coordinates of a system alongside its allocated weight and coordinates. This reveals where weight trades can be made and allows calculation of a trade's effect on the overall weight and the location of the center of gravity.

11.15 Tools

Three special tools should be designed and fabricated: an array stand to support the array during static charging (unless, as shown in Fig. 11.4, the array is hinged from one side of the car), a jack for changing tires, and a wheel chock. Making the latter is simple and could be done from plastic pipe, as is the chock under the right front wheel in Fig. 11.3, or wood. Figure 11.36 shows a jack in the "up" position. The bar at the top fitted into the socket built into each wheel strut. A lever placed into the horizontal tube was used to pivot the car up or down around the hinge at the bottom.

Fig. 11.37 Finished! (M. Griffin)

11.16 Completion

After many months of meetings, calculations, choices, spending, fund raising, planning, trips, public relations, attention to detail, missed holidays, arguments, and hours of devoted work, your car has been built—at last. Your team has learned a priceless lesson: the meaning of commitment. In this they have become professionals. Figure 11.37 shows the proud moment.

References

Kaw, A. K. (1997). *Mechanics of composite materials*. Boca Raton: CRC Press LLC.

Kyle, C. R. (1990). The sunraycer: Wheels, tires, and brakes, lecture 3.3. In P. MacCready et al. (Ed.), *Sunraycer case history*. Warrendale: Society of Automotive Engineers.

Marshall, A. C. (1998). *Composite basics* (5th ed.). Walnut Creek: Marshall Consulting.

McCarney, S., Olson, K., & Weiss, J. (1987). *Photovoltaics, a manual for design and installation of stand-alone photovoltaic systems*. Colorado: Appropriate Technology Associates.

Photon. (1995). *Soldering techniques, photon technologies, P.O. Box 790*. Maryland: Severna Park.

Photon. (1997). *High performance solar cells for sunrayce '97, Photon Technologies, P.O. Box 790*. Maryland: Severna Park.

Rutan, B. (1983). *Moldless composite homebuilt sandwich aircraft construction* (3rd ed.). Mojave: Rutan Aircraft Factory, Inc.

Storey, J. W. V., Schinckel, A. E. T., & Kyle, C. R. (1993). *Solar racing cars*. Canberra: Australian Government Publishing Service.

Chapter 12
Testing

12.1 Introduction

This chapter discusses ways of measuring the characteristics of a vehicle and that of some of its components. Emphasis has been placed on methods that are straightforward and economical. Wind tunnel testing of models is not economical unless the team can obtain the wind tunnel time as part of a sponsorship. Nevertheless, a section documenting the construction and test of a quarter-scale model of the Table Top concept has been included. It will be useful, should an opportunity to test in a wind tunnel arise.

12.2 Center of Gravity

Refer to Fig. 12.1. The objective of this measurement is to find the coordinates of the center of gravity of a vehicle: $x=a$, $y=c$, and $z=h$. The vehicle front points in the positive x-direction.

The upper drawing of Fig. 12.1 shows the setup for measuring a and c. Scales are placed under each tire to measure the wheel reactions. The wheel base, L, and track, T, must be known. Then:

$$a = \frac{N_B}{W} L \tag{12.1}$$

where, $N_B = N_3 + N_4$, the total reaction of the back (rear) tires (or tire), and taking moments about the x-axis.

$$c = \frac{N_R - N_L}{W} \frac{T}{2} \tag{12.2}$$

Thus, $c > 0$, if $N_R > N_L$, where:

$N_R = N_2 + N_4$ and $N_L = N_1 + N_3$. Note that for a three-wheeled car, the single back (or front) tire reaction does not appear explicitly as its moment about the x-axis is zero.

E. F. Thacher, *A Solar Car Primer*, DOI 10.1007/978-3-319-17494-5_12

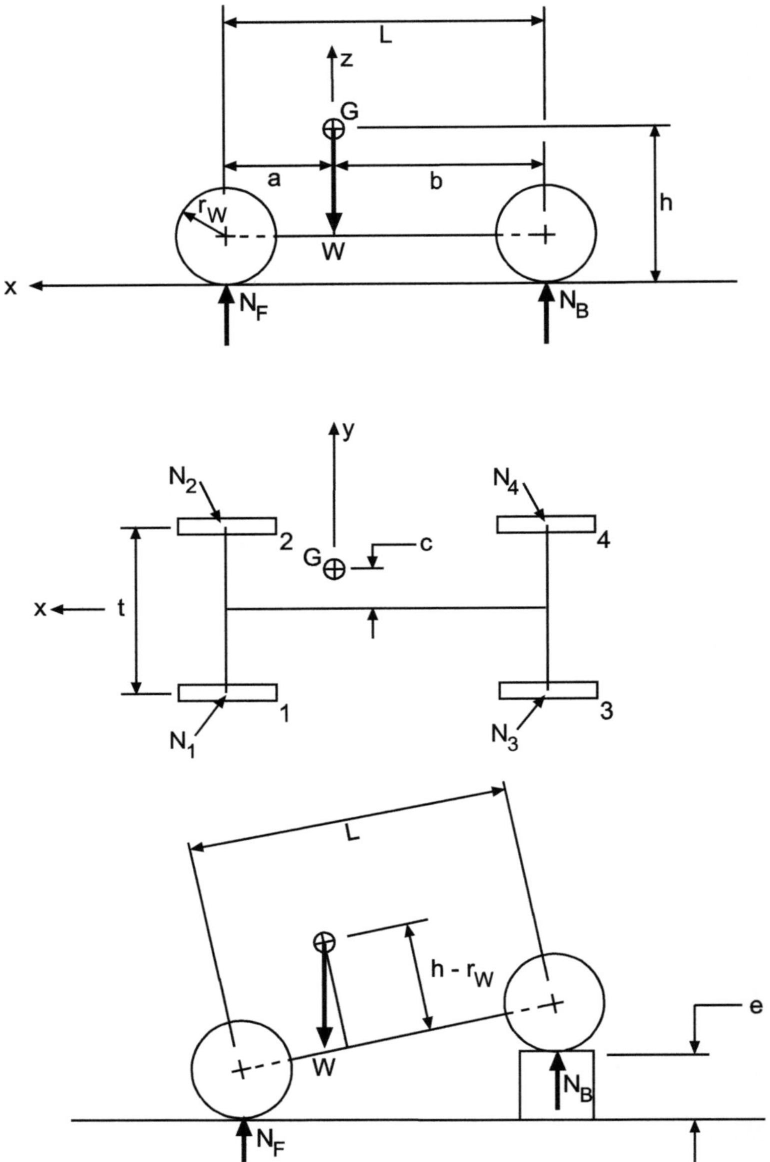

Fig. 12.1 Finding the center of gravity

The lower drawing in Fig. 12.1 shows the setup for finding h for either a three-wheeled or a four-wheeled car. This test must be done after the measurements of a and c as a and W must be known. N_R, e, and the wheel radius, r_W, must be measured.

The wheel radius must be measured at the contact patch. The height of the center of gravity is given by:

$$h = r_W + a\cot\Theta \tag{12.3}$$

where, Θ is the tilt angle, positive up, and

$$\cot\Theta = \sqrt{\left(\frac{L}{e}\right)^2 - 1} \tag{12.4}$$

12.3 Wind Tunnel Testing

Model Testing If wind tunnel services are available, the drag coefficient can be measured on a scale model. The drag area of the full-scale car may then be estimated using the drag coefficient of the model. However, as we have mentioned in earlier chapters, the drag of the as-built car will usually be higher than that of the model. Kurtz (1980) suggests this value to be 10–20 % higher.

Scaling Proper scaling of model measurements requires geometric similarity between the model and the full-scale car, and equality between the Reynolds number at which the scale model is to be tested and at which the full-scale car is to be operated. If the temperature and pressure of the air is the same in both cases, this equality means that the relationship between the air speeds must be:

$$V_m = V_{car}\frac{L_{car}}{L_m} \tag{12.5}$$

where, L is the characteristic length (often the wheelbase) in the Reynolds number. Therefore, if the model is quarter-scale, the airspeed in the tunnel must be four times that at which the full-scale car operates: a 200-mph quarter-scale model test represents a 50-mph car.

Figure 12.2 shows a quarter-scale model solar car under construction. To save weight and to allow easier shaping, the model was mostly made from the foam insulation placed for strength on a plywood frame. The model must be strong enough to take the air loads in the tunnel. The outer surface was sealed with a body filler material, sanded very smooth, and covered with yellow high gloss paint.

Errors The maximum possible number of features of a car should be incorporated in the model. This will yield the most realistic drag coefficient. A corollary to this rule is that the model must be carefully made as the effect of seemingly small flaws will be in a sense amplified by the small scale. For instance, the surface of a quarter-scale model must be quite smooth relative to its dimensions as the height of a small

Fig. 12.2 Constructing model of table top

protuberance on the model is equivalent to one four times as high on the full-scale car.

Figure 12.3 shows the Table Top model mounted in the test section of the Canadian National Research Center's Advanced Aeronautical Laboratory (CNRC-AAL) wind tunnel. The car rests upon a stationary artificial ground plane. When the tunnel is operated, there will be a boundary layer at the ground plane's surface. In the actual case of still air with the car moving, there is no boundary layer on the road surface under the car. The ground plane-boundary layer will alter the pressure distribution about the model, and consequently, the force and moment measurements

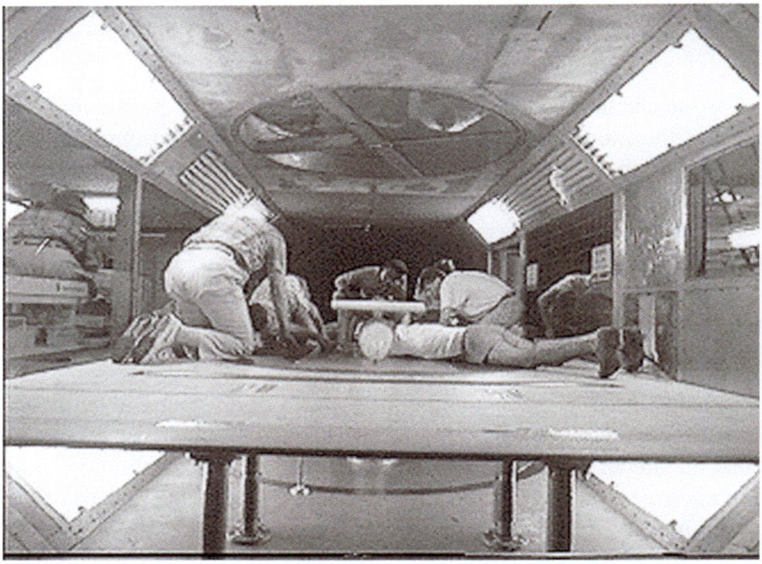

Fig. 12.3 Table top in wind tunnel

(Beauvais et al. 1968). A moving ground plane formed by a moving belt is a solution to this problem. However, such a system is complicated, expensive, not widely available, and not free of problems caused by the belt.

Flow blockage caused by the presence of the model between the tunnel walls increases the velocity past the model above that of the upstream flow and consequently alters the pressure distribution over it. These effects can be kept small by keeping the ratio of the model frontal area to the tunnel cross-sectional area less than about 5 % (Hurst et al. 1983). The tunnel's data reduction scheme also incorporates blockage correction factors.

When on the road, the vehicle rarely travels through still air; that is, the relative wind is usually from a non-zero yaw angle, may be unsteady, may contain large-scale turbulence, and has a velocity increasing from zero at the surface.[1] The first condition can be simulated and should be incorporated into the model test plan. Notice in Fig. 12.3 that the Table Top model rests on a circular plate set on the ground plane. This plate can be rotated to create yaw angles. The latter two conditions are not usually simulated in wind tunnels.

Models and full-scale cars, due to their cost and convenience, are usually tested with stationary wheels. Hucho (1987) asserts that, when the wheels are integrated into the body, the effect of wheel rotation on the forces and moments acting on the vehicle is small. Cogotti (1983) showed, for isolated wheels, that it is more important to maintain the wheel's contact with the ground than to reproduce its rotation, say by elevating it slightly so that it can spin. However, Cogotti also found that streamlining a stationary wheel with a fairing produced a much smaller drag reduction than similarly streamlining a rotating wheel.

Design teams, obtaining testing services for a model at different tunnels, should not be surprised to find different results reported for the same conditions. Hucho (1987) summarized results from a systematic comparison of tests on the same car[2] in ten different tunnels in the USA, Europe, and Canada. The standard deviation about the mean drag coefficient obtained was $\pm 2.2\,\%$, which corresponded to a drag change of ± 0.009. The largest error was 9.3 %, showing that, although the agreement was good on average, large differences could also exist. Moreover, the percentage drag change from modifying the same detail can differ by roughly the same percentage from tunnel-to-tunnel.

12.4 Flow Visualization

"Flow visualization" refers to techniques for making the flow around the car observable. We would like to see where the flow separates from the car, or where interfering flows exist. Then, we can better understand how to reduce or eliminate these flow features.

---

[1] The vehicle operates in an atmospheric boundary layer.

[2] A Volkswagen 1600 notchback, type 31.

Fig. 12.4 Stagnation point

In this chapter, three flow visualization methods will be discussed: colored oils, yarn tufts, and streaming smoke. These meet the criteria of straightforward application. The smoke stream requires the purchase of a smoke generator and a "wand," to create the stream. Thus, it is the most expensive of the three.

Colored Oil The mixture of oil and pigment should be of such consistency that it will, ideally, only run when the desired test conditions exist and leave behind streaks of pigment showing the direction of the flow. There must be sufficient contrast between the streak pattern and the color of the model surface so that the significant details of the flow may be examined and photographed easily.

Maltby and Keating (1960) list (in order of increasing viscosity) kerosene, light diesel oil, and light transformer oil as suitable[3] for low-speed tests. Properly combining one of these oils with a pigment makes a mixture that will start to flow at about 75 % of the test air speed and produce a completely developed pattern in about 2 minutes.

Lampblack mixed with about three times its weight of oil is a suitable pigment for use on light-colored surfaces. The mean carbon particle size should be between 1 and 10 microns. Titanium oxide, a white pigment, may be used for dark colored models. Mix it in about the same proportion as the lampblack. Make a test run on a sample of the model's surface. If the paint does not run at the test speed, use oil with a lower viscosity or dilute it with linseed oil. The wind speed could also be increased if the Reynolds number is not crucial.

Figures 12.4, 12.5, and 12.6 show streaks on the Table Top model and ground plane. The streaks were made by placing small dots or lines of dark-pigmented oil

[3] Note that all three oils are flammable.

Fig. 12.5 Separated flow

Fig. 12.6 Array edge vortex

Fig. 12.7 Smoke and tufts

on the model's surface and the tunnel floor. In Fig. 12.4, the streaks reveal the stagnation region on the pod and the flow near the floor.

The oil streaks cannot be blown through a region where separation occurs. Separated flow is, therefore, shown by the unstreaked areas on the rear surface of the pod and the strut in Fig. 12.5. These areas show where improved streamlining is required. Oil was neither applied to the clear area underneath the pod nor beneath the array.

Figure 12.6 illustrates the kind of detail obtainable with the pigmented oil. It clearly shows the swirling pattern of the vortex on one side of the array near the leading edge.

Tufts Yarn tufts taped to the surface of the car show the pattern of flow (see Fig. 12.7). The kind of flow may also be inferred. In laminar flow, the yarns lie against the surface, pointing downstream, and almost still. Turbulence produces rapid vibrations, but the yarns lie against the surface and point downstream. Zones of separated flow are revealed by large, rapid oscillations of the tufts; some tufts may even point upstream, indicating backflow. The locations of the beginning of turbulence and separation can be observed, approximately, if the yarns are spaced closely enough.[4] Drag measurement should not be done when tufts are attached to the car because the drag of the tufts will alter the measurement. Also, if tufts are placed on a model, the size of the yarn should be reduced by the scale of the model. Two- or three-inch yarn might be used on a full-scale car, and one half to three quarter inch thread on a quarter-scale model.

[4] It is possible that the yarns, if too closely spaced or too large, could trip the flow to turbulence.

The flow about the as-built car, rather than the model, may be visualized during actual driving conditions using tufts as shown in Seal (1991). Drive the car at a steady speed on a day with no wind. Videotape the tufts from another vehicle moving with the test car, if possible, or from a stationary video camera. It would also be possible to use pigmented oil in this way. The viscosity of the oil would have to be reduced to allow it to flow properly at the lower relative air speeds.

Smoke A smoke generator connected to a discharge nozzle, or "wand" produces a stream of white smoke from, for example, propylene glycol. Figure 12.7 shows such a smoke stream revealing the flow over Clarkson's 1995 Sunrayce car, Helios. The smoke stream shows some turbulence in the wake of the car. Notice the yarn tufts taped to the solar array. This was a baseline test, part of the test program described in the next section, that Clarkson's team was allowed to do in the wind tunnel at the General Motors Aerodynamics Laboratory.

12.5 Test Plan

A test plan was prepared in order to maximize the information gleaned from time in the wind tunnel. The major objective was to examine ways of reducing the drag area of the car while improving cockpit visibility. The four objectives for the test were: (1) to determine the pitch angle giving minimum drag with the car in a benchmark configuration; (2) to measure the reduction in the drag by successive smoothing of certain features (the wheel fairing roots, the ventilation intake and exit, the wheel fairings, wheel wells, and the solar array); (3) to measure the drag with each of three bubble canopy designs; and (4) to measure the drag of the car when yawed with respect to the airflow. The drag, lift, side force, pitching moment, yawing moment, and the rolling moment were measured during each run. Objective (5) was to look, during each run, for dangerous conditions such as large pitch, roll, or yaw moment, or a wheel reaction going to zero. After the optimum pitch (0.5 down) was found, the remaining tests were made at that pitch. The test plan was specific, spelling out the configuration, airspeeds, yaw angles, etc., for each test.

Special apparatus supported the test. To carry out objective (1), shims were made that fit into the suspension and changed the pitch of the car and a simple method was devised for measuring the pitch. For objective (2), fillets were made that could be attached to the wheel fairing roots with duct tape, the ventilation system sealed, etc., but in succession, so that both the cumulative and marginal effects of each smoothing could be observed. Dummy canopies were made for objective (3), which could be tested either in the baseline or smoothed condition as only the additional drag would be of interest.

Helios Test Results Figure 12.8 shows the drag coefficient of Helios at three yaw angles. Notice the reduction in drag as yaw increases. This is "sailing."

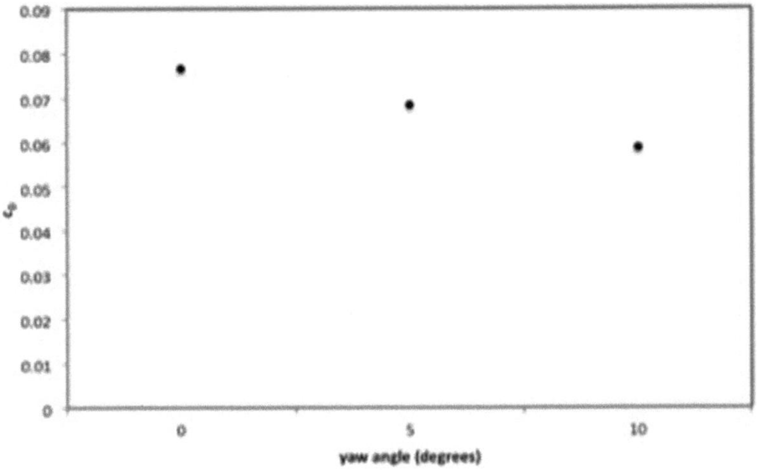

Fig. 12.8 Helios drag vs. yaw angle

12.6 Table Top Model Test Results

Figures 12.9 and 12.10 show wind tunnel data for the Table Top. Note that the drag coefficients shown in the graphs are averages of many readings taken rapidly by the data gathering system of the tunnel. Although the mean flow is steady,[5] at any instant the aerodynamic force on the car is not steady, but oscillates rapidly about a mean value. The oscillations are caused by turbulence, separation, and vortex shedding on the vehicle. Averaging the measurements for a sufficiently long period causes cancellation of the small fluctuations because they are random.

In Fig. 12.9, the filled circles mark drag coefficient values measured at different airspeeds with the model at zero pitch and zero yaw. The decrease in the drag coefficient until about 55 m/s may be caused by the transition from laminar to turbulent flow. After 55 m/s, the drag coefficient increases, but slowly. This is typical of behavior in the transition region; compare the curves in Fig. 2.6 with Fig. 12.9.

Figure 12.10 shows the result of a yaw angle sweep at an airspeed of 73 m/s (equivalent to 18.25 m/s or 65.7 km/h for the full-scale car), at zero pitch. Note that between −5 and 5 ° yaw the drag coefficient hovers about 0.14. At larger yaw angles (only positive yaw angles were recorded above 5° as the model was symmetric) it jumps up as high as 0.16 before decreasing steadily with yaw angle. At 30 ° yaw it is negative, indicating thrust, rather than drag. The Table Top is an eclectic collection of shapes. Subsequently, the jump in the drag coefficient between 5 and 10 ° yaw may indicate that separation occurred on a portion of the car, maybe the pod, but probably not on the large rear wheel fairings. As the yaw angle continued

[5] The flow approaching the model will itself contain some turbulence, although a small amount, in well-made tunnels. It is probably impossible to eliminate it entirely.

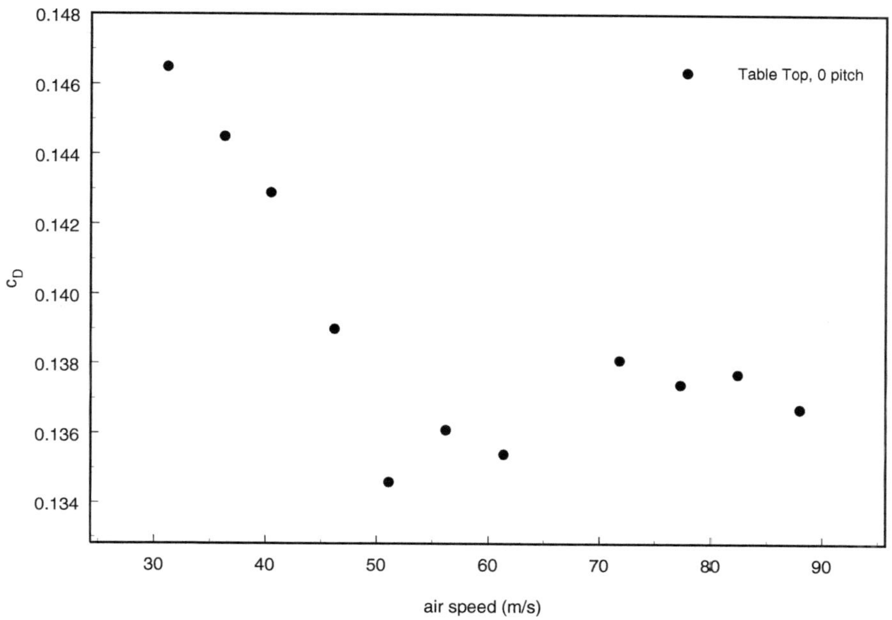

Fig. 12.9 Table top model drag

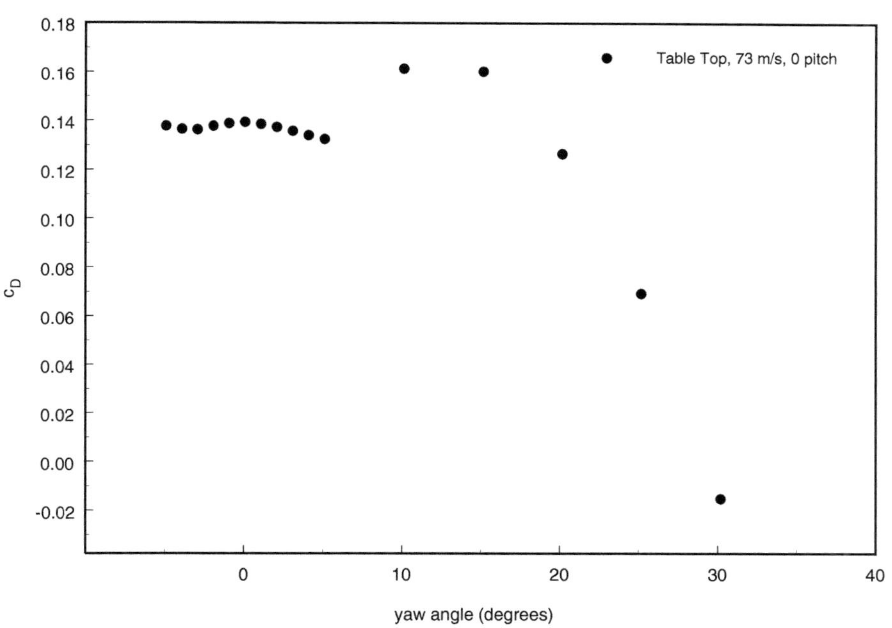

Fig. 12.10 Table top yaw sweep

to increase, the sailing action of the fairings (horizontal "lift" force on the vertical airfoils pointing increasingly in the direction of motion) perhaps overcame the drag increase from pod separation.

12.7 Coast Down Tests

Introduction A coast-down test is a very economical means of measuring the drag area and rolling resistance coefficients of the solar racing car. For many teams, coast-down testing will be the only way to acquire the data they need to characterize the design of their car or to manage it during the race.

Coast-down testing will be especially helpful during the design phase if the new car will be a modification of the preceding one. Then, the effect of shape changes can be ascertained by tests with temporary smoothing appliances made of cardboard, balsa wood, or some other convenient way, attached to the car by duct tape. The drag of the ventilation system can be assessed by noting the drag area difference between a test with the ventilation inlet, and one of the outlet taped closed and the other open.

The rolling resistance will decrease with increasing temperature of the tires, and increase in the roughness of the road surface. When racing, the solar car has no opportunity to warm up its tires before it leaves the starting line each morning. Therefore, the rolling resistance should be measured, both when the tires are cold and when they have been warmed. Warming requires operating the car for at least 20–30 min before the first test, or if the tires have had an opportunity to cool. The rolling resistance should be measured on each of the surfaces of the road course, if possible.

Description The car and driver are carefully weighed. Then, the effective mass is computed from Eq. (22.15).

The ideal operating conditions of a flat, dry road surface, no traffic, and still air will be at first assumed. Some ways of dealing with non-ideal conditions will be presented later. The atmospheric pressure and temperature are recorded before each run. As the name implies, the car is brought to an initial speed of 55 mph, or at least 40 mph, the motor is disconnected from the driven wheel and the car coasts straight ahead until it reaches 15–25 mph. This may take more than a minute. During the coast-down, the vehicle's speed is recorded as a function of time, and the deceleration, a, is computed from the speed–time data.

A relatively low-budget way of obtaining the speed–time data[6] is to temporarily install a video camera in the car. The camera should be able to record the time of each video frame, and be positioned to record the speed indicated on the car's speedometer. If the camera cannot, however, record the time, place a watch with a sweep second hand in the camera's field of view. An inexpensive global positioning system (GPS) receiver can serve quite well as a speedometer. A cheaper option is a

[6] Buckley (1974) contributes non-low-budget ideas for other measurements.

digital bicycle speedometer. Korst and Funfsinn (1977) used a small electric generator attached to the axle of one wheel such that it produced a voltage proportional to speed.

SAE (1996) gives a detailed discussion of a standard coast-down test procedure that includes a listing of a computer program for reducing the data. The SAE and Korst and Funfsinn (1977) use a rolling resistance model that depends on the square of the speed, rather than the linear model employed herein.

Equation (2.19), with the tractive force and the gravity term (T and G) set to zero, and solved for the rate of change of momentum, when equated to a quadratic, gives:

$$M_e a = R + D = c_0 + c_1 V + c_2 V^2 = -\mu_1 W - \mu_2 WV - \frac{1}{2}\rho c_D A_D V^2 \tag{12.6}$$

The deceleration values and the corresponding speeds are used to select c_0, c_1, and c_2 so that the sum of the squared error

$$e^2 = \sum_{i=1}^{i=N} \left[M_e a_i - \left(c_0 + c_1 V_i + c_2 V_i^2 \right) \right]^2 \tag{12.7}$$

from each data point is minimum. This is called "curve fitting" because the constants in the quadratic are being chosen to make that expression predict $M_e a$ as closely as possible along the speed–time curve. Any spreadsheet program can be used to perform such a curve fit.[7] Once the constants are found, they are given meanings based on Eq. (12.6), as follows:

$$\mu_1 = -\frac{c_0}{W} \tag{12.8}$$

$$\mu_2 = -\frac{c_1}{W} \tag{12.9}$$

$$c_D A_D = -\frac{2c_2}{\rho} \tag{12.10}$$

12.8 Deceleration

A good way to calculate the deceleration is to fit a polynomial function of time[8] to the speed–time data. There will be scatter in this data caused by, among other things, pairing runs in opposite directions to minimize the influences of wind and grade. The best-fit polynomial will smooth this scatter, making it easy to find the

[7] The spreadsheet program probably calls it "regression analysis" or something similar.

[8] If the speed is represented by the "Nth-order" polynomial $a_0 + a_1 t + a_2 t^2 + \ldots + a_N t^N$, the deceleration would be $a_1 + 2a_2 t + \ldots + N a_N t^{N-1}$

deceleration: it is the slope of the graph of the polynomial at any time. Second-, third-, or may be fourth-order polynomials should be tried to find the best one. Compute the mean squared error and see when it is the smallest.[9] Always plot a graph of the polynomial and the data together. If the graph of the polynomial passes through, or near, each data point but wiggles in between, use the next lower-order polynomial, even if its *correlation coefficient,* R^2, is smaller. Keep reducing the order until you get the best-fitting polynomial having a smooth graph. Also, note that all the "c" coefficients must be negative.

12.9 Number of Tests

Plan to make at least 12 good runs, but make more, if possible. The more data collected, the better the fluctuations from run-to-run in the calculated coefficients average out. Then the mean of the run results for a parameter will better reflect the actual value. Runs in opposite directions should be made, in case the fluctuations are biased in some way by the direction of travel. So of the 12 runs, 6 should be one way and 6 the other. Another reason for making lots of runs is that the data of a run may have to be discarded for a reason that appears later.

12.10 Profile Area

The profile area, A_D, must be known before c_D can be calculated. A method of measuring A_D is to photograph the car or model from the front and the side and find A_D by scaling it from the photograph. The camera must be far enough away to reduce the distortion caused by perspective. Pershing and Masaki (1976) enlarged their 35 mm black and white negatives to 8×10 before taking measurements. The vehicle or model should be photographed against a contrasting background. A square reference surface of known area must be located on the car at a known spacing. The square is photographed with the front view.

The side view is used to aid in correcting the front view for the apparent narrowing of the parts of the car farther from the reference plane (the plane containing the reference square).

No perspective correction is necessary[10] for the parts of A_D visible in the front view photograph. Otherwise, all parts of the profile of A_D, not in the reference plane, should be orthographically projected onto it using their spacing from the plane. The

[9] Or else, note when the square of the correlation coefficient is closest to 1.0 as calculated by the spread sheet.

[10] It is possible that the widest and tallest parts of the car may not be at the same axial location. Select one or the other as the reference location and project into that plane.

side view assists in identifying these features and their spacing. This is illustrated in Fig. 12.11. Length scales photographed with the car can assist in this correction.

After correcting for perspective, the corrected outline of the car and the reference square are very carefully cut out of the photograph and accurately weighed. The profile area of the car is calculated from:

$$A_D = (area\ of\ square)\,\frac{corrected\ image\ weight}{square\ image\ weight} \tag{12.11}$$

Tracing the corrected image with a planimeter, if available, is an alternate method of measurement.

Pershing and Masaki assert that if the uncertainty in A_D is to be not more than $\pm 2\,\%$, and the uncertainty in the location of A_D is no more than $\pm$ 1 m, the spacing of the camera from the reference cross section should be at least 100 m. Kurtz (1980) recommended a spacing of 200 m for full-scale cars and a telephoto lens of 1000 mm focal length. For models, the spacing may be reduced by the model scale, i.e., to 50 m for a quarter-scale model.

Low-Budget A_D This is a more accurate version of the method used in Chap. 9. Print the engineering drawing of the front view of your car on paper which has a rectangular grid on it. Count the number of squares contained within the front view. Use the scale of the drawing to calculate the profile area. Using very small squares, compared to the size of the front view, makes the calculation more accurate. Again, a planimeter could also be used.

12.11 Grade

Constant Grade If the course is not flat, the next best condition is one with small, constant slope. The slope should be small so that the car will still decelerate to 15–25 mph when coasting the downhill way. For a constant up-slope (positive, weight retards motion), add $-W \sin c$ to the right-hand side of Eq. (12.6). Reverse the sign for a down-slope. Multiply the rolling resistance terms by $W \cos \alpha$ rather than W. A constant slope changes the interpretation of c_0 and c_1 to:

$$\mu_1 = -\frac{c_0}{W\cos\alpha} \pm \tan\alpha, \quad \mu_2 = -\frac{c_1}{W\cos\alpha} \tag{12.12}$$

where, the positive sign denotes a down slope. However, if runs are made in each direction, the $\tan\alpha$ term tends to cancel because its sign reverses for up-hill and down-hill runs.

Varying Grade If the test track is of mixed grade, the grade must be known as a function of distance. Then, the motion of the car over the test course must be

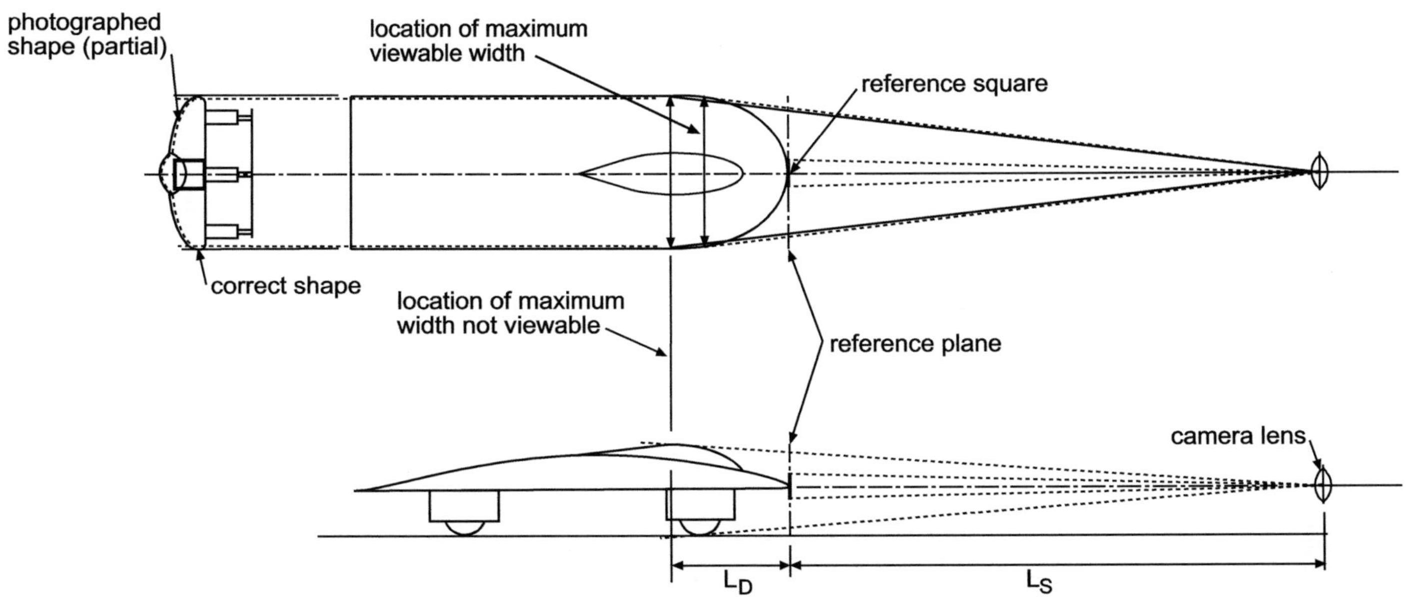

Fig. 12.11 A_D imagery (after Pershing and Masaki 1976)

simulated numerically and an optimization procedure used to select the drag area and rolling resistance coefficients that give the best fit to the measured data. This procedure is explained by Swift (1991).

12.12 Sample Results

Figures 12.12a and b show results for one run of an actual car coasting down a slight constant grade. Note the scatter in the data, for the most part, is small and that many points were taken. These two characteristics greatly improve the results.

12.13 Wind

Measurement The relative wind speed and direction may be measured by placing an anemometer on the solar car, elevated sufficiently to remove it from the influence of the flow about the car. This is complicated, expensive, and may increase the drag of the car by 4–6% (Passmore and Jenkins 1988). An alternative procedure is to measure the wind by a relatively cheap, battery-powered anemometer (which could be put in a temporary, home-built stand) located about the midpoint of the coast-down course. The wind speed and direction should be recorded about every 5 s during the coast-down, but at least as frequently as the speed. Find the relative wind and yaw angle at each recorded solar car speed from:

$$V_R = \sqrt{V_W^2 + V^2 - 2V_W V \cos\left(C_W - C_V\right)} \tag{12.13}$$

where, V and C_V are the speed and true course of the car at any time, and V_W and C_W are the average wind speed and true course of the wind, respectively. The wind direction, D_W, is always measured as the direction from which the wind is coming, measured clockwise from true north; whereas, the course of the car is the direction in which it is traveling, also measured clockwise from true north. To convert D_W to C_W follow this rule:

$$C_W = D_W + \begin{Bmatrix} -180^\circ, & D_W \geq 180^\circ \\ +180^\circ, & D_W < 180^\circ \end{Bmatrix} \tag{12.14}$$

The yaw angle, β, may be found from:

$$\beta = \sin^{-1}\left|\frac{V_W}{V_R}\sin\left(C_W - C_V\right)\right| \tag{12.15}$$

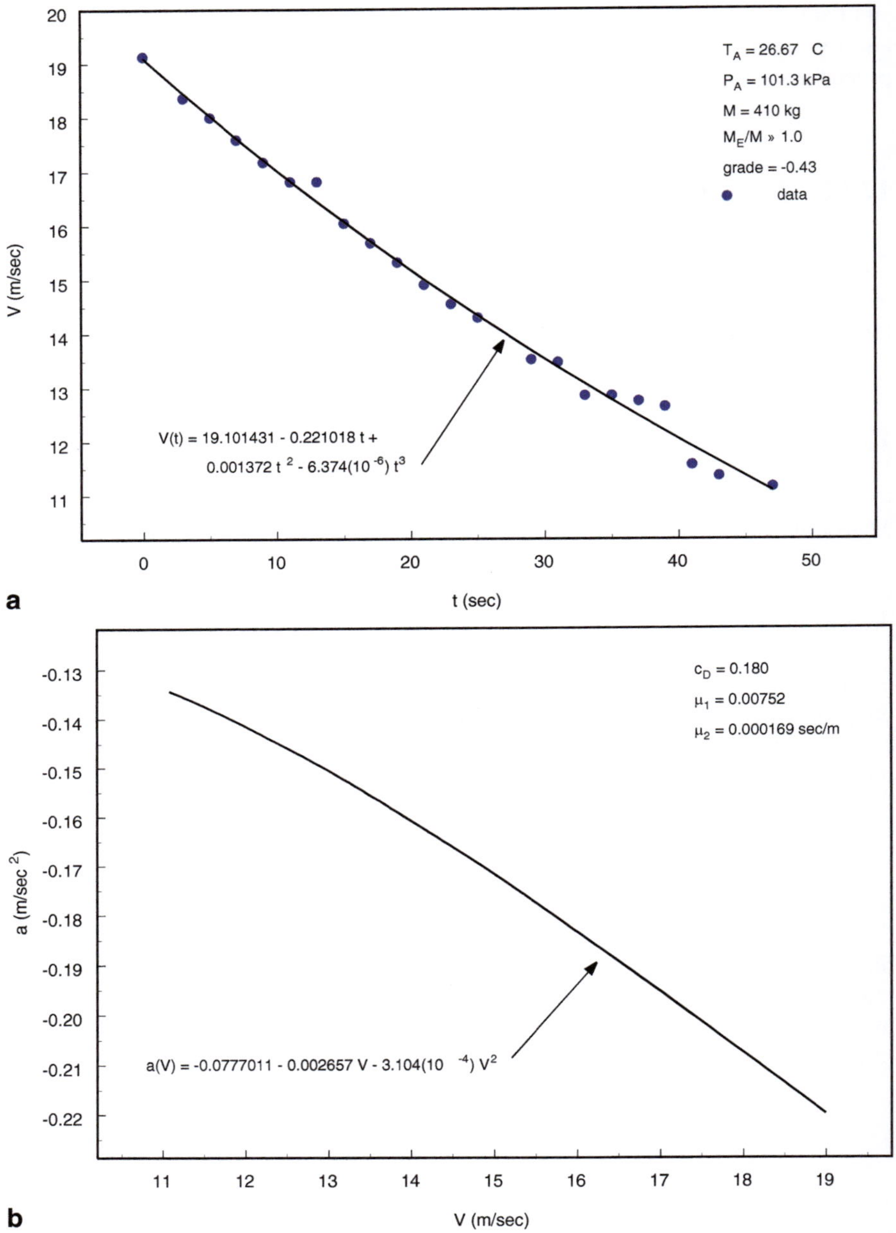

Fig. 12.12 **a** Coast-down data and curve fit. **b** Results from coast-down test fit

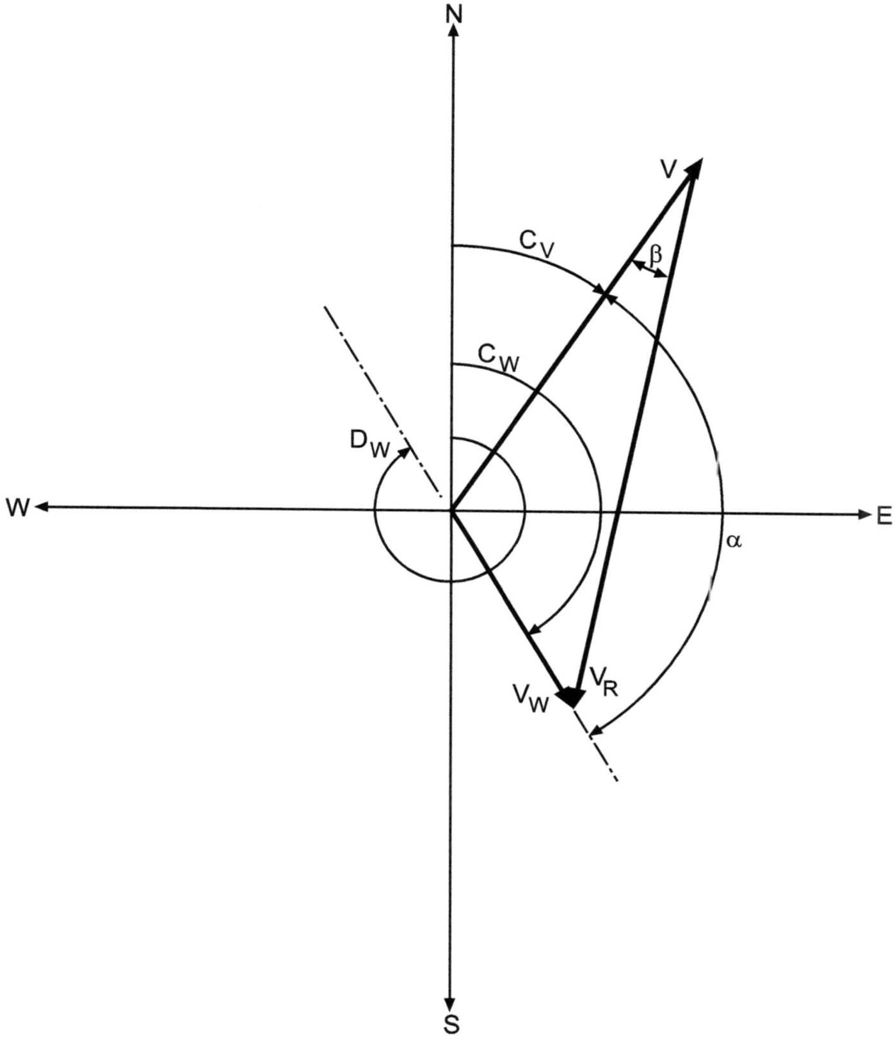

Fig. 12.13 Relative wind

If $C_W > C_V$, the relative wind will be from the left of the course of the car. Figure 12.13 shows these relationships.

Yaw Effects A non-zero yaw angle means there is a side force. This causes a slip angle and the driver must compensate; thus, increasing rolling resistance. This is a small effect, and following Passmore and Jenkins (1988), will be neglected. However, the effect of yaw on the drag cannot be neglected. We adopt the model used by Swift (1991) that was based on the test of Bowman (1966).

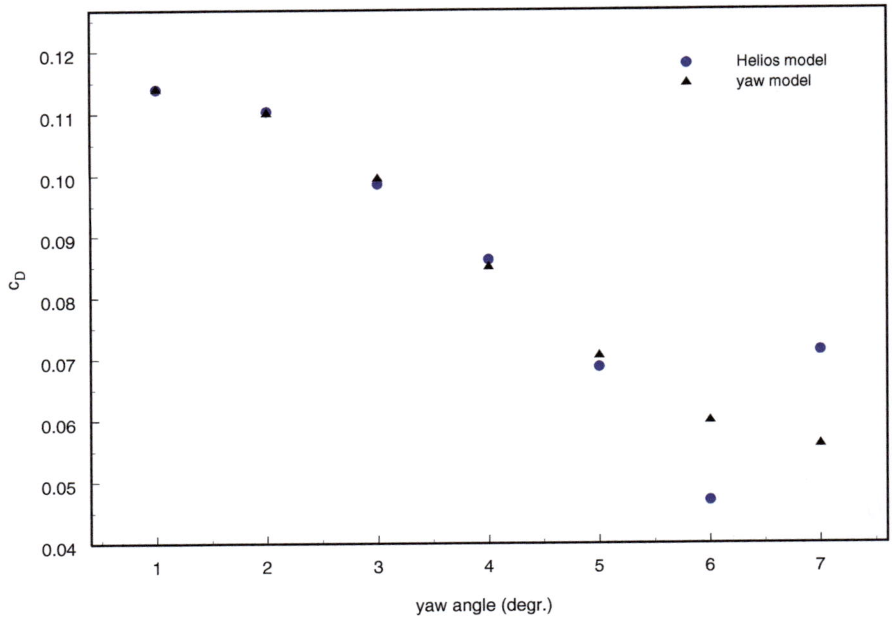

Fig. 12.14 Yaw model comparison

$$c_D(\beta) = c_{D_0} + k_D \sin^2(2\beta) \tag{12.16}$$

Figure 12.14 shows Eq. (12.16), with k_D set to -0.0581/degree, plotted with yaw data from the test of a model of the Shark in the Glenn L. Martin wind tunnel at the University of Maryland. The agreement is good out to a yaw angle of 20 °.

Lift Effects Lift may reduce or increase the ground reaction on a tire hence altering the rolling resistance force.

The lift coefficient must be known beforehand to compensate for this. Passmore and Jenkins (1988) recommend neglecting this effect, which is usually small. Lift is usually always present to some extent, so this will introduce an error in the rolling resistance coefficients. If the net lift is up, the measured rolling resistance coefficients will be smaller. The reverse case occurs when the net lift is negative.

12.14 Ambient Temperature

As the static rolling resistance is function of the tire temperature, it can be correlated with the ambient temperature. Passmore and Jenkins (1988) used the following relation to model this correlation:

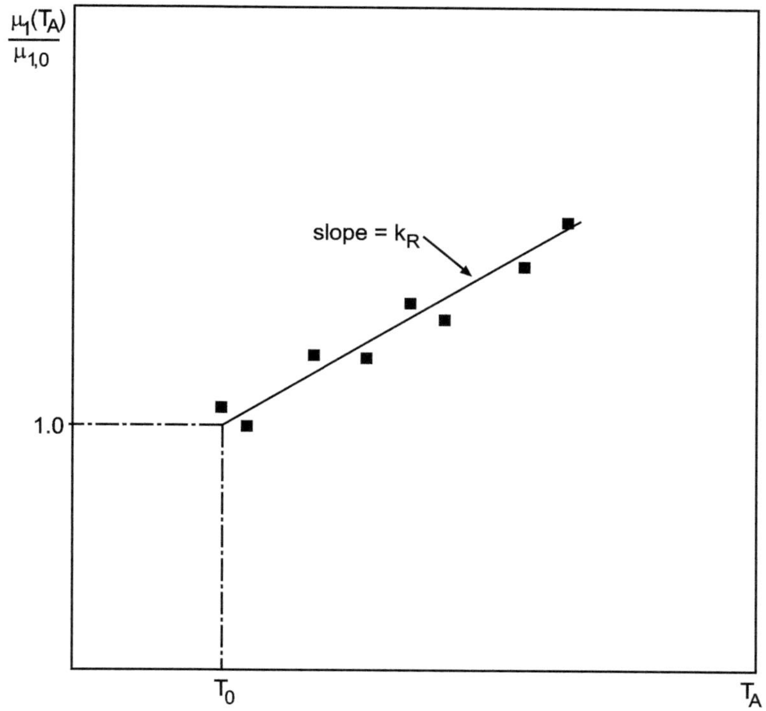

Fig. 12.15 Finding k_R

$$\mu_1(T_A) = \mu_{1,0}\left[1 + k_R(T_A - T_0)\right] \tag{12.17}$$

where, T_A is the ambient temperature, $\mu_{1,0}$ is the coefficient at the reference temperature T_0, and k_R is a constant, which Passmore and Jenkins set at 0.013 1/C°. It is found from coast-down test data taken at different ambient temperatures, as Fig. 12.15 illustrates (synthetic data). Other test conditions should be the same.

12.15 Problems

Safety Roll-down testing may involve operating the solar car on public roads in traffic. In such cases, the car should be under the control of a chase vehicle at all times. Radio communications between the driver of the solar car and the test conductor must be established for the entire period of the test.

The car must be legally entitled to operate on a public road.

Clutch A solar racing car may not have a clutch to disengage the drive from the driven wheel after the initial speed has been reached. Or, the car may be fitted with a wheel-mounted motor directly driving the wheel. If for either reason the motor cannot be disconnected from the wheel, the rotational inertia and mechanical losses of the drive will apply an additional retarding torque to the driven wheel. The motor should be electrically disconnected.

One could tow or push the car to the test speed and then cast-off. In the first case, the car would coast-down in the wake of the towing vehicle, which would introduce error in the drag measurement. In either case, the car should have convenient and safe towing or pushing features built-in from the outset. If a hill of sufficient slope or length followed by a flat stretch of sufficient length is available, the car could use the hill to attain an initial speed in the desired range.

Parasitic Torque The drive's mechanical parasitic torque is, in general, a function of rotational speed; so, separating it from drag and rolling resistance requires an independent measurement of it. More on this is given in the Sect. 12.18.

12.16 Low Speed Test

The objective of the low-speed coast-down test is to measure μ_1 and μ_2. Aerodynamic drag on the rotating wheel makes a small contribution to the rolling resistance, and this test will virtually ignore it, so the speed-dependent rolling resistance coefficient will be slightly less. With the rolling resistance coefficients independently measured, and the no-load losses of the drive measured as described later, the car's motor can be used to reach the initial speed required for the high speed coast-down test.

The time, t_S, to coast-down to zero speed when drag is negligible is:

$$t_S = \frac{M_e}{W\mu_2} ln\left(\frac{\mu_1 cos\alpha + \mu_2 V_0 + sin\alpha}{\mu_1 \, cos\alpha + sin\alpha}\right) \tag{12.18}$$

where, V_0 is the speed entering the coast-down course. The time to zero speed may be recorded and used to find the rolling resistance coefficients. However, the exact moment when the speed is zero may be less precise than a distance-traveled measurement. The distance traveled to zero speed, S, is:

$$S = \frac{M_e\left(\mu_1 \, cos\alpha + sin\alpha\right)}{\mu_2^2 W} ln\left(\frac{\mu_1 cos\alpha + \mu_2 V_0 + sin\alpha}{\mu_1 \, cos\alpha + sin\alpha}\right) + \frac{M_e v}{\mu_2 W} \tag{12.19}$$

Suppose that several runs are made, m of them. For each, S, V_0, and v (the coast-down distance, initial velocity, and the kinematic viscosity of the air) are recorded. The values of μ_1 and μ_2 sought are those minimizing the sum of the squared errors.

$$e^2 = \sum_{i=1}^{i=m} \left[S\left(\mu_1, \, \mu_2, \, V_{0,i}, \nu_{0,i}\right) - S_i\right]^2 \tag{12.20}$$

Preparation The preparation for the low-speed test is similar to that for the high-speed test. Except that a temporary mechanism connecting the driving wheel to the chassis should replace the motor, if it is wheel-mounted. The test can be done in an area sheltered from the wind and of relatively small extent because the car will not coast very far.

Push the car straight ahead and fast enough so that it coasts a few car-lengths. Initial speed should not be more than about 2 m/s. The initial maximum instantaneous drag will then be less than about 2.3 % of the total force on the car, and of course, decreases from that value. The initial speed and coasting distance must be recorded for each test. If the initial speed cannot be recorded, record the time to coast-down and the distance, and use Eq. (12.18) to calculate (or eliminate) V_0. The as-tested weight of the car must of course be known.

12.17 Correlation of Wind Tunnel and Coast-Down Tests

In a wind tunnel, the vehicle is at rest on the ground plane (some suspension vibration will be present, but it will be less than that on the road), all rotating parts are still, the test chamber walls are present, and the velocity distribution and turbulence level of the air will usually be different from that experienced during coast-down tests. However, the correlation between the drag coefficients measured by wind tunnel tests and coast-down tests can be good. The coast-down test results must be corrected for important effects not present in a wind tunnel test. Tunnel results are routinely corrected for blockage.

Morelli et al. (1981) reported negligible errors introduced by energy absorbed in suspension motions, and by the effect of braking the air mass in the coast-down vehicle's wake. They found the effect of wheel rotation to not be negligible, but this effect is included in the rolling resistance measured during coast-down. The corrected drag coefficients of three identical vehicles, excluding ventilation drag measured by coast-down, agreed with wind tunnel results to within an average error magnitude of 0.71 %. The authors assumed a V^2 speed-dependence for the rolling resistance. Consequently, the speed-dependent part of the rolling resistance had to be measured independently so that it could be separated from the drag.

The results summarized above increase confidence in coast-down tests as a source of potentially accurate drag coefficient and rolling resistance coefficient information.

12.18 Drive Losses

Model The contact patch force equivalent to the windage and friction in the drive at no load was modeled by Passmore and Jenkins (1988) as a linear function of speed.

$$F(V) = W\,(b_0 + b_1 V) \tag{12.21}$$

Measurement A method for measuring the constants b_0 and b_1 was outlined by Hoerner (1950). The driven end of the car is elevated on a jack. The driven wheel is spun up using the motor, then the power is shut off and the drive allowed to coast to zero rotational speed. The initial rotational speed and the time to zero rotational speed are recorded. The test is repeated several times to gain statistical confidence. The moment of inertia of the entire drive (motor, sprockets, and chain) and the driven wheel must be known. A solution for the coast-down time, t_S, is:

$$t_S = \frac{I_E}{r_W^2\, W\, b_1}\, ln\left(1 - \frac{\omega_{W,0}\, r_W\, b_1}{b_0}\right) \tag{12.22}$$

where, I_E is the moment of inertia of the drive, r_W is the wheel radius, and $\omega_{W,0}$ is the initial angular rotation rate. An error function, similar to Eq. (12.20), is defined to express the sum of the squared error between the measured and predicted coast-down time. The constants b_0 and b_1 are chosen to minimize this sum. Note that these constants will include wheel-bearing torque and aerodynamic drag on the wheel. However, these are small compared to the effects of the drive.

12.19 Correction Summary

To summarize the preceding discussions, Eq. (12.6) below is corrected for wind yaw angle, constant up-slope, ambient temperature, and drive losses.

$$M_e a = -\frac{1}{2} c_D\,(\beta)\; A_D \rho V_R^2\,(\beta) - [\mu_1\,(T_A)\; cos\,\alpha + \mu_2 V\,]\; W - W sin\alpha - F\,(V) \tag{12.23}$$

where, $c_D(\beta)$ would be given by Eq. (12.16), $V_R(\beta)$ by Eq. (12.13), $\mu_1(T_A)$ by Eq. (12.17), and $F(V)$ by Eq. (12.21). The relative wind and the yaw angle would be measured during the coast-down. Note that if yaw data available was zero, the yaw coefficient, k_D, could be found from coast-down data.

12.20 Battery Energy Rate

The coast-down testing allows the energy per unit distance delivered to the driven wheel to be found at any speed by Eq. (2.17). Another important piece of information about the car is the battery energy it consumes per unit distance, e_{BATT}, when unaided by the sun. Measurement of this parameter may be done as outlined below.

First, locate and survey a standard road course. The course should have a mixture of terrain features typical of a race. It should be about 80 km long so that a significant amount of battery energy will be consumed when traversing it. Document the road surface, grades, and direction, and the distances to each such feature.

Disconnect the solar array from the battery bus and drive the course at some average speed. Record the time, the battery bus voltage, and the corresponding ampere-hour reading (and other potentially useful data such as the battery and motor currents). Measure the barometric pressure, wind speed, and wind direction during the test. Repeat this procedure several times at different average speeds (perhaps by changing climbing speeds on certain hills) and compute the battery energy reduction for each run. Compute the battery energy consumed per unit distance as:

$$\bar{e}_{\text{BATT}} = \frac{\Delta E_{\text{BATT}}}{S} \tag{12.24}$$

where, ΔE_{BATT} is the energy removed from the battery and S is the total distance driven.

This procedure could be adapted to separate the effect of the driver from the performance of the car or to determine the energy consumption differences between drivers.

12.21 Solar Radiation Measurement

Intensity on a Surface To measure the global solar radiation on the surface of the array requires an instrument called a *pyranometer*. Commercially-made pyranometers are expensive. However, a solar cell may be made into a pyranometer using the upper circuit of Fig. 3.14 at a far less expense. When a cell is purchased, the information furnished with it should include the short circuit current at standard temperature and irradiance conditions. To make it into a pyranometer, obtain a small, battery-operated, digital multimeter. Mount the cell on a piece of perforated circuit board and connect its positive (rear) and negative (front) buses to flexible, insulated wires terminated with plugs that will fit into the receptacles on the multimeter. Set the multimeter to read the short circuit current from the cell, which is proportional to the irradiance normal to its surface.

$$G = K_{\text{G}} I_{\text{SC}} \tag{12.25}$$

where, the symbols have the same meaning as in Chap. 3.

$$K_{\text{G}} = \left(\frac{1000}{I_{\text{SC}}}\right)_{rated} \tag{12.26}$$

The instrument's reading will change with the cell's temperature due to the small temperature-dependence of I_{SC}, and with changes in the spectral content of the irradiation from the rated conditions. Nevertheless, it will measure the irradiation on the array to within a few percent. To improve this, K_G should be recalculated using measurements from a commercial instrument ("source"), if the opportunity arises:

$$K_{\mathrm{G}} = \left(\frac{1000}{I_{\mathrm{SC}}}\right)_{rated} \left(\frac{G_{source}}{G_{cell}}\right) \tag{12.27}$$

Solar cells are *very* fragile and expensive. Even small cells purchased at electronics parts stores cost about $ 10.00. Allow for thermal expansion by the cell when gluing it to the board. The wires from the cell should be connected to binding posts, not directly to the wires to the meter, and should have bends in them to allow for thermal expansion. Use no larger than number 24 braided copper wire. Solid copper wire soldered to the cell can *easily* crack it when the wire is manipulated to attach it to the binding post. Make a slot or hole in the PC board for the positive wire. The slot allows the cell to lie flat on the PC board; thus, preventing cracking by contact with the positive wire.

Angle of Incidence Figure 12.16 shows a device for measuring the angle of incidence on any flat surface. Knowing the height of the threaded rod, measure the length of the rod's shadow, *L* (or rule the disk's surface in concentric rings) and find the angle of incidence as:

$$\Theta = tan^{-1}\frac{L}{H} \tag{12.28}$$

The device will not work well when there is a high percentage of diffuse irradiance as the shadow of the rod will not be well defined.

There should be provisions (pads, for example) for preventing damage to the surface upon which the device rests (which could be covered with solar cells). Special care must also be taken to make the rod perpendicular to the disk's surface.

12.22 Solar Cell Tests

Cell I–V Curve and Efficiency The best environment for testing solar cells is a clear, calm day, an air temperature of 25 °C, and a time within 1 h of solar noon. Use the lower circuit of Fig. 3.14. Place the cell on a flat surface away from other objects or buildings that could reflect significant light on the cell. Connect the cell to a "decade box." A decade box contains precision resistors that can be selected by switches to give values of resistance ranging from fractions of an ohm up to several decades higher. Connect a voltmeter across the decade box. Record the irradiation on the cell and the air temperature. Change the resistance from nearly a short to effectively an open circuit. Record the voltage and resistance at each setting of the box. After the last resistance setting, record the irradiance and air temperature again

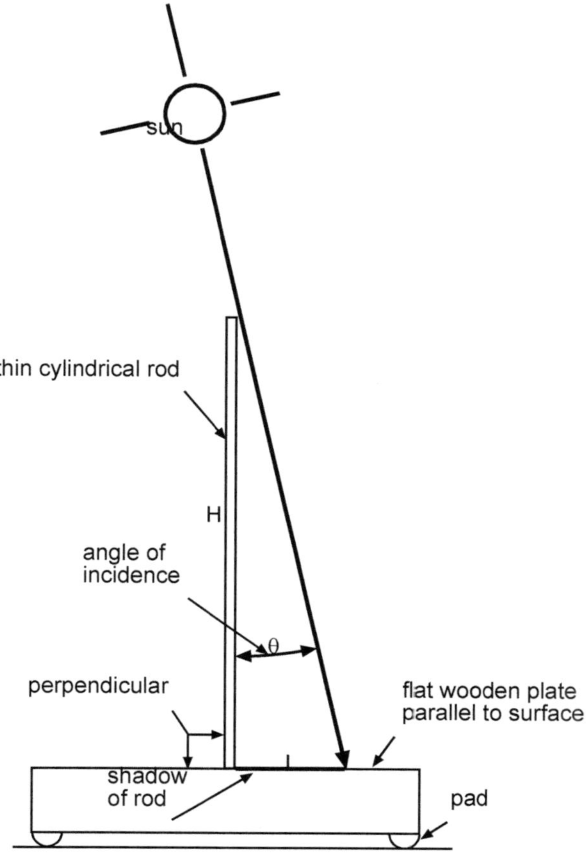

Fig. 12.16 Incidence angle measurement

and average them with the first readings. Calculate the efficiency of the cell at each data point. Plot the current, power, and efficiency as functions of the voltage across the decade box.

Indoor Testing Testing indoors confers the advantage of control of the test environment. The problem that comes with indoor testing is that artificial light only approximates the spectral distribution of solar radiation. Quartz-halogen projection lamps are an economical light source that approximate the spectral distribution of solar radiation and come equipped with a reflector. They are intended for use in equipment, such as overhead projectors, and may require non-standard voltages.

The electrical setup is similar to outdoor testing. The projection lamp is powered through a variable transformer, a "variac," to not only provide the non-standard voltage[11] but to also allow the irradiance level to be adjusted. A wide range of

[11] Be sure not to exceed the rated voltage of the lamp; they are quite easily burned out by this. Have some replacements in hand.

adjustment should be avoided, as major reductions in the lamp's operating voltage will cause its light to redden. To avoid stray light, the apparatus should be put into a test enclosure, such as a cardboard box, with blackened walls. In this case, air flow should be provided to cool the lamp, which runs very hot, and to keep the cell under test at a constant temperature.

Solar Array Efficiency The pyranometer described above can be used to find the total radiation striking the array at some particular array orientation. The array should be operating with a load, such as when charging the battery bank. Or, the load could be a decade box, rated for the power of the array. In this case, the array I–V characteristic could be obtained.

Work efficiently so that measurements are taken before the solar conditions change enough to introduce significant error. On a mostly clear day, the best time of day for this is within 1 h of solar noon. However, the characteristics of the array when charging in the morning or evening are of interest.

Record the voltage and current on the main bus and calculate the power being delivered ($V_B I_B$). Then, using the pyranometer, measure the irradiance normal to each array facet. Multiply a facet's irradiance by the gross area of the facet. The sum of the results from all the facets is the rate of solar energy delivery to the array, P_{sun}. As soon as all the facets have been surveyed, record the bus voltage and current again and average it with the first values taken. The efficiency in percentage at about the midpoint of the measurement time interval is:

$$\eta_{array} = 100\frac{V_B I_B}{\bar{P}_{sun}} \tag{12.29}$$

where, the overbar on P_{sun} denotes the average value.

12.23 Constant Radius Steering Test

This test determines at what lateral accelerations, a_y, the solar car understeers, has neutral steering or oversteers (Wong 1978).

Procedure Drive the car along a circular path of radius R at different constant speeds, V, in succession. Record the speed, the angle, δ, of the steering wheel required. The lateral acceleration is $a_y = V^2/R$. Plot steering wheel angle versus lateral acceleration in g's (a_y/g). If, over a range of a_y's, the steering wheel angle is the same, this is neutral steer. If the steering wheel angle increases with a_y, this is understeer. And, if the steering wheel angle decreases as a_y increases, it shows oversteer. All three may appear on the plot; see Figure 12.17 for an illustration. See the discussion on steering in Chap. 21.

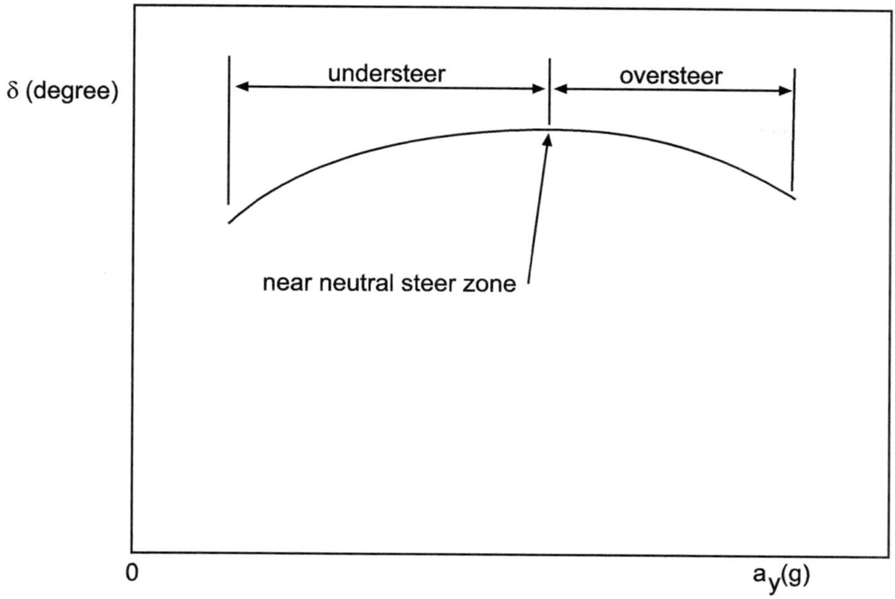

Fig. 12.17 Qualitative steering characteristic

12.24 Cross-Wind and Roll Stability

Dominy (1991) presents a general procedure for cross-wind testing of models if wind tunnel facilities are available. A basically similar, but highly qualitative, test was performed by Clarkson University prior to the first Sunrayce. Clarkson's car was driven through the propeller wash of an aircraft and its tendency to deviate from a straight path was noted. The velocity distribution in the wash was sampled using a hand-held anemometer.

Van Valkenburgh et al. (1982) describe roll stability testing of three-wheeled cars. The techniques used would be applicable to four-wheeled vehicles.

References

Buckley, B. S. (1974). Road test aerodynamic instrumentation. *Proceedings of Automobile Engineering Meeting*, Toronto, Canada, October 21–25, 1974, SAE paper 741030.

Cogotti, A. (1983). Aerodynamics of car wheels. Technological advances in vehicle design series, SP3. In M. A. Dorgham (Ed.), *Impact of Aerodynamics on Vehicle Design*, (pp. 173–196). Interscience Enterprises Ltd: La Motte Chambers. (Int. J. of Vehicle Design).

Dominy, R. G. (1991). A technique for the investigation of transient aerodynamic forces on road vehicles in cross winds. *Journal of Automobile Engineering, 205*, 245–250. (Proceedings of the Institution of Mechanical Engineers vol *205, Part D*).

Dyke, M. (1960). *Flow visualization*. Library Bib'y. No. 222. Royal Aircraft Establishment, Ministry of Aviation, London.

Hoerner, S. F. (1950). Determining drag of an automobile under actual road conditions. *Automotive Industries*, Sept 15, pp. 43, 96, 100.

Hucho, W.-H. (1987). Wind Tunnels for Automobile Aerodynamics. In W.-H. Hucho (Ed.), *Aerodynamics of Road Vehicles* (pp. 398–435). Boston: Butterworths.

Hurst, D. W., Allan, J. W., & Burgin, K. (1983). Pressure measurements on a full-scale tractor-trailer combination and comparison with data from wind-tunnel model tests. *International Journal of Vehicle Design*. Technological advances in vehicle design series, SP3. In M. A. Dorgham (Ed.), *Impact of Aerodynamics on Vehicle Design* (pp. 471–479).

Lucas, G. G., & Britton, J. D. (1970). Drag Data from deceleration tests and speed measurement during vehicle testing. In A. J. Scibor-Rylski (Ed.), *Proceedings of First Symposium on Road Vehicle Aerodynamics, The City University, 6 & 7 November, 1969*. London.

Korst, H. H., & Funfsinn, M. A. (1977). Determination of Effective Rolling Resistance by Coastdown Experiments on Smooth and Rough Roads. In R. H. Snyder, (Ed.), *Tire Rolling Losses and Fuel Economy, SAE Conference Proceedings. P-74*, (pp. 133–141). Society of Automotive Engineers, Warrendale.

Kurtz, D. W. (1980). *Aerodynamic design of electric and hybrid vehicles: A guidebook*. Jet Propulsion Laboratory: JPL Publication. (80–69 (NASA-CR-163744 and NTIS N81-12943).

Maltby, R. L. & Keating, R. F. A. (1960). *Flow visualization in low-speed wind tunnels: Current British practice*. Tech. Note Aero.2715, Royal Aircraft Establishment, Ministry of Aviation, London.

Merzkirch, W. (1987). *Flow visualization*. Orlando: Academic Press.

Morelli, A., Nuccio, P., & Visconti, A. (1981). *Automobile aerodynamic drag on the road compared with wind tunnel tests*. SAE 810186, SAE International Congress and Exposition, Cobo Hall. 23–27 Feb 1981.

Passmore, M. A., & Jenkins, E. G. (1988). *A comparison of the coastdown and steady state torque methods of estimating vehicle drag forces*. International Congress and Exposition, Detroit. 29 Feb–4 March 1988, SAE 880475.

Pershing, B., & Masaki, M. (1976). *Estimation of vehicle aerodynamic drag*. EPA-460/3-76-025 (NTIS PB 275 948). Washington, DC: National Technical Information Service.

SAE. (1996). *Road load measurements and dynamometer simulation using coastdown techniques*. (SAE J1263 Feb 1996) *Society of Automotive Engineers Handbook, 2*, 26.531–26.550.

Seal, M. R. (1991). Design for the ultimate Viking car 1972–1991. *Proceedings S/EV 91 Solar and Electric Vehicle Symposium and Trade Show 26–27 Oct, Boxborough Massachusetts* (pp. 26–44). Northeast Sustainable Energy Association, Greenfield.

Swift, A. (1991). Calculation of vehicle aerodynamic drag coefficients from velocity fitting of coastdown data. *Journal of Wind Engineering and Industrial Aerodynamics, 37*, 167–185.

Stanbrook, A. (1960). *The surface oil flow technique as used in high speed wind tunnels in the United Kingdom*. Tech. Note Aero.2712, Royal Aircraft Establishment, Ministry of Aviation, London.

Van Valkenburgh, P. G., Klein, R. H., & Kanianthra, J. (1982). Three-Wheel Passenger Vehicle Stability and Handling. (SAE 820140). *SAE Transactions*, vol. 90.

Chapter 13
Energy Management

13.1 Introduction

This chapter emphasizes energy management. But this skill is not practiced alone; it is practiced within the context of race management, and it is influenced by the preparations for the race. The fastest racing *system* wins. Therefore, to put energy management in its proper context, these topics will be touched upon as well. For an alternative view of energy management, readers may wish to study the variational calculus-based method reported in MacCready et al. (1990). A complete description of the theory of this method is beyond the mathematical scope of this book. Wright (1997) describes an energy management strategy used by a 1995 Sunrayce™ team. Shimizu et al. (1998) give a description of the energy management strategy and its supporting energy management system used by the winner of the 1993 World Solar Challenge.

13.2 Energy Management

Energy management refers to the way the solar energy available during the race is expended. The goal is to win the race, not necessarily a particular race day. To do this, the car must maintain the highest average speed of any entry over the period of the race. Energy management has two parts: planning period management and micromanagement.

The *planning period* is the longest period for which reliable hourly weather data are available. Probably this will be the next race day plus the day after. However, forecasts of general daily conditions are available for periods as long as a week. These should be used to guide long-term strategy. *Micromanagement* is the moment-by-moment adjustment of the car's energy consumption to keep it within the planning-period strategy. Micromanagement also includes strategies for passing through shadows, climbing and descending hills, regeneration, and city driving.

E. F. Thacher, *A Solar Car Primer*, DOI 10.1007/978-3-319-17494-5_13

We begin with planning period management. Study the following example for the first 2 days of a race.

13.3 Example 13.1

Cornflower University's (C.U.) solar car uses Ni-Cd batteries and has the characteristics in Table 13.1. Also, the motor controller in C.U.'s car will shut down at a battery bus voltage of 47 V.

C.U. measured the rolling resistance coefficients in Table 13.1 on dry asphalt pavement.

Rain or overcast clouds have been predicted for both day 1 and day 2, but day 3 should be mostly sunny. The average air pressure and temperature will be 101.3 kPa and 27 °C, respectively, for both days, and negligible head winds are expected. The travel distance will be 115.5 miles for day 1 and 198 miles for day 2. The net change in elevation for each day will be zero. Is there an average speed for each of the first 2 days that will maximize the average speed over both days, that is, minimize the total travel time?

Solution We construct a simplified model of the situation in order to highlight the important issues. Suppose that:

1. The average acceleration of the car will be small.
2. The drag area and rolling resistance coefficients are constant.
3. The drag area is independent of the relative wind (note that this is conservative if c_D has the falling characteristic with increasing yaw angle shown in Fig. 2.14, but not otherwise).
4. The array efficiency, battery efficiency, and drive efficiency are constant.
5. The average speed may be used to compute the drag.
6. The car is in the cruise condition.
7. Race route information shows that the pavement will be almost exclusively asphalt and that the average grade is about zero.

The foregoing are the major simplifications; other assumptions will be introduced as the need arises.

The total time to travel distances S_1 plus S_2 miles is:

Table 13.1 Data for example car

$c_D A_D$	0.15 m^2
μ_1	0.004
μ_2	0.0002 s/m
M	408 kg, gross
M_e/M	1.01

$$\Delta t = \frac{S_1}{\overline{V}_1} + \frac{S_2}{\overline{V}_2}, \tag{13.1}$$

where the over-bars denote the average speeds for a day.

The average speed over both days is:

$$\overline{V} = \frac{S_1 + S_2}{\Delta t}. \tag{13.2}$$

This is to be maximized (Δt minimized) subject to energy balance equations and the following two constraints:

$$\overline{V}_1 \geq \frac{S_1}{\Delta t_\mathrm{M}}, \qquad \overline{V}_2 \geq \frac{S_2}{\Delta t_\mathrm{M}}, \tag{13.3}$$

where Δt_M is the maximum travel time allowed before C.U. must trailer its car, taken to be 8 h. For this example, the average speeds for days 1 and 2 must be at least 14.44 mph and 24.75 mph, respectively.

Write an energy balance on the power bus between the start and finish times for day 1. The initial and final speeds are zero. The array and battery each feed the power bus. The balance is:

$$(W_{12})_1 = (Q_{12})_1 + (E_{\mathrm{B1}} - E_{\mathrm{B2}})_1. \tag{13.4}$$

The subscript "12" refers to the time between start and finish times, t_1 and t_2. The additional subscript "1" appended to each term denotes the day.

W_{12} is the total work done by the drive (or energy supplied to the drive) to overcome the opposing forces of drag and rolling resistance. Hence, the division by η_D, the drive efficiency, in Eq. (13.5):

$$(W_{12})_1 = \left[c_\mathrm{D} A_D \frac{1}{2} \rho \overline{V}_1^2 + Mg\left(\mu_1 + \mu_2 \overline{V}_1\right) \right] \frac{S_1}{\eta_\mathrm{D}}. \tag{13.5}$$

For either day, the solar energy delivered to the drive wheel is:

$$Q_{12} = \eta_\mathrm{A} Q_{\mathrm{S}12}, \tag{13.6}$$

where η_A is the array efficiency and $Q_{\mathrm{S}12}$ is the total solar energy intercepted by the array,

$$Q_{\mathrm{S}12} = (\overline{GR})_{12} A_\mathrm{A} (t_2 - t_1). \tag{13.7}$$

G is the irradiance on a horizontal surface, and R is the tilt correction factor. Hence $(\overline{GR})$is the average irradiance on the array surface. A_A is the array area and t_2-t_1 is the time spent racing.

For the battery,

$$\eta_B(F_1 - F_2)E_{B0} = E_{B1} - E_{B2}, \tag{13.8}$$

where η_B denotes the battery transaction efficiency, taken as the same constant for charge and discharge, F represents the fractional state of charge, and E_{B0} is the fully-charged energy of the battery.

Combining the foregoing with the drive work, Eq. (13.5), and the energy supply on the left and right, respectively, gives for day 1

$$(W_{12})_1 = \frac{\eta_A \eta_D (\overline{GR_{12}})_1 A_A}{\overline{V}_1} + \frac{\eta_B \eta_D (F_1 - F_2)_1 E_{B0}}{S_1}. \tag{13.9}$$

The time difference has been replaced by $S_1 / \overline{V}_1$. The energy equation for day 2 is similar.

The state of charge at the starting time, t_1, on day 2, depends on the state of charge at the finishing time, t_2, on day 1. This, in turn, depends on the average day-1 speed, $\overline{V}_1$.

Take the average values of E_{B0}, F_1 (day 1), η_A, η_B, and η_D to be 5.0 kW·h, 1.0, 0.1, 0.9, and 0.85, respectively. Set the beginning-charge time, t_1, and the impound time to 6 a.m., to 10:15 a.m., and 8:45 p.m., respectively. The latter two times may of course vary, but this variance will not be significant for the purpose of the example. The average irradiance will be taken as 300 W/m^2 for both days as typical of high overcast conditions.

First, examine the solution to Eq. (13.9) as a function of the fractional state of charge at the finish line of day 1. Note that the equation is arranged so that the energy loss per mile must be balanced by the energy available per mile at the average speed. In Fig. 13.1, the energy loss and the energy supply per mile[1] for each of three values of the finishing charge fraction, F_2, are plotted separately as functions of the average speed. The intersections of the three energy supply curves with the energy loss curve give the average speed for each F_2.

The lowest value of F_2, 0.1, corresponds approximately to a bus voltage of 47 V, when C.U.'s motor controller will stop operating. The average speed could be

[1] Because it is presented on an energy per mile basis, the result in Fig. 13.1 is independent of the distance to be traveled.

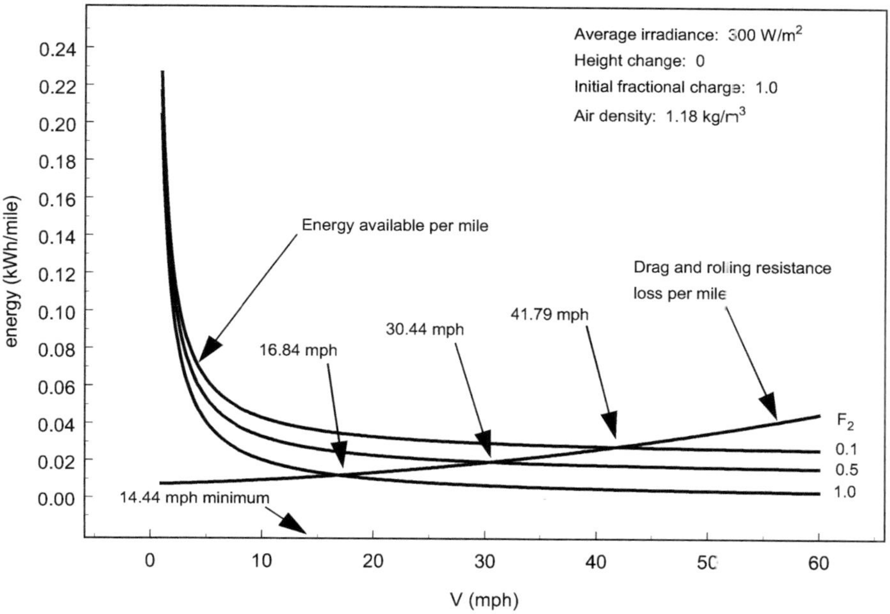

Fig. 13.1 Solution of Eq. (13.9)

nearly 42 mph if C.U. were willing to discharge to this level. This could place the team high in the day's standings and would maximize the time available for battery charging at the day 1 destination. But the next day's operation could be quite slow (recall that the irradiance will remain about 300 W/m²) and might include trailering. At the other extreme, an F_2 of 1.0 corresponds to a net zero battery discharge. The speed indicated in the figure for this case (16.84 mph) is the speed attainable using direct solar energy alone. Fortunately for C.U., it is greater than the minimum speed (14.44 mph). This means that it would be possible to travel the entire distance of 115.5 miles without a net battery discharge. This would maximize the energy available to run the next day but could reduce the average speed for the 2-day period. Charging in the afternoon of day 1 and the morning of day 2 would not be necessary (or possible).

With other quantities regarded as constant, the optimum speed would be a function of the fractional state of charge at the finish lines of day 1 and day 2. We will search for the optimum speed for each day by a trial-and-error process. Successive values of $(F_2)_1$ between 0.1 and 1.0 will be set and $\bar{V}_1$, $(F_1)_2$, and $\bar{V}_2$ found by trial and error until the 2-day average speed is a maximum. The irradiance for day 2 will be the same as for day 1, and the other quantities used to develop Fig. 13.1 will also be the same. Because day 3 will be sunny, we will set $(F_2)_2$, the fractional state of charge at the finish line of day 2, at 0.1.

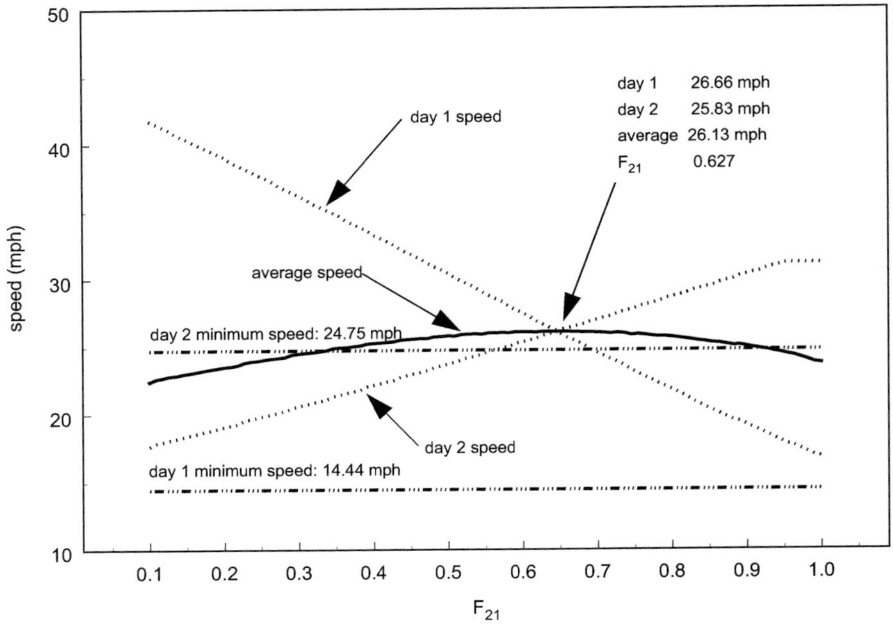

Fig. 13.2 Optimal vehicle speed

The results of the solution process are shown in Fig. 13.2. This figure shows the average speeds for day 1, day 2, and the 2-day average speed as functions of $(F_2)_1$. Also shown are the minimum speeds for each day.

The maximum average speed for the 2-day period was 26.13 mph, giving a total running time of 12 h (4.33 h on day 1 and 7.67 h on day 2). Note that the maximum is broad, and it would seem that C. U. has wide latitude in expending energy on day 1. However, the day-2 minimum speed line intersects the day-2 speed line at an $(F_2)_1$ of about 0.57. Thus, traveling at, or above, about 28 mph (estimated from Fig. 13.1) on day 1 will risk, or cause, trailering on day 2. However, the maximum is broad enough so that the team could hedge its bet a bit (in case the weather prediction for day 3 turns out to be optimistic) by keeping $(F_2)_1$ greater than 0.57 but less than perhaps 0.75. This would mean an average speed of less than about 28 mph and greater than about 22 mph during day 1.

The minimum speed for day 1 is so low that it does not intersect the day-1 speed curve over the range of fractional charge employed. Thus, as mentioned above, the team could run even more conservatively on day 1. However, the reader should keep in mind that the speed of the car is influenced by traffic and road conditions. In order to average about 17 mph (rounded from 16.84 mph), it might be necessary to travel much faster at times to make up for the instances when the car was stopped in traffic or slowly climbing a hill.

The greatest margin between the minimum speed for day 2 and the optimum speed for day 2 is just over 1 mph. In this situation the preparation of the team, especially the drivers, is critical. Energy wasted by poor driving technique or by unfamiliarity with the car could cause trailering on day 2.

The results above were predicated on a target $(F_2)_2$ of 0.1. If this were set at 0.2, to provide additional energy for day 3, what would the results be? The best average speed for the 2 days drops slightly to 25.06 mph. The total travel time consequently increases to 12.51 h, with 4.51 h expended in traveling on day 1 and just under 8 h on day 2.

With a detailed car, route, and weather simulation, the optimum speed could be found using the "little tap" method.

Attaining an average speed high enough to win does not necessarily mean that the team will finish first on any given day. If a given day is overcast and raining, the team must be disciplined and patient and keep its energy expenditure low, as shown in the example. This may mean squelching the desire to speed up if another team is closing; the team must be willing to finish low in the pack to conserve energy so that its average speed can be higher than those of its perhaps more aggressive competitors over the expected period of cloudy days.

The charging and discharging characteristics of the battery bank are vital parts of the energy management data base. The team must be able to estimate well the state of charge of the battery, the battery transaction efficiency, and must know the lower limit of the battery bus voltage. In the example, this limit was not set by the state of charge, as it might be with Pb-acid batteries, but by the motor controller shut-down voltage.

13.4 Micromanagement

Energy Dead Reckoning Ships navigate by "dead reckoning." The navigator predicts the next position by estimating the effects of weather and ocean currents. Micromanagement is energy dead reckoning, predicting the energy state from the current state and the predicted supply and use. Prior to the run, the average speed should be selected to leave the battery at a certain state of charge at the finish line. This speed would be the optimum vehicle speed (OVS), or possibly a more conservative speed. These choices were brought out in Example 13.1. During the day's run, the speed of the car is adjusted so that energy use remains in the proper relation to the energy supply to achieve these ends. Tools to assist in this are discussed in the section on the energy information system (EIS). Strategies for hills, cloud shadows, and the use of regeneration will now be suggested.

Hills Climbing a hill is analogous to acceleration. Each results in higher stored energy state: the higher potential energy at the top of the hill or the higher kinetic energy at the end of acceleration. The energy stored in either case is proportional to the vehicle's mass.

Climbing a hill usually costs additional battery energy depending on the climbing speed. Electrical losses proportional to the square of the current are part of this cost. Thus, climbing at low motor torque keeps the losses in the battery, power wiring, and motor smaller. A transmission, or an axial flux motor with an adjustable air gap, allows the drive to keep the shaft torque higher to lift the car up the hill, but at a lower motor current. This improves the car's efficiency while climbing, but the speed is necessarily less.

A low state of charge will dictate a low-speed climb. The additional current draw caused by higher speed could pull the battery voltage below an operational limit, such as the motor controller shut-down voltage. However, in other situations we may ask: Is there a climbing speed that minimizes the charge removed from the battery? Consider the following example:

13.5 Example 13.2

The car of Table 13.1 climbs a 1-mile-long hill at various steady speeds between 1 and 55 mph in still air under battery power only. Its battery consists of 60, 56-A h, SAFT STX600 Ni-Cd cells in series. The car is driven by an NGM-SC-M100 wheel motor (efficiency curves in Fig. 5.11) with its air gap set to 2.5 mm. The wheel radius is 0.241 m. The air temperature and pressure are as for Example 13.1. The car starts up the hill at the designated speed and fully charged (Fig. 13.3: H=0), and thus its battery voltage is at a maximum.

Plot the charge removed from the battery and the final, that is, minimum battery bus voltage as functions of the climbing speed for grades of 1, 2, and 3° (1.75, 3.49, and 5.24 %, respectively). For comparison, show the charge removed for a horizontal run of 1 mile under the same conditions.

Solution Figure 13.3 shows the open circuit voltage and the effective internal resistance (mΩ) of the STX600 at 23 °C as a function of its fractional discharge, H (or $1-F$). A curve fit (solid lines) to each data set has also been plotted. The terminal voltage is given by Eq. (4.3), with the current positive for charge and negative for discharge.

Using the data of Table 13.1 and the atmospheric conditions and wheel radius specified above, Eqs. (2.15) and (2.17) were used to find the shaft torque and consequently the shaft power, P_S. The motor characteristics then gave the battery current, I_B, as

$$I_B = \frac{P_S}{V_B \eta_D}. \tag{13.10}$$

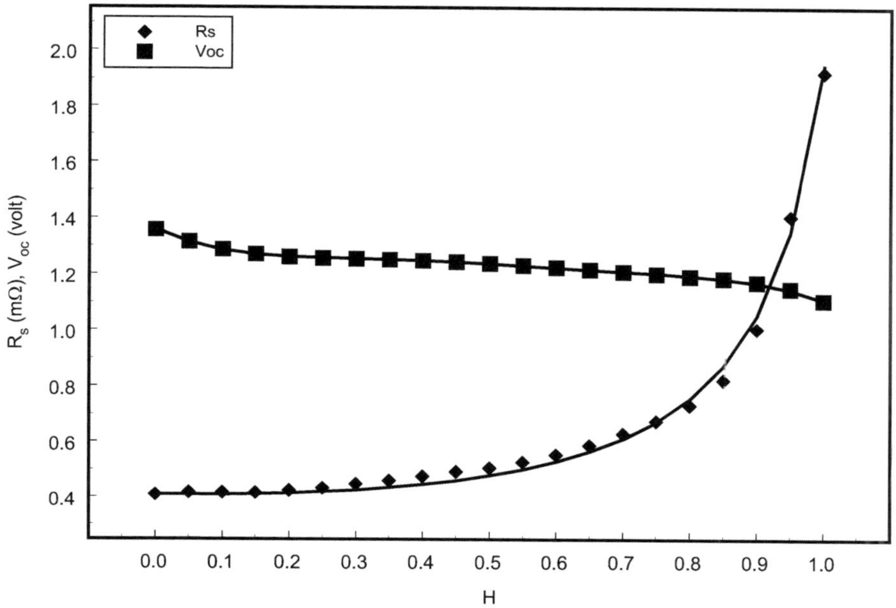

Fig. 13.3 SAFT STX600 characteristics. (Courtesy SAFT engineering)

The battery voltage was found from the curve fits of Fig. 13.3 by calculating the fractional state of discharge as

$$H = 1 - \frac{Q(t)}{Q(0)} = 1 - F(t), \tag{13.11}$$

where $Q(t)$ is the present state of charge and $Q(0)$ the initial charge. The "little tap" method was used: during each small time step Δt, the charge $I_B \Delta t$ was subtracted from the charge at the end of the previous step and the voltage of the battery calculated. The total charge removed and the final battery voltage were then plotted in Fig. 13.4 and 13.5, respectively.

Figure 13.4 shows that at each grade above zero there was a speed at which the charge removed was a minimum. The removed charge at first decreased as the speed increases because the time to reach the top of the hill rapidly decreased; therefore, the time to remove charge was less. However, this was eventually overcome by the increase in the current. The current increased both because the shaft power increased as a cubic function of the speed and because the battery bus voltage decreased as charge was removed from the battery.

Figure 13.5 shows that the minimum bus voltage, which occurred at the top of the hill, passes through a maximum at a particular speed. Following the removed charge, the fractional state of discharge, H, passed through a minimum.

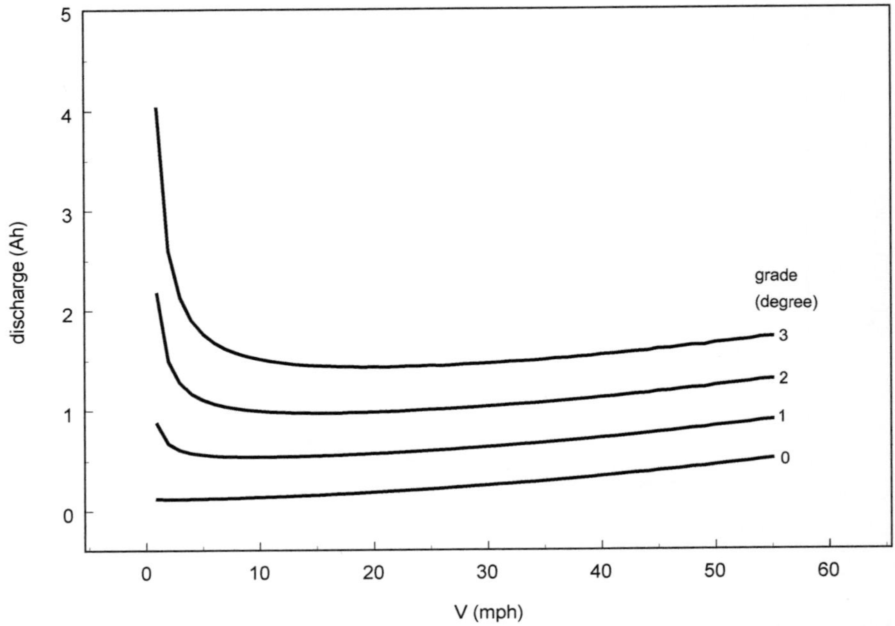

Fig. 13.4 Battery discharge vs. speed on four hills

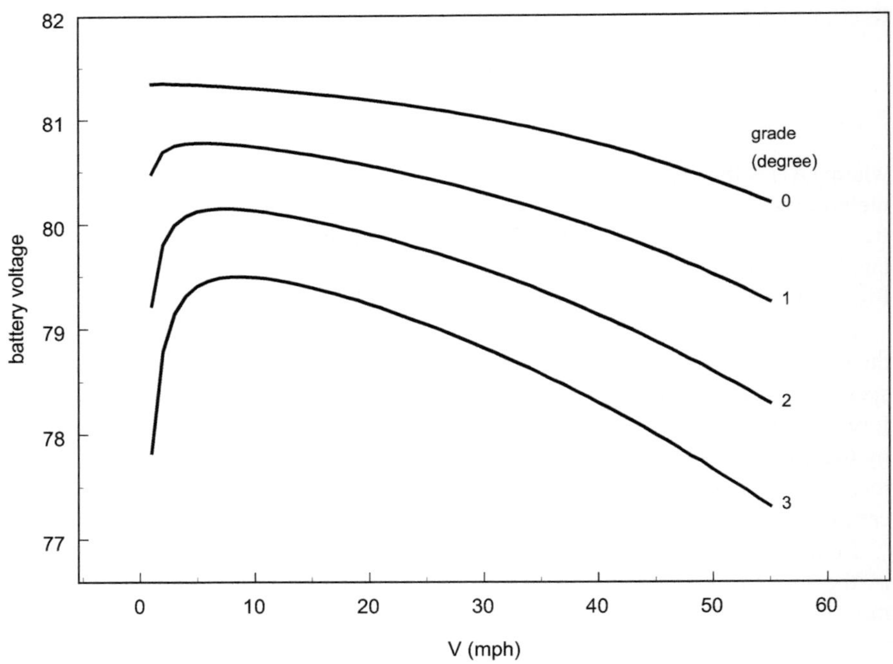

Fig. 13.5 Minimum battery voltage vs. speed on four hills

Consequently, so did the battery equivalent resistance, R_S. This caused the minimum battery voltage to be maximized, but at a lower speed because of the influence of the increasing current.

The climbing speed which minimizes the battery discharge is 10–20 mph, depending on the grade.

Rather than climbing at constant speed, the hill could be climbed at constant motor current. For example, suppose a motor current of 10 A is required when at cruising speed just before the hill. The car will slow to the climbing speed dictated by 10 A. Alternately, the battery current draw could be held constant. The car will then slow to a climbing speed that depends upon the array current available. The advantage of this latter method is that the battery current is directly controlled. This may be quite important when the remaining charge is low.

Regeneration A strategy for descending hills is to use the gravity assist to reach the local speed limit. The conversion of the potential energy to kinetic energy provided by gravity costs the vehicle no battery energy. Then regeneration should be used to maintain the local speed limit while descending. The kinetic energy thus attained is used to coast up the next hill until the car reaches its desired climbing speed; then the motor is used to climb the rest of the way. Regeneration should always be used for braking and for controlling the speed while descending hills in towns, where speed is restricted anyway. Then at least some of the car's precious kinetic energy will be recovered, rather than dissipated in heating the brakes.

Cloud Shadows Intermittent cloud shadows are best handled by keeping up speed and to move out of them as soon as possible.

13.6 Energy Information System

The term energy information system (EIS) refers to the equipment and information sources used to estimate the OVS for each day of the planning period and to select the highest, practical current speed considering the available solar radiation, the state of charge of the battery, and the projected energy requirement for the next hour needed to attain the OVS.

The EIS has six parts:

1. A weather and solar radiation forecasting facility
2. Present-time weather and solar radiation monitoring
3. Race vehicle telemetry
4. Energy management software
5. Road data
6. Global positioning information

13.7 Weather and Solar Radiation Forecasting

Equipment Radio, cellular telephone, and probably Internet communications are required in order to query various weather and solar energy prediction services.

From Weather Forecasts The hourly clearness index may be predicted using weather parameters that are given by readily available hourly forecasts (Accuweather 2013, for example). This avoids the uncertainties arising when attempting to predict hourly global radiation from the daily total. The hourly total extraterrestrial radiation, I_0, can be calculated by Eq. (3.8). Then the hourly total radiation on a horizontal surface may be calculated from Eq. (3.13).

Weather records going back several years for many locations can be obtained from the National Oceanic and Atmospheric Administration (NOAA). Free typical meteorological year (TMY) data (more than 200 locations are available, see Chap. 3) can be obtained from the Solar Energy Laboratory of the University of Wisconsin, Madison, and from the National Renewable Energy Laboratory (NREL). The following outlines the process of developing the clearness-index correlation for Atlanta, GA, for June and July.

13.8 Example 13.3

Outline the process of developing the clearness-index correlation for Atlanta, GA, for June and July.

Solution The set of weather parameters for the correlation will be based on those used by Jensinius (1983),[2] except that day length would have no bearing. And, although TMY files contain a large number of weather variables besides hourly total solar radiation, they do not contain the cloud probabilities used by Jensinius, but simply the total cloudiness in tenths. The work reported by Davies and McKay (1988, 1989) suggests that the clearness index is a nonlinear function of this cloud cover measure. Suppose we assume that the hourly clearness index, k_T, may be given by

$$k_T = f\left(\overline{c}^2, \overline{c}, \overline{T}_{\mathrm{db}}, \overline{T}_{\mathrm{dp}}, \overline{m}, I_0\right), \tag{13.12}$$

where f denotes an unknown formula using the variables in parentheses, c denotes cloud cover in tenths, T_{db} the dry bulb temperature (°C), T_{dp} the dew point temperature (°C), \$\$\$\$ the air mass, and I_0 the total, global, horizontal, extraterrestrial radiation (W h/m^2). The overbar denotes an average for the hour. We obtain the TMY2 data file for Atlanta (13874.tm2, 1,232 KB) and a copy of NREL (1999), a free handbook describing the TMY2 data files. The result, Eq. (13.13), may be

2 See also Jensinius in Hulstrom (1989), Chap. 7.

obtained by a specially written program or by using the built-in correlation facility invariably supplied with spreadsheet programs.

$$k_T = 0.09645 - 0.32143\bar{c}^2 + 0.11614\bar{c} + 0.01514\bar{T}_{db} - 0.007986\bar{T}_{dp} - 0.002056\bar{m} + 0.0004356 I_0. \quad (13.13)$$

Figure 13.6 shows the global radiation predicted by this correlation plotted against the corresponding TMY2 global radiation. The average root-mean-squared error was about 21 % with the least error evident at low irradiances.

Weather The energy management software must calculate the drag on the car. For this it requires the dry bulb temperature, the atmospheric pressure, and the wind speed and direction at the ground. These must be known at least hourly at the current location of the car. This information is available from sources such as Accuweather (2013). When it was not feasible to use such sources, teams have done their own forecasting using student meteorologists from their own or another school.

Satellite Images Solar radiation measurements derived from images received from earth-orbiting satellites can be used to predict the solar radiation by predicting the cloud cover motion from the images. Methods of predicting the cloud cover from satellite images were presented by Muench and Hawkins (1979) and Muench (1979). The methods used to measure solar radiation using geostationary satellite images were presented in the same papers. The methods used to measure solar radiation using geostationary satellite images were surveyed in Noia et al. (1993).

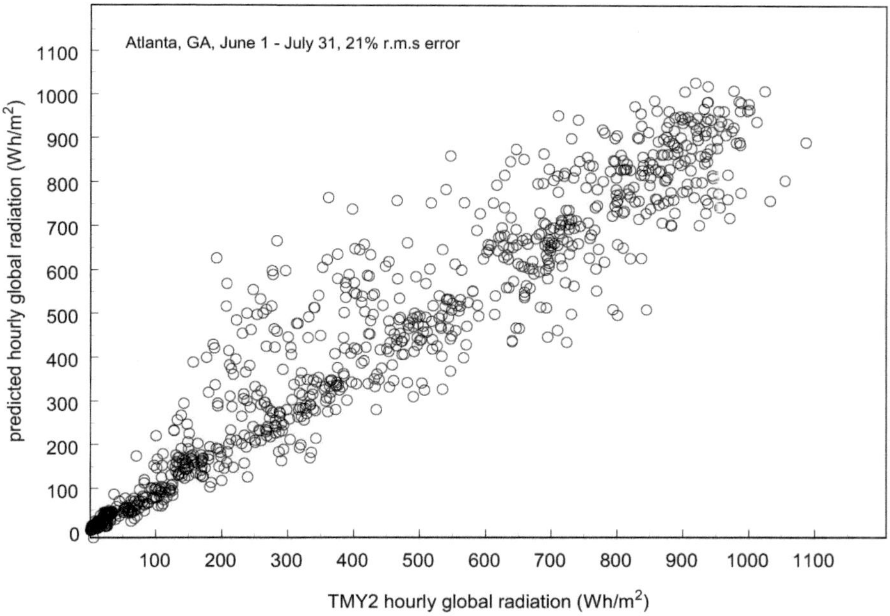

Fig. 13.6 Predicted hourly global irradiance

Justus, et al. (1986) presents results for the USA, Mexico, and South America. The statistical method discussed in Cano et al. (1986), Diabaté et al. (1989), and Perez et al. (1994) has been tested against ground measurements. Zalenka et al. (1992) reported a study of the error in extrapolating daily total radiation from a ground-based measurement compared to using satellite-derived radiation.

Satellite-based forecasting methods are complex and probably impractical for most solar racing teams. However, Kyle et al. (1994) report the superposition of the race route on satellite images, thus clearly showing the relation of the route to clouds and clear air.

13.9 Software

Components The software must accomplish two purposes: assist in estimating the OVS and assist in micromanaging the energy. A simulation code, if it is user-friendly and conveniently integrated into the EIS, could serve the first purpose. A spreadsheet program would probably be more convenient for micromanagement.

Simulation The operation of the simulation code must be convenient. Push button or point-and-click operations should be programmed to calculate a new solution from the current location and time, to edit the road data file (so "what if" can be played) or select a new one, to create or select new weather files, and to change the parameters of the car model.

The code should be configured to interpolate in space and time between the locations for which weather correlations have been prepared in order to estimate the solar and weather parameters at the current location of the simulated car. This was done by Craparo and Thacher (1995), for example.

The route information used by the simulation may be obtained by surveying the route, or it may be constructed by the use of commercial map software. Some commercial map codes can be used to find the grade and direction of segments of the day's race route as functions of the distance traveled. This method is easier and cheaper than traveling the route. But the resolution of the resulting topographical data will probably not be as good as that of a survey. Start several months before the race to allow time to garner the data and set it up to be read by the simulation code.

Surveying the route provides information about traffic conditions, road surfaces, stop lights, etc. This is not provided by the map software. The surveyors also become valuable guides to the team for navigating through unfamiliar cities and in choosing convenient hotels. A surveying trip would be done during the summer before the race or perhaps as late as the spring break before the summer of the race. The final route selected by the race organizers may deviate from that followed by the surveyors, requiring corrections to route data. By the time these are known, the only practical method of making these corrections is by map software.

Day's run	169.42 miles	
Minimum time	8 hours	
Minimum speed	21.178 mph	
Optimum speed	29.2 mph	
Start time	9 h	9 min
Current time	13 h	0 min
Elapsed time	3.85 h	
Odometer	111.86 miles	
Average speed	29.055 mph	
Distance to go	57.559 miles	
Arrival time	14 h	58.86 min
Current F	0.5256	
Predicted F2	0.4109	
Battery energy/mile	0.0204 kWh/mile	
Extrapolated F2	0.2815	

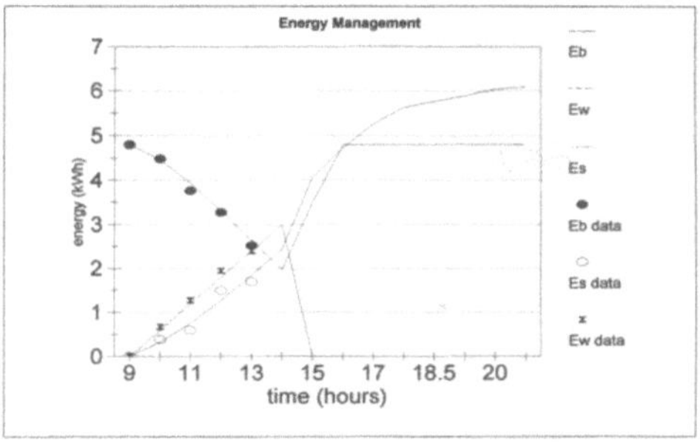

Fig. 13.7 Example of an energy display

Map software that can accept data from a Global Positioning System (GPS) receiver can be used to show the position of the car on the route and relative to upgrades and downgrades. This is valuable information for micromanagement.

Energy Display The spreadsheet program would receive the telemetry from the solar car, current solar irradiance measurements, and the predictions from the simulation. Figure 13.7 shows such a display as it might appear at 1 p.m. (13:00). The total solar energy received since starting the run, $E_S(t)$, the energy in the battery, $E_B(t)$, and the energy that has been expended to move the car, $E_W(t)$, both actual (symbols) and those predicted by the simulation (lines) are shown. These are referenced to the driving wheel. The simulation should be set up to allow the operator can play "what if" by changing the speeds over the remaining course at any time. The actual values for E_S, E_W and E_B would be produced by the spreadsheet using telemetry data.

The energy manager can now estimate the effect of hills and speed changes and also see at a glance how the energy usage and predicted usage compare.

13.10 Preparations

Preparations should be carried out against the background of the realization that, from the racing system viewpoint, racing begins the moment the team decides to enter the race.

Preparations culminate in the testing of the racing system. This has two components: thorough car testing and thorough team training. The car testing, methods for which are discussed in detail in Chap. 12, has two objectives: to measure the drag area and rolling resistance coefficients of the car (on both wet and dry pavement) and to drive the car at least as many miles as the total length of the race and under various conditions—particularly over hilly terrain and in inclement weather. This driving tests the durability of the racing system and provides opportunities to train

Fig. 13.8 Racing in the rain

the team. The team must learn how to race the car when it is overcast and raining, as in Fig. 13.8, and in heavy traffic, as in Fig. 13.9. There must be strategies for every eventuality. There should be a mix of training and testing trips, some short and some long. Short trips focused on testing should be made over the same route

Fig. 13.9 Racing in heavy traffic

so that the route itself will be removed as a variable affecting the energy consumption of the car.

If the car has accumulated many hundreds of miles before the race, probably all the wheel bearings should be replaced. Leave enough time so that you can run 50–100 miles with the new bearings, then remeasure the rolling resistance coefficients. When racing, carefully inspect the car after each day's run, and especially during and after trailering, including all welds and suspension elements. Do not assume that such vital parts cannot fail, even if they have shown no wear during the training period. Come to the race with spare parts, such as shock absorbers. Bring welding equipment. *Take nothing for granted.*

13.11 Race Management

The average speed is influenced by the performance of the race team as race managers. Race management comprises solar car energy management and management of ancillary race activities. There are a multitude of these.

Flats Flat tires will result from improperly adjusted wheel spokes or poor matching of tires to rims as well as running over broken glass. Flats cause lost time. Rapid tire changes are a function of good wheel design, but they are also the result of having replacement wheels on hand, balanced and pressurized. This means that the wheels and tools must always be in a designated place in the chase or lead vehicle. Members of the team must be preassigned to do the change and should know exactly what to do so that no extra time is lost milling around. This means tire-change training must be accomplished during the prerace period.

Organizational Scope As the tire-changing example above shows, a racing team must be efficient. That is, every activity must be carried out in the smallest possible time. This implies rapid action but also minimum error; error requires repeated action. Rapid and error-free action requires organization, training, and reliable communications.

The race organization must account for both large and seemingly small activities. The large activities include feeding the team, sleeping accommodations, and fuel for the support vehicles. The seemingly small activities include awakening the team in sufficient time each morning so that all may be fed, the car removed from impound and put on charge, etc. This activity seems small, but it can loom large if time elapses and the team is not ready to start at its appointed time.

Communications Radio communications between the driver and the lead and chase vehicles must be available at all times when racing. Reliable communication is a function of proper equipment and propagation conditions, but also of making sure the batteries in the radio handsets are charged. Radio procedure must be disciplined. There must be no unnecessary chatter and only those designated to give information to the solar car driver should speak. This prevents confusion, which is unsafe.

Under the rules in Chap. 16, support vehicles must not travel on the race route. Therefore, at times during a race day, the suggested route for the off-route crew operating the vehicle towing the trailer may be separated by many miles from the race route. Therefore, communication with the off-route crew must be reliable over these distances. The personnel operating the towing vehicle must try to remain within the radio communication range, in case its services are needed.[3] This procedure is best worked out during the prerace period. However, remaining within the radio communication range is not always possible. Cell telephone communication is usually more reliable than radio communication over longer distances. Cell coverage during the 1999 race was excellent, but during earlier races there were regions with no cell coverage. Another drawback of this method of communications is that it is expensive.

References

Accuweather. (2013). http://www.accuweather.com.

Cano, D., Monget, J. M., Albuisson, M., Guillard, H., Regas, N., & Wald, L. (1986). A method for the determination of the global solar radiation from meteorological satellite data. *Solar Energy, 37*(1), 31.

Davies, J. A., & McKay, D. C. (1988). Estimating solar radiation from incomplete cloud data. *Solar Energy, 41*(1), 15.

Davies, J. A., & McKay, D. C. (1989). Evaluation of selected models for estimating solar radiation on horizontal surfaces. *Solar Energy, 43*(3), 153.

Diabaté, L., Moussu, G., & Wald, L. (1989). Description of an operational tool for determining global solar radiation at ground using geostationary satellite images. *Solar Energy, 42*(3), 201.

Hulstrom, R. L. (Ed.). (1989). *Solar Resources*. Massachusetts Institute of Technology.

Jensenius, J. S. (1983). *Sunshine and Solar Energy Guidance*, National Weather Service Technical Procedures Bulletin 334, U.S. Department of Commerce, National Oceanic and Atmospheric Administration, National Weather Service Program Requirements and Planning Division, Silver Spring, MD, September 15.

Justus, C. G., Paris, M. V., & Tarpley, J. D. (1986). Satellite-measured insolation in the United States, Mexico, and South America. *Remote Sensing of Environment, 20,* 57.

MacCready, P., et al. (1990). *Sunraycer case history*. Warrendale: Society of Automotive Engineers.

Muench, H. S. (1979). *Short-range forecasting through extrapolation of satellite imagery patterns*. Part II: Testing Motion Vector Techniques, Air Force Geophysics Laboratory, Hanscom, Mass. AFB, Technical Report AFGL TR-79/0294.

Muench, H. S., & Hawkins, R. S. (1979). *Short-range forecasting through extrapolation of satellite imagery patterns*. Part I: Motion Vector Techniques, Air Force Geophysics Laboratory, Hanscom, Mass. AFB, Technical Report AFGL TR-79/0096.

NREL. (1999). User's manual for TMY2s, National Renewable Energy Laboratory, Golden, Colorado.

Noia, N., Ratto, C. F., & Festa, R. (1993) Solar irradiance estimation from geostationary satellite data: I. Statistical models. *Solar Energy, 51*(6), 449.

[3] The off-route team cannot delay because they leave later than the race car (because of packing up the trailer) and they may have to drive twice as far as the race team. And, if the day is sunny, they will be hard pressed to keep up.

Perez, R., Seals, R., Stewart, R., Zelenka, A., & Estrada-Cajigal, V. (1994). Using satellite-derived insolation data for the site/time specific simulation of solar energy systems. *Solar Energy, 53*(6), 491.

Shimizu, Y., Komaatsu, Y., Torii, M., & Takamuro, M. (1998). *Solar car cruising strategy and its supporting system*. JSAE Review, vol. 19, JSAE9830181, Society of Automotive Engineers of Japan, Inc. pp. 143–149.

Wright, G. S., (1997). *Optimal energy Management for solar car race*. Proc. 39th Midwest Symposium on Circuits and Systems, August 18–21, 1996, Iowa State, Ames, Iowa, IEEE, Piscataway, NJ, pp. 1011–1014.

Zelenka, A., Czeplak, G., D'Agostino, V., Josefsson, W., Maxwell, E., & Perez, R. (1992). *Techniques for Supplementing Solar Radiation Network Data. Final Report of* International Energy Agency Solar Heating and Cooling Program, Task 9, subtask 9D, IEA, Paris, France.

Chapter 14
Fund Raising and Public Relations

14.1 Introduction

Experience shows that developing a racing system from scratch and competing in a cross-country race, such as the American Solar Challenge, costs $ 50,000–190,000, in cash and in kind, over a 2-year period.[1] Most schools cannot afford this expense. Therefore, the solar car team must seek sponsorships from individuals and agencies, principally businesses, outside the school. However, this is on balance a good thing because it provides an opportunity to involve students in yet another phase of real work. And if the university has a business school, then seeking sponsorships is made to order for that school's marketing majors.

Public relations work is coupled to raising funds. Publicity for their participation is the principal *quid pro quo* for the team's major sponsors. Also, the school will want to publicize the solar car team to attract students, and solar car teams *do* attract students. High school and grade school teachers and service organizations frequently request visits by the solar car and talks on solar energy by the team. This builds community support and may result in donations.

14.2 Fund Raising

Budget The first step in fund raising is to estimate how much money must be raised. This requires a budget, which in turn requires a project plan; these two must be developed together. The first try at either will probably not be well informed so revisions will be necessary as the project proceeds. The major categories of the budget might be: solar vehicle, race logistics, energy management, communications, training, administration, and fund raising and public relations. A brief discussion of each of these categories follows. Bear in mind that the figures in the tables that

[1] 2013 dollars; teams have spent far more, and much less. However, this range represents an approximate threshold for fielding a competitive team.

E. F. Thacher, *A Solar Car Primer*, DOI 10.1007/978-3-319-17494-5_14

Table 14.1 Solar vehicle costs

Component	Cost range ($)
Solar cell array	2500–25,000
Maximum power point trackers	2500–4700
Battery	2800–40,000
Motor and controller	3000–21,000
Chassis and body shell	2000–20,000
Wheels, tires, and brakes	2000–14,000
Steering and suspension	2600–10,000
Auxiliary power and instrumentation	2500–7200

follow may represent either cash or in-kind donations. Most of the construction was assumed done by the solar car team, with the principal exception of the body mold. Note also that costs for some of the reusable equipment would be one-time expenditures.

Solar Vehicle Table 14.1 shows the components of this budget category and typical cost ranges for each. Each entry should be taken to include all elements of the component. For example, the array entry includes the solar cells, tabbing, diodes, underlayment, and encapsulate. Only the maximum power point trackers are listed separately. The high end of the range is based on ordering two times the number of cells actually needed. This provides ample replacements for breakage. The battery entry ranges from lead-acid modules at the low end to nickel-metal hydride modules at the high end. The high end of the chassis and body shell range assumes a shell made from pre-impregnated composite material and a commercially made mold. (The mold is very expensive and should be a prime sponsorship target.) The running gear includes wheels, tires, suspension, and steering. The auxiliary power and instrumentation cost includes numerous small components such as wire ties, wire, DC-DC converters for low-voltage DC, running lights, circuit bakers, fuses, cockpit instruments, etc.

Race Logistics Table 14.2 shows logistics costs for a solar race including the Qualifier. The Qualifier is usually about a month and a half before the team must leave for the race. The Qualifier for the 1999 Sunrayce™ was held at the General Motors

Table 14.2 Logistics costs

Item	Cost or cost range ($)
Fuel and oil	600–1900
Vehicle rental	3000–5500
Food	700–3000
Accommodations	2500–5000
Spare parts	1000–1500
Uniforms	500–1000
Fees	8000

[a] The fees include insurance cost and race entry. These costs are now higher than in Chapter 16.

Table 14.3 Energy management

Item	Cost range ($)
Chart route	2000–2500
Computer	1500–2000
Software	200–500
Telemetry	1000–2500

Table 14.4 Communications

Item	Cost range ($)
Radio equipment	500–700

Proving Ground in Milford, Michigan, and required a 5-day round trip by eight people. This was the basis for the budget item cost. The race itself typically requires a 3-week round trip of 4000–5000 miles by about 12 people operating four vehicles (lead, chase, solar, and trailer towing.) The race costs in Table 14.2 includes the rental of two of the service vehicles and fuel. The remaining vehicle was assumed to be obtainable from the school.

Energy Management Two or three persons could travel down the proposed race route during the summer prior to the race. This survey team would collect information about the route, such as road surface conditions, grades, and so on. This information will be invaluable in managing the energy of the car during the race. The cost for this trip follows the "chart route" item in Table 14.3.The computer and software would be used to reduce the telemetry data from the car and to operate a vehicle simulation code. A cell telephone is useful for obtaining weather predictions and of course for communications when on the road.

Communications Amateur radio gives the team great flexibility in choice of communication frequency and in equipment. This may be important in races with many teams. The drawback is that, race team members who will use it must obtain their technician-class radio amateur licenses. Citizen's Band (CB) radio equipment, while more frequency-restricted, is cheaper and easier to obtain.[2] Licenses are not required. Cell telephones may also be used. But cell service may not be reliable over the entire race route. The costs are based on CB equipment in the solar, lead, chase, scout, and trailer-puller (Table 14.4).

Training A major goal of the project should be to finish the car early enough so that vehicle testing and race team training can be done before the race.[3] The cost shown in Table 14.5 shows three or four 1-day trips in full racing configuration.

The race organization holds a workshop in the spring of the year before the race. It is important to send some team members to this event.

[2] Teams must monitor the designated official race CB channel in the chase vehicle.

[3] This may seem obvious. Nevertheless, accomplishing it often turns out to be quite challenging.

Table 14.5 Training

Item	Cost or cost range ($)
Training trips	600–2000
Workshop	1000
First aid training	50–100

Table 14.6 Administration

Table 14.6 Administration

Item	Cost or cost range ($)
Telephone	1000–2000
Local trips	500–1000
Paper	100–200
Photocopying	500–500
Postage	300–400
Printing	200–400

Administration The telephone costs in Table 14.6 include costs incurred for fund raising and public relations as well as those associated with design and procurement.

Fund Raising and Public Relations The major expenses in this category arise from preparing and printing the proposal sent to corporations. This proposal, like the race proposal and the structural report, is a pivotal document. The experience of the advisors should be brought directly to bear on its preparation to bring out the best in the students. They strongly desire to produce an excellent, professional proposal. And they are capable of so doing. Often a trip to formally present the proposal at the corporate headquarters is required. This is excellent experience, of course.The "Adopt-A-Cell" program is discussed in more detail in the following section (Table 14.7.)

Methods There are four means of fund raising: donations from individuals and service organizations, sponsorships by corporations, grants by foundations, and grants from government agencies.

To solicit individuals and service organizations, Clarkson University's team produced an "Adopt-A-Cell" program brochure. The brochure contained information about the team, information about the cars used in previous races, and information about the Adopt-A-Cell program. This latter information included four levels of monetary support. Each level included certain rewards, such as a team T-shirt and a

Table 14.7 Fund raising and public relations

Table 14.7 Fund raising and public relations

Item	Cost range ($)
Press kits	100–700
Adopt-a-cell	500–700
Corporate proposals	1000–3000
Presentation trips	500–1000

subscription to the team's newsletter. A coupon with blanks to be filled with information about the donor and the level of support was also included.

Corporate sponsorships are obtained by submitting a formal proposal to corporations that could conceivably have some interest. This would include corporations employing a few, but highly placed alumni of the school, corporations that hire large numbers of the school's graduates, and corporations whose business interests might be served by an association with solar car racing. Benefits, such as whether the sponsor's trademark would be placed on the race car, should depend upon the level of support. Less-prestigious locations for the trademark could be the support vehicle, team uniforms, and other promotional materials. Other benefits could include making the race car available for display at events, such as trade shows or technical conferences, in which the sponsor participates. The team's marketing group should work out a list of such benefits to place in the proposal. Of course it is absolutely necessary to deliver the promised benefits once a sponsorship has been received.

The procedure for soliciting foundation support is similar, except that a formal presentation is usually not necessary. There are literally hundreds of foundations in the USA representing a great variety of interests. Some fund projects only within a particular city; others have national scope. The school's Development Office (see next section) will have publications that give details about the foundations' objectives and what they have supported in the past. A foundation proposal must show how the requested support would serve the foundation's objectives.

Government support is also possible. For example, if the state has an organizational unit that fosters energy research, this unit may be receptive to a proposal. However, this unit will probably find it must support all the solar race teams in the state or none of them.

Liaison Every school has an organizational unit that is charged with raising money for new buildings, scholarships, etc. This unit was referred to above as the "Development Office." Some of the same corporations and foundations to which the solar car team may contemplate sending sponsorship proposals for $ 25,000 may also be on the Development Office's "ask" list for millions. It is therefore imperative that the Solar Car Team's Marketing Group work through and with the Development Office, obeying its policies and procedures at all times. This may result in some corporations being off-limits for solar car team sponsorship proposals. However, establishing good relations with the fund raising professionals in the Development Office will result in long-term benefits for both the team and the office. The professionals cultivate corporate contacts over time and can therefore be very helpful to the solar car team. Furthermore, the mentoring effect of a close liaison with the development office is valuable training for the students. Finally, the public relations value of the solar car is very high. This is appealing to potential donors to the school and consequently may help the development office raise money.

The team should also maintain good liaison with the school's legislative representative. This person will have cultivated contacts in the legislature and in units of the state bureaucracy. These contacts may be receptive to sponsorship proposals.

14.3 Public Relations

At first, the engineers on the solar car team may feel that public relation is an annoying interference with their work. However, the contrary is true: it is part of their work. It benefits the school and therefore gains the support of the school. It interests high school and grade school children in building solar cars and therefore helps educators raise the level of knowledge and achievement in secondary and in grade schools. It garners public support; people begin to think, "This is our solar car team."

This work is done in different venues. Fig. 14.1 shows Clarkson's 1995 car, Helios, on display in a shopping mall. This particular trip required trailering Helios for 75 miles and was arranged by a service organization from the town containing the mall. Display boards behind the car show pictures of the design and construction sequence, information about solar energy, and Clarkson University. A steady stream of adults and children peppered the team members with questions. This points to an important aspect of the team's training: all members of the team, business school students not excepted, must acquire a certain level of understanding of the technical aspects of the solar car. This level must be such that when they speak to the public or the press their statements will be correct.

Perhaps the most frequent activity is visits to schools to speak about solar cars and to demonstrate the car. The most inquisitive students are often those in grade school. Their questions are good and sometimes come so rapidly that one can scarcely finish an answer before another question is asked. This interest is most gratifying to the participating team members.

Fig. 14.1 A solar car at a public relations venue

Opportunities to show the car at alternative energy events occasionally occur. These events may involve both static displays of the car and parades. One of most enjoyable features of parades is the obvious pleasure people of all ages take in seeing the car whir by. A solar car is a true, locally nonpolluting vehicle. As such it is an icon of what many hope our transportation system will become.

Liaison Contact should be maintained with the unit charged with public relations for the school. Let us call this unit the "Public Relations Office." This office will usually be staffed by energetic, articulate folks who recognize the public relations value of the solar car project. Keep them informed of the team's activities and of opportunities for photographs and stories. The public relations office will often initiate ideas for public activities and serves as a contact point for persons interested in learning more about the team. Most of these initiatives will be practical. An "unveiling," however, should be most carefully considered.

An unveiling is an official presentation of the solar car to the sponsors and the public. Requests for it will naturally occur in the spring before the race because the car must be finished in this period. Do not agree to participate in such an event unless the car is in all respects race-ready and unless there is no possibility of interference with the preparation schedule. The spring before the race is a very crowded time. Unless the car has been completed in the preceding fall or early in the spring, there will be no time, energy, or money to spend on an unveiling. Once such an event is scheduled, invitations issued, and plans set in motion, it will be nearly impossible to stop. Even if the production schedule should slip, the unveiling must go forward. It will control the project schedule because of the public commitments that have been made. This could lead to public embarrassment, rather than public acclaim, and loss of precious time.

Remember also that student volunteers do not necessarily share the school's public relations goals and have their own heavy scholastic obligations. Extra pressure put on them may cause a loss of workers during the most crucial period of the project. On the other hand, having a non-slippable milestone may help motivate the students. Both outcomes have occurred in the writer's experience.

The best scenario is to finish the car by early spring and to abide by the thumb rule: "substance first; public relations later."

14.4 Newsletter

14.4.1 Public Relations

A newsletter is a way of informing the public, sponsors, and Adopt-A-Cell program participants about the team's activities. It should contain articles about the progress of the car and preparations for the race. Leavening this strictly technical mix should be articles about members of the team. Parents should be particularly pleased about this. It tends to compensate them for giving up the company of their sons and daughters during holidays and summers spent working on the car. Also, it puts information about the folks on the team before potential employers.

14.5 Management of Funds

It is imperative to train the team to manage its money (and other resources) professionally. The project is too large and diverse for the advisors to maintain detailed management control. Proper accounting records must be kept by the team and reports made at administrative meetings. Otherwise there will be chaos and many headaches for the advisors. Competent student managers have to be found, trained, and held accountable. This of course also serves the purpose of the project, whereas detailed management by the advisors would work counter to it. Nevertheless, final control of funds must rest with the advisors. This should be exercised through the school's purchasing system. Usually, this system will have the advisors listed as the managers of the accounts used by the team. This means that the signature of an advisor will be required for purchases above a certain cost. However, local purchases below this cost may be handled by a voucher recognized by local merchants. The voucher may not require an account manager's signature. The advisor must work with the students to make sure that such a system is carefully supervised.

Choosing student managers must be carefully done, but the talent will be there. The challenge of the project will attract students who want to learn to be professionals and who are willing to make the sacrifices of time that are requisite.

Chapter 15
A Solar Car-Based Learning Community

15.1 Introduction

Most of the ideas in this chapter are originally from Thacher and Compeau (1999). That material is copyrighted by Roskilde University Press and is used with permission. I gratefully acknowledge the contributions of Dr. Larry Compeau, my collaborator for several years, and co-advisor of Clarkson University's Solar Car Team.

Chapter 1 asserted that acquiring the art of solving ill-posed problems should be the primary objective of education. This chapter explores four questions raised by that assertion. What should be the means for acquiring the problem-solving art? What knowledge and skills facilitate problem solving? What should be the general structure of a curriculum designed to teach ill-posed problem solving? Into what form might an institution evolve when it centers itself on teaching problem solving?

15.2 Acquiring the Art

The PBLC The ill-posed problem-solving art includes the art of collaboration. Hence, it must be learned in a collaborative setting. Formal collaborative associations, teams, are not necessarily the easiest or most efficient means of solving a problem. But ill-posed problems are often complex enough so that collaboration is the only practical alternative. On the other hand, the concept of a lone creator generating a solution in isolation is probably a myth. Collaboration, that is, idea exchange and complementary work, even if carried on only through the technical literature and between persons unknown to each other and separated by space or time, is part of the solution of every ill-posed problem.

A team of students and mentors working on a project constitutes a project-based learning community (PBLC). PBLCs have the defining characteristics given below.

1. Collaboration must not be artificial, but driven by the demands of the problem to be solved. A PBLC should therefore be centered on a project that is multidisciplinary, thus requiring collaboration between students and faculty from different

E. F. Thacher, *A Solar Car Primer*, DOI 10.1007/978-3-319-17494-5_15

academic departments. It should be functionally demanding so that all of the students are challenged in their complementary work.

2. A PBLC is not necessarily destroyed and created as semesters come and go. Its schedule is shaped by the requirements of its project, and not by the regular academic year schedule.
3. Students of all grade levels, freshman through graduate, may participate.
4. The project goal, not grades, provides the motivation.
5. The project forces engineering and business cooperation with industry and other professionals outside of the school.
6. Students in the PBLC have real responsibility and authority, and the possibility of success or the risk of failure that accompanies them.
7. Learning takes place primarily through mentored experience. The mentoring may come from faculty, from other professionals on- or off-campus, and from students in the PBLC.

Solar car-based PBLCs have all of the preceding characteristics. They operate continually. Developing and racing a solar car includes *many* technical and business issues, as the reader has seen. Beyond this, the relatively small amount of energy available drives the PBLC to seek the lowest weight, most efficient racing system. However, the possibility of failure, the safety requirements, the concurrent demands of the academic schedule, and the need to sell the project in order to acquire resources, provide constraints. These conflicts force trade-offs and tend to balance the project on the sharp edge of reality. Students must mature and grasp the requirements of professional work.

Motivation The characteristics of a PBLC motivate students to accomplish tasks that would be impossible in ordinary courses. This motivational engine is the PBLC's principal advantage. It unlocks the door to professional-level experience. This high-level experience, properly mentored, can produce graduates who are actual entry-level professionals, that is, junior engineers, not proto engineers.

Knowledge originally comes from the need to solve problems: it is created out of necessity or curiosity. After the problems are solved, the knowledge of them is written in books for students to study. In a PBLC, the need for solutions is reunited with learning and, therefore, tends to motivate study.

Integration Large multidisciplinary projects, such as a solar car, force students to apply integrated knowledge. For example, the design and manufacture of a solar cell array on a car is a balance of electrical, mechanical, and aerodynamic requirements. Students are often required to learn beyond their specialty. An engineering major may have to work with marketing majors to develop and present proposals to prospective sponsors, or to speak about the project to the local Rotary Club. Also, skills and knowledge already acquired, but untapped by conventional courses, are often utilized. The SCP makes use of machining and welding expertise, and radio communications experience. This characteristic gives students avenues for making contributions which are valued, even if not academic.

Business A PBLC is essentially a small business, embedded in an academic institution. It is product driven. The profit it returns to its members is experience, which cannot be conventionally measured. Its costs are direct, and financing them is part of the experience the PBLC supplies. The monetary benefits to the school are indirect, but real.

A PBLC may require office space, computers, service vehicles, manufacturing facilities, extensive travel, and insurance. If it is properly operated, a PBLC will continually seek to improve its plant and equipment to provide an ever-higher level of experience. There is no apparent reason why a PBLC should not rise to operate at the highest professional level, with state-of-the-art equipment. The institution must provide a supporting framework in the form of administration and design space, manufacturing facilities, schedule and organizational innovation, and flexibility. When a particular community's members realize that the potential of their PBLC is really in their hands, this will further fuel the motivational engine. So, most of the cost of its operation and growth should be born by the PBLC itself.

Mentoring The relationship between students and faculty in a PBLC tends toward that between senior and junior colleagues. This has proven much more fruitful than the somewhat adversarial relationship characteristic of conventional courses.

If asked to define teaching, many would consider it to be imparting knowledge. Further, they might describe it as taking place in a classroom and from a specially-trained teacher. However, following the definition of "education" adopted herein, teaching may be pragmatically defined as any activity that facilitates the acquisition of the art of the application of knowledge by the students in the PBLC. A mentor's intent must be to create an environment in which such activities may go on.

Mentors must be willing, and able, to relinquish some real authority to the students—and sometimes to allow them to fail, or nearly fail—while still maintaining overall control so that one can maintain safety, solvency, legality, and continue to progress. This is difficult and requires seasoning. It requires irregular working hours because to keep fully informed and to gain the trust of the students one must, to a certain extent, be present when they are working.

Faculty are used to being consulted and to pontificate. But to be an effective mentor, one must know when and how to speak, and when to keep silent. This is not always easily learned. One must always keep in mind that discovery is the best method of learning; what a student discovers is her own and has a vivid relevance. A remark by motion picture producer-director Steven Spielberg succinctly defines the art of mentoring (Corliss and Ressner 1997): "The delicate balance of mentoring someone is not creating them in your own image, but giving them the opportunity to create themselves."

Although one may have a Ph.D., one may still not be well trained in all areas of a large, multidisciplinary project. One must be willing to endure the loss of student-perceived omniscience to show them how research proceeds from the known into the unknown. Not every student is mature enough to retain her respect for the mentor when the perception of omniscience is lost. Respect must sometimes be built, or rebuilt, on shared accomplishment, rather than credentials.

Portfolio A portfolio provides a way of recording and giving structure to the experiential learning of the PBLC. It can help to create a symbiosis between the PBLC and the laboratories and classroom courses that relate to it, provide the basis for the mentoring staff to certify the experience credits of the students, establish cross-training requirements between the disciplines of a PBLC's project, and provide the requirements for training peculiar to the PBLC, such as solar race-team training. Portfolio also provides a place to record special qualifications and reward excellent work, say by inserting letters of commendation.

An element of the symbiosis, mentioned above, is to have portions of the portfolio incorporated as assignments in relevant courses and laboratories. Selling this idea to the relevant departments is a job for the mentoring staff.

The Solar Car Project (SCP) portfolio might have three elements: satisfactory written answers to a set of questions, passing oral examinations on the solar car's systems, and satisfactory performance of a set of operations. These three elements would be divided into basic and specialized levels of knowledge. Each student in the PBLC would complete both the basic portion of the portfolio and the specialization section appropriate to her position in the team. She could also be allowed to complete as many other specialization sections according to her ability.

The questions, oral examinations, and operational qualification requirements might be focused on topics or areas such as vehicle dynamics, body system, suspension and steering, wheels and brakes, power system, instrumentation, energy-management system, race operations, marketing and public relations, and project administration. A few sample questions, oral examinations, and operational qualifications follow. These are intended to suggest the level and scope of the material that should be considered.

15.2.1 Vehicle Dynamics

Basic Explain the origin and characteristics of each force acting on a solar car in the cruise condition, and how these forces are influenced by the characteristics and speed of the car.

Mechanical Specialist Using a mathematical model, predict the energy required by a solar car to traverse mixed (hilly and flat) terrain at different average speeds.

15.2.2 Power System

Basic How do solar cells convert solar energy into electric energy?

Electrical Specialist Design, construct, and test a small solar cell array to supply a direct current (DC) load. Compare its actual and predicted performance. Explain any differences.

15.2.3 Marketing and Public Relations

Basic Write an essay describing an event that required you to speak to the public about the SCP.

Business Specialist Develop and deliver a presentation to solicit a corporate partnership. Include visual, audio, and scripted elements. Present in a convincing and professional manner.

15.2.4 Project Administration

Basic Construct an organization chart of the SCP. Fill in the names, telephone numbers, and e-mail addresses of each manager and mentor. Show your position on the chart.

Mechanical or Electrical Specialist Participate in the development of a plan, a set of milestones, and a budget for the design, construction, and test of your portion of the car.

15.2.5 Examinations

Oral Each oral exam would be conducted at the car and consist of both theoretical and operational questions. The questions would be keyed to the section of the portfolio the student desired to complete. For example, a basic question from the wheels and brakes area would be to explain how the brakes work while pointing to and naming each component. The examiner (a qualified SCP member) would be free to take the examinee as far as she could go in the subject area.

Skill The operational skills demonstration would also be keyed to a subject area. For example, as part of the power system basic subject area, the student could be asked to demonstrate how to set up the solar array for charging (other team members would of course be needed here). Or, the student could be asked to show how to turn on and shut off the car, operating each switch in its proper sequence. Students specializing as race-team drivers would demonstrate race-qualification maneuvers, highway operation (highway convoy procedures, passing), urban operation, blowout procedure, and minimum energy operation.

15.3 Attainments that Promote Problem Solving

Listed below are the knowledge, characteristics, and skills that graduates should have in order to effectively participate in the solution of ill-posed problems.[1] Comments are included that support or amplify each item, especially as it relates to experiential learning in mentored learning communities.

1. Competence in employing the "design process."

The term "design process" refers to the process used to work through the solution of an ill-posed problem. This is the heart of problem-solving art. The parts of this process were illustrated by the house-building example in Chap. 1. The process cannot be effectively taught academically, that is, in the abstract. A student may study a book about design, receive an "A" on a test covering the book, but be unable to solve an actual ill-posed problem. The process must be taught by an experienced mentor.

The design process should be the unifying theme in education because it is the process used by all the persons who solve ill-posed problems, and because real problems are ill-posed. Then what would differentiate one student's educational path from the other would be the kind of problem being solved, its context, or the form the solution must take. An author produces a book; an engineer produces a machine; an artist produces a quilt.

If this unifying theme were woven through all education, it would revolutionize its results: Students would become creatively mature much earlier, and in larger numbers, because they would understand and utilize the design process routinely. Such is not the case now, of course; there is no such unifying theme. In grade school and high school, students take an eclectic set of courses in separate subject areas. In college, students enter "disciplines." These disciplines, such as mechanical engineering, are rubrics for sets of related specialty areas in which the faculty are studying, such as heat transfer or chemistry. But even though the specialty areas are related, there is generally no unifying theme.

Furthermore, well-posed problems, like the distance–time example described in Chap. 1, are pervasive in our technical-education system. Students are quite accustomed to, and expect, them. Thus, when students are given an ill-posed problem to solve for the first time, the problem's lack of definition is disconcerting. What should be done first? Where are the formulas? What specific result is required? What information should be used? It is as if an implied contract between the students and the educational process had been violated. Their task had always been: To study analytical techniques and facts, do well-posed homework problems, and then take well-posed tests. This was the definition of "learning," to them and to their teachers.

This culture must be changed to one that embodies Whitehead's idea. Examples from the other creative arts[2] confront engineering educators. A composer does not

[1] Based partly on a list in Redish (1988).

[2] Yes, engineering is *quintessentially* a creative art. Can anyone who has seen a well-executed solar car doubt this? Before you is a concrete expression of an ideal: transportation with minimum

simply study music theory, he composes. Painters and sculptors do not just study abstract techniques, they create objects of art. Who would think of awarding a music degree to a trumpet player who had neither performed a solo nor played in a band? In other professions, a heavy emphasis is placed on professional-level experience under tutelage.

2. An integrated understanding of mathematics, science and the engineering sciences, and applied disciplines built upon them.

"Integrated" means that the student should grasp the connections between these subjects, not just view them as disconnected specialties. A very effective way of forcing the integration of knowledge is to give students multidisciplinary problems to solve; the knowledge is integrated in its application, which is the natural state of things.

3. A strongly-developed physical intuition and an ability to apply a variety of insights to evaluate the plausibility of a proposed solution.

The design process is a decision-making process, whatever the problem to which it is applied. The basis of a decision can often be physical intuition, which brings insight. This is more effective than constructing an elaborated computer model for certain decisions, particularly in early phases of the design process when alternative concepts are being generated and selected. Competency in the application of intuition and insight is developed over time by mentored experience.

4. Competence in transforming a physical system into a mathematical model and in extracting meaning from that model by numeric or analytic solutions, approximation, or estimation.

Model-making is an important design-process tool. Students often can be quite expert at getting answers when the model has been constructed. Teachers tell them "Given that…" and "Assume that…" but this does not exercise them in the most important skills: To make the model and to understand the limits of the model and the errors it introduces. When students are given a physical object to model they are often quite reluctant to make modeling assumptions.

Partly because of this reluctance, there is a tendency to jump directly to sophisticated computer packages, and then to accept their results uncritically. Skepticism should be taught; naiveté in the design process leads to errors. Make a simple model first and find out which football field, or which part of the field, the solution is in, and to provide a check on the sophisticated model. Also, in the design process, time–money is often important. If a decision can be made with a calculation on the back of an envelope, it should be that The philosophy of modeling should be taught.

Confidence and skill in modeling comes with modeling experience, especially experience in which the predictions of the model can be compared to the actions of the object modeled. Mentored project experience creates many opportunities to teach and to learn modeling skills and philosophy.

---

waste. Thus it's beauty derives from the intelligence it expresses and because it is a realization of how our hearts know things should be, but are not.

5. Clear communication in written, oral, or graphical form.

Communication on a project team requires effort; it is not the casual conversation on familiar subjects or the quick exchange of greetings to which people are accustomed. It means discussing design choices, writing technical specifications, reporting calculations, keeping clear and complete records, and finally making drawings of the design that contain the information necessary to build it. This kind of communication was illustrated in the house-building example of Chap. 1.

This more intense, detailed, and focused communication is often difficult for students to achieve at first. It needs to be taught and practiced. Mentored experience, again, is requisite.

6. Competence in collaboration with professionals of differing backgrounds.

This characteristic cannot really be separated from characteristic 5, communication: The essence of collaboration is communication. However, there are other issues bearing on collaboration, or teamwork, that arise when training students to work in teams.

Students are not accustomed to working in teams. They have different academic schedules and thus different overall objectives and priorities, are accustomed to working at their own pace, and have widely varying levels of knowledge and maturity. Consequently, the work load is often nonuniformly distributed over the team, consistent attendance at meetings is difficult to achieve, and assignments made by the team to its members are often not completed when required.

Overcoming these obstacles and learning teamwork is a maturing process requiring time and consistent mentoring. The time required to thoroughly learn these skills is probably much longer than one semester. Joining a learning community as a first-year student would provide four years to learn the skills.

7. An understanding of the effects of a proposed action on the social, economic, man-made, and natural environments.

The interactions of the device being designed with the social, economic, man-made, and natural environments enters forcefully and specifically into the design process. It is these interactions that shape the design. As Chap. 8 illustrates, the entrance begins with the design specification where weight, cost, appearance, and other characteristics that constrain the design are listed. The limits of these are quantified, otherwise there is no ruler against which the design can be measured. The quantification is developed by benchmarking against other products, market studies, researching environmental regulations, experimental results, etc.

More generally, it is important for professionals not to be simply producers of technology for sale. The work they do affects, in some way, the life on this planet. So they must be aware of the social, economic, and environmental context of their work also from a philosophical perspective. Life is an ill-posed problem; humanity is the ultimate design team. What are the guides we have to solve this problem?

8. Knowledge of the range and history of ideas.

An engineer, or other professional, should know what ideas have been proposed and explored in both the narrow "literature search" sense and also in the broad cross-disciplinary and historical senses. What is the history of the effects of the genre of the technology that engages him? What is its relation to other areas of thought? It seems almost dangerous for an individual's intellect to be only narrowly informed. Would such an individual be too easily manipulated to be trusted with higher responsibilities?

9. Ethical professional behavior.

Struggling to solve a real complex problem in close collaboration with others of differing views tends to remove any abstract, academic sense about ethics. It replaces this sense with direct experience of their relevance to the problem at hand. There is an ethics of collaboration. It is difficult to collaborate with someone, one does not trust.

Ethical issues play an important role, sometimes in the design process, and frequently in engineering practice, as the Challenger disaster and other cases show. Profitability can be a false guide in the engineering business, but it is not easy to recognize when this is so. Experience in actual work, but with mentors present to point out ethical issues when they arise, strengthens the student's ability to navigate ethically in professional life.

10. A philosophy of continuous intellectual growth.

New ideas are constantly being uncovered. Some computer equipment retailers say that about every two months much of what they have for sale becomes obsolete. Global competition forces companies to be flexible. Consequently, they look for employees who are flexible, and can move from project-to-project with a minimum of retraining. The employee must adopt a policy of continuous learning to avoid obsolescence.

Ideally, experience in a PBLC forces students to recognize that they are responsible for their own learning; that in professional life it will not be prepackaged and handed to them. A properly-selected, posed, and mentored problem should force the members of the PBLC to study beyond their current academic training. However, not all students are mature enough to appreciate the opportunity to acquire this element of professionalism. Weaning these individuals from their dependency is one of the important activities of the mentor.

Beyond this, intellectual growth is necessary to life, to a sense of progress and unfoldment. In the long run, perhaps this is its strongest justification.

15.4 Curriculum Structure

PBLC Core Learning ill-posed problem solving is the primary goal of a curriculum, the project work (the application of knowledge to ill-posed problems) in the curriculum, rather than the accumulation of knowledge, becomes the most important part. It becomes the trunk of the tree, instead of just one of the branches because that is where the "art of the utilization of knowledge" is acquired. This experience would be provided by the integration of a set of PBLCs into the curriculum. The combination of the PBLCs, which emphasize mentored, but professional-level, design process experience, and the traditional courses, which emphasize knowledge, will the meet the educational objective. Both components would be present and support each other as an integrated system. Learning the art of the application of knowledge cannot be separated from learning the knowledge. The roots, branches, and trunk of the tree are one; they are related symbiotically.

Organization No organizational feature or rule should compromise the defining features of the PBLCs. Such compromise will limit or destroy the motivational engine. This engine is indispensable.

A line and task organization is required. The task groups are the PBLCs; the line organizations are the regular academic departments and schools. Each PBLC would be continuously supported by several academic departments, depending on the technical, social, and economic demands of the project. The support would have two levels: "embedded" and "ancillary." Embedded support from a department means the department would supply a teacher to the PBLC, academic support to its portfolio, and a basic level of financial support. Ancillary support means the department would only give academic support to the PBLC's portfolio. For example, the SCP would be embedded in the (local equivalents of the) Mechanical Engineering, Electrical Engineering, and Marketing departments. The Physics, Mathematics, Civil Engineering and Chemical Engineering departments would furnish ancillary support.

"Academic support" means that the essays and projects in the PBLC's portfolio would be integrated into the courses taught by the department in some combination of the following points: by direct assignment; by instruction in the subject areas supporting the portfolio; by consulting with individual students.

Each PBLC would have its own budget. Each school would support the PBLCs via the embedded departments' budgets. This would be a floor, seed money, and not intended to meet the entire expense of the PBLC; the PBLC must make industrial, and other outside contacts and raise support.

Let us call the PBLCs, the departments supporting them, and the facilities assigned to them the Experiential Learning Project (ELP). Some representative body, a Deans' Council perhaps, should have general oversight of the ELP and help to guide its growth. The objective would be to promote the evolution of the ELP into a "super-learning community," explained below. The presence of all the Deans would be necessary because of the integrated nature of the ELP.

Staff The staff of a PBLC should include faculty mentors from the embedded departments. It may also include advisors from other organizational units from inside, or from outside, the school. The more genuine connections the PBLC makes to the expertise present in the University and external communities, the more effective it will be in accomplishing its objective

The PBLC faculty mentors, with the student managers, set up the portfolio requirements for the PBLC, specific to the PBLC but based on the general rules herein, set standards and certify their attainment, insure safety, the proper management of funds, and establish a framework so that the product of the PBLC is properly designed, manufactured, marketed, and operated.

Mentors would be assigned to the project by their departments as part of the teachers' regular course load, and not a voluntary add-on. By agreement between the embedded departments, one teacher would be designated as in charge of the PBLC. When the schedule of a PBLC requires summer work, faculty would be compensated for summer time spent on the project.

A successful PBLC requires continuity of faculty presence and effort. Therefore, faculty assignments should last at least for two periods over which a PBLC goes through a characteristic cycle, such as the 2-year race cycle of the SCP. Departments would be required to insure that replacement faculty are trained, and not just simply rotated in with no preparation. If the success of the ELP depends upon the continued presence and energy of particular persons, the ELP will eventually fail. A goal of the Deans' Council should be to continually broaden the faculty base of ELP. Mentor training and rotation will help to do this.

Registration and Credit Students of any grade level could register for the PBLC as a course. Ideally, each PBLC should constitute a single course, not segmented by grade level. These might be called "Multi-disciplinary Project" courses, or MP courses. The enrollment should be limited to keep the PBLC to a practical size. Students not registered may earn credits if they execute independent study topics based on their PBLC work while volunteering, or use part of their PBLC work for a conventional, department-centered, or design course (PBLCs will exist side-by-side with conventional design courses at first). Graduate projects can sometimes be based on PBLC work.

15.5 Evolution of the School

Industrial and governmental "in kind" donations include, in effect, training for the learning community members. For example, Clarkson University's SCP members obtained a donation of wind-tunnel time at the Advanced Aerodynamic Laboratory in Ottawa and tested a quarter-scale model of their car there (Chap. 12). The wind tunnel engineers trained the team members. Team members have also helped high school students with solar car projects and made presentations at high schools and grade schools on the SCP. Building up such outside contacts and alliances incorporates them into the school. The "campus" becomes enlarged and decentralized. The school plus the outside organizations becomes a *super-learning community* linked by project-team visits and modern communications: electronic mail and perhaps video conferencing and groupware.

At the core of the super-learning community, would be the project teams both continuing and noncontinuing. The continuing PBLCs would be engineering competition teams or teams doing nationally-recognized work, such as designing and flying space shuttle payloads. Projects of this magnitude tend to attract sponsor support and are multifaceted enough to keep many students involved. So the actual number of continuing, core projects, would not be large. There would be a mechanism for establishing new projects, which projects would not necessarily be continuing.

Each evolutionary step should be managed such that the ELP always builds on the foundation of ideas that work, and that preserve the amazing driving engine of the PBLC: The thirst of the students for real experience. The evolution will probably never quite reach an end; the school must always be considering the next step of improvement of community building.

The connections made between academic departments by the necessity to support multidisciplinary projects at levels commensurating with national competitions will tend to unify the campus around this curriculum. The boundaries between departments will fade as the campus reorganizes to support multidisciplinary projects, rather than disciplines.

At first, this evolution may be most developed among the professional schools of Engineering and Business.

15.6 Educational Outcome

Business and engineering undergraduates would leave the school as junior professionals with a certain level of knowledge certified by course grades and a certain level of project experience certified by an approved PBLC portfolio.

As the super-learning community develops, the boundary between the world of its industrial and governmental members and the world of the school will blur. Companies will find the school's graduates require far less in-house training before becoming contributing professionals. Graduates will find the transition from mentored experience at the school to professional experience in industry to be an extension or continuation, rather than a discontinuous jump.

Note that the benefit of the ELP is not primarily the raising of grades, although that it will probably take place because of the motivation for study supplied by the PBLC. The primary benefit will be to supply an element that may be mostly missing in the school's curriculum: training and experience in the solution of multidisciplinary, ill-posed problems at a professional, or near-professional level.

References

Corliss, R., & Ressner, J. (1997). Peter Pan grows up. *Time*, May 19, 1997, p. 75.

Redish, E. F. (1988). From here to the future: The impact of the computer on college physics teaching. *Academic Computing,* 18.

Thacher, E. F., & Compeau, L. D. (1999). *Project-based learning communities at Clarkson University, in project studies—a late modern university reform?* (pp. 194–207). Denmark: Roskilde University Press.

Chapter 16
American Solar Challenge™ Regulations

Official Version April 30, 2001.

16.1 Purpose

16.1.2 The Fundamental Missions

The fundamental missions of the American Solar Challenge™ are to promote and celebrate educational excellence and engineering creativity. Fuelled by the spirit of friendly competition and teamwork, the American Solar Challenge (ASC) champions the creative integration of technical and scientific expertise across a range of exciting disciplines.

Our mission includes:

1. The support and encouragement of bright young minds to succeed in the fields of engineering, sciences, mathematics, in multidisciplined learning and in subsequent careers.
2. The creation of public awareness and enthusiasm, both for education excellence itself, and for the technologies that emerge from that excellence.

16.2 Administration

16.2.1 Application of Regulations

These regulations will apply to the Formula Sun-American Solar Challenge (the "Event"), which includes the selection of teams, registration of teams, the inspection of solar cars ("Scrutineering"), the qualification of solar cars (the "Qualifier"), and the cross-country competition (the "Rayce").

Used by permission. Intended as sample rules only. See http://americansolarchallenge.org for the latest version. See also http://americansolarchallenge.org/about/formula-sun-grand-prix/ for FSGP rules.

E. F. Thacher, *A Solar Car Primer*, DOI 10.1007/978-3-319-17494-5_16

16.2.2 Supplemental Documents

Additional documents may be distributed to all teams entered in the event to supplement these regulations. These documents will clearly state that they are a supplement to the regulations, and they will have the same force and effect as these regulations. If there is a conflict between a supplemental document and these regulations, the document having the later date shall take precedence. Supplemental documents specifically referenced in these regulations include the Formula Sun newsletter, official interpretations, and instructions for scrutineering.

16.2.3 Acceptances of Regulations

All persons or groups selected to participate in the event are assumed to know these regulations. Their participation in the event will constitute acceptance of them.

16.2.4 Interpretation of Regulations

Prior to scrutineering all interpretations must be published in the Formula Sun newsletter or posted to the Internet under "Official Interpretations" on the ASC page in order to become official. During and after scrutineering, all official interpretations will be announced at briefings and posted at headquarters and on the Internet. The only group authorized to interpret the regulations is the regulations committee.

16.2.5 Advertising, Promotion, and Publicity

All advertising, sales promotion, and publicity material produced by the teams or their sponsors concerning or referring to the event will refer prominently to the event as the Formula Sun-American Solar Challenge. All teams, by entering the event, specifically agree to abide by this regulation. By entering the event, all teams and team members agree to the use of their names and their likenesses in any publicity materials (brochures, magazines, videos, photographs, etc.) that may be issued by the event's sponsors or organizers.

16.2.6 Headquarters

During scrutineering, the Qualifier, and the Rayce, a headquarters will be established at the site of each function and will assume the management functions for the event.

16.2.7 Officials

A team of officials to conduct registration, scrutineering, the Qualifier, and the event will be selected by American Solar Challenge organizers. Officials having specific duties shall be announced to the teams through the Formula Sun newsletter and briefings.

16.2.8 Jury

A jury will be formed to evaluate protests on conformity with these regulations, to resolve team disputes, and assign penalties. In addition, the jury is empowered to decide cases not specifically covered by these regulations. The jury will be available to teams during the Rayce.

16.3 Entries

16.3.1 Entry Registration

The event is open to all to participate. Each team wishing to participate in the event must submit an entry package consisting of a team information sheet and a signed participation agreement. No team will be officially registered until the team information sheet and entry fee are submitted to headquarters. Thef schedule is as follows:

16.3.1.1 Open Class—US$ 2000

16.3.1.2 Stock Class—US$ 1000

16.3.2 Registration Deadlines

Registration opens May 1, 2000 and closes June 1, 2001

16.3.3 Number of Entries

In the interest of safety, the number of solar cars entered in the Rayce will be limited to 60 (sixty), and (with the exception of seeded teams—see **Sect. 16.3.4**) will be registered on a first come, first served basis.

16.3.4 Seeded Entries

The top two finishing vehicles in any International Solar Racing Federation (ISF) sanctioned event worldwide since 1996 will be awarded seeded status in American Solar Challenge provided they submit an intent to register before January 1, 2001 and complete their entry for American Solar Challenge before the registration cutoff date of June 1, 2001. These "seeded" teams must still comply with the requirements of these regulations to ensure their entry in the Rayce.

16.3.5 Faculty Advisor

Teams representing an educational institution must have at least one faculty advisor who will provide guidance as needed throughout the solar car design, building, and testing process. The advisor will be responsible for signing documents representing the school.

16.3.6 Technical Documents

Technical documents describing the solar car's structure, batteries, and solar cells must be submitted to ASC Headquarters by April 1, 2001 for approval. Early submissions will receive prompt review by headquarters. The technical information provided in these documents will not be made public. The information contained in each team's final submission must match the solar car presented at scrutineering.

16.3.6.1 Structural Report

Safety should be the primary concern with regard to the structural development and fabrication of the solar cars. The structural report must present and address the design issues involved in impact, roll over, and suspension scenarios. Particular attention should be paid to the roll over and impact protection systems for the driver. Document with calculations and/or testing, photos, drawings, and anecdotal references are acceptable. The entire document including appendices shall not exceed fifty (50) pages (not sheets) in length.

16.3.6.2 Battery Approval

All storage batteries used in the solar car must be approved by ASC Headquarters. Each team must provide a copy of the manufacturer's battery specification sheet, the material safety data sheet (MSDS) obtained from the battery manufacturer, and the following battery information:

a. Manufacturer's name and contact information
b. Stock number, type, or description
c. Module voltage (e.g., 6, 12, or 24 V)
d. Buss voltage
e. Number of modules to be used in the solar car
f. Manufacturer's specifications, including capacity (kWh), weight (kg), and cost (US$)
g. Spill/damage protocols and procedures (if these are not provided in the MSDS then the team must obtain this information from the manufacturer and submit it to headquarters with the MSDS)

16.3.6.3 Solar Cell Approval

All solar cells must be approved by ASC Headquarters. Each team must provide a copy of the manufacturer's solar cell specification sheet, and the following solar cell information:

a. Manufacturer's name and contact information
b. Stock number, type, or description
c. Manufacturer's quote for cell area (cm^2)
d. Manufacturer's quote for performance
e. Cost (US$) per cell
f. Cell area (cm^2) after trimming or cutting or placement on the solar car

16.3.7 Team Data

Each team must submit team photos and data sheets to American Solar Challenge Headquarters by May 1, 2001. The photo and data will be publicly released and used in event brochures. Late submissions will be omitted. Early submissions will not be made public prior to June 15, 2001 without permission of the team representative.

16.3.7.1 Team Photo

The team photo shall be submitted as a color print measuring approximately 20 by 25 cm. The photo must clearly show the solar car and the team members. Team members in the photo must be identified by name and by their company or institution when there is more than one company or institutional sponsor. The photos will be used in the ASC programs and other publications.

16.3.7.2 Data Sheets

The data sheet must include solar car weight (with battery but no driver), solar car dimensions, motor type and rating, solar cell type and manufacturer, estimated peak solar array power in both Raycing and charging configuration (overhead sun, clear sky), battery weight and estimated capacity, chassis description, braking system, and wheel type and size. All specifications must be provided in metric units (SI). The team leader, crewmembers, designated drivers, and faculty advisor(s) must also be listed.

16.3.7.3 Team Data Changes

Teams may change specifications of the solar car and crew up to the scheduled time of scrutineering, with the exception that solar cell and battery specifications may not change after April 1, 2001 without specific approval from ASC Headquarters. Any changes submitted after June 1, 2001 may not appear in print.

16.3.8 Registration

All people taking part in the event must be registered with headquarters. This includes team members, sponsors, officials, guests, and the media. Badges will be issued and used to obtain access to restricted areas. These badges must be visible at all times.

16.3.9 Crew Requirements

All team members involved in the ("crew") must present themselves at registration to complete all required forms. Team members will be required to complete and sign liability waivers and emergency medical information forms.

16.3.10 Driver Requirements

Only registered solar car drivers will be allowed to drive solar cars during the event. A team shall have a minimum of two (2) drivers available at all times. In addition to meeting the crew requirements, solar car drivers must be 18 years or older, present a valid driver's license, and must supply their own ballast container and ballast (sand or metal shot only). The official weight of each driver, including driving clothes, helmet, and shoes, will be 80 kg. If the driver weighs less than 80 kg, ballast will be added to make up the difference. If the driver weighs more than 80 kg, no credit will be given.

16.3.11 Insurance

All teams must purchase the liability insurance provided by the organizers or show a certificate of commensurate purchased insurance or self-insurance.

16.4 Event Components

16.4.1 Scrutineering

Each team registered for the event must submit their solar car for inspection prior to the Qualifier to verify compliance with these regulations. In addition, spot checks for regulation compliance may take place during and immediately after the Qualifier and Rayce, and the top five overall finishing cars will be impounded immediately following the Rayce for a final inspection.

16.4.1.1 Scrutineering Time and Location

The date and location of scrutineering for the American Solar Challenge is yet to be determined. The order of inspection will be determined by drawing. Teams that fail to present their solar car at their designated time will drop to the back of the queue, and will risk not having enough time to complete the scrutineering process.

16.4.1.2 Scrutineering Format

Scrutineering will involve inspection stations for sizing, body, electrical, and mechanical; plus dynamic tests to verify handling and braking performance. Instructions for scrutineering and a detailed description of the scrutineering tests will be distributed in advance to all registered teams.

16.4.2 Qualifier

Each team must successfully participate in the Qualifier, a 2-day track rally for solar cars, before they will be allowed to compete in the Rayce. The date and location of the Qualifier for American Solar Challenge is yet to be determined. A maximum of 60 (sixty) Qualified teams will be permitted to participate in the Qualifier. The team with the most officially logged laps will be declared the winner and will gain pole position for the Rayce.

16.4.3 The Rayce

A maximum of 60 (sixty) qualified teams will be permitted to participate in the Rayce, a cross-country rally for solar cars. Solar cars must Rayce in the same configuration used at the Qualifier. The team with the shortest official elapsed time will be declared the winner of the Rayce or class thereof.

16.4.4 Safety

Each team is responsible for the road-worthiness of its solar car. Passing scrutineering or implementing changes suggested in comments on the team's structural report does not relieve the team of any liability. All solar cars and support vehicles must be maintained in a safe, road-worthy condition and be operated safely at all times. A team may be disqualified and withdrawn from the event at any time if it is judged to be operating their solar car in an unsafe manner.

16.4.4.1 Team Safety

Each team is required to have at least one member trained in basic first aid, including CPR. Proof of training will be required.

16.4.5 Withdrawals

Any team wishing to withdraw must notify American Solar Challenge Headquarters in writing. All written withdrawals signed by the team representative are final. American Solar Challenge Headquarters may withdraw teams that do not meet the technical document deadlines or fail to present a solar car at scrutineering or the Qualifier.

16.5 Solar Car Regulations—Electrical

16.5.1 Power

Global solar radiation received by the solar car without artificial external augmentation is the only source of energy that can be used for propulsion, except for energy stored in the solar car's battery system at the beginning of the first day of Raycing. Wind energy as well as direct and diffuse radiation is the considered form of global solar radiation. With the exception of the effects of wind on the basic shape of the car, all components used to convert global solar radiation for propulsion shall be considered part of the solar array described below.

16.5.2 Solar Array

At any given moment, the solar array comprises all components that are involved in the conversion of solar energy for use by the vehicle. In addition to direct energy conversion components (such as photovoltaic cells), the solar array includes any reflective surfaces, refractive lenses, or thermal-cooling systems employed to increase power output. Components that carry or process the energy after conversion are not considered part of the solar array, nor are structural members whose sole function is to support the solar array. The entire solar array must fit within an imaginary right rectangular parallelepiped ("box") of limited size whenever the solar array is connected to the solar car's motor or battery. The "box" may not exceed:

16.5.2.1 Stock

For Classic Dimension Solar Cars

Up to 5 m in length, 2 m in width, 1.6 m in height. Furthermore, the product of the length and width, less any single rectangular region not occupied by solar array components, may not exceed 8 m^2.

For New International Standard (NIS) Solar cars

Up to 5 m in length, 1.8 m in width, 1.6 m in height

16.5.2.2 Open

For Classic Dimension Solar Cars

Up to 5 m in length, 2 m in width, 1.6 m in height. Furthermore, the product of the length and width, less any single rectangular region not occupied by solar array components, may not exceed 8 m^2.

For New International Standard (NIS) Solar cars

Up to 5 m in length, 1.8 m in width, 1.6 m in height. (The reason for the two sets of dimensions is to allow bodies and arrays built for previous ISF events to compete.

It is the intent of the organizers of ASC and other solar car organizers to phase out the "Classics" over a period of 3 or 4 years.)

16.5.3 Raycing Configuration

Whenever the solar car is moving under its own power, the solar array must be in its Raycing configuration. In Raycing configuration, the "box" must be defined such that the length and width lie parallel to the ground. Furthermore, all portions of the solar array must remain fixed with respect to the solar car chassis, in the same orientation and configuration used when the solar car was inspected during scrutineering.

16.5.4 Charging Orientation

Whenever the solar car is stationary, the solar array may be reoriented to maximize solar exposure for charging. Reconfiguration of the array is not allowed. In charging orientation, the "box" can have any orientation relative to the ground. Charging configuration will be demonstrated as part of scrutineering.

16.5.5 Electrical Connection

All connections between the solar array and the solar car must be carried by the solar car.

16.5.6 Water Spray

Ambient-temperature water from an external source may be applied to the solar array using hand-pumped sprayers if the water is applied while the solar car is stationary and the application does not present a shock hazard. This is a unique exception to the general requirement that cooling systems must be considered part of the solar array.

16.5.7 Solar Cell Technology Limitation

16.5.7.1 Stock Class

If photovoltaic technology is used, only solar cells that are listed on the ASC Suppliers list will be allowed. These will have been determined to be available to all registered teams at a price not exceeding US$ 10/watt for bare cells; teams may pay extra for cutting, tabbing, or lamination of the cells. Substantial modification of the crystal structure, junction, or metallization constitutes manufacture of a new cell. Teams or suppliers wishing to make an addition to the list must submit all appropriate data to ASC Headquarters by January 15, 2001.

16.5.7.2 Open Class

There are no limitations on cells that may be used.

16.5.8 Storage Batteries

All solar cars are allowed to store solar-generated energy in a battery system composed of individual modules having a weight determined by the technology used (see **Sect. 16.5.8.2**).

For both Stock and Open Class entrants, the battery weights will include all charge control devices that were packaged with the batteries. The actual pack to be used during the Rayce must be weighed. Sample batteries may not be used for weighing.

Adherence to weight limitations does not imply automatic battery approval. Battery approval forms must be submitted to headquarters before official approval may be issued. ASC officials reserve the right to refuse approval of modules.

16.5.8.1 Stock Class

May use up to 165 kg of sealed (non-spill) lead-acid battery. Battery size shall be based on the manufacturer's published specifications submitted by the team. The solar car storage battery may be composed only of rechargeable, commercially produced lead-acid modules. Batteries must be available in sufficient quantities to be accessible to all participating teams. The battery modules may not be modified in any manner, including the addition of electrolyte additives; case modification; or plate addition, removal, or modification.

16.5.8.2 Open Class

May use up to:

- 165 kg of sealed PB-acid battery
- 60 kg of NiMH battery (see section "Grandfathering of Ovonics NiMH" for exception)
- 100 kg of NiCad battery
- 30 kg of Li Ion battery
- 30 kg of Li Ion Polymer/Li Ion Alloy (polymer) battery

(This weight will be determined using scales provided by the organizers).

Grandfathering of Ovonics NiMH

Open Class entrants may use the Ovonics NiMH battery that was approved for Sunrayce 99 only if the batteries were purchased prior to January 1, 2000. Teams must submit the model number and proof of purchase date to Headquarters for approval. ASC officials reserve the right to refuse approval of modules considered under this category.

Hybrid Battery Packs

Allowances for hybrid packs will be based on percentages of the weight allowances for the types of modules used, i.e., if a NiMH/Lead acid hybrid pack comprises 50 % of the allowable weight for NiMH, then the lead acid allowance for that pack will be 50 % of the 165 kg allowance for lead acids. The total of the percentages used in the pack may not exceed100 %.

16.5.8.3 Supplemental Batteries

Supplemental, replaceable batteries carried in the solar car may be used to power only the following accessories: radios, electronic panel meters, driver ventilation fans (if solely for driver ventilation), main disconnect relay, horn, and data telemetry.

16.5.8.4 Other Storage Techniques

If a power condenser is used, the electric charge must be proved to be zero before the start of each day of the Rayce. If a flywheel is used, it must be proved not to be rotating before the start of each day of the Rayce. Fuel cells may not be used.

16.5.9 Battery Enclosures

All battery modules must be fully contained in enclosures that are electrically isolated from the solar car. The enclosures must be constructed from nonconductive, electrolyte-resistant material. The battery enclosure covers must be constructed from the same material used in the fabrication of the rest of the enclosure. The cover must be firmly secured. The resistance measured between the battery terminals and any portion of the solar car chassis shall be greater than 1 MΩ for applied potentials up to 500 V. The battery enclosures must be secured to the solar car chassis so as to prevent them or the modules within from coming loose in the event of an accident or rollover. Velcro fasteners/straps will not be approved. All sides of each battery enclosure, including top, must be marked using 10-mm-high letters with "Caution: Chemical Hazard" and "High Voltage" and any other standard hazard markings specific to the type of battery enclosed.

16.5.9.1 Battery Removal

Battery enclosures must be designed such that the entire set may be removed and placed in impound overnight. Cell/module level removal is allowed but discouraged.

16.5.9.2 Battery Stacking

Stacking the batteries is discouraged. If it is necessary to stack the batteries, a battery rack must be used. The rack must be made of nonconductive, electrolyte-resistant material that is strong enough to support the weight of the entire battery system. The rack shall meet the same electrical isolation requirements as the battery enclosures.

16.5.9.3 Battery Ventilation

Battery enclosures must be equipped with a forced ventilation system rated at a minimum of 280 L per min. It must operate whenever the battery system is electrically connected to the solar car or to the solar array. Such ventilation systems must exhaust to the exterior of the solar car and must be powered by the battery system.

16.5.10 Main Fuse

A separate fuse (not a circuit breaker) must be placed in series with the battery system and the rating must not exceed 200 % of the maximum expected current draw. All low-voltage taps from the battery system must be separately fused. All fuses must be placed first in series with the battery starting at the positive connection.

16.5.11 Battery Switch

The battery system must be equipped with a manually operated, high-current switch to quickly disconnect the battery from the electrical system. This switch must be capable of interrupting the full load current. The switch must be located within easy reach of the driver.

The switch must be plainly marked in letters at least 10-mm high as the “Battery Switch” with “ON” and “OFF” designations. These markings must be clearly visible to the driver inside the solar car and to rescue personnel outside the solar car (canopy removed); use two sets of markings if necessary. Relays for this purpose must be normally open, and power for the relay may be supplied by auxiliary batteries

16.5.12 Motor Switch

All solar cars must have a motor switch wired to disconnect all power to the motor from either the battery or the solar array. The switch must be able to interrupt full load current, and it must be separate from the battery switch. It must be within easy reach from the driver’s position and clearly marked in letters at least 10-mm high as the “Motor Switch” with “ON” and “OFF” designations. These markings must be clearly visible to the driver inside the solar car and to rescue personnel outside the solar car (canopy removed); use two sets of markings if necessary.

16.5.13 Cable Sizing

All electrical cables must be properly sized to expected system currents.

16.5.14 Electrical Shock Hazards

All exposed or easily exposed conductors, junction boxes, solar cells, etc., operating at greater than 36 volts must be protected from inadvertent human contact and must be marked “High Voltage” in letters at least 10-mm high.

16.5.15 Lighting

Solar cars must have amber front indicators, red or amber rear turn indicators and red brake lights which must all be clearly visible from 30 m in full sunlight. Turn signals must be located at the front extremity of the vehicle with a 1.5-m minimum left to right separation. Turn signals and brake lights must be located at the rear extremity of the vehicle with a 1.5-m minimum left to right separation. The geometric visibility of each light shall be 30° from center and 15° up and down. Additional brake lights may be centrally located if desired.

16.5.16 Horn

Solar cars must be equipped with a horn that can be heard at a sound power level between 75 and 102 dBA at a distance of 15 m in front of the solar car. The horn must be permanently mounted and must be acoustically coupled to the air outside the solar car.

16.5.17 Accelerator

Accelerator mechanisms on solar cars must be free moving, and when released, must return to the zero current position. If the solar car is equipped with cruise control, it must be designed with an automatic shut-off when the brake is activated.

16.5.18 Control

Acceleration, braking, and steering must be under the sole control of the driver.

16.6 Solar Car Regulations—Mechanical

16.6.1 Solar Car Dimensions

The solar car (including solar array) will have the following maximum dimensions when moving under its own power. These define both open and stock Classes (see **Sect. 10.10** for array dimensions)

16.1.1.1 Classic

Length = 6 m, height = 1.6 m, width = 2 m. When turning corners, wheels and wheel fairings may exceed these dimensions.

16.1.1.2 NIS

Length = 5 m, height = 1.6 m, width = 1.8 m. When turning corners, wheels and wheel fairings may exceed these dimensions.

16.6.2 Tire and Wheel Requirements

The solar car shall have a minimum of three tires in contact with the ground at all times. The wheels and tires shall be designed for the intended application.

16.6.3 Tire Ratings

Tires in contact with the ground shall be loaded and inflated within the manufacturer's rating at all times during vehicle operation. Each wheel and tire on a single axle must be rated for the full weight applied to that axle.

16.6.4 Dynamic Stability

All wheels and their suspensions, steering linkage and geometry will be inspected for safe operation in normal and adverse conditions.

16.6.5 Driver Cockpit

The driver's cockpit may not subject the driver to excessive strain during normal operation, and must be designed to protect the driver from injury in the event of an accident. The driver must be clear of moving parts and linkages, so as to provide adequate space for safe operation of the vehicle.

16.6.5.1 Seating Position

The normal driving position must place the driver's entire head higher than the highest point of his or her legs. No headfirst positioning is allowed for the driver.

16.6.5.2 Belly Pan

The cockpit must be equipped with a full belly pan to isolate the driver from the road. The belly pan must be strong enough to support the full weight of an 80-kg driver.

16.6.5.3 Roll Cage

All solar cars must be equipped with a roll cage that encompasses the entire driver. The roll cage shall be a fixed, integral part of the solar car structure. The protection provided for the driver in a collision must be documented in the team's structural report. In addition to providing collision and rollover protection, the roll cage must be designed so as to deflect body/array panels of the car away from the driver in the event of an accident. There must be 5 cm of clearance in all directions between the roll cage and the helmet of the driver seated in the normal driving position. The roll cage must be of steel tubing having a minimum carbon content of 0.18 %. The roll cage tubing must have a minimum outside diameter of 2.5 cm and minimum wall thickness of 2 mm. Alternate materials which afford equivalent protection for the driver are permitted, provided they are fully documented in the team's structural report.

16.6.5.4 Padding

The roll cage must be padded with energy-absorbing material wherever it may come into contact with the driver's helmet. This energy-absorbing material may be included within the required 5 cm of clearance. In addition, a headrest of at least 2 cm thick resilient material must be mounted behind the driver's head.

16.6.5.5 Crush Space

The driver, when seated, must have a minimum of 15 cm of horizontal distance between his or her shoulders, hips, and feet and the car's outer body surface.

16.6.5.6 Safety Belts

All solar cars must be equipped with a minimum of a five-point lap and shoulder belt (harness system). The use of safety belts is mandatory. The safety belts must be attached securely, as recommended by the manufacturer, to a strong component connected to a main frame member, or to a main frame member itself in the solar car. The harness must be attached with bolts and nuts; bolts threaded into a structural member or "insert" are not allowed. If a hammock-type seat is used, the safety belts must remain functional in the event of a structural failure in the driver's seat. Only commercially manufactured safety belts are allowed. They must bear the manufacturer's emblem, and they must not be modified in any way from the condition in which they were received from the manufacturer.

16.6.5.7 Fresh Air Circulation

Fresh intake air from vents or wheel openings must be provided for the solar car's driver.

16.6.5.8 Egress

The driver's cockpit must provide for the driver's unassisted exit within 10 s. Driver's doors and/or canopies may not be taped shut at any time.

16.6.6 Visibility

16.6.6.1 Eye Height

In the normal driving position with ballast on board, the driver's eyes must be at least 70 cm above the ground.

16.6.6.2 Windshield

All solar cars must have a windshield made of shatter-resistant material. The windshield must be free of excessive distortion. This will be tested by having the driver identify 2-cm high letters at a distance of 3 m through any of the required viewing angles referenced below. Solar cars must have a method to clear at least 0.1 m^2 of the windshield of rain. The clearing method must operable at all times and must be in use when it becomes necessary to use the windshield wipers on the team's support vehicles.

16.6.6.3 Forward Vision

From the normal driving position, the driver must be able to see at all times without artificial assistance: (1) a point on the ground 8 m in front of the solar car, (2) a minimum of 17° above the horizon on level ground, and (3) a full 100° to either side of center. To provide an "encompassing" roll cage, some elements of the roll cage may obstruct a portion of the forward vision. However, this view must be essentially unobstructed by the solar car structure so the driver can easily see the road and traffic.

16.6.6.4 Rear Vision

All solar cars must be equipped with a rear view system that will allow the driver at all times to see a vehicle 15 m directly behind the solar car and up to 30° off center.

The system must provide the driver with a single reflex type image. Having the driver identify 20-cm high letters at a distance of 15 m will test this.

16.6.7 Fasteners

All fasteners must be of suitable type, strength, and durability for their application, with the following minimum requirements:

16.6.7.1 Bolts

Bolts used in the steering, braking, suspension, seat mounts, safety harness, drive train, and battery box systems must at minimum meet SAE grade 5, metric grade M 8.8 and/or AN/MS specifications. Bolts must be of the correct length, and extend at least two threads beyond the nut. Bolts in tension must not have shaved or cut heads

16.6.7.2 Securing of Bolts

The bolts described above must be secured from unintentional loosening by safety wire, cotter pins, and nylon lock nuts. In difficult areas only, Inspectors may allow Loctite, or other means deemed appropriate. Lockwashers may not be used.

16.6.7.3 Hose Clamps

Hose clamps must not be used to secure any structural or critical members of the car. Their use to secure ducting or wire cables is allowable.

16.6.8 Covers and Shields

All moving parts must be suitably covered to prevent accidental human contact when the solar car is fully assembled. The driver must be shielded from contact with all steering linkage and other moving parts.

16.6.9 Steering Stops

The steering system must include steering stops to prevent dangerous or damaging steering travel.

16.6.10 Clearance

Interference or rubbing of the wheels with the solar car's body, wheel well, or structure at full steering lock or suspension travel is not permitted. Movement of rod-end bearings may not be obstructed in any axis throughout the full travel of suspension and steering. Other moving parts, such as the motor shaft, must not contact stationary parts except through properly designed bearings.

16.6.11 Ballast

Any solar car driver weighing less than 80 kg will require ballast to bring his or her weight to 80 kg.

16.6.11.1 Ballast Carrier

Teams must supply a single sealable container for the purpose of ballasting each driver. The container itself will be loaded to the weight required for the driver. This container must be shown to securely fasten to a structural member of the solar car and/or be demonstrated to remain fixed in the event of an impact.

16.6.11.2 Ballast Access

The ballast container and its identification and security markings must be visually accessible during driver changes.

16.6.12 Brakes

Solar cars must have a balanced, co-reactive, dual braking system so that if one system should fail, the solar car can still be stopped. All wheels must have at least one component of this dual system. The two systems must be operationally independent and may be either front/rear or redundant front and rear (one-sided systems, left or right, are not permitted). Hydraulic systems must have separate master cylinders. Regenerative brakes may not be considered as one of the braking systems.

16.6.12.1 Braking Performance

Solar cars must be able to repeatedly stop from speeds of 50 kph or greater with an average deceleration on level-wetted pavement exceeding 17 kph per second.

The time interval over which the deceleration is averaged shall be from the first indication that the driver should stop until the solar car comes to a complete halt. When braking, the solar car must not veer excessively to the left or right, or exhibit structural instability. The tire pressure and mechanical systems settings used in this test will be considered Raycing configuration.

16.6.13 Handling Performance

Solar cars must be able to negotiate a figure-8 course (of which the center circle of each half of the figure-8 has a radius of 4 m) with a 5-m-wide-lane without knocking over any of the cones or exhibiting signs of structural instability in less than 11 s per side.

16.6.14 Turning Radius

Solar cars must be able to make a U-turn in either direction, without backing up, such that all wheels remain within a 16-m-wide lane.

16.6.15 Graphics

Solar cars must prominently display their assigned number, Institution name, and the event logo such that they are clearly visible from a roadside vantage point. Additional graphics related to the team's Institution(s) or sponsors are permitted, provided they are neither offensive nor disruptive.

16.6.16 Solar Car Numbers

Each team registered for the event will have a unique number approved by American Solar Challenge Headquarters (positive integer, three digits maximum). This number must be clearly displayed on both sides of the solar car. Each number must have a minimum of 5 cm of unobstructed background color on all sides. These colors can be black on white, white on black, or another high-contrast color approved by American Solar Challenge Headquarters. The numerals themselves must be a minimum of 25 cm high, 12 cm wide (except the numeral one), and have a minimum brush stroke of 4 cm. Numbers containing more than one digit must have a minimum of 2.5 cm spacing between them.

a. Teams fielding a vehicle that has participated in previous ISF events and registering on time for the by American Solar Challenge have the right to retain the number they used formerly.

b. If a conflict arises, American Solar Challenge Headquarters will determine the numbers assigned.

16.6.17 Institution/Company Name

The name of the institution(s) or organization sponsoring the team must be displayed on the solar car. American Solar Challenge Headquarters must approve the use of abbreviations or initials. The institution's name shall be larger and more prominent than any team sponsor's logo or name.

16.6.18 Event Logo

The event logo must be applied on both sides of the solar car. The logo will be provided by American Solar Challenge Headquarters and will measure no more than 20 cm in height by 30 cm in width. The logo must be mounted with 5 cm of unobstructed background color on all sides.

16.7 Raycing Regulations

The Rayce will start in Chicago, IL on July 15, 2001 and will finish in Claremont, CA on July 25, 2001.

Any location where all of the competitors are released from the same point is termed a "Staged Start." The first stage of the Rayce will run from Chicago, IL to Rolla, MO, where all the vehicles will stage again according to rank at the time of staging. The second Staged Start will be in Barstow, CA.

The distance between the two staging areas is such that it is unlikely any vehicle, falling within these regulations, will complete that distance in one Raycing day. At the end of any Raycing day during which an entry has not reached the next planned staging area (and does not trailer to the staging area), the team will stop where they are. They will begin the next morning at the same point, following regulations established for non-staged starts.

16.7.1 Traffic Laws

During the course of the Rayce, all state and local traffic laws must be obeyed. Solar cars must observe a maximum speed limit of 65 mph (Note: while event organizers may or may not be aware of or enforce specific local regulations, under no circumstances does this imply that jurisdictions will not enforce local ordinances, laws, or regulations).

16.7.2 Team Uniforms

On Rayce days, from 6:00 a.m. to 9:00 p.m. team members shall wear uniforms representing their institution(s)/company(s). The only information or graphics approved to appear on the front of the team uniform (jacket, shirt, hat, or other wearable) shall be the institution/company name/logo, team name/logo, car name/number, and ASC logo.

Team sponsors may also be displayed, but only on the back of the team uniform shirt or jacket. If team sponsors are displayed, the event sponsor must also appear in a similar manner on the back of the team uniform. Artwork for ASC logo and for the event sponsors may be obtained from ASC Headquarters.

16.7.3 Rayce Time

Official clock time for each team each day of the Rayce will be based on the local time at that day's start line, as displayed by the officials. The same official time ("Rayce Time") will remain in effect for each team for the entire day (until midnight), even though that day's route may cross into a different time zone.

16.7.4 Drivers

Only one person, the authorized driver, may ride in the solar car at any time.

16.7.4.1 Driver Helmets

Drivers must wear a helmet while operating the solar car. The helmet must meet or exceed the Snell M95 or DOT motorcycle standard. Bicycle helmets will not be allowed.

16.7.4.2 Driver Shoes

Drivers must wear closed-toed shoes in the solar car. Sandals are not permitted.

16.7.4.3 Driver Ballast

Drivers and ballast will be identified with unique identification tags. The tags on the ballast carried by the solar car must match the tags on the solar car driver at all times.

16.7.4.4 Driving Time

Drivers may not drive more than a total of 6 h in a given Rayce day.

Water/Fluids The driver must have sufficient quantities of water/fluids in the cockpit to stay properly hydrated. *(Minimum 1 L for each driver switch).*

16.7.5 Briefings

A briefing will be held each staged morning and in the case of emergency. Attendance at this meeting by a team representative and driver(s) is required. Briefing notes and other daily updates will be available at checkpoints, posted to the Internet, broadcast by e-mail and available by phone. All official statements, rule interpretations, and special instructions will be contained in these postings.

16.7.5.1 Official Statements

These include starting order official statements, rule interpretations, and special instructions are announced at briefings.

16.7.5.2 On Non-staged Days

It will be the responsibility of the team to check available outlets for updates and instructions.

16.7.6 Starting Line

16.7.6.1 Staged

Each team will be assigned a start time each day, which will be distributed to the teams at the briefing. If the start of the Rayce is delayed then all assigned start times for that day will be adjusted accordingly. If the team leaves the starting line at their assigned time, then that becomes their official start time for that day. If the team leaves before their assigned time because they were moved forward in the queue by the start line officials, then the team's official start time is their actual start time. If the team leaves after their assigned time because they were not ready, then the team's official start time will remain their assigned time. The solar cars will be released from the official starting line at 60 s intervals beginning at 9:00 a.m. All solar cars must report to their starting position by 8:45 a.m. Each team's lead and chase vehicles must merge with their solar car after it leaves the starting line. The movement of all vehicles in the start-line area is under the control of the start line officials.

16.7.6.2 Non-staged

Solar cars will be released from their start point at 8:00 a.m.

16.7.7 Starting Order

The starting order for the first day of the Rayce will be determined at the Qualifier. On all other staged days, the order is based on the solar cars' official total elapsed time available at 7:00 a.m. of that morning, from shortest to longest. In case of a tie on any day, the first of the two teams to cross the previous stage's finish line will precede the other in the starting line-up.

16.7.7.1 Teams Not Ready

If a team's solar car, lead, and chase vehicles are not in their assigned starting positions at 8:45 a.m., the start line officials may, at their discretion, move all of the following cars up one slot, and the tardy team must move to the end of the starting queue.

16.7.8 Delayed Start

The start of the Rayce, at any stage, may be delayed if inclement weather or other hazardous conditions appear likely to pose a threat to the solar cars or their drivers.

16.7.9 Rayce Route

An American Solar Challenge Route book will be distributed to each team that qualifies for the Rayce. The Route book will contain information to direct the team along the official route. It will specify days, distances, directions, route numbers, maps, and points of reference. For a team to receive an official time, they must follow the official Rayce route.

16.7.9.1 Route Revisions

Due to unforeseen events, it may be necessary to detour. When advance warning is available, Rayce headquarters will correct the official route accordingly and provide revisions to the Route Book to all Rayce teams, or provide written revisions at the briefing, at checkpoints, by e-mail, and on the Internet.

16.7.9.2 Teams Departing from the Rayce Route

Any team leaving the Rayce route must rejoin the route at the same intersection where they left the route, or they will receive no credit for distance driven beyond that point.

16.7.9.3 Checkpoints ("Media Stops")

A checkpoint, otherwise referred to as a media stop, is a mandatory stop. Checkpoints will be established along the Rayce route. Checkpoints will remain "active" for a specified number of days (or portions of days) and "open" from 7:45 a.m. to 6:00 p.m. within those days. After the specified number of active days (or portions of days), checkpoints will be permanently shut down and will be referred to as "Closed." Failure to stop at an active checkpoint will result in no credit for distance driven beyond that point. A "missed Checkpoint" is one that has closed before a team arrives at it, not an active checkpoint that a team failed to stop at. The distance between consecutive checkpoints will be referred to as an "interval".

Checkpoint stops are mandatory for all solar cars. The length of the checkpoint stop time will be specified by Rayce officials (typically this will be 30 min). Checkpoints may be added or subtracted as needed by Rayce officials. Within the checkpoint area, the movement of all team vehicles shall be under the control of checkpoint officials.

Solar charging of solar car batteries and solar car maintenance are allowed during the mandatory checkpoint time. However, teams must not interfere with or block any other team's passage through the checkpoint. Teams unable to leave the checkpoint area after the mandatory time must move their solar car and support vehicles elsewhere. Mandatory time spent in an active/open checkpoint will not be factored into a team's "Interval Time."

Late Arrival at Active Checkpoints

If a team fails to arrive at an active checkpoint during open hours, they must remain at the checkpoint until open hours to officially check-in at that stop.

Arrival at Closed Checkpoints (Missed Checkpoints)

Teams that arrive at a checkpoint after it has closed (i.e., a missed checkpoint), must reach the next checkpoint while it is active and check-in during open hours.

If a team misses one checkpoint and has not trailered any section of that interval, the team may drive without trailering to the next checkpoint while it is active. They will then receive credit for the entire distance driven to that active checkpoint.

Teams that miss two consecutive checkpoints must arrive at the next checkpoint while it is active and check-in during open hours. Teams must then trailer directly to the end of that stage and will receive no credit for the distance to the end of that stage, or for the distance driven past the first missed checkpoint.

Teams that miss three consecutive checkpoints will be disqualified from the Rayce and their observer will be withdrawn.

16.7.10 Trailering

Should it become necessary to load the solar car onto a trailer for transport, it may be pushed onto the trailer. Battery charging from the solar array while trailering is allowed during non-impound hours.

16.7.10.1 Trailering Penalties

Teams will be assessed a prepublished driving time for each interval trailered, plus a penalty per uncompleted (driven by solar car) mile of the interval trailered of 1.5 min per mile. Once a team has decided to trailer, they must trailer to the next active checkpoint before they can drive again.

If a team trailers any section of an interval, they must check-in at the next active checkpoint. The team will receive a trailering penalty for the portion of the interval trailered. Credit will then be given for distance driven after the team has checked-in at an open checkpoint.

If a team drives through an interval but misses the checkpoint, and trailers for any portion of the interval to the next active checkpoint, they will receive a trailering penalty for that entire second interval.

16.7.11 Support Vehicles

All vehicles and trailers associated with a team other than the solar car itself, are support vehicles. These vehicles must be registered with ASC Headquarters.

16.7.11.1 Support Vehicle Graphics

All support vehicles, including trailers, must be marked with the team's solar car number (at least 15-cm tall) on both sides and the rear. The name of the team's sponsoring institution(s)/company must also be displayed prominently on each vehicle. Additional graphics are permitted provided they are neither offensive nor disruptive.

16.7.11.2 Scout Vehicle

Each team will be permitted to include a "scout vehicle" in their convoy for the purpose of investigating road and traffic conditions ahead of the solar car. The scout vehicle must meet US Federal Motor Vehicle Safety Standards. The scout vehicle must display the team's solar car number on its front windshield (at least 15-cm tall), in addition to both sides and the rear. The scout vehicle shall not be larger in height or length than a standard 15-passenger, full-size van. The scout vehicle may not tow a trailer, and must maintain at least a 500 m separation from the solar car caravan. The scout vehicle should not obstruct traffic or other solar car convoys.

16.7.11.3 Lead Vehicle

Each team must provide a support vehicle meeting US Federal Motor Vehicle Safety Standards to alert oncoming traffic to the presence of the solar car. This "lead" vehicle must travel within 500 m ahead of the solar car, with its headlights on and with roof-mounted flashing amber lights. The lead vehicle may not tow a trailer. The lead vehicle must display the team's solar car number on its front windshield (at least 15-cm tall), in addition to both sides and the rear. The lead vehicle shall not be larger in height or length than a standard 15-passenger, full-size van.

16.7.11.4 Chase Vehicle

Each team must provide a support vehicle meeting US Federal Motor Vehicle Safety Standards to protect the solar car from the rear. This "chase" vehicle must follow directly behind the solar car, with roof-mounted, flashing amber lights. The chase vehicle may not tow a trailer. The chase vehicle must display the team's solar car number on its front windshield (at least 15-cm tall), in addition to both sides and the rear. A sign provided by ASC Headquarters must appear on the rear of the chase vehicle to warn overtaking traffic of the solar car caravan. The chase vehicle shall not be larger in height or length than a standard 15-passenger, full-size van.

16.7.12 Other Support Vehicles

Other support vehicles may travel on the Rayce route, but must maintain at least a 500 m separation from the solar car caravan. (The intent of this rule is to allow support vehicles to be close but not to obstruct other traffic.)

16.7.13 Radios

The chase vehicle must be in two-way radio communication with the solar car at all times. All two-way radio channels must be registered with ASC Headquarters. All teams must also have a separately monitored CB radio in the chase vehicle tuned to a designated channel to communicate with other nearby teams and officials.

16.7.14 Passing Traffic

When six or more vehicles are lined up behind a team's chase vehicle, the team must pull over as soon as safely possible to allow the traffic to pass.

16.7.14.1 In Traffic

Teams need not disrupt their own progress to permit other vehicles to pass when they themselves are traveling at the posted speed limit or trapped behind other traffic.

16.7.15 Passing Teams

In the event that one team is overtaken by another, the overtaking team can signal their intention to pass by flashing the headlights of their lead vehicle between high and low beam. The overtaking team must also attempt to make CB radio contact with the team being passed to coordinate the pass. Once the overtaking team has signaled their intention to pass, the team being passed must facilitate the pass at the first available safe opportunity, either by slowing down by at least 8 kph (5 mph) in a zone where passing is permitted and feasible, or by pulling completely out of the traffic lane.

16.7.16 Drafting

Drafting by a solar car is prohibited. A solar car will be considered to be drafting if it continuously follows behind another vehicle at less than a three-second interval. The only exception to this is in congested traffic at speeds of 40 k/hr (25 mph) or less.

16.7.17 Pushing

Except for the following situations, solar cars may not be pushed or pulled from the time they are moved into their starting position for the daily start until they reach the finish line later that day. In no case shall regenerative braking be engaged while pushing or pulling the solar car.

16.7.17.1 Checkpoint

Solar cars may be pushed within the confined area of the checkpoint.

16.7.17.2 Emergency

In an emergency or breakdown situation, the solar car must be removed from the road. In this circumstance, the car may be pushed or lifted off the roadway. The solar car may then be pushed or lifted back onto the roadway at the same location where it left the roadway.

16.7.17.3 Weather

The solar car may be pushed onto and off of a trailer to protect it from the weather, provided the solar car is moved back to its original location after it is unloaded from the trailer.

16.7.18 Accidents and Reinspection

All accidents involving either solar cars or support vehicles must be reported immediately to ASC Headquarters. In the case of an accident involving personal injury, notification of the appropriate emergency medical services and public safety officials shall take priority. If a solar car is involved in an accident it must:

16.7.18.1 Team Inspection

Stop and be visually inspected by team members and the observer.

16.7.18.2 Reinspection

Be reinspected by an inspector at or before the next checkpoint. The inspector may require repairs prior to resuming the Rayce.

16.7.19 Timing

Timing and distance determinations for the event will be the responsibility of ASC timing officials. ASC Headquarters will recognize no other timing or distance information.

16.7.20 Raycing Hours

16.7.20.1 Staged start days

These are 9:00 a.m. to 6:00–6:30 p.m. (depending on official start time—i.e., an entry with an official start time of 9:00 a.m. may officially Rayce until 6:00 p.m. where an entry with an official start time of 9:32 a.m. may officially Rayce until 6:32 p.m.)

16.7.20.2 Non-staged start says

These are 8:00 a.m. to 6:00 p.m. (see **Sect. 16.7.22.2** for allowable stop time window)

16.7.21 Elapsed Time

Interval time

It will be based on the actual Rayce time that elapsed during the interval. Rayce time will begin from the official start time or the end of the last checkpoint time and will continue until the team enters the checkpoint at the end of the interval.

Teams Off Course

If a team departs from the Rayce route but then returns properly to the route and continues, their Interval time will be determined in the normal manner; no credit will be given for the time the team was off-course.

Official Interval Time

Official interval time will be the team's interval time plus any penalties and any protest filing fees. Note that protest filing fees are counted against the interval on which the protest is filed, whereas penalties are counted against the interval in which the infraction occurred. Thus, the official interval time is not final until after the end of the Rayce.

Official interval time = Interval time + Penalties + Protest filing fees

Official Elapsed Time

Each team's official elapsed time for the Rayce will be the sum of the team's official interval time for all of the intervals of the Rayce.

16.7.22 Overnight Stops

16.7.22.1 Finishes in a Staged Area

Once a team's solar car crosses the finish line of each stage, the movement of that team's vehicles shall be under the control of finish line officials. Specific areas will be designated for solar charging, impound, support vehicle parking, Rayce Headquarters, and food service. These areas will become the staging area for the start of Raycing the following morning. Solar cars may be pushed within and between these areas, but regenerative braking may not be used during such times.

16.7.22.2 Finishes in a Non-staged Area

Official Rayce time ends at 6:00 p.m. each day in the interest of safe bivouac teams may stop as much as 15 min before or as late as 30 min after their official stop time without penalty. The following day they must start as much early or late as they ended the night before. Example: Team A elects to drive 17 min late to find an appropriate over night venue. The following morning they may not begin Raycing until 8:17 min. Conversely, Team B elects to stop Raycing 7 min early. The next morning they may start as early as 7:53 a.m. Teams that elect to stop Raycing more than 15 min early will gain no additional credit. Teams that Rayce beyond 30 min past their official stop time will be penalized 2 min for every minute beyond that limit. These time adjustments do not apply to staged starts.

16.7.23 Impound

All registered and sealed batteries must be removed from the solar car and kept overnight in battery storehouses or under the direct supervision of the observer. Headquarters should be appraised of special issues for impound (i.e., other than ambient temperature, super ventilation needs, propensity to explode, etc.)

16.7.23.1 Impound Times

Batteries must be impounded by 9 p.m. each evening and will be released from Impound at 6 a.m. the following morning.

16.7.24 Accommodations and Lodging

All teams are responsible for team accommodations and food during the Rayce. Teams are responsible for their own reservations.

16.7.25 Charging Area

16.7.25.1 Staged

A charging area will be provided for the teams. Internal combustion generators will not be permitted within the charging area. Solar cars must be charged within this designated area.

16.7.25.2 Non-staged

Teams may choose appropriate charging areas.

16.7.26 Observers

Trained observers, selected and sponsored by ASC Headquarters, will travel with each team to alert the inspectors to possible infractions of these regulations, and to help teams deal with unforeseen events. The observer has the authority to warn teams when they believe that a rule infraction is imminent. Observers will determine official start times for non-staged starts. Observers may not interpret these regulations or give advice on Rayce strategy.

Observers will be rotated in their team assignments at checkpoints.

16.7.26.1 Observer Access for Inspection

Observers will be assigned to keep each solar car in sight from the release from impound to the time of impound each day. The observers shall witness any and all work done on the solar cars during this period. The observers must be allowed access to the solar cars for inspection of ballast during all driver changes.

16.7.26.2 Observer Record of Performance

The details of the activities of a team will be recorded in a logbook carried by the Observer. The team leader will be permitted to review the book each day; however, failure to do so does not make any record invalid. The records kept by the Observer include the Official Start Time, stopping times (including Checkpoint), the distances traveled, and any apparent rule infractions either by their assigned team or by any other team.

16.7.26.3 Observer Accommodations

During Raycing Hours

Teams must allow the observer the seat of his or her choice behind the driver in the chase vehicle. The observer must be able to see the solar car and read the chase vehicle's speedometer from this location, and must also be able to determine, at least periodically, how many vehicles are following behind the team.

Before and After Raycing Hours

Teams must provide a secure shelter for the observer and the battery impound box.

Meals and Lodging

Observers should be considered another team member for whom the team will supply adequate food, drink, shelter, and amenities.

16.8 Penalties

Any team failing to comply with these regulations during scrutineering, the Qualifier, or the Rayce will be penalized. Penalties range from official warnings to disqualification from the event. It is the responsibility of the chief inspector, with input from the other inspectors and the observers, to determine whether an infraction occurred, the severity of the incident, and the appropriate penalty. All time penalties will be submitted by the chief inspector to Rayce Headquarters for subsequent posting. Disqualification of a team from the event requires concurrence of the director. Penalties will generally be applied to total elapsed time on the official elapsed time sheet (posted by 7 a.m.) on staged days, at the start of non-staged days, or at checkpoints.

16.8.1 Posting of Penalties

Except for the last day, all compiled time penalties will be posted and broadcast by Rayce Headquarters by 8 a.m. each morning. On the last day of Raycing, time penalties will be posted no later than 30 min after the finish of the Rayce.

16.8.2 Conduct

Penalties, including disqualification from the event, may be imposed for improper conduct or the use of alcohol or illegal substances. Improper conduct may include, but is not limited to, improper language, unsportsmanlike conduct, unsafe behavior, or cheating.

16.8.3 Non-solar Charging of Batteries

After the start of the Rayce until the official finish, teams will be disqualified from the event for charging their solar car's storage batteries from any source of energy other than the solar car's solar array, without specific written instruction from Rayce officials. Such charging of a solar car's storage battery will constitute replacement and is subject to regulation **Sect. 16.8.4.**

16.8.4 Replacement of Batteries

Decisions to exchange (or externally recharge—see **Sect. 16.8.3**) all or part of a battery must be communicated formally to the team's observer or an Inspector. The penalty will be computed as follows:

Time penalty (minutes) $= 480 * (n + S)/N$, where:
n = number of replacement modules
S = sum of all modules previously replaced
N = total number of modules in solar car battery pack

16.8.5 Disturbing Official Battery Seals

Solar car batteries will be marked with an official seal. Disturbing these seals in a manner that prevents proper identification by Inspectors will be penalized as though all of the battery modules affected had been replaced.

16.8.6 Traffic Violations

Any solar car committing a traffic violation will be penalized up to 15 min for each violation. Any solar car driver who commits three traffic violations over the course of the Rayce will be individually disqualified from the event.

16.8.7 Failure to Allow Other Traffic to Pass

Any team failing to properly facilitate passing by traffic or other teams will be penalized up to 15 min for each offense.

16.8.8 Drafting

A penalty of up to 1 min will be assessed for each minute that a solar car drafts behind another vehicle.

16.8.9 Pushing

A penalty of up to 30 min will be assessed each time it is necessary for a team to push or pull their solar car in order to advance along the Rayce route. Teams pushing or pulling their solar car along the Rayce route for more than 15 s (except as in **Sect. 16.6.11.2**) will be assessed a time penalty. The penalty may be as great as if the team had made no further progress beyond that point on that day.

16.8.10 Improper Ballast

A penalty of up to 60 min will be assessed each time a team operates their solar car with ballast that does not match the solar car driver.

16.8.11 Failure to Impound

A penalty of up to 3 min will be assessed for every minute between 9 p.m. and 6 a.m. that a solar car's Raycing batteries are not in Impound.

16.8.12 Exceeding Size Specifications

Oversized solar arrays will be penalized up to 10 min per Rayce day per excess centimeter in each dimension beyond the allowed size specification. Oversized solar cars will be penalized up to 5 min per Rayce day per excess centimeter in each dimension. If both the array and car are oversized, both penalties will be applied.

16.8.13 Protests

Any team desiring to file a protest must do so by submitting an official protest (signed by the team leader) to Rayce Headquarters. Protests may be filed for any reason, including disputing a penalty levied against any team, correcting timing errors, or protesting the actions of another team. A "filing fee" of 10 min will be assessed against the team's official elapsed time for the day on which the protest is filed. The jury will hear all protests.

16.8.14 Protest Judgments

The decision of the jury is final and no further appeals are allowed. The jury will notify Rayce Headquarters of their decision, and Rayce Headquarters will then inform the affected teams. The jury may refund some or the entire filing fee, which will be credited to the day the filing fee was assessed.

16.8.15 Opportunity to Be Heard

Protests will normally be heard by the jury at the earliest possible jury sitting. It may be necessary in some instances for the jury to postpone the hearing on a protest.

16.8.16 Time Limit

Except for the last day, all protests against penalties must be filed by 8:30 p.m. the day the penalty is posted. Protests that do not directly relate to a penalty must be filed by 8:30 p.m. on the day after the offence occurred. On the last day of Raycing, protests for any purpose must be filed within 60 min after the finish of the Rayce.

These regulations are copyrighted by New Resources Group (NRG) April 2001 and may not be sold or used without written consent of NRG.

"American Solar Challenge," "Formula Sun," and the Formula Sun Logo are all trademarks of NRG.

Chapter 17
The Drag Build-up Method

17.1 Description

A way of making an estimate of the drag area is to model the car as a composite of shape elements which have known drag coefficients. The drag areas of these shape elements are added to give the drag area of the car at a particular speed. This is called the *drag build-up* method.

Figure 17.1 shows two views of a solar car. An arrow representing the drag of each shape element is shown attached to that element. The front view shows the profile areas of each element. The profile area of the car is the sum of the element profile areas; likewise, the total drag force is the sum of the element drag forces. But each element drag force can be represented by Eq. (2.1), and dividing out the dynamic pressure, q, which is the common factor, gives an equation for the drag area of the composite shape. Therefore, if there are n_E shape elements, the drag area estimate would be, including ventilation drag,

$$c_D A_D = \sum_{i=1}^{i=n_E} (c_D A_D)_i + c_V A_D. \tag{17.1}$$

Both viscous friction and separation are included in the drag coefficient measurements of the individual shape elements. Chapter 18 explains how to estimate the ventilation drag, represented by c_V in Eq. (17.1).

Corrections When shapes are assembled into a car or brought near the ground, their drag contributions are altered by the change in the airflow caused by the nearness of the other shapes or the ground. Also, the surface of an actual car will usually be rougher than that of a more idealized mathematical model and the smoothed shape elements tested in wind tunnels.

Furthermore, objects such as rearview mirrors and antennas must be attached to some of the shape elements. The drag contributed by these protuberances must also be included.

E. F. Thacher, *A Solar Car Primer*, DOI 10.1007/978-3-319-17494-5_17

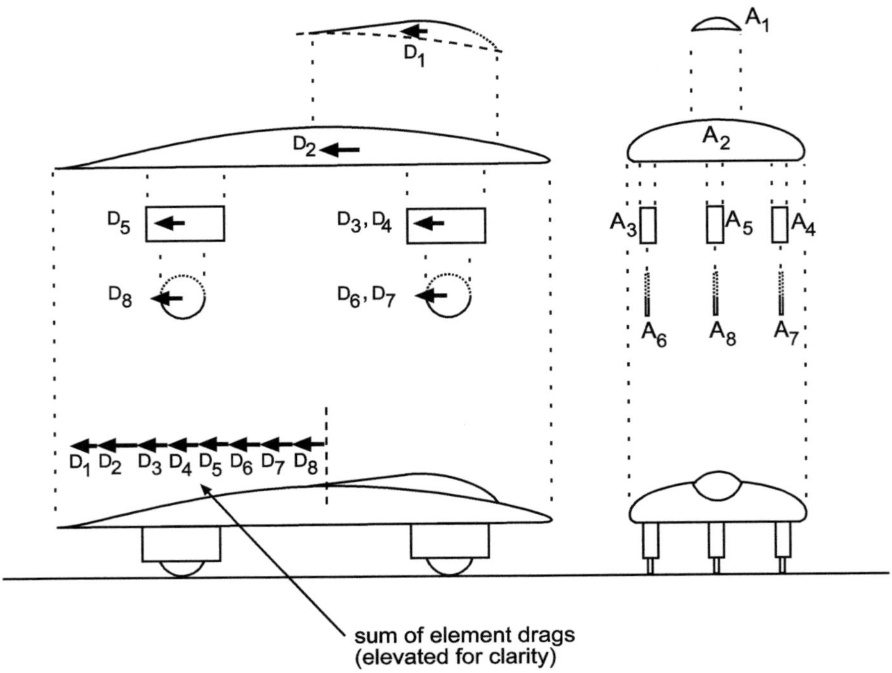

Fig. 17.1 Drag elements

The drag measurement of a shape element that has been taken free of the influence of the other elements, the ground, the roughness that comes from fabrication, and protuberances is called a *free air* measurement. It is denoted by $c_{D\infty}$. Corrections for interference, ground effect, protuberance, and roughness must be added to the free-air drag to produce an estimate of the actual drag. When all drag corrections have been referred to the element's profile area, A_{Di} in Eq. (17.1), the corrected drag coefficient for the a shape element may be found from

$$c_D = c_{D\infty} + \Delta c_{DG} + \Delta c_{DI} + \Delta c_{DP} + \Delta c_{DR}. \tag{17.2}$$

The subscript *"G"* denotes ground effect, *"I"* interference effect, "P*"* a protuberance, and *"R"* refers to extra roughness. The corrections, each beginning with "Δ", increase (usually) the drag from the free-air measurement.

Flow Regime The discussion in Chap. 2 pointed out that most of the operation of the solar car will be in turbulent, separated flow with Reynolds numbers in the $0.5(10^6)$–10^7 range. In this range the drag coefficient increases slowly with Reynolds number, i.e., essentially the vehicle's speed, as Fig. 2.6 shows. So it is possible to speak, in a shorthand way, of "the" drag coefficient, as if it were a constant. By this is meant the slowly increasing drag coefficient in the characteristic Reynolds number range mentioned above. Note also that the drag coefficients of very blunt bodies, like the disk shown in Fig. 2.6, are nearly constant once separation occurs and quite insensitive to the flow regime in the boundary layer.

When combining shape elements and applying corrections, as described earlier, we must use data often taken at different Reynolds numbers. So we rely on the insensitivity of the drag coefficient to Reynolds number changes when the flow regime is separated on a small, blunt body, such as a rearview mirror, or turbulent on a more streamlined body, such as a basic body in the shape of an ellipsoid with a large fineness ratio. As their individual drag coefficients are insensitive to Reynolds number, combining these elements produces a vehicle drag coefficient estimate insensitive to Reynolds number and this is what we observe in testing.

Shape Geometry Figure 17.2 shows a side and front view of a wheelless, basic-body shape. The length, L, and profile area, A_D, already defined, are shown. The *camber line* is the mean line of the shape; that is, it passes through the centroid of every cross-section between the nose and tail. The *chord* is the straight line connecting the points where the camber line intersects the boundary at the nose and tail. The *camber*, b, is the perpendicular distance between the camber and chord lines at any location along the chord. The *pitch angle*, π, is the angle between the chord line and the horizontal, positive up. The chord length, L_C, and the body length, L, are related by $L/L_C = \cos\pi$. The ratio of the minimum ground clearance to the width, h_{min}/W, is called the *clearance ratio*. The ratio of the maximum camber to the chord, b_{max}/c, is called the *camber ratio*. The ratio of the height of the profile area to the width, h_D/W, is called the *aspect ratio*. The ratio of the length to the diameter of a circle of area A_D, L/d, is called the *fineness ratio*.

Steps The drag build-up method may be divided into eight steps.

1. Set the design ambient temperature, pressure, and airspeed relative to the car. Calculate the air density from the ideal gas law.
2. Make at least side- and front-view drawings of the vehicle to scale on cross-hatched paper (see Fig. 9.2).

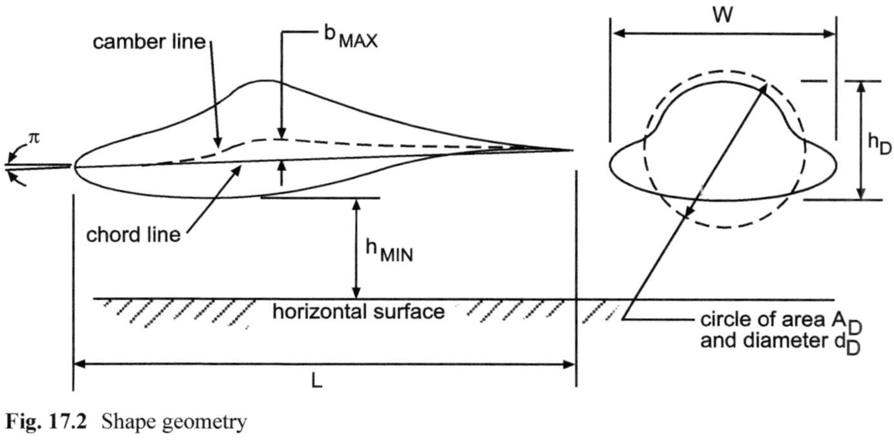

Fig. 17.2 Shape geometry

3. Divide the car into shape elements for which free-air drag coefficients are available.
4. Calculate the profile area of each shape element and that of the car by counting the scaled squares or partial squares their respective front-view drawings include, allowing for overlap.
5. Estimate the free-air drag coefficient of each shape element for the design conditions and the characteristic length of the shape element.
6. Correct the free-air coefficients for the effects symbolized in Eq. (17.2).
7. Sum the element drag areas as shown in Eq. (17.1) and include the ventilation drag, using the method of Chap. 18 .
8. The drag coefficient is the quotient of the drag area and the vehicle's profile area found in step 4.

Two examples of the application of steps 1 through 8 are presented in Chap. 9. Hoerner (1965) uses this method to predict the drag of an ME-109-G, a German fighter aircraft used in World War II.[1]

17.2 Shape Elements

The remainder of this chapter is devoted to presenting free-air drag coefficient information for shape elements and protuberances, and means of correcting for interference and ground effect. Some information is presented as graphs. These were constructed by carefully measuring points from the data in the referenced source and then fitting interpolating curves through these points. Each such curve has a correlation coefficient[2] of at least 0.99. This information is limited to that required for the examples in Chap. 9. However, a very large amount of information is available in the literature. Consult the references at the end of this chapter.

Ellipsoid Table 17.1, from White (1986), presents drag coefficient data for ellipsoids of various fineness ratios and circular cross-sections. The data are for both laminar separated flow and turbulent separated flow. The reference area is $\pi d^2/4$. To interpolate in Table 17.1, use $c_D = 0.4311(L/d)^{0.3994}$ for laminar flow and $c_D = 0.1835(L/d)^{-0.4590}$ for turbulent flow.

Table 17.1 Free-air drag coefficient of ellipsoid ($R_d \geq 10^4$)

L/d	Laminar	Turbulent
0.75	0.5	0.2
1	0.47	0.2
2	0.27	0.13
4	0.25	0.1
8	0.2	0.08

[1] Page 14–3

[2] As previously explained: a measure of the goodness of the fit, with 1.0 indicating an exact fit.

Airfoils The zero-lift drag coefficient data in Table 17.2 was read from the airfoil section data in Abbott and von Doenhoff (1959) and that in Table 17.3 from Bullivant (1941). "R_C" is the Reynolds number based on the section chord. The reference area used to define c_D is the unit planform area, that is, the product of the chord and the unit span. Hence, to convert to a profile area reference,

$$c_{D_{profilearea}} = c_{D_{planformarea}} \frac{c}{t}. \tag{17.3}$$

The tables show both four-digit and six-digit airfoil designations. The meaning of the four- and six-digit designations is given in the next two Tables (17.4 and 17.5).

Wheels Adding wheels to a basic body shape can give the largest increase in drag. For example, Morelli (1983) reported a drag coefficient increase of about 0.09 from adding four wheels to a basic body which had an initial drag coefficient of about 0.06. Thus the wheels alone increased the drag by more than the drag of the entire basic body.

A wheel is essentially a short, rotating cylinder. In free air, friction and separation cause drag on the wheel. When the wheel is rolling on the ground, ground interference comes into play. Finally, when the wheel is installed on a car inside a wheel housing, additional effects are induced by the housing and the external flow about the car. Clearly this is in general a complex situation not easily reduced to simple rules.[3]

However, Cogotti (1983) pointed out that according to Morelli (1969), the drag coefficient of an isolated wheel, when based on the profile area of its unshielded

Table 17.2 Free-air airfoil section drag coefficients (Lift = 0)

R_C	$3(10^6)$	$6(10^6)$	$9(10^6)$
NACA section			
0006	0.0043	0.0050	0.0053
0009	0.0052	0.0055	0.0055
0012	0.0057	0.0057	0.0057
2415	0.0070	0.0065	0.0065
2418	0.0074	0.0066	0.0065
2421	0.0078	0.0070	0.0070
2424	0.0087	0.0081	0.0075
63_2–015	0.0053	0.0051	0.0048
63_3–018	0.0057	0.0052	0.0049
63_4–021	0.0064	0.0056	0.0053

Table 17.3 More free-air airfoil section drag coefficients (Lift = 0)

$R_C(10^{-6})$	1.75	2.00	2.50	3.00	4.00	5.00	6.00	7.50
NACA section								
0018	0.0080	0.0079	0.0077	0.0076	0.0075	0.0073	0.00715	0.0070
0025	0.0089	0.0088	0.0082	0.0081	0.0081	0.0081	0.0081	0.0081

[3] Fortunately, solar racing cars can take advantage of some options that passenger cars usually cannot. Solar racing cars can use wheels with smaller thickness-to-diameter ratios, and thus smaller drag areas. They can partially enclose the wheels in streamlined fairings or in nearly-sealed wheel housings.

Table 17.4 NACA four-digit airfoil designation

Digit	Meaning
1	Maximum distance of mean line from chord (%c)
2	Distance from leading edge to maximum camber ($c/10$)
3 & 4	t/c (%)

Table 17.5 NACA six-digit airfoil designation

Digit	Meaning
1	Series designation
2	Position of minimum pressure from leading edge (c/10)
3	c_L range ± design c_L of low drag (10ths)
4	Design c_L (10ths)
5 & 6	t/c (%)

portion, is approximately unchanged when the wheel is partly enclosed by the fairing. These results suggest the thumb rule that the drag of the wheel is proportional to the profile area of its unshielded portion,

$$A_{DW} = hb_W, \tag{17.4}$$

where, A_{DW} is the unshielded profile area, h is the height of this area above ground, and b_W is the width of the wheel.

Figure 17.3 shows the drag coefficient of an isolated, stationary wheel as a function of yaw angle. Accounting for the forgoing discussion, the drag of a wheel (the jth one) would be

$$D_{W_j} \approx qc_{DW_{isolated}} A_{DW_j}, \tag{17.5}$$

Where, $c_{DW_{isolated}}$ would be taken from Fig. 17.3, and q is the dynamic pressure defined in Chap. 2. The equation of the curve in the figure is

$$c_{DW_{isolated}} = 0.5272 - 3.2289(10^{-4})\beta + 4.8399(10^{-4})\beta^2, \tag{17.6}$$

Where, β is the yaw angle in degrees. Interpret this as the "free-air" drag coefficient of the wheel.

The airflow changes direction at the front of the car in order to pass around the car. Therefore, the wheel yaw angles will probably differ from zero for vehicles with two wheels in front.[4] Figure 17.3 shows that this will probably cause the drag to increase.[5] The yaw angle will depend upon the direction of the relative wind near the wheels, which will be difficult to estimate.

[4] The flow would be symmetric with respect to a centered single wheel.

[5] The drag coefficient passes through a maximum at $\beta \approx 15^\circ$ (not shown).

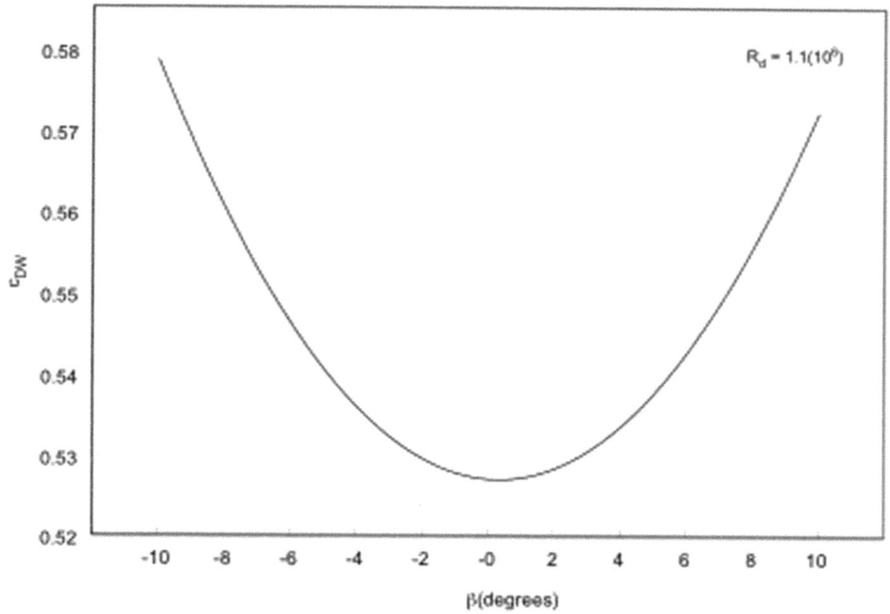

Fig. 17.3 Drag coefficient of an isolated wheel. (adapted from Cogotti 1983)

Cogotti's data show that the rotation of the wheel may slightly increase or decrease the drag of wheels in housings. Which effect occurs depends upon the inner and outer housing shape and the location of the wheels relative to the front of the car. We will make no attempt to account for wheel rotation.

Cogotti also found that placing a flat or nearly flat fairing over standard wheel rims decreases the drag of the car (not the wheel alone) by an amount

$$\Delta c_{D_{rf}} = -0.009 \pm 0.003. \tag{17.7}$$

The subscript "*rf*" denotes "rim fairing."

Protuberance This category includes items such as rearview mirrors and antennas which are small enough not to cause significant interference drag. They can therefore be included using only their free-air drag characteristics. In addition, as Pershing and Maskai (1976) pointed out, their characteristic lengths are small enough to ensure that the flow about the protuberance is in the laminar-separated region. Based on information in Hoerner (1965) the drag coefficients for cylindrical protuberances, such as flat mirrors and antennas, are all about 1.1 and for cup-like shapes which are convex in the upwind direction, are about 0.4.[6]

[6] Pershing and Maskai also point out that such items may be located in regions of accelerated or locally-separated flow. This will change their drag contributions.

17.3 Interfering Flows

For most of the plots in this section, the original data has been normalized by the tested $c_{D\infty}$. This helps to minimize dependence on the airfoil shapes tested.

Parallel Airfoils Consider the two identical symmetric airfoils used as fairings over the rear wheels of the car in Fig. 17.1. Figure 17.4 shows first the flow around the airfoil shape when tested independently at zero angle of attack to the wind.[7]

There is no lift force on the airfoil because the angle of attack is zero. Thus the free-air drag coefficient could be selected from Tables 17.2 and 17.3. Figure 17.4 shows that free-air drag exists when y_m/t is greater than 5.

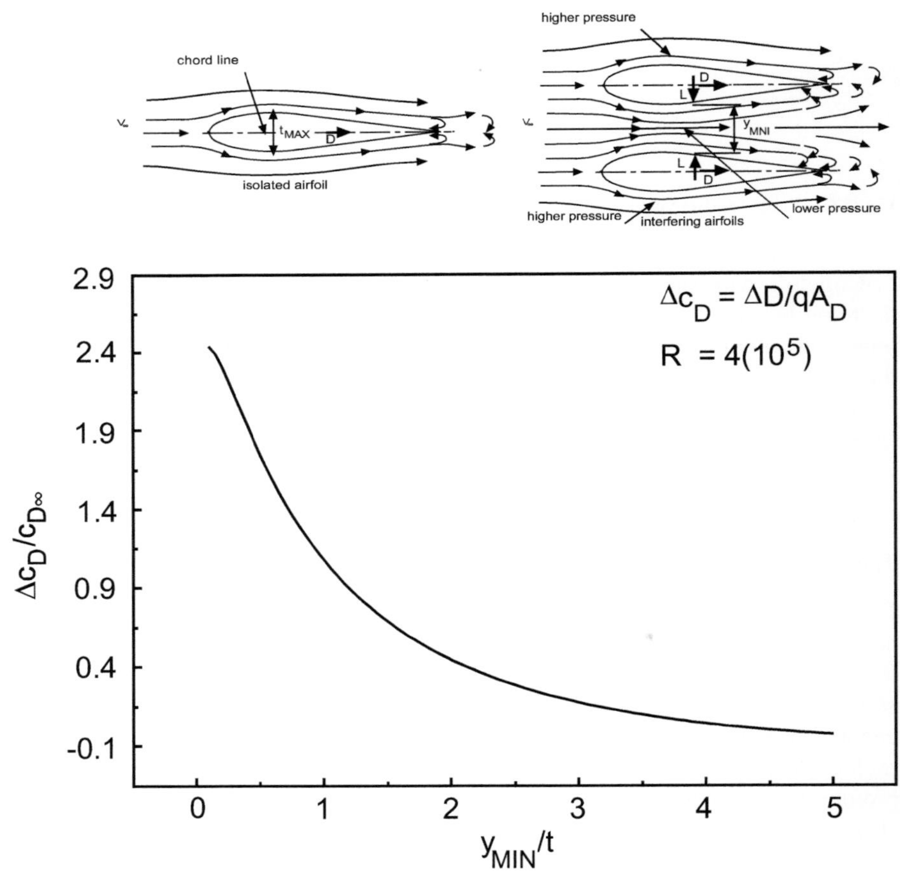

Fig. 17.4 Interference between parallel airfoils

[7] The angle of attack is the angle between the chord line and the relative wind.

When the airfoils are brought near each other on the car, the symmetry of the flow is disturbed because of the Bernoulli effect (Chap. 2). The figure shows that the lower pressure region created between the fairings by the acceleration of the flow creates an inward-directed "lift" force with an associated increase in drag. The increase in the drag coefficient over the free-air drag has been plotted in Fig. 17.4 as a function of y_{min}/t, the minimum spacing divided by the airfoil thickness. The curve was based on the data reported in Biermann and Herrnstein (1933).

$$\frac{\Delta c_{DI}}{c_{D\infty}} = 2.819929 - 2.084828\frac{y_m}{t} + 0.614002\left(\frac{y_m}{t}\right)^2 - 0.084466\left(\frac{y_m}{t}\right)^3 + 0.00450\left(\frac{y_m}{t}\right)^4. \tag{17.8}$$

The interference is zero for y_{min}/t more than about 5.

The drag increase is defined as

$$\Delta D = (D_1 + D_2)_{interfering} - (D_1 + D_2)_{free}. \tag{17.9}$$

In the parallel airfoil case, the symmetry of the geometry implies that $D_1+D_2=2D_1$ or $2D_2$, and that the total profile area will be twice the profile area of one airfoil. Such is not the case for airfoils in tandem.

Tandem Airfoils Figure 17.5 shows two identical symmetrical airfoils in tandem. Cogotti also found that placing a flat or nearly flat fairing over standard wheel x/c is greater than 4. The free-air drag coefficients could be selected from Tables 17.2 or 17.3. The upstream and downstream curves are, respectively,

$$\frac{\Delta C_{D_1}}{C_{D\infty}} = 0.000143\exp\left[\begin{array}{l} 26.395047\frac{x}{c} - 24.782729\left(\frac{x}{c}\right)^2 \\ +10.118397\left(\frac{x}{c}\right)^3 - 1.911910\left(\frac{x}{c}\right)^4 + 0.13672\left(\frac{x}{c}\right)^5 \end{array}\right] \cdot \tag{17.10}$$

and

$$\frac{\Delta C_{D_1}}{C_{D\infty}} = 2.630656 - 0.575883\ln\frac{x}{c} + 21.487285\left(\ln\frac{x}{c}\right)^2 - 37.823911\left(\ln\frac{x}{c}\right)^3 + 25.489452\left(\ln\frac{x}{c}\right)^4 - 6.118458\left(\ln\frac{x}{c}\right)^5. \tag{17.11}$$

Hoerner (1965) assumes that the increase in the drag of the rear airfoil is caused by separation induced by the wake of the upstream airfoil. The decrease in the drag of the upstream airfoil is explained by the increased static pressure between the airfoils when they are close together. Note that these two effects tend to offset one another so that the sum of the interference corrections is approximately zero.

Airfoil-Wall Figure 17.6 shows an airfoil attached to a wall. According to Hoerner (1965) the interference drag is

$$\Delta c_{Dt} = \frac{\Delta D_I}{qt^2} = 0.75\left(\frac{t}{c}\right) - \frac{0.0003}{\left(\frac{t}{c}\right)^2}. \tag{17.12}$$

Hoerner writes that this interference drag is independent of the span of the airfoil. Hence, it is referenced to the "thickness area," t^2, instead of the profile area. Figure 17.6 shows Δc_{Dt} going to zero below a t/c of 0.1. Hoerner points out that the data upon which Eq. (17.12) was based suggests this, but does not actually extend to zero beyond 0.1.

Equation (17.2) requires that $c_{D\infty}$ and corrections to it be referenced to the shape element's drag area. This is $A_D = tS$, where S is the airfoil's span. Using this formula, the reader can show that $\Delta c_{DI} = \dfrac{\Delta D_I}{qA_D} = c_{Dt}\dfrac{A_D}{S^2} = c_{Dt}\dfrac{t}{S}$.

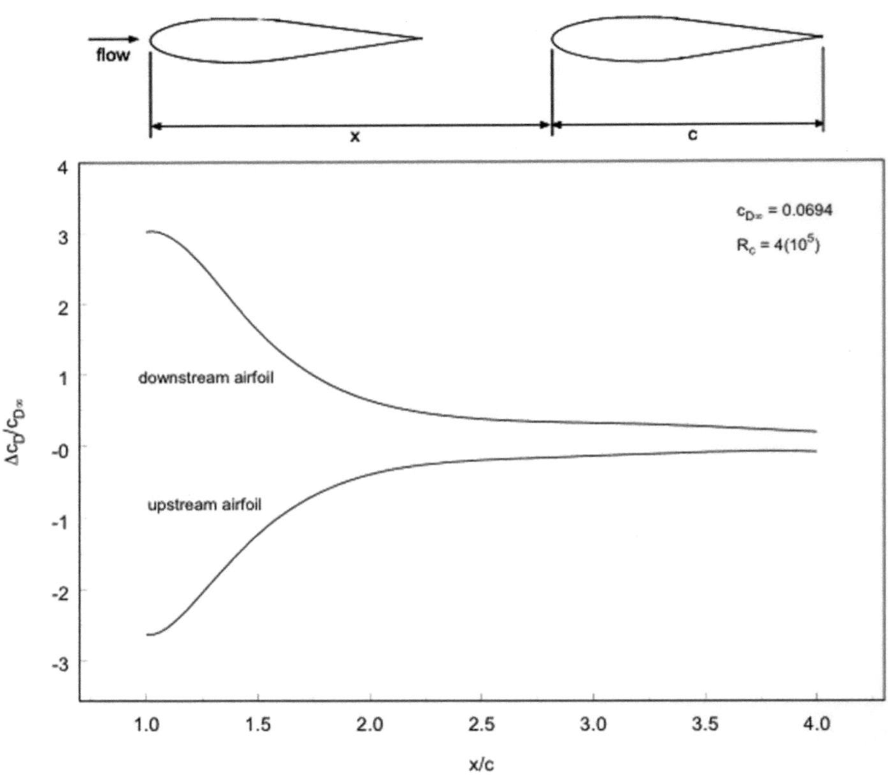

Fig. 17.5 Interference between tandem airfoils. (Adapted from Bierman and Herrnstein 1933)

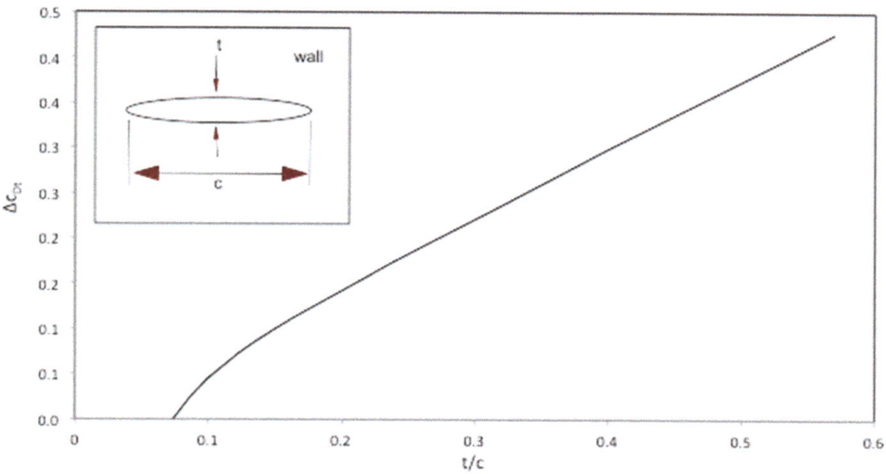

Fig. 17.6 Airfoil-wall interference

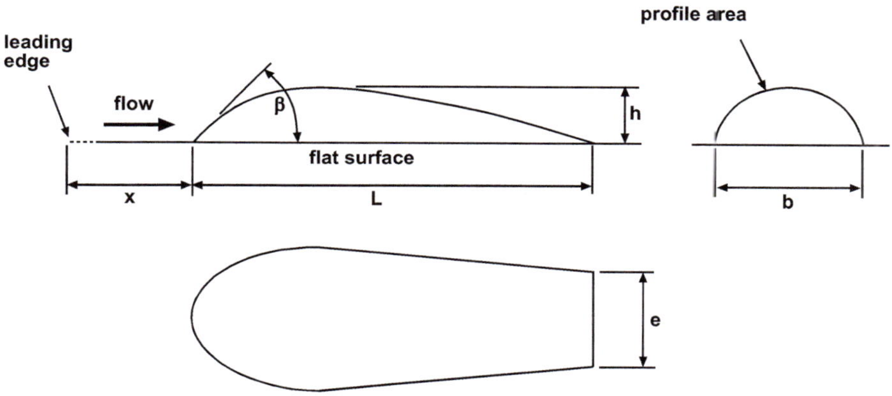

Fig. 17.7 Canopy dimensions

Fairing Figure 17.7 shows a fairing (which could be a canopy) placed on a plane surface. Hoerner (1965) summarizes the results of a number of tests. The drag of the fairing or canopy is a function of its height ratio, h/x, and (specially defined) fineness ratio, L/h, width ratio, e/b, and the angle, β, between the tangent to its leading edge and the plane wall.

Following Hoerner, we assume a $c_{D\infty}$ of 0.03. As the height ratio becomes less than about 0.3, $\Delta c_{DI}/c_{D\infty}$ rises, reaching a maximum of about 1.0 at about 0.1. At height ratios larger than 0.3, $\Delta c_{DI}/c_{D\infty}$ is about zero. All tests that Hoerner summarizes were done at R_L greater than the critical Reynolds number. Hence, he attributes the influence of h/x to be the interference drag associated with junction between the surface and the fairing.

The shape of the canopy, described by β, L/h, and e/b, also affects the drag. Although, Hoerner gives no measurements of the tangent angle, he shows that reducing β to considerably less than 90 can reduce $\Delta c_{DI}/c_{D\infty}$ by as much as 43 %. The effect of increasing L/h is generally to increase the interference drag.

$$\frac{\Delta c_{D_I}}{c_{D_\infty}} = \begin{cases} 0.4, & 5 \le \dfrac{L}{h} \le 15 \\ 0.0667\dfrac{L}{h} - 1, & 15 < \dfrac{L}{h} < 30 \end{cases}. \tag{17.13}$$

17.4 Ground Effect

The discussion in Chap. 2 explained that when a shape element is moved toward the ground from free air, lift and drag forces are created by the alteration in the pressure distribution caused by the Bernoulli effect between the element and the ground. This is similar to the parallel airfoil interference effect, with the plane of symmetry between the foils representing the "ground." As we shall presently see, ground effect is dependent upon the shape and pitch angle of the element.

We present curves that have been derived from the results of Morelli (1983). This paper presented data from wind tunnel tests of the lift and drag forces on cambered, wheelless, basic-body shapes (similar in appearance to Fig. 17.2) designed to minimize the drag near the ground by adjusting the camber ratio. Call these "shark" shapes, as in Chap. 9. Three $\Delta c_D/c_{D\infty}$ curves for these shapes, are plotted in Fig. 17.9. They display $\Delta c_D/c_{D\infty}$ for zero pitch and camber ratios of 3.7, 5.3, and 6.9 % as a function of clearance ratios from 0.05 to 1.0[8]. Morelli (1983) shows that the zero pitch data approximately represents small pitch angles, say between $\pm 1^\circ$. At larger pitch angle magnitudes the drag coefficient is higher.

Morelli (1983) also displayed c_D data for a family of teardrop shapes of different camber ratios as a function of clearance ratio at zero pitch angle. The results for $\pm 1^\circ$ pitch were nearly the same. The 0 and 3 % camber ratio curves from this family are also shown in Fig. 17.8. The data were taken at a chord Reynolds number of $4.7(10^6)$.

The free-air drag coefficients of all the shapes are given in the figure.

The curves show the strong effect of clearance ratio on the drag. For the teardrop shapes, as the clearance ratio increases the drag decreases, with the lowest value occurring in free air. However, for the shark shapes at the two higher camber ratios, the drag coefficient passes through a minimum in the clearance ratio range 0.1–0.2, followed by a shallow maximum.

The general form of the equation of the curves for the shark shapes is

$$\frac{\Delta c_{D_G}}{c_{D\infty}} = a_0 + a_1 ln\frac{h_{min}}{W} + a_2\left(ln\frac{h_{min}}{W}\right)^2 + \rightleftharpoons + a_n\left(ln\frac{h_{min}}{W}\right)^n. \tag{17.14}$$

[8] According to Morelli (1983), most automobiles lie between clearance ratios of 0.05 and 0.2.

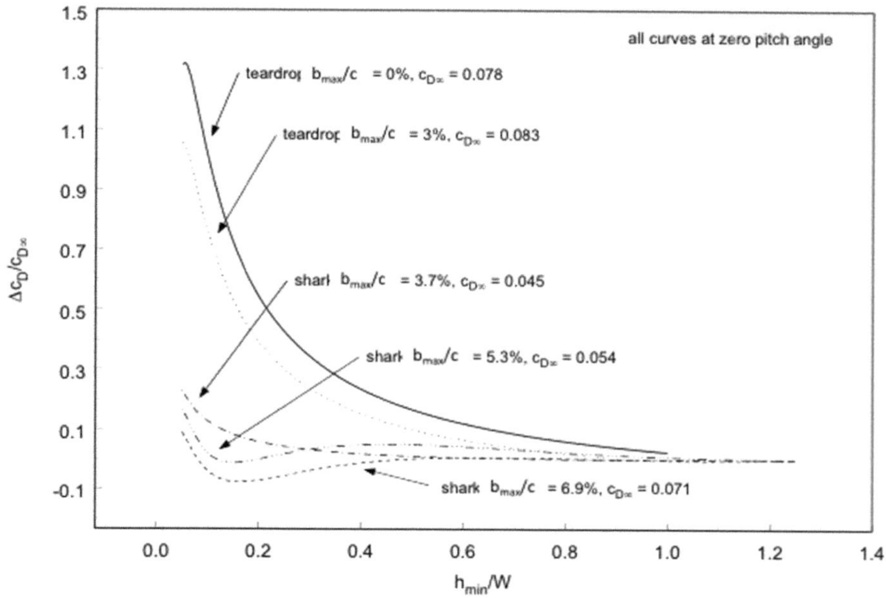

Fig. 17.8 Ground effect for two shapes. (Adapted from Morelli (1983))

Table 17.6 gives the set of a's for each of these curves.

The equation for the teardrop curves has the form

$$\frac{\Delta c_{D_G}}{c_{D\infty}} = a_0 + a_1\left(\frac{h_{min}}{W}\right)^{-1} + a_2\left(\frac{h_{min}}{W}\right)^{-2}. \qquad (17.15)$$

Table 17.7 gives the a's for each teardrop camber ratio case

17.5 Flat Plate

The critical Reynolds number for a flat surface with the flow parallel to it is

$$R_{x_c} \approx 5(10^5) - 3(10^6). \qquad (17.16)$$

Table 17.6 Shark shape curve constants

Camber ratio (%)	a_o	a_1	a_2	a_3	a_4	a_5	a_6
3.7	0.00031	−0.00132	0.01741	−0.00254	0	0	0
5.3	0.00776	−0.05589	0.08372	0.12379	−0.00117	−0.02897	−0.00599
6.9	−0.0041	−0.0069	−0.0928	0.2371	0.1303	0.0203	0

Table 17.7 Teardrop shape curve constants

Camber ratio (%)	a_0	a_1	a_2
0	−0.1256	0.1541	−0.0041
3	−0.1218	0.1168	−0.0029

The upper limit corresponds to a very smooth plate. The symbol "x_c" is the distance from the leading edge of the plate to the flow transition. The friction drag coefficient is defined as

$$c_{DF} = \frac{D_F}{qA_W}, \tag{17.17}$$

where, A_W is the wetted area of the surface and the subscript "F" denotes "friction." The following expression gives the friction drag coefficient of one side of the plate when the laminar—turbulent transition occurs on the plate and the width of the transition region in the flow direction is ignored.[9]

$$c_{DF} = \frac{1.328}{R_{x_x}^{\frac{1}{2}}}\frac{x_c}{L} + \frac{0.031}{R_L^{\frac{1}{7}}}\left[1 - \left(\frac{x_c}{L}\right)^{\frac{7}{8}}\right]. \tag{17.18}$$

Bear in mind that Eq. (17.18) applies to flow over a smooth plate.

17.6 Roughening

Joints occur around access doors or around other portions of the body that must be repositioned. There may be isolated defects such as cracks or pits, or there may be wires and fasteners that project into the airflow. The distinguishing parameters differ in each case. Hoerner (1965) contains methods for accounting for many of these relatively isolated defects.

Now, suppose the roughness is approximately uniformly distributed over the surface. Then it makes sense to characterize the smoothness of the surface by the ratio L/ε, where ε(Greek "epsilon") is the mean roughness height. The larger this *relative roughness*, the smoother the surface is relative to its length. For example, the 800-odd cells of the solar array are often applied directly to the array substrate and thus are raised above it by perhaps 1–2 mm. In addition, there may be 1.0 mm thermal expansion gaps left between the cells. If the glazing does not smooth the array sufficiently, the cells will contribute additional surface roughening which could be characterized by a relative roughness.

[9] D_F was calculated over the laminar and turbulent regions (x=0 to x_C and x_C to L, respectively). The results were added and (17.17) applied.

We will consider the flat surface of the previous section. (The surface of an actual vehicle will be curved. But for practicality we will ignore this when considering roughness as long as we judge that the flow is strongly one-dimensional.) Suppose the flow is turbulent over the entire surface ($x_c = 0$ in Eq. (17.18)). If the Reynolds number continues to increase, the flow will enter the *fully rough* regime. Then the friction drag coefficient no longer depends upon the Reynolds number, but only on the relative roughness. Also, the Reynolds number at which this condition occurs becomes smaller as the relative roughness is reduced. If we were to measure the friction drag coefficient on a smooth surface, L/ε about 10^6 perhaps, in laminar flow at a Reynolds number of $5(10^5)$, we would find a value of about 0.002. Suppose we were to paint that surface with a roughening agent that reduced L/ε to 500. At the same Reynolds number the flow would move into the fully rough regime and the friction drag coefficient would rise to 0.01, a 5-fold increase.

The *critical relative roughness* is that which precipitates fully rough flow. From information in White (1986),

$$\left(\frac{L}{\varepsilon}\right)_c = -654963 + 335446.3 logR_L - 57477.5(logR_L)^2 + 3299.78(logR_L)^3. \quad (17.19)$$

17.7 Other Methods

Drag Rating White (1967) presented a method in which features of vehicles were assigned to nine categories and then awarded a score, with the lowest score (one) assigned to the feature contributing the least to the drag. The sum of the category ratings is the drag rating, R, of the vehicle. The method does not incorporate ventilation drag or ground and pitch angle effects. The correlation relating the drag rating to the drag coefficient is

$$c_D = 0.16 + 0.0095R. \quad (17.20)$$

This relation was developed from drag data for 141 vehicles and it correlates that data is within $\pm 7\,\%$.

Note that even if a vehicle received a drag rating of zero, Eq. (17.20) would predict a drag coefficient of 0.16, higher than most solar racing cars. However, the method would be useful in the initial stages of design when selecting shape elements, or when modifying a stock body to reduce its drag.

Feature Build-up Pershing and Masaki (1976) reported a method that is well adapted to predicting the drag of automobile shapes similar to that of the typical sedan. Equations for the drag contribution of the elements of the body's shape were developed from the available test data. These equations were in terms of the details of the shape element. For example, the front hood drag coefficient is to be calculated from

$$c_{D_{hood}} = 0.707 \frac{\left(\frac{A_h - A_F}{L_h}\right)^2}{A_R}. \tag{17.21}$$

A_h is the projected area of the body below the hood-windshield intersection, L_h is the length of the hood, A_F is the front end projected area, and A_R is the projected area of the car (less protuberances). If the difference A_h—A_F is negative, it is to be set to zero.

The method does not account for the effects of lift, side wind, and ground clearance, and it assumes a fixed wheel drag coefficient of 0.14. It accounts for ventilation drag.

Smalley and Lee (1978), reported that this method agreed with wind tunnel data to within ±10 % for 13 of 17 vehicle tests. Five of which were duplicate tests using a different tunnel blockage correction method. Three of the outliers were 12–18 % below the tunnel tests and one was 16.5 % higher.

Due to its correlation with details of the vehicle's design, this method will be very useful for setting shape details, that is, actual dimensions rather than qualitative features, when starting with stock bodies. Its major drawback is the use of a large, constant, wheel drag coefficient. As in the drag rating method case, this value alone is larger than the drag coefficient of most solar racing cars.

Ashley and Carr (1982); Rose (1986); and Calkins and Chan (1998) have further developed the feature build-up method. Calkins and Chan have incorporated a refined version of the method in a drag prediction tool called CDaero. CDaero is a component of the Automobile Design Support System reported by Calkins et al. (1996). The refined feature build-up method estimates the drag of the wheels from their exposed projected area, as in this chapter, instead of assuming a constant value. The calculation is for zero yaw and does not account for ground effect.

References

Abbott, I. H., & von Doenhoff, A. E. (1959). *Theory of wing sections*. New York: Dover Publications, Inc..

Ashley, C., & Carr, G. (1982). A Prediction Method for the Aerodynamic Drag of Cars, *Energy and Mobility: XIX International FISITA Congress Proc., Melbourne, Australia, Nov. 8–12, 1982*, SAE paper 82068, Parkville, Victoria, Australasia.

Biermann, D., & Herrnstein, Jr., W. H. (1933). Interference Between Struts in Combinations, NACA-TR-468.

Bullivant, W. K. (1941). Tests of the NACA 0025 and 0035 Airfoils in the Full-Scale Wind Tunnel, NACA-TR-708.

Calkins, D. E., & Chan, W. T. (1998). 'CDaero'-A Parametric Aerodynamic Drag Prediction Tool, *Developments in Vehicle Aerodynamics (SP-1318), International Congress and Exposition, Detroit, Michigan, February 23–26, 1998*, SAE paper 980398, SAE Warrendale, PA.

Calkins, D. E., Guan, L., & Koeser, L. (1996). Automobile Conceptual Design Support System (AutoDSS), SAE paper 960551, *Proc. SAE International Congress and Exposition, Detroit, Michigan, 26–29 February, 1996*, SAE Warrendale, PA.

Cogotti, A. (1983). Aerodynamics of Car Wheels, *Int. J. of Vehicle Design*, Technological Advances in Vehicle Design Series, SP3, *Impact of Aerodynamics on Vehicle Design*, Dorgham, M. A. ed., Interscience Enterprises Ltd., La Motte Chambers, United Kingdom, pp. 173–196.

Hoerner, S. F. (1965). *Fluid-Dynamic Drag*, Hoerner Fluid Dynamics. Bakersfield.

Kurtz, D. W. (1979). *Aerodynamic Characteristics of Sixteen Electric, Hybrid, and Subcompact Vehicles, Complete Data*, JPL Publication 79–59, (NASA-CR-158814 and NTIS N79-29107), Jet Propulsion Laboratory, Pasadena, California.

Kurtz, D. W. (1980). *Aerodynamic Design of Electric and Hybrid Vehicles: A Guidebook*, JPL Publication 80–69 (NASA-CR-163744 and NTIS N81-12943), Jet Propulsion Laboratory, Pasadena, California.

Morelli, A. (1983). Aerodynamic Basic Bodies Suitable for Automobile Applications, *Int. J. of Vehicle Design*, Technological Advances in Vehicle Design Series, SP3, *Impact of Aerodynamics on Vehicle Design*, Dorgham, M. A. ed., Interscience Enterprises Ltd., La Motte Chambers, United Kingdom, pp. 70–98.

Pershing, B., & Masaki, M. (1976). Estimation of Vehicle Aerodynamic Drag, *EPA-460/3-76-025 (NTIS PB 275 948)*, National Technical Information Service, Washington, DC.

Rose, M. J. (1986). Appraisal and Modification of the Empirical Method for Predicting the Aerodynamic Drag of Cars, report no. 1986/1, Motor Industry Research Association (MIRA), England.

Schlichting, H. (1979). *Boundary-layer theory* (7th ed.). New York: McGraw-Hill Book Company.

Smalley, W. M., & Lee, W. B. (1978). *Assessment of an Empirical Technique for estimating Vehicle Aerodynamic Drag from Vehicle Shape Parameters*, Aerospace Corporation, El Segundo, CA, NTIS accession number PB-292 160/9, NTIS issue number 7913.

White, R. G. S. (1967). A Rating Method for Assessing Vehicle Aerodynamic Drag Coefficients, Motor Industry Research Association, United Kingdom.

White, F. M. (1986). *Fluid mechanics* (2nd ed.). New York: McGraw-Hill Book Company.

Chapter 18
Ventilation System Analysis

18.1 Ventilation Design

Purpose The ventilation system influences the cockpit comfort conditions and removes battery-evolved gases and battery-heated air from the battery compartment at a specified minimum rate.

Discussion Restrictions There may be separate cockpit and battery ventilation subsystems. But we consider only the case in which each subsystem (or entire system) has one inlet and one outlet. This will keep the analysis for the flow through the system simple and focused on the basic ideas.

Battery Ventilation The race rules specify the minimum flow requirements for the battery compartment. Chapter 16 gives this requirement as a fan-forced flow of at least 280 L/min. This flow must be present whenever the battery is connected to the main bus, and it must exhaust to the exterior of the car. The designer's task then is to insure that this flow is present under the most limiting condition. This will occur when the car is stationary and the battery fan, or fans, must supply the flow unaided by the motion of the car.

Driver Comfort Figure 18.1 shows the summer human *comfort zone* (adapted from ASHRAE (1981)) defined by lines of constant temperature (T) and relative humidity (Φ).

We would like the cockpit conditions to remain within this zone. But if the outside air temperature is typical of a late June day, say an air temperature (*dry bulb* temperature) of 30 °C and relative humidity of 70 %, the figure shows that this cannot be done without air conditioning. The weight and power penalties that accompany air conditioning preclude its use in solar racers

Therefore, the temperature and humidity of the air entering the cockpit will be those of the outside air. Our objectives then become to prevent the air temperature in the cockpit from exceeding that of the outside air and to minimize the direct heating of the driver by solar radiation.

E. F. Thacher, *A Solar Car Primer*, DOI 10.1007/978-3-319-17494-5_18

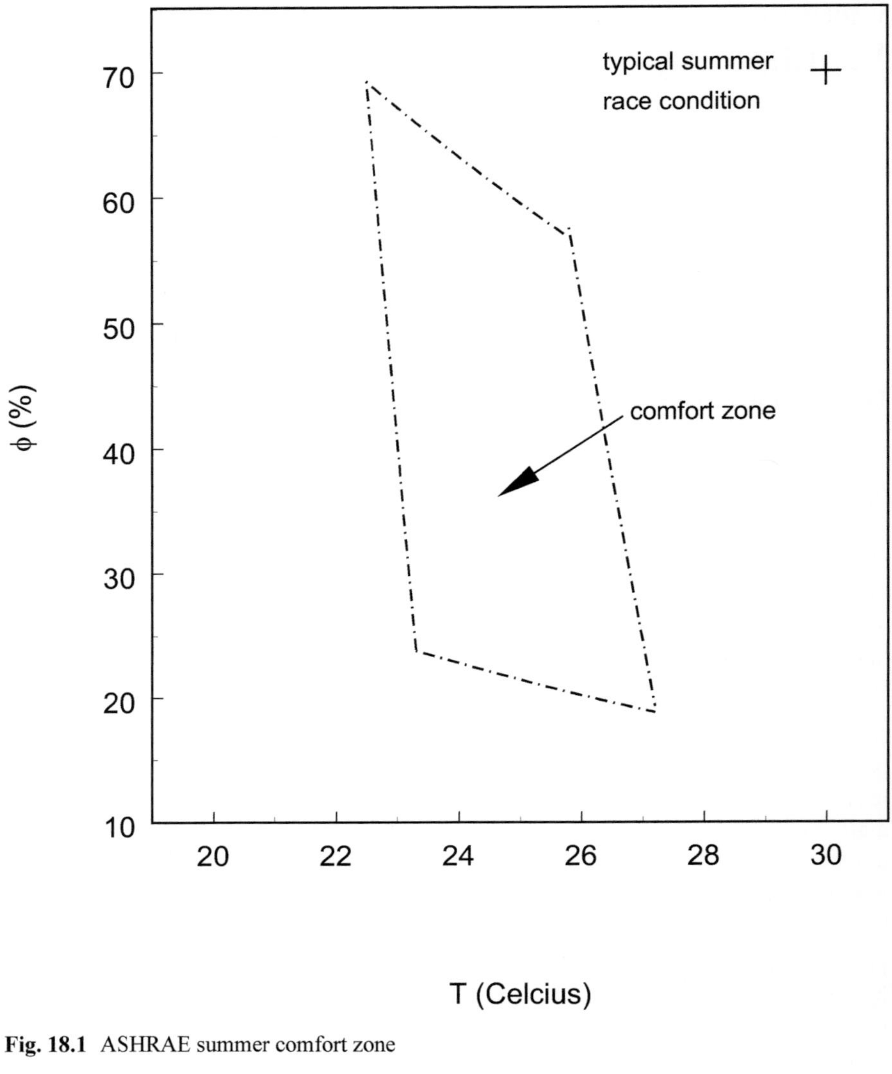

Fig. 18.1 ASHRAE summer comfort zone

The options available are: reducing the solar gain through the canopy, reducing the absorption of solar radiation by surfaces in the cockpit, and changing the air in the cockpit. The latter can be done by providing forced circulation of outside air through the cockpit. Making the cockpit surfaces more reflective reduces absorption. The former can be done by shading, tinting or coating the glazing, and reducing the amount of glazing, subject to meet the visibility requirements of the race rules and to provide sufficient light to read the instruments.

Analysis Outline We would like to know the static pressure at the inlet and outlet of the ventilation system so that the ventilation drag and the ventilation drag coefficient can be calculated. We would also like to know the flow through the battery box, to see if it is sufficient, and the temperature in the cockpit, to see if we might be in danger of frying the driver.

The first problem we will consider, then, is how to get an approximate pressure distribution around the car. This will be followed by an explanation of how to calculate the flow through the ventilation system, once the pressure at its inlet and outlet is known. Finally, a method for estimating the cockpit temperature will be explained. An example of each of these calculations is in Chap. 10.

18.2 Inlet and Exit Pressures

Inlet and Outlet Locations The pressure distribution in the external flow is a function of the vehicle's shape (see the list of assumptions below). This flow is approximately frictionless and therefore, may be analyzed by Bernoulli's equation. Thus if the velocity at a point in the flow can be found, then the static pressure at that point can also be found.

Bear in mind that the above statements do not apply to regions of separated flow. Also, near the inlet or outlet, the flow is deflected inward or outward, respectively, and this alters the static pressure at that point because the "shape" of the car at that point is effectively changed. For example, suppose an inlet for cockpit air is installed just ahead of the canopy. Some flow enters the car, so there is more flow just upstream of the inlet than just downstream of it. Velocities upstream of the inlet are therefore higher and the pressure is lower. Figure 18.2, adapted from Wallis (1971), illustrates this situation. Wallis observed that at a particular vehicle speed, the average inlet static pressure remains constant for any inlet flow, even though the pressures upstream and downstream of the inlet shift, as previously described, as the inlet flow changes.

Locating inlets at or near stagnation points will not only maximize flow but may also increase drag. Figure 18.2 shows an inlet near the stagnation point in front of the canopy. However, wind tunnel measurements of ventilation flow have shown that if the inlet is small and is located at the stagnation point in the nose of a highly streamlined car, the ram flow may be sharply reduced by small pitch changes which shift the stagnation point away from the opening. Wallis (1971) recommends placing a small outlet in a region where $p-p_\infty$ is negative or a large, low-loss outlet where $p-p_\infty$ is zero or slightly positive.

Approximate Pressure Distribution We now describe a simple method for estimating the pressures at the ventilation inlet and outlet using Bernoulli's equation. However, some idealizations must be made. These are:

1. The air outside the boundary layer behaves as if it had no viscosity.
2. The thickness of the boundary layer may be neglected.
3. There is no heat transfer to the air.

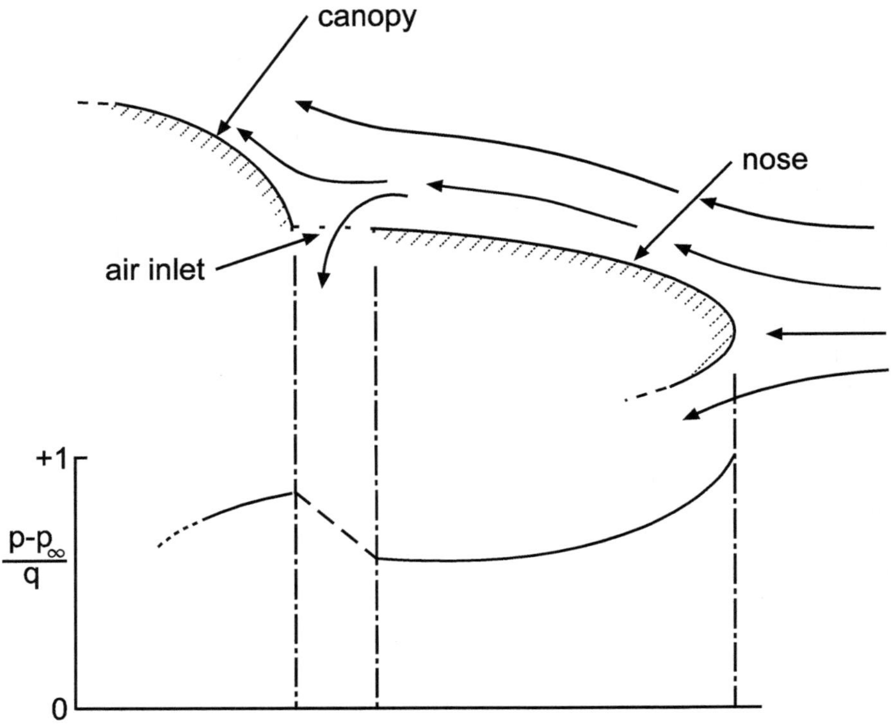

Fig. 18.2 Qualitative pressure distribution

4. There are no regions of separated flow in the area of the car's surface that is of interest.
5. The flow is steady.
6. The flow about one car length upstream and downstream of the car is uniform.
7. The velocity distribution transverse to the vertical plane through the car's longitudinal axis is uniform.
8. The air density is constant.

The fundamental assumption of boundary layer theory is that viscous effects may be thought of as residing solely in the boundary layer. This implies assumption 1. Assumptions 1, 3, and 4 greatly simplify the description of the flow external to the boundary layer, allowing Bernoulli's equation to be used.

The boundary layer thickness is small compared to the dimensions of the car. The average error introduced by assumption 2 will be small. The boundary layer thickens as the flow moves toward the rear of the car, thus deflecting the stream surfaces away from the surface of the car. This displacement causes the velocity estimate near the car to be lower than the real flow. Therefore, by Bernoulli's equation, the pressure estimate will be higher than in the real flow.

Assumption 5 further simplifies the problem by removing time dependence from the variables. Assumption 6 allows the inlet and outlet properties of the flow to be calculated.

Assumption 7 relies on the observations that, in view of assumption 6, the flow is symmetric with respect to the car's centerline and, near the centerline, the transverse slope of the surface is usually small. Assumption 7, therefore, allows the flow to be approximated as two-dimensional (2D). That is, to be treated as if the velocity depended only on the coordinates perpendicular and parallel to the car's long axis.

The air density has been taken as approximately constant. This assumption will be used throughout this chapter. The error it introduces is small, but the simplification it allows is large.

Imagine the car placed in a 2D wind tunnel (H, height, and length, x, only: no flow perpendicular to the plane defined by H and x) that supplies the uniform upstream flow. A rule of thumb used in designing actual wind tunnels is that the tunnel cross-sectional area should be at least five times the profile area of the car. This makes the blockage effect that arises when the flow is confined between the car and the tunnel walls negligible. The tunnel is of width W and height H. But because the flow is 2D, W may be set to unity. Therefore the height, H, of the tunnel must be such that

$$H \times 1 \geq 5A_D \tag{18.1}$$

Because there are no boundary layers and no flow separation, the tunnel walls and the car's surface are stream surfaces. Using this fact and the known entrance and the exit conditions, we can sketch a plot of the stream surfaces (or streamlines in our 2D representation). From this information we can estimate the pressure distribution.

Flow Net A *flow net* is a net of streamlines and intersecting lines, called constant *velocity potential* lines. The flow net is constructed such that the streamlines and the lines intersecting them form, approximately, curvilinear squares. This means that the lines intersect at right angles and the sides of the figure thus constructed are approximately equal, as Fig. 18.3 shows.

First, the car is carefully drawn on crosshatched paper. Then the flow net is drawn by trial-and-error. The smaller the squares, the more accurate the approxima-

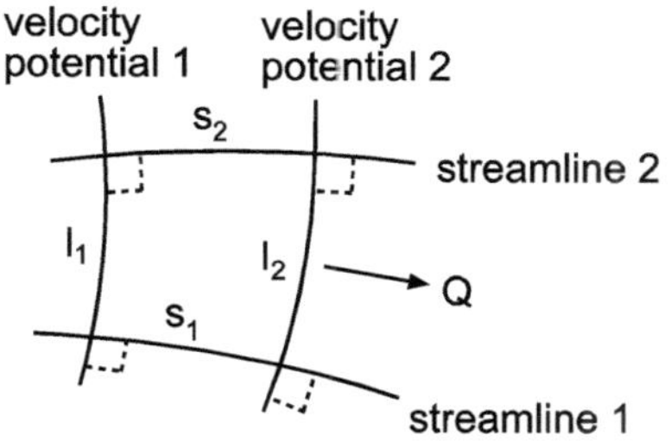

Fig. 18.3 Curvilinear square

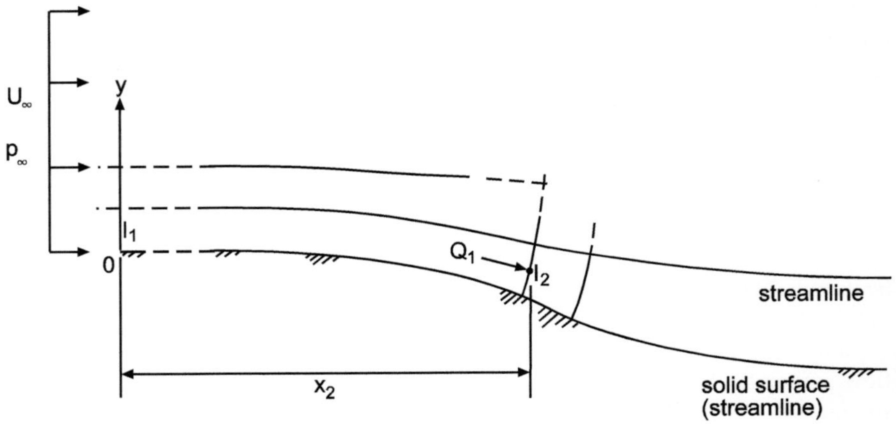

Fig. 18.4 Partial flow net near a surface

tion. Figure 18.4 shows a streamline next to a surface and a curvilinear square with the center of its upstream side at a distance x_2 from the origin. We wish to estimate the pressure at this position.

Because of assumptions 1, 3, 4, 5, and 6, Bernoulli's equation is a constant throughout the flow. Therefore

$$p_\infty + \frac{1}{2}\rho U_\infty^2 = p_2 + \frac{1}{2}\rho V_2^2 \tag{18.2}$$

We solve Eq. (18.2) for the pressure difference $p_2 - p_\infty$ and divide by the dynamic pressure at the origin, q_∞. This quotient is called the pressure coefficient, c_P.

$$c_{P_2} = 1 - \frac{V_2^2}{U_\infty^2} \tag{18.3}$$

The flow has a unit depth into the page, and so we may substitute

$$V_2 = \frac{Q_1}{l_2}, \quad U_\infty = \frac{Q_1}{l_1} \tag{18.4}$$

in Eq. (18.3). Note that the first of Eq. (18.4) gives the *average* velocity between the streamlines. As the flow net is made finer, this average approaches the true, ideal-flow velocity. The substitution produces

$$c_{P_2} = 1 - \frac{l_1^2}{l_2^2} \tag{18.5}$$

after canceling common factors. Measurement of l_1 and l_2 from the flow network gives the pressure coefficient. Measurement of the lengths, which are curved, by a ruler, which is straight, will only be approximate. However, this approximate measurement is consistent with the approximation to V_2 just discussed.

The pressure at point 2 is then

$$p_2 = p_\infty + c_{P_2} q_\infty \tag{18.6}$$

Flow Around a Car Fig. 18.5 shows a vehicle body in a 2D flow. (Some of the features that make the actual flow three-dimensional (3D), such as the wheels, are shown in dashed lines.) Note that unlike the flow in Fig. 18.4, the air must flow on both sides of the body.

The flow between the streamline at height l_1, Q_1 must divide at the front stagnation point such that

$$Q_1 = Q_U + Q_L = U_\infty l_1 \tag{18.7}$$

The previous method cannot be applied unless Q_U and Q_L are known. One way of estimating them would be to locate the stagnation streamline. The streamline should bisect the angle between the velocity potential lines labeled "l_{1U}" and "l_{1L}." In Fig. 18.5, the streamlines were drawn such that $Q_U \approx Q_L$.

A more analytical approach, which could be used as a check on the flow net near the car, is as follows. Locations 1 and 2 are imagined to be at pressures p_1 and p_2, respectively, and each to be "at" the point of flow division. Applying the Bernoulli equation to the flow along the upper surface of the car gives

$$p_2 - p_1 = \frac{1}{2}\rho(V_{1U}^2 - V_{2U}^2) \tag{18.8}$$

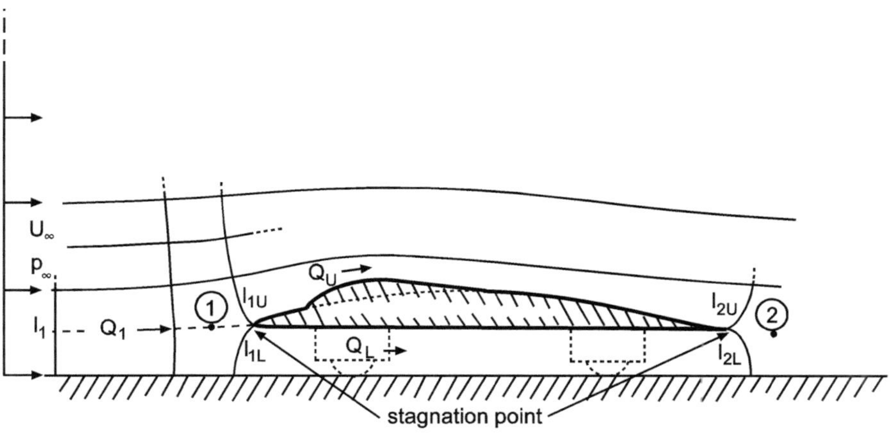

Fig. 18.5 Divided flow

Applying Eq. (18.4) at each end of the surface gives

$$p_2 - p_1 = \frac{1}{2}\rho\left(\frac{1}{l_{1U}^2} - \frac{1}{l_{2U}^2}\right)Q_U^2 \tag{18.9}$$

The same procedure is now applied to the lower surface and the pressure changes are equated, giving

$$Q_U = \left(\frac{k_L}{k_U}\right)^{\frac{1}{2}} Q_L \tag{18.10}$$

where k_L and k_U are defined as

$$k_L = \frac{1}{2}\rho\left(\frac{1}{l_{1L}^2} - \frac{1}{l_{2L}^2}\right) \tag{18.11}$$

$$k_U = \frac{1}{2}\rho\left(\frac{1}{l_{1U}^2} - \frac{1}{l_{2U}^2}\right) \tag{18.12}$$

Using Eq. (18.10) with Eqs. (18.11) and (18.12) gives

$$Q_L = \frac{1}{\left(\frac{k_L}{k_U}\right)^{\frac{1}{2}} + 1} Q_1 \tag{18.13}$$

$$Q_U = \frac{1}{\left(\frac{k_U}{k_L}\right)^{\frac{1}{2}} + 1} Q_1 \tag{18.14}$$

Note that these relations require the flow net to be drawn such that

$$\frac{k_L}{k_U} > 0 \tag{18.15}$$

Therefore, k_L and k_U both must be positive or negative.

18.3 Design Strategy

Design Operating States The vehicle operational states that factors controlling the ventilation design are zero speed operation and the cruise condition. In the former state, drag is zero but so is the ram flow. The latter state not only causes the most drag but can also provide the most ram flow. Therefore, the design strategy will be to locate the inlet and outlet, and design the ducting to minimize ventilation drag at the cruise condition and to size the fan, or fans, to provide the minimum required flows at zero speed.

Equation 2.8, repeated below as Equation (18.16), shows that the ventilation drag is the sum of the pressure force difference and the reaction force from the momentum change. Subscripts 1 and 2 denote the inlet and outlet, respectively. The subscript "x" denotes the component of the force or air speed parallel to the direction of motion. The bar over the gauge pressure symbol, p_G, denotes the average over the inlet or outlet. The symbol "A" denotes the area of the opening normal to the entering flow or pressure. If there is more than one subsystem, the drag contribution of each one must be calculated using Eqn (18.16). The total ventilation drag will be the sum of the subsystem drags.

$$D_V = (\overline{p}_{G1}A_1)_X - (\overline{p}_{G2}A_2)_X + \dot{m}_1(V_{1X} - V_{2X}) \tag{18.16}$$

As for flow between streamlines, the cross-sectional area and speed for a given steady, constant-density flow in a duct (no leakage) are related by

$$Q = A_1V_1 = A_2V_2 = constant \tag{18.17}$$

The subscripts refer to any two stations in the flow path, and Q is the volumetric flow rate (m^3/s). The mass flow rate (kg/s) is

$$\dot{m} = \rho Q \tag{18.18}$$

18.4 Component Pressure Losses

The energy used to force the flow through components such as inlets, lengths of duct, sudden expansions and contractions, and valves causes a pressure drop in the flow direction. The net pressure drop (i.e., including pressure increases caused by fans) through the ventilation system must be equal to that imposed by the external flow. If this flow is zero, then the pressure rise caused by the fans must exactly cancel the total pressure loss at the desired internal flow.

The pressure loss in a component may be expressed by the product of a *head loss coefficient*, K, and a reference kinetic energy per unit mass.

$$h_L = \frac{\Delta p}{\rho g} = K_{12}\frac{V_{ref}^2}{2g} \tag{18.19}$$

The pressure loss would then be $\rho g h_L$ (N/m^2). The location of the reference kinetic energy will be identified for each loss coefficient case presented below.

The units of $\Delta p/\rho$ are $N \cdot m/kg$, or J/kg. This quotient represents the work per unit mass of air necessary to drive the flow through the component. Division by g, the acceleration of gravity, converts the units to meters. The power, or the rate of doing this work, would be

$$\dot{W} = \dot{m}\frac{\Delta p}{\rho} = \dot{m}gh_L \tag{18.20}$$

The head loss coefficients of the components are approximately independent of Reynolds number, that is, independent of the speed of the internal flow, as long as the flow is fully turbulent. However, in the case of components such as the cockpit or the battery box, the flow cross-sectional area is much larger than in the ducting connecting them, so the reference flow speed will be smaller. We must be sure to check the Reynolds number in such components.

We begin a flow calculation by assuming that the flow is turbulent and that the Reynolds number at each component is within the limits associated with the loss coefficient data for that component. The flow-dependent loss coefficients for the next iteration would be estimated from the flow just calculated. This procedure would be continued until the changes in the flow from the previous calculation are satisfactorily small.

Coefficient Sources Handbooks such as Idel'chik (1966) and Crane (1969) present information on loss coefficients for many different devices. Textbooks such as White (1986) also contain such information, but for a smaller variety of devices. The loss coefficient cases presented below may be found in any of these references. The loss coefficient curve for each component has been fitted with an interpolation formula.

Rounded Inlet from Reservoir Suppose the flow enters a pipe from an approximately infinite reservoir, such as the atmosphere, energy is lost from the flow at the entrance. This loss can be reduced by rounding the edge of the entrance. Figure 18.6 shows how K_{12} for the rounded inlet changes as a function of the rounding radius. V_{ref} is V_2, the velocity at the exit of the inlet. The equation for Fig. 18.6 is

$$K_{12} = a_0 e^{a_1\frac{r}{d} + a_2\left(\frac{r}{d}\right)^2 + a_3\left(\frac{r}{d}\right)^3 + a_4\left(\frac{r}{d}\right)^4} \tag{18.21}$$

where r is the rounding radius, and d is exit diameter. Table 18.1 gives the constants.

Any Exit to a Reservoir Because all the kinetic energy of the flow is lost by dissipation into the reservoir,

$$K_{12} = 1.0 \tag{18.22}$$

Sudden Expansion and Contraction When the flow from a duct enters a larger duct through an abrupt transition, that is, no gradually sloping duct connecting the two

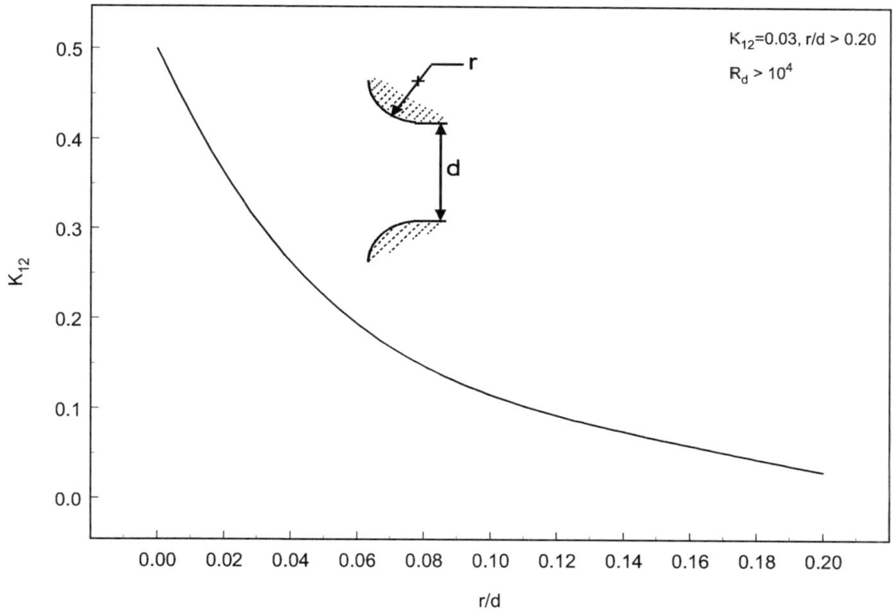

Fig. 18.6 Rounded inlet loss coefficient

Table 18.1 Constants for Eq. (18.21)

a_0	a_1	a_2	a_3	a_4
0.499540	−15.415485	−40.244830	721.990759	−2432.856432

conduits, energy is lost in the sudden expansion. When the flow leaves a larger duct and abruptly enters a smaller, energy is lost in the sudden contraction. The reference velocity in both cases is that in the smaller pipe, denoted by the subscript "1."

The loss coefficient for the sudden expansion is

$$K_{12} = \left(1 - \frac{d_1^2}{d_2^2}\right) \tag{18.23}$$

and for the sudden contraction is

$$K_{12} = 0.42\left(1 - \frac{d_1^2}{d_2^2}\right) \tag{18.24}$$

Diffuser A diffuser is a conical-shaped duct that widens in the flow direction. Its purpose is to slow the flow and thus increase the static pressure. Figure 18.7 shows a diffuser and its associated loss coefficient. Note the minimum loss coefficient at a cone angle of about 5°. Below this angle, the length of the diffuser becomes

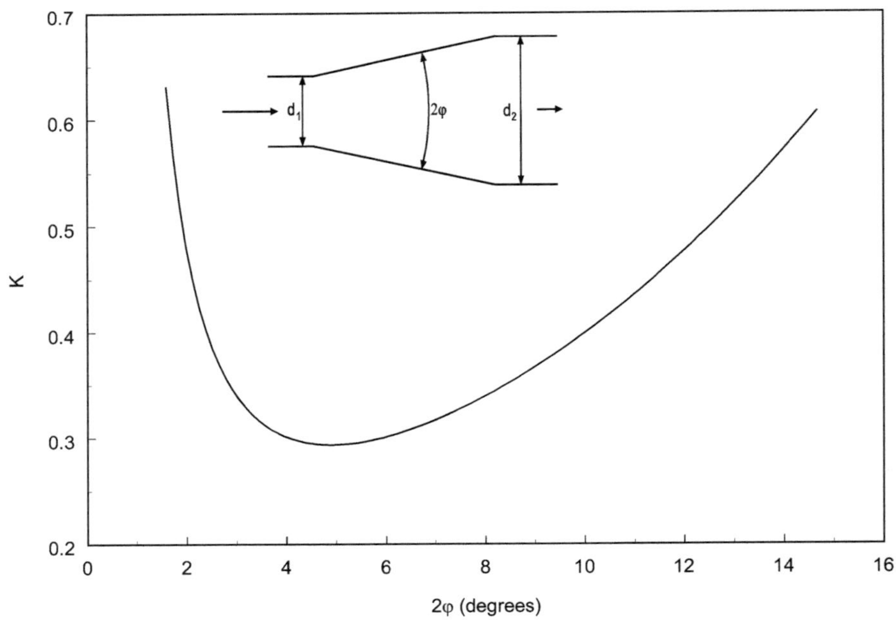

Fig. 18.7 Diffuser loss coefficient

extreme and frictional losses increase. Above this angle flow separation occurs, beginning near the discharge end. As the cone angle increases the separation loss likewise increases until at about 40° the loss coefficient exceeds that of the sudden expansion.

The loss coefficient curve presented is for the so-called fully-developed[1] entrance flow. In this condition, the developing boundary layer inside the pipe has thickened until it reached the centerline of the pipe. This inlet condition yields the largest loss coefficient at a particular cone angle.

The diffuser pressure recovery coefficient, c_p, is the dimensionless pressure increase.

$$c_P = \frac{p_2 - p_1}{q_1} \tag{18.25}$$

The pressure and loss coefficients are related by

$$c_P = 1 - \frac{d_1^4}{d_2^4} - K_{12} \tag{18.26}$$

[1] If an initially uniform flow enters a duct, the boundary layer begins to thicken, just as on a flat plate.Eventually, it meets itself at the centerline.After that, the velocity profile no longer depends on the distance down the duct, but only on the radius from the centerline. Thus it is called *fully-developed.*

Table 18.2 Diffuser constants

a_0	a_1	a_2
1.359836	−1.926919	0.605670

The equation of the loss coefficient curve is

$$K_{12} = a_0 \left(2\varphi\right)^{a_1 \ln(2\varphi) + a_2 \ln\left(4\varphi^2\right)} \tag{18.27}$$

Table 18.2 gives the constants for Eq. (18.27).

Pipe, Tubing, and Ducting The loss coefficient of duct, pipe, and tubing is

$$K_{12} = \frac{f L_{12}}{d} \tag{18.28}$$

where f is the *friction factor*. The friction factor depends upon whether the flow is laminar or turbulent and how rough the surface of the duct is.

The roughness is measured by the *relative roughness*, e/d, (or d/e as in Chap. 17) where e is the mean height of the bumps and d is the pipe's inside diameter. The roughness of commercial steel pipe is about 0.046 mm. Three-inch, schedule 40 pipe has an actual inside diameter of 3.068 in. or 77.95 mm. The relative roughness of this pipe would then be 0.00059. When the flow is turbulent, the loss coefficient increases with the relative roughness. However, when the flow is laminar, the roughness plays no role in the friction factor.

Figure 18.8 shows how the friction factor depends on e/d and the Reynolds number.[2] Laminar flow ends at Rd≈2300 and a transition region begins; the flow is fully turbulent by Rd≈4000. Note that in much of the turbulent region the friction factor is nearly independent of Rd, except for very smooth surfaces, whereas in the laminar region it varies inversely with Rd. Note the similarity in this respect between internal pipe flow and flow over a flat plate (Chap. 17). There are no reliable friction factors in the transition region.

In the case of a noncircular duct use the *hydraulic diameter*, d_h, defined as

$$d_h = \frac{4A_F}{P_W} \tag{18.29}$$

where A_F is the cross-sectional flow area, and P_W is the wetted perimeter of the duct. This approximation gives "reasonable accuracy," according to White (1986), who recommends the use of $2d_h/3$ for "extreme accuracy."

The friction factor in Fig. 18.8 is given by

[2] This is a simplified *Moody Chart*, named after L.F. Moody, who published a more elaborate chart in 1944.

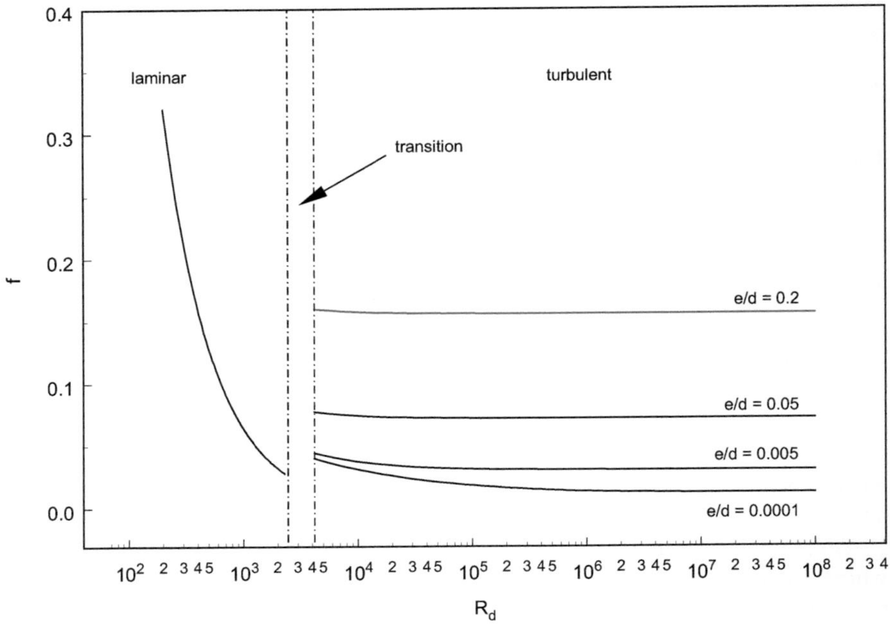

Fig. 18.8 Friction factor

$$f = \begin{cases} \dfrac{64}{R_d}, & R_d \le 2300 \\ \dfrac{0.3086}{\log^2\left[\dfrac{6.9}{R_d} + \left(\dfrac{e/d}{3.7}\right)^{1.11}\right]}, & R_d \ge 4000 \end{cases} \tag{18.30}$$

The first result, due to Prandtl, may be found in many textbooks, for example, White (1986). The second equation was given by Haaland (1983).

18.5 System Characteristic

Consider the ventilation system diagrammed in Fig. 18.9. Let us suppose that we have estimated the pressure distribution around the car. We would now like to develop a composite loss coefficient for the system and use it to find the flow under the cruise condition and to size the fan so that we get at least the required minimum flow when the car is stopped.

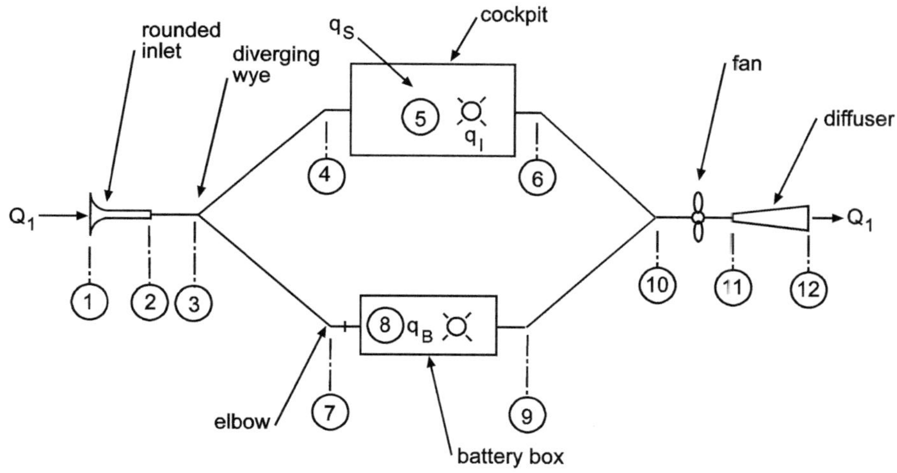

Fig. 18.9 A ventilation system

The system has an inlet, a cockpit branch, and a battery branch, and then the ducts join and connect to the fan. The fan discharges through a diffuser to the external flow. Also shown are heat inputs to the cockpit from the sun (q_s, arrow) and from the driver and the instruments (q_I, rayed circle).

We will ignore the fan for now and discuss only the system external to it. Note that some of the components are in series with the flow so that the pressure drops to an additional amount as each component is traversed by the air. So the rule for series components is to add the pressure (or head) loss for each. The flow is the same through each series component, although the velocity may vary because of flow area changes. We can use Eq. (C.16) to convert the velocity to the volumetric flow. This can then be factored from the sum. For example, suppose the flow area of duct length $l_{2,3}$ is A_3 and the discharge area of the rounded inlet is A_2.

$$\Delta p_{2,3} = \frac{\rho K_{2,3}}{2A_3^2} Q_1^2 = k_{2,3} Q_1^2 \tag{18.31}$$

The lower case k represents a pressure loss coefficient, rather than a head loss coefficient, referenced to the entering flow.

But how about the parallel cockpit and battery branches? Proceeding in a fashion similar to the analysis of the divided flow around the car, we can develop an equivalent flow resistance, $k_{E_{3,10}}$ to represent the pressure loss of those two paths together.

$$\Delta p = k_{E_{3,10}} Q_1^2 = k_{U_{3,10}} Q_U^2 = k_{U_{3,10}} Q_L^2 \tag{18.32}$$

Substituting this into Eq. (18.7) gives

$$k_{E_{3,10}}^{-\frac{1}{2}} = k_{U_{3,10}}^{-\frac{1}{2}} + k_{L_{3,10}}^{-\frac{1}{2}} \tag{18.33}$$

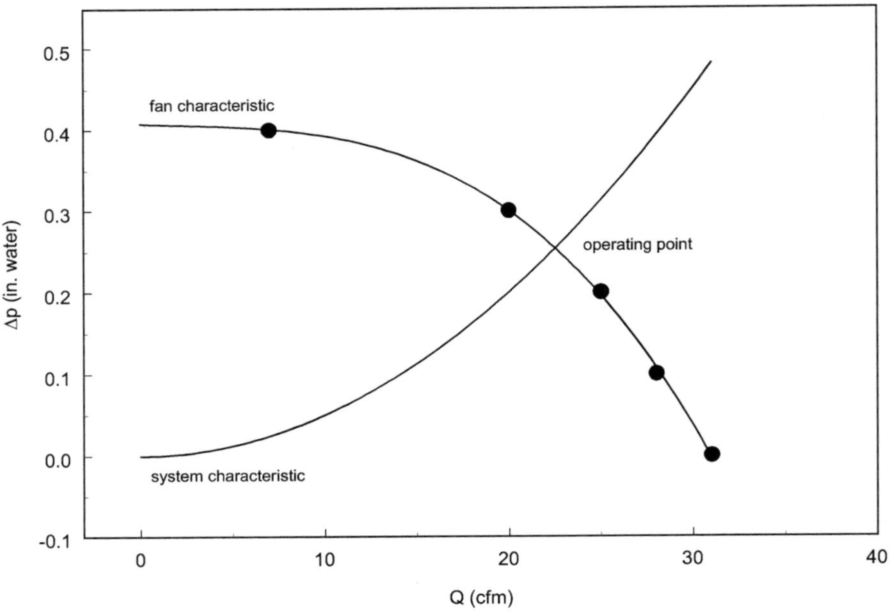

Fig. 18.10 Fan and system characteristic

Where $k_{E_{3,10}}$ is the equivalent loss coefficient of the parallel branches and $k_{U_{3,10}}$ and $k_{L_{3,10}}$ are the series loss coefficients of the upper and lower branches, respectively. Now, the equivalent loss coefficient can be summed with the series loss coefficients for the components between stations 1 and 3 and stations 7 and 12.

$$k_S = k_{1,3} + k_{E_{3,10}} + k_{10,12} \tag{18.34}$$

The pressure loss for the system is

$$\Delta p = k_S Q_1^2 \tag{18.35}$$

This equation is plotted as the "system characteristic" in Fig. 18.10.

18.6 Fan

The purpose of the fan is to add energy to the air flow to offset the head losses. This head gain is a function of the flow rate and the rotational speed of the fan. We will consider herein a single-speed fan because this represents the usual case. Figure 18.10 shows a fan head-flow characteristic. The filled circles indicate that this curve must be obtained from the data furnished by the manufacturer. Note that the head produced is a function of the flow rate.

The intersection of the fan and system characteristics determines the flow through the system. This point usually must be calculated by trial-and-error, although it may be easily read from a graph (to within the scale) like Fig. 18.10.

18.7 Heating

Design Heat Load The rate of solar heating of the cockpit, plus the contribution of the instruments and the driver's body, is the load that must be used to design the cockpit's ventilation system. The radiation entering the cockpit causes negligible direct heating of the air through which it passes. However, surfaces in the cockpit absorb some of the radiation. This energy input tends to cause the temperatures of the absorbing surfaces to rise, thereby heating the air in contact with them. If the air flow through the cockpit cannot remove this energy at the rate it is absorbed, the temperatures in the cockpit will increase. Chap. 7 mentions that an air temperature of 120°F was measured in a solar racer cockpit when the racer was in slowly-moving traffic on a hot day in summer.

Battery The heat given off by the battery can be estimated with the knowledge of the battery's efficiency. For instance, suppose at the ventilation design condition, the battery was discharging at 120 V and 10 A, that is, at 1200 W. Then, if the discharge efficiency was 80%, 240 W would have been dissipated in the process. The battery box air temperature is estimated by assuming that the walls of the box are essentially insulators and therefore all of the battery heat must be carried off by the air.

Solar Gain The fraction of the solar energy incident on the cockpit's glazing and absorbed in the cockpit surfaces is

$$\tau_c \alpha_{eff} = \tau_c \frac{\alpha_i}{\alpha_i + (1 - \alpha_i)\tau_d \dfrac{A_a}{A_i}} \tag{18.36}$$

(Duffie and Beckman 1991). Where τ_c is the transmittance of the glazing for the incident solar radiation, τ_d is the transmittance for isotropic diffuse solar radiation (calculate as for beam but with an effective angle of incidence of 60°), A_a is the area of the cockpit's glazing, A_i is the area of the surfaces inside of the cockpit (not including the glazing), and α_i is the mean absorbance for diffuse radiation of these inner surfaces.

Siegel (1973) presents a more elaborate model.

Instruments The heat dissipation of each instrument or cockpit light is equal to its operating power. For example, if an instrument requires 12 V at 1 mA, then it dissipates 12 mW of heat.

Temperature We will assume that the ventilation and heat flows are steady and that the air temperature is uniform across the inlet and outlet of the cockpit. Accounting for all energy flows gives T_{outlet}, the cockpit outlet temperature, as

$$T_{outlet} = \frac{\dot{Q}_{sun} + \dot{Q}_{inst} + \dot{Q}_{driver}}{\dot{m}c_P} + T_{inlet} \tag{18.37}$$

Where the symbol $\dot{Q}$ denotes the heat inputs previously discussed, $\dot{m}$ is the mass flow rate, c_P now represents the specific heat at constant pressure of the entering air. This is a conservative calculation because it assumes that all the absorbed radiation will be transferred to the ventilation air.

18.8 Relative Humidity

The design environmental conditions should include a relative humidity and an air temperature. These should be assigned to the incoming ventilation air.

References

ASHRAE. (1981). *Thermal environmental condition for human occupancy*. Atlanta: American Society of Heating, Refrigerating and Air-Conditioning Engineers, Inc.

Crane. (1969). *Flow of fluids through valves, fittings, and pipe*. Technical Paper no. 410. New York: Crane Company.

Haaland, S. E. (1983). Simple and explicit formulas for the friction factor in turbulent pipe flow. *Journal of Fluids Engineering, 105,* 89–90.

Idel'chik, I. E., (1966). *Handbook of hydraulic resistance coefficients of local resistance and of friction*, AEC-TR-6630, NTIS. Springfield: Virginia.

Siegel, R. (1973). Net radiation method for enclosure systems involving partially transparent walls, NASA TN D-7384, August

Wallis, S. B. (1971). Ventilation system aerodynamics–A new design method. SAE paper 710036, Automotive Engineering Congress, Detroit, Michigan, January 11–15, 1971.

White, F. M. (1986). *Fluid mechanics* (2nd ed.). New York: McGraw-Hill Book Company.

Chapter 19
Performance Simulation

19.1 Purpose

This chapter suggests a method by which the performance of a solar–electric car over a route may be simulated. The motivation for writing such a simulation is to study the battery energy consumed as a function of the speed, the road course, or the parameters of the car.

19.2 Simulation Methods

A practical method of achieving a realistic solar car simulation is to move the car over short intervals of time during which the quantities entering in to the acceleration may be assumed approximately constant. The intervals may be as short as a second. A short time step is required to simulate speed changes without excessive error. This is the "little tap" method discussed in several places in this book.

The simulation should be written for the cruise condition, at least. This will capture the forces that make the largest contributions to the energy consumption of the car. One technique is to specify the speed over different segments of the route. The tractive force needed to achieve the specified speed may then be computed. From this information and from models of the solar array, motor and controller, and the battery, the net battery energy consumed may be estimated. Alternatively, a specified motor torque may be supplied and the resulting speed computed. Torque specification gives direct control of the energy consumption per unit distance. Specifying the speed is nearer the usual actual method of operation, so this will be discussed herein.

The simulation will consist of six parts: a means of entering information, a model of the car, information about the road course, information about the solar radiation and weather along the course, instructions for logically connecting the previous three parts as a function of time, and a means of communicating the results.

E. F. Thacher, *A Solar Car Primer*, DOI 10.1007/978-3-319-17494-5_19

19.3 Entering Information-Reporting Results

Spreadsheet Building the simulation with a spreadsheet program makes data entry convenient. Several sheets constitute a "notebook" that may be given a unique name. Each sheet may also be given a unique name. These features help to organize the calculation. The data may be entered directly into a sheet designated and organized to receive it. Some programs provide dialog boxes that may be used to prompt for and accept entries. These are handy for making sure that all data is entered and entered correctly. Data files created elsewhere, suitably formatted, may be imported into the sheet. Data on that sheet may be accessed from any other sheet. Formatting features allow the appearance of each sheet in the notebook to be modified to suit the needs of the calculation.

Results are always available in tabular form. However, flexible plotting abilities allow the user to graph the simulation results in almost any form desired.

Languages Programming environments of great power have been developed for time-based simulations. These environments do not provide the same ready-made data entry convenience as the spreadsheet. But they provide libraries of functions that may be used to create a convenient interface with the user. Their strengths lie in their calculation engines, specialized for forward-time-marching analysis, in the relative ease with which a simulation can be constructed, and in their function libraries. The user may write functions to perform calculations not available in these libraries.

Plotting capabilities are extensive. And many language commands are available to customize the appearance of the plots. In addition, printed output may be formatted and controlled in various ways.

19.4 Car

The model of the car will have parts that account for the car's interactions with the road and the atmosphere, for the motor and controller, for the battery, and for the solar array.

Interactions The interactions with the road and the atmosphere were summarized in Eq. (19.1)

$$T = M_e a_X + D + R + W \sin\alpha. \tag{19.1}$$

The design route should be divided into segments on each of which the speed, course angle, grade, and certain other information has been specified. A new segment begins whenever one of these characteristics changes. The section on road data contains an example.

From the discussions in Chap. 2 and 13, we know that the tractive effort represents the energy required at the driving wheel per unit distance traveled. A handy unit for this quantity is kilowatt-hours per kilometer (kW·h/km) because battery energy is customarily expressed in kW·h. So the energy consumed (kW·h) at the wheels while traversing a distance S (km) at constant speed is

$$\Delta E_W = \frac{T}{3600}\Delta S = \frac{T}{3600}V_X \Delta t = \frac{P_W}{3600}\Delta t.$$

The "wheel energy" is of interest because it is a measure of the efficiency of the design. Keep a running total like this:

$$E_W^{(n)} = E_W^{(n-1)} + \Delta E_W. \tag{19.2}$$

Where P_W and E_W are the power (kW) and energy (kW·h) required at the driven wheel.

Note that if the segment is downhill, the grade angle α will be negative. If the magnitudes of the drag and rolling resistance terms in (19.1) are equal or less than the magnitude of the gravity term, the tractive force, that is the energy consumed per unit distance, will be zero or negative. If zero, gravity supplies the required tractive force and no battery or solar energy is consumed for propulsion. If less than zero, gravity not only maintains the steady motion, but regeneration converts gravitational potential energy into electrical energy that is stored in the battery.

Motor The power the motor must supply to the drive shaft is

$$P_S = \frac{P_W}{\eta_G}. \tag{19.3}$$

Where η_G is the efficiency of the transmission. The power demanded from the main bus will then be

$$P_M = I_M V_B = \frac{P_S}{\eta_D} = \frac{P_W}{\eta_D \eta_G}. \tag{19.4}$$

Where η_D is the drive (motor plus controller) efficiency and V_B is the main bus voltage.

Chapter 5 explains that the efficiency of the motor will depend upon its rotational speed and torque. The rotational speed (rpm) of the motor is related to the speed of the car by

$$N_M = n_G N_W = n_G \frac{30}{\pi}\frac{V_X}{r_W}. \tag{19.5}$$

Where n_G is the ratio of the rotational speed of the motor to that of the wheel (the reduction ratio of the transmission). The flattening of the wheel at the contact patch has been neglected. The shaft torque, τ_S (N·m), required of the motor (or supplied to the motor, in the case of regeneration) is

$$\tau_S = \frac{P_W}{\eta_G n_G \dfrac{V}{r_W}}. \tag{19.6}$$

The efficiency may be found by entering the motor efficiency characteristics (supplied by the motor manufacturer) with the torque and speed. These characteristics should be tabulated for convenience in interpolation. Or, interpolating functions, written in terms of torque and rotational speed, may be developed from samples taken from the original curves.

Loads Chapter 6 points out that some of the electrical loads, such as the turn signals, will be switched on intermittently, and others, such as the battery fan, will be switched on continuously. It is possible, especially when the simulation is built in a programming environment, to simulate each of these loads. But the added precision is probably not worth the extra effort. An alternative would be to represent the average load by a suitably-chosen resistor connected in parallel with the battery bus.

Battery Equation (19.4) gives the motor current once the bus voltage is known. Chapter 4 explains that the bus voltage is determined by the battery's charge, Q, and shows how the voltage is related to the charge through an equivalent resistance, R_S. Chapter 13 shows RS as a function of Q for a particular battery (Fig. 13.3). If this information is available it is a simple and effective way to model the effect of charge on the bus voltage. Other, more complicated, methods exist.[1] The charge, Q, must be found by keeping a running tab (IB denotes the magnitude of the battery current).

$$Q^{(n)} = Q^{(n-1)} + \Delta t \begin{cases} I_B, \text{charge} \\ -I_R\ , \text{discharge} \end{cases} \tag{19.7}$$

The usual convention is that I_B is positive when charging the battery and negative when discharging it. Note that (19.7) does not account for charge loss during charging or discharging: the Coulomb efficiency is taken to be 100 %.

On discharge, the energy change in one time step equals that delivered by the battery plus that consumed in the equivalent internal resistance, R_S. The running tab for the energy in the battery is

$$E_B^{(n)} = E_B^{(n-1)} + I_B \Delta t \begin{cases} -V_B - I_B R_S, \text{discharge} \\ V_B - I_B R_S, \text{charge} \end{cases}. \tag{19.8}$$

Array The solar array supplies current to the main bus through maximum power point trackers. Let us suppose the array is flat, so that the angle of incidence of each sub-array is the same. Faceted arrays are more complicated to model (and are not as efficient)[2]. And let us also suppose that each sub-array has the same number of

[1] See model and references in Craparo and Thacher (1995).

[2] $V_M \approx$ constant and $I_A \approx G_S I_{A0} / G_{S0}$, where G_{S0} is a reference irradiance. Then $h_A = I_M V_M / A_A G_S = (G_S / G_{S0}) I_{M0} V_M / A_A G_S = I_{M0} V_M / A_A G_{S0}$.

cells in series and that the temperatures of corresponding cells in different strings are equal. The array power (kW) would be

$$P_A = \frac{G_T}{1000} A_A \eta_A \eta_{MT}. \tag{19.9}$$

Where G_T (W/m^2) is the total solar irradiance normal to the array's surface, A_A the total area included in the array (cells plus unused space), η_A is the array efficiency (including the area efficiency), and η_{MT} is the maximum power point tracker efficiency. The area efficiency is the ratio of the total cell and total array areas. The array efficiency is the product of the area efficiency and the cell efficiency.

The efficiency, η_A, should be that at the maximum power point. It may be estimated from the I–V curves furnished by the cell manufacturer. These curves should show the current as a function of voltage at different irradiances normal to the array and at the standard cell temperature of 25 °C. The efficiency at the maximum power point is approximately independent of irradiation. These efficiencies should be slightly reduced using the methods of Chap. 3 to allow for the thin glazings applied during construction. Chapter 3 showed that the array will be hot when in operation, often well above the cell rating temperature of 25 °C. This will reduce the maximum power point efficiency. Duffie and Beckman (1991) point out that the change in V_M with temperature is approximately the same as the change in the open circuit voltage with temperature. The temperature dependence of V_{OC} may be derivable from the cell manufacturer's data. Some manufacturers supply I–V curves at different cell temperatures at a particular irradiance. Hu and White (1983) give −2.0 mV/K for silicon-based solar cells and state that this value agrees well with measured temperature sensitivities. Call this value s_T. Duffie and Beckman (1991) establish η_A as a linear function of temperature

$$\eta_A = \eta_{A0} + s_{TM}\left(T - T_0\right). \tag{19.10}$$

Where η_{A0} and T_0 are the reference maximum efficiency and its corresponding temperature (25 °C would be convenient), s_{TM} is the sensitivity of the efficiency to temperature, and T denotes the average temperature of the array[3]. These authors give s_{TM} as

$$s_{TM} = \eta_{A0} \frac{s_T}{V_M}. \tag{19.11}$$

Chapter 5 discussed the three modes of operation of the power system: battery charge, battery float, and battery discharge. The logic of the model must account for these scenarios and calculate the bus voltage and currents properly. For example,

[3] A better simulation would be obtained by breaking the array into "thermal pieces" and estimating the temperature of each piece, then combining them. This is a much more complicated task, however.

if the motor current calculated for the current time step exceeds the available array current, the logic must force the battery to discharge and reduce the main bus voltage and stored charge accordingly.

Speed Changes The discussion of Fig. 2.16 of Chap. 2 explained two methods of changing speed. In the first, the tractive force at the new steady speed was computed and held until the new speed was reached. This, as the figure showed, could take a relatively long time. The second method was to multiply the new tractive force by some factor greater than one, and then to reduce the tractive force to the steady speed value as the new speed was approached.

19.5 Predicting Solar Radiation

Chapters 3 and 13 presented methods for predicting the total solar irradiance and its beam and diffuse components during an hour. This chapter will present the list of weather stations used in the National Weather Services (NWS) daily total irradiance predictions (mentioned in Chap. 13) followed by a means for estimating the total beam and diffuse components of the solar irradiance over intervals less than an hour.

NWS Daily Total Table 19.1 gives the weather reporting stations used by the National Weather Service.[4] The locations of the stations may be used to interpolate in the charts shown in Chap. 13 for other locations. These values may be broken into hourly total beam and diffuse components using the techniques of Chap. 3. The WBAN station designator is an acronym for "Weather-Bureau-Army-Navy," "Lat," "Lon," and "Hte" are abbreviations for "Latitude," "Longitude," and "Height" (above mean sea level). Finally, "TZ" is the difference in hours between Greenwich (England) Mean Time and the time zone of the station.

Approximating G(t) From Hourly Data Solar radiation values approximating $G(t)$, the instantaneous solar irradiation on a horizontal surface, and its beam and diffuse components must be supplied to the simulation. Only hourly total global radiation data are generally widely available, so these will be used.

Figure 19.1 shows a synthetic $G(t)$ (solid line) for a day beginning at 6 a.m. and ending at 6 p.m., hour 18. The values of G at dawn and sunset have been set to zero, thus neglecting the pre-dawn glow and evening twilight. The jagged parts of $G(t)$ represent cloud cover. This figure will be used to illustrate techniques for estimating $G(t)$.

The $I(n)$ values derived from this $G(t)$ for each hour are plotted as "data" points in Fig. 19.2 (filled squares). Also plotted in this figure is the accumulated solar energy at the end of each hour of daylight, $E(t^{(n)})$ (kJ/m^2, solid line).

$$E\left(t^{(n)}\right) = E^{(n)} = \sum_{i=1}^{i=n} I_i \qquad (19.12)$$

[4] Check with the NWS for changes to this and for international weather station locations and data.

Table 19.1 National weather service reporting stations. (Courtesy National Weather Service)

Station	Name	State	WBAN	Lat	Lon	Hte	TZ
ABE	Allentown	PA	14,737	40.65	75.43	385	−5
ABI	Abilene	TX	13,962	32.42	99.68	1790	−6
ABQ	Albuquerque	NM	23,050	35.05	106.62	5314	−7
ABR	Aberdeen	SD	14,929	45.45	98.43	1300	−6
ACT	Waco	TX	13,959	31.62	97.22	508	−6
ACY	Atlantic City	NJ	93,730	39.45	74.57	67	−5
AGS	Augusta	GA	3820	33.37	81.97	148	−5
AHN	Athens	GA	13,873	33.95	83.32	803	−5
ALB	Albany	NY	14,735	42.75	73.80	292	−5
ALO	Waterloo	IA	94,910	42.55	92.40	878	−6
AMA	Amarillo	TX	23,047	35.23	101.70	3604	−6
Y62	Sault Ste Marie	MI	14,847	46.48	84.36	721	−5
APN	Alpena	MI	94,849	45.07	83.57	693	−5
AST	Astoria	OR	94,224	46.15	123.88	22	−8
ATL	Atlanta	GA	13,874	33.65	84.43	1034	−5
AUS	Austin	TX	13,958	30.30	97.70	621	−6
AVL	Asheville	NC	3812	35.43	82.55	2170	−5
AVP	Scranton	PA	14,777	41.33	75.73	948	−5
BDL	Hartford	CT	14,740	41.93	72.68	179	−5
BDR	Bridgeport	CT	94,702	41.17	73.13	17	−5
BFD	Bradford	PA	4751	41.80	78.63	2150	−5
BFF	Scottsbluff	NE	24,028	41.87	103.60	3958	−7
BFL	Bakersfield	CA	23,155	35.42	119.05	492	−8
BGM	Binghamton	NY	4725	42.22	75.98	1629	−5
BGR	Bangor	ME	14,606	44.80	68.82	192	−5
BHM	Birmingham	AL	13,876	33.57	86.75	630	−6
BIL	Billings	MT	24,033	45.80	108.53	3570	−7
BIS	Bismarck	ND	24,011	46.77	100.75	1660	−6
BKW	Beckley	WV	3872	37.78	81.12	2514	−5
BNA	Nashville	TN	13,897	36.12	86.68	605	−6
BNO	Burns	OR	24,134	43.58	118.95	4170	−8
BOI	Boise	ID	24,131	43.57	116.22	2868	−7
BOS	Boston	MA	14,739	42.37	71.03	29	−5
BRL	Burlington	IA	14,931	40.78	91.12	702	−6
BRO	Brownsville	TX	12,919	25.90	97.43	20	−6
BTR	Baton Rouge	LA	13,970	30.53	91.15	76	−6
BTV	Burlington	VT	14,742	44.47	73.15	340	−5
BUF	Buffalo	NY	14,733	42.93	78.73	706	−5
be	Boothville	LA	12,884	29.33	89.40	12	−6
BWI	Baltimore	MD	93,721	39.18	76.67	155	−5
CAE	Columbia	SC	13,883	33.95	81.12	225	−5

Table 19.1 (continued)

Station	Name	State	WBAN	Lat	Lon	Hte	TZ
CAK	Akron Canton	OH	14,895	40.92	81.43	1236	−5
CAR	Caribou	ME	14,607	46.87	68.02	628	−5
CDC	Cedar City	UT	93,129	37.70	113.10	5618	−7
CHA	Chattanooga	TN	13,882	35.03	85.20	688	−5
CHS	Charleston	SC	13,880	32.90	80.03	48	−5
CLE	Cleveland	OH	14,820	41.40	81.85	805	−5
CLT	Charlotte	NC	13,881	35.21	80.95	769	−5
CMH	Columbus	OH	14,821	40.00	82.88	833	−5
CNK	Concordia	KS	13,984	39.55	97.65	1484	−6
CON	Concord	NH	14,745	43.20	71.50	346	−5
COS	Colo. Springs	CO	93,037	38.82	104.72	6170	−7
COU	Columbia	MO	3945	38.82	92.22	898	−6
CPR	Casper	WY	24,089	42.92	106.47	5290	−7
CRP	Corpus Christi	TX	12,924	27.77	97.50	44	−6
CRW	Charleston	WV	13,866	38.37	81.60	982	−5
CVG	Covington	KY	93,814	39.05	84.67	877	−5
CVS	Cannon Afb	NM	22,008	34.38	103.32	4295	−7
CXY	Harrisburg	PA	14,751	40.22	76.85	351	−5
CYS	Cheyenne	WY	24,018	41.15	104.82	6141	−7
DAB	Daytona Beach	FL	12,834	29.18	81.05	41	−5
DAG	Daggett	CA	23,161	34.87	116.78	1929	−8
DAY	Dayton	OH	93,815	39.90	84.20	1003	−5
DBQ	Dubuque	IA	94,908	42.40	90.70	1080	−6
DCA	Washington	DC	13,743	38.85	77.03	65	−5
DDC	Dodge City	KS	13,985	37.77	99.97	2592	−6
DEN	Denver Intl	CO	3017	39.87	104.67	5382	−7
DFW	Dallas-Ft. Worth	TX	3927	32.90	97.03	596	−6
DLH	Duluth	MN	14,913	46.83	92.18	1417	−6
DRT	Del Rio	TX	22,010	29.37	100.92	1027	−6
DSM	Des Moines	IA	14,933	41.53	93.65	963	−6
DTW	Detroit	MI	94,847	42.23	83.33	664	−5
EAU	Eau Claire	WI	14,991	44.87	91.48	895	−6
EKN	Elkins	WV	13,729	38.88	79.85	1997	−5
EKO	Elko	NV	24,121	40.83	115.78	5077	−8
ELP	El Paso	TX	23,044	31.80	106.40	3916	−7
ELY	Ely	NV	23,154	39.28	114.85	6262	−8
ERI	Erie	PA	14,860	42.08	80.18	737	−5
EUG	Eugene	OR	24,221	44.12	123.22	373	−8
EVV	Evansville	IN	93,817	38.05	87.53	388	−6
EWR	Newark	NJ	14,734	40.70	74.17	30	−5

Table 19.1 (continued)

Station	Name	State	WBAN	Lat	Lon	Hte	TZ
EYW	Key West	FL	12,836	24.55	81.75	21	−5
FAR	Fargo	ND	14,914	46.90	96.80	899	−6
FAT	Fresno	CA	93,193	36.77	119.72	327	−8
FCA	Kalispell	MT	24,146	48.30	114.27	2973	−7
FLG	Flagstaff	AZ	3103	35.13	111.67	7018	−7
FMN	Farmington	NM	23,090	36.75	108.23	5502	−7
FMY	Fort Myers	FL	12,835	26.58	81.86	18	−5
FNT	Flint	MI	14,826	42.97	83.73	766	−5
FSD	Sioux Falls	SD	14,944	43.57	96.73	1427	−6
FSM	Fort Smith	AR	13,964	35.33	94.37	463	−6
FWA	Fort Wayne	IN	14,827	41.00	85.20	828	−5
GEG	Spokane	WA	24,157	47.63	117.53	2365	−8
GGW	Glasgow	MT	94,008	48.22	106.62	2298	−7
GJT	Grand Junction	CO	23,066	39.12	108.53	4839	−7
GLD	Goodland	KS	23,065	39.37	101.70	3688	−7
GRB	Green Bay	WI	14,898	44.48	88.13	702	−6
GRI	Grand Island	NE	14,935	40.97	98.32	1856	−6
GRR	Grand Rapids	MI	94,860	42.88	85.52	803	−5
GSO	Greensboro	NC	13,723	36.08	79.95	886	−5
GSP	Greenville-Spart.	SC	3870	34.90	82.22	971	−5
GTF	Great Falls	MT	24,143	47.48	111.37	3657	−7
HLN	Helena	MT	24,144	46.60	112.00	3898	−7
HON	Huron	SD	14,936	44.38	98.22	1289	−6
HSE	Hatteras	NC	0	35.23	75.62	11	−5
HSV	Huntsville	AL	3856	34.65	86.77	644	−6
HTL	Houghton Lake	MI	94,814	44.37	84.68	1160	−5
HTS	Huntington	WV	3860	38.37	82.55	838	−5
HVR	Havre	MT	94,012	48.55	109.77	2599	−7
IAD	Wash-Dulles	VA	93,738	38.95	77.45	323	−5
IAH	Houston	TX	12,960	29.97	95.35	108	−6
ICT	Wichita	KS	3928	37.65	97.42	1340	−6
ILG	Wilmington	DE	13,781	39.67	75.60	80	−5
ILM	Wilmington	NC	13,748	34.27	77.92	38	−5
IND	Indianapolis	IN	93,819	39.73	86.28	808	−5
INL	International Falls	MN	14,918	48.57	93.38	1183	−6
INW	Winslow	AZ	23,194	35.02	110.73	4883	−7
IPT	Williamsport	PA	14,778	41.25	76.92	525	−5
ISN	Williston	ND	94,014	48.18	103.63	1905	−6
JAN	Jackson	MS	3940	32.32	90.08	331	−6

Table 19.1 (continued)

Station	Name	State	WBAN	Lat	Lon	Hte	TZ
JAX	Jacksonville	FL	13,889	30.50	81.70	31	−5
JFK	New York-Kennedy	NY	94,789	40.65	73.78	22	−5
LAN	Lansing	MI	14,836	42.78	84.60	874	−5
LAS	Las Vegas	NV	23,169	36.08	115.17	2180	−8
LAX	Los Angeles	CA	23,174	33.93	118.40	104	−8
LBB	Lubbock	TX	23,042	33.65	101.82	3241	−6
LBF	North Platte	NE	24,023	41.13	100.68	2787	−6
LCH	Lake Charles	LA	3937	30.12	93.22	32	−6
LEX	Lexington	KY	93,820	38.03	84.60	989	−5
LFK	Lufkin	TX	93,987	31.23	94.75	316	−6
LGA	New York-Laguardia	NY	14,732	40.77	73.90	31	−5
LGB	Long Beach	CA	23,129	33.82	118.15	40	−8
LIT	Little Rock	AR	13,963	34.73	92.23	257	−6
LND	Lander	WY	24,021	42.82	108.73	5558	−7
LOL	Lovelock	NV	24,172	40.07	118.55	3904	−8
LSE	La Crosse	WI	14,920	43.87	91.25	663	−6
LYH	Lynchburg	VA	13,733	37.33	79.20	937	−5
MAF	Midland	TX	23,023	31.95	102.18	2862	−6
MCI	Kansas City	MO	3947	39.32	94.72	1025	−6
MCN	Macon	GA	3813	32.70	83.65	362	−5
MCO	Orlando Intl	FL	12,815	28.43	81.32	105	−5
MCW	Mason City	IA	14,940	43.15	93.33	1225	−6
MDW	Chicago-Midway	IL	14,819	41.78	87.75	623	−6
MEI	Meridian	MS	13,865	32.33	88.75	310	−6
MEM	Memphis	TN	13,893	35.05	90.00	284	−6
MFR	Medford	OR	24,225	42.37	122.87	1329	−8
MGM	Montgomery	AL	13,895	32.30	86.40	202	−6
MIA	Miami	FL	12,839	25.82	80.28	12	−5
MKE	Milwaukee	WI	14,839	42.95	87.90	693	−6
MKG	Muskegon	MI	14,840	43.17	86.23	633	−5
MLI	Moline	IL	14,923	41.45	90.52	594	−6
MOB	Mobile	AL	13,894	30.68	88.25	221	−6
MOT	Minot	ND	24,013	48.27	101.28	1714	−6
MSN	Madison	WI	14,837	43.13	89.33	866	−6
MSO	Missoula	MT	24,153	46.92	114.08	3189	−7
MSP	Minneapolis	MN	14,922	44.88	93.22	838	−6
MSS	Massena	NY	94,725	44.93	74.85	214	−5
MSY	New Orleans	LA	12,916	29.98	90.25	30	−6
OAK	Oakland	CA	23,230	37.73	122.20	7	−8

Table 19.1 (continued)

Station	Name	State	WBAN	Lat	Lon	Hte	TZ
OKC	Oklahoma Cty	OK	13,967	35.40	97.60	1304	−6
OLM	Olympia	WA	24,227	46.97	122.90	200	−8
OMA	Omaha	NE	14,942	41.30	95.90	982	−6
ORD	Chicago-O'hare	IL	94,846	41.98	87.90	674	−6
ORF	Norfolk	VA	13,737	36.90	76.20	30	−5
OTH	North Bend	OR	24,284	43.42	124.25	17	−8
PBI	W Palm Beach	FL	12,844	26.68	80.12	21	−5
PDT	Pendleton	OR	24,155	45.68	118.85	1495	−8
PDX	Portland	OR	24,229	45.60	122.60	39	−8
PHL	Philadelphia	PA	13,739	39.88	75.25	28	−5
PHX	Phoenix	AZ	23,183	33.43	112.02	1107	−7
PIA	Peoria	IL	14,842	40.67	89.68	662	−6
PIH	Pocatello	ID	24,156	42.92	112.60	4478	−7
PIR	Pierre	SD	24,025	44.38	100.28	1726	−6
PIT	Pittsburgh	PA	94,823	40.50	80.22	1225	−5
PNS	Pensacola	FL	13,899	30.47	87.20	118	−6
PUB	Pueblo	CO	93,058	38.28	104.52	4720	−7
PVD	Providence	RI	14,765	41.73	71.43	62	−5
PWM	Portland	ME	14,764	43.65	70.32	63	−5
RAP	Rapid City	SD	24,090	44.05	103.07	3168	−7
RBL	Red Bluff	CA	24,216	40.15	122.25	353	−8
RDM	Redmond	OR	24,230	44.27	121.15	3084	−8
RDU	Raleigh-Durham	NC	13,722	35.87	78.78	441	−5
RFD	Rockford	IL	94,822	42.20	89.10	734	−6
RIC	Richmond	VA	13,740	37.50	77.33	177	−5
RKS	Rock Springs	WY	24,027	41.60	109.07	6745	−7
RNO	Reno	NV	23,185	39.50	119.78	4400	−8
ROA	Roanoke	VA	13,741	37.32	79.97	1176	−5
ROC	Rochester	NY	14,768	43.12	77.67	555	−5
RSL	Russell	KS	93,997	38.87	98.82	1869	−6
RST	Rochester	MN	14,925	43.92	92.50	1320	−6
SAC	Sacramento	CA	23,232	38.52	121.50	25	−8
SAN	San Diego	CA	23,188	32.73	117.17	28	−8
SAT	San Antonio	TX	12,921	29.53	98.47	794	−6
SAV	Savannah	GA	3822	32.13	81.20	51	−5
SBN	South Bend	IN	14,848	41.70	86.32	773	−5
SBP	San Luis Obispo	CA	93,206	35.23	120.63	207	−8
SCK	Stockton	CA	23,237	37.90	121.25	27	−8
SDF	Louisville	KY	93,821	38.18	85.73	488	−5

Table 19.1 (continued)

Station	Name	State	WBAN	Lat	Lon	Hte	TZ
SEA	Seattle-Tacoma	WA	24,233	47.45	122.30	450	−8
SFO	San Francisco	CA	23,234	37.62	122.38	18	−8
SGF	Springfield	MO	13,995	37.23	93.38	1270	−6
SHR	Sheridan	WY	24,029	44.77	106.97	3968	−7
SHV	Shreveport	LA	13,957	32.47	93.82	259	−6
SJT	San Angelo	TX	23,034	31.37	100.50	1908	−6
SLC	Salt Lake City	UT	24,127	40.77	111.97	4227	−7
SLE	Salem	OR	24,232	44.92	123.00	201	−8
SPI	Springfield	IL	93,822	39.83	89.67	613	−6
SPS	Wtchita Falls	TX	13,966	33.97	98.48	1030	−6
STL	St. Louis	MO	13,994	38.75	90.38	564	−6
SUX	Sioux City	IA	14,943	42.40	96.38	1103	−6
SYR	Syracuse	NY	14,771	43.12	76.12	407	−5
TCS	Truth Or Consq	NM	93,045	33.23	107.27	4858	−7
TLH	Tallahassee	FL	93,805	30.38	84.37	68	−5
TOL	Toledo	OH	94,830	41.60	83.80	692	−5
TOP	Topeka	KS	13,996	39.07	95.63	885	−6
TPA	Tampa	FL	12,842	27.97	82.53	11	−5
TPH	Tonopah	NV	23,153	38.07	117.08	5434	−8
TRI	Bristol	TN	13,877	36.48	82.40	1525	−5
TUL	Tulsa	OK	13,968	36.20	95.90	676	−6
TUS	Tucson	AZ	23,160	32.12	110.93	2555	−7
TVC	Traverse City	MI	14,850	44.74	85.57	630	−5
TYS	Knoxville	TN	13,891	35.82	83.98	980	−5
UIL	Quillayute	WA	94,240	47.95	124.55	205	−8
VCT	Victoria	TX	12,912	28.85	96.92	117	−6
WMC	Winnemuca	NV	24,128	40.90	117.80	4314	−8
YKM	Yakima	WA	24,243	46.57	120.53	1066	−8
YNG	Youngstown	OH	14,852	41.27	80.67	1186	−5
YUM	Yuma	AZ	23,195	32.67	114.60	206	−7

The real $E(t)$ would be a continuous function, but the lines joining the E_j values are straight and change slope abruptly at each hour. Thus they describe a "faceted" approximation to $E(t)$. Note the changes in the slope caused by the clouds.

The slope of $E(t)$ at any time is $G(t)$, the function sought. The slope of $E(t)$ can be computed by numerical approximation. Equation (19.12) is the simplest form of this approximation. Taking the time interval as 3600 s and $I^{(n)}$ in J/m^2, it gives the average slope for the hour in W/m^2.

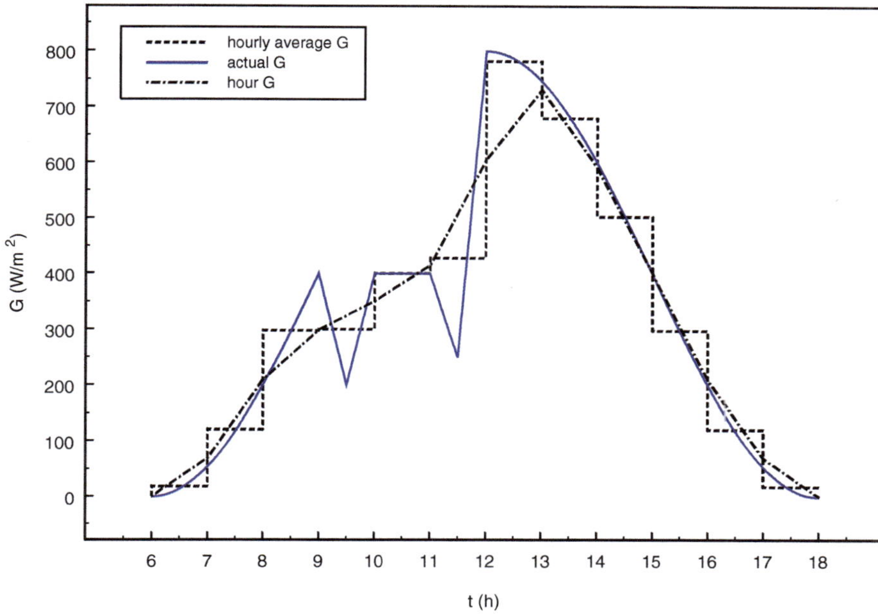

Fig. 19.1 Artificial G(t) and estimates

$$\bar{G}^{(n)} = \frac{E^{(n)} - E^{(n-1)}}{t^{(n)} - t^{(n-1)}} = \frac{I^{(n)}}{\Delta t} \tag{19.13}$$

This average is the same for each moment in the hour and is shown superimposed on the synthetic *G(t)* in Fig. 19.1. During each hour, *E(t)* may be estimated from

$$E(t) = E^{(n-1)} + \bar{G}^{(n)}\left(t - t^{(n-1)}\right). \tag{19.14}$$

To find $\bar{G}_B$ and $\bar{G}_D$ on a horizontal surface, first find I_D from I as in Example 3.5 and then apply the preceding method to give $\bar{G}_D$. $\bar{G}_B$ is equal to the difference between $\bar{G}$ and $\bar{G}_D$. Figure 19.3 shows $\bar{G}_B$ and $\bar{G}_D$ for the Massena, NY, on June 11 derived using this method and the cumulative hourly solar energy values of Fig. 19.2.

Discussion Some improvement over the simple approximation may be obtained by using an approximation to *G(t)* at each hour:

$$G^{(n)} = \frac{I^{(n)} + I^{(n+1)}}{2\Delta t}. \tag{19.15}$$

$G^{(n)}$ may be transformed into $G_D^{(n)}$ by defining an instantaneous diffuse fraction $f_{GD}^{(n)}$ such that

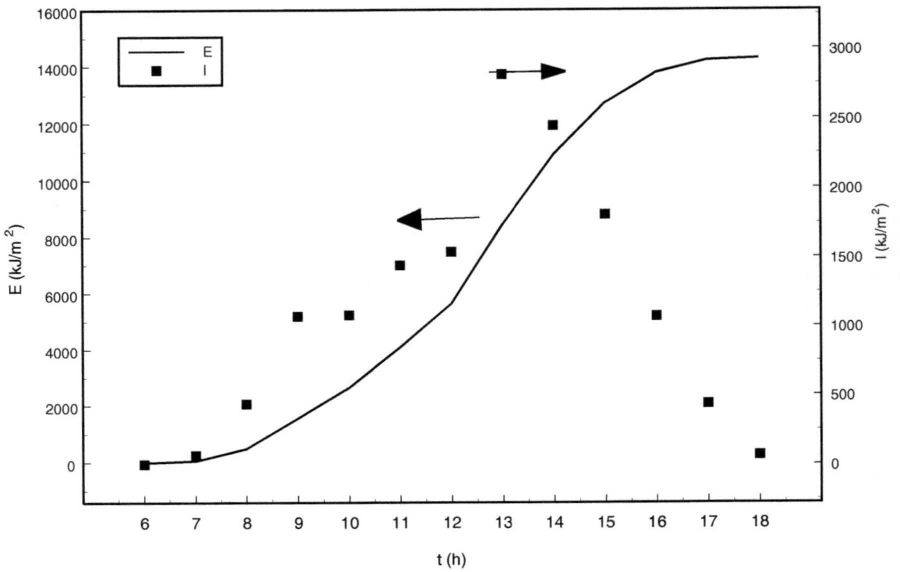

Fig. 19.2 Synthetic hourly data and corresponding E(t)

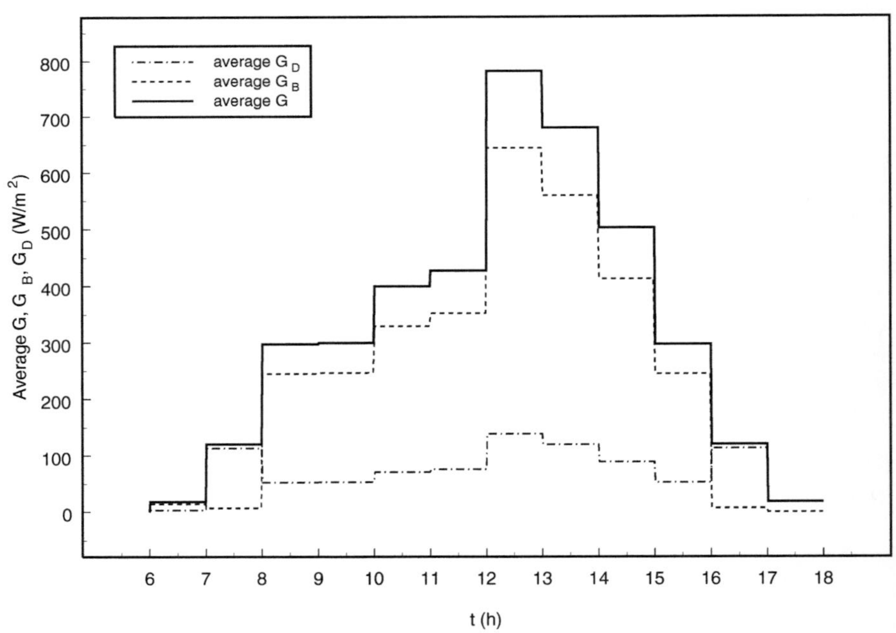

Fig. 19.3 Synthetic data on June 11 for Massena, New York

$$G_D^{(n)} = f_{GD}^{(n)} G^{(n)}. \tag{19.16}$$

Where

$$f_{GD}^{(n)} = \frac{f_D^{(n)} I^{(n)} + f_D^{(n+1)} I^{(n+1)}}{I^{(n)} + I^{(n+1)}} \tag{19.17}$$

$G_B^{(n)}$ can be found by subtraction, as before. An interpolating polynomial can then be fitted through the sets of $G^{(n)}$, $G_B^{(n)}$ and $G_D^{(n)}$ to find values at intermediate times. The results of this method using a straight line interpolating polynomial are also shown superimposed on the original *G(t)* in Fig. 19.1. Notice the smoothing of the sharp points of *G(t)*. However, the approximation is better than using the average, $\bar{G}$, during the periods of clear sky. Figure 19.1 also suggests that the average values could be assigned to the midpoint of each hour during clear periods, at least. This procedure could double the number of estimated values available for interpolation and improve the approximation at the peaks at 9 a.m. and noon.

19.6 Road Data

Road data is created by first decomposing the route into segments and assigning certain properties to each segment. These are: distance traveled to the start of the segment (km), grade (percent, positive for climb, negative for descent), direction of travel (degrees true), road tilt angle (degrees), road surface (represented by the static rolling resistance coefficient), ground reflectivity, event code, and event duration (minutes). A new segment begins whenever one or more of these properties changes.

Figure 19.4 shows a 40 km road course. Table 19.2 shows how the course's segments could be entered road data file. Observe from the segments beginning at 30 and 35 km that a stop, because it is a change, should be entered at the beginning of a new segment. Other properties that must be changed are given their new values. Then when the stop time expires, the car will be moved into the new segment and use these new values.

The final entry in Table 19.2 is 41 km, not 40 km. The road data file must always end with an extra distance entry for the "Driver" (next section) to read. Otherwise an error will result. The extra, dummy, entry will play no role because the car will already be stopped.

$$s^{(n)} = s^{(n-1)} + V^{(n)}\Delta t. \tag{19.18}$$

Battery charging with sun tracking and stops for some specified interval may be initiated with event codes. These are segment properties, too. Table 19.3 shows

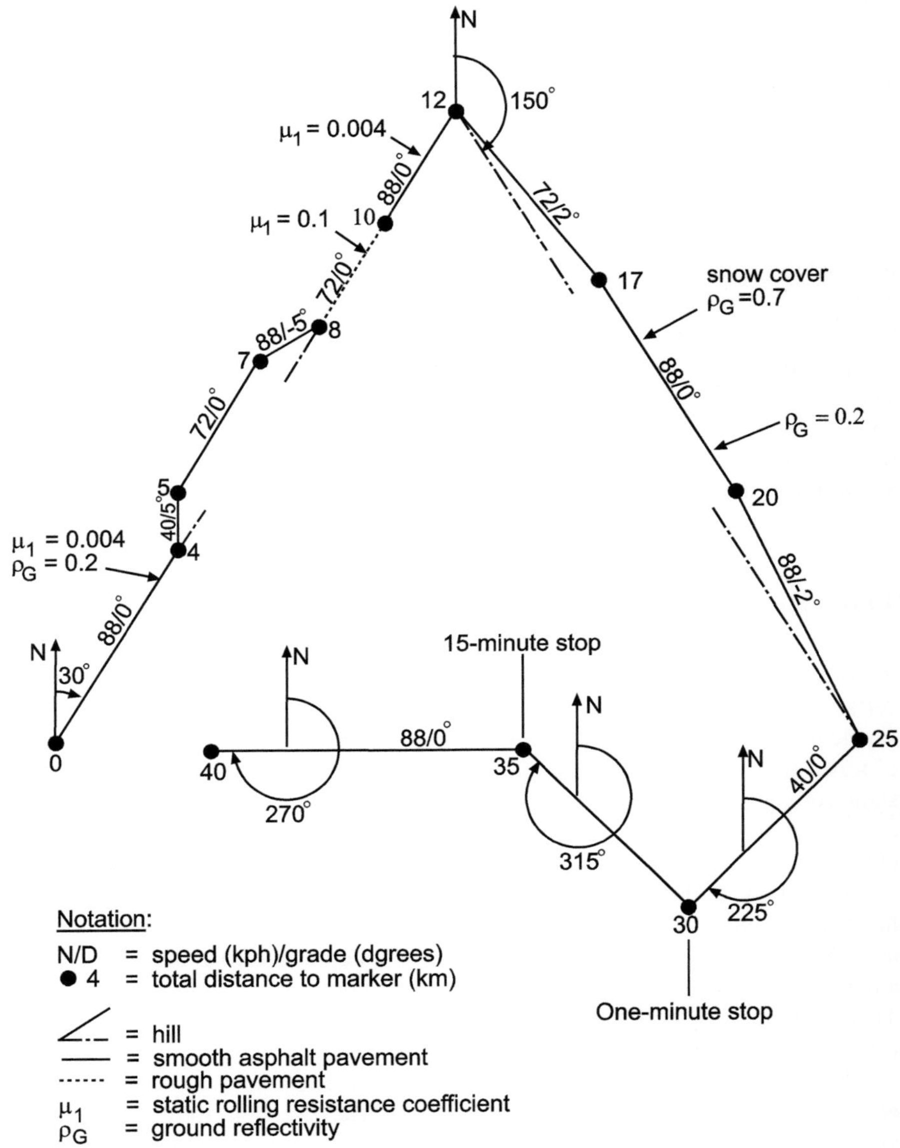

Fig. 19.4 Sample road course. (Craparo and Thacher 1995)

possible codes and their meanings. In the road data file, event code 1 should have an accompanying time interval during which the car is to be stopped.

Driver The "driver" is the logic that connects the parts of the simulation. In a spreadsheet-based simulation, this logic will be embedded in the various sheets. In

Table 19.2 Sample road characteristics

Distance (km)	Grade (deg)	Course (deg)	Speed (kph)	Tilt (deg)	Road surface	Ground refl.	Event code	Event interval (h)
0.0	0	030	88	0	0.004	0.2	0	0
4.0	5	030	40	0	0.004	0.2	0	0
5.0	0	030	72	0	0.004	0.2	0	0
7.0	−5	030	88	0	0.004	0.2	0	0
8.0	0	030	72	0	0.100	0.2	0	0
10.0	0	030	88	0	0.004	0.2	0	0
12.0	2	150	72	0	0.004	0.2	0	0
17.0	0	150	88	0	0.004	0.7	0	0
20.0	−2	150	88	0	0.004	0.2	0	0
25.0	0	225	40	0	0.004	0.2	0	0
30.0	0	315	40	0	0.004	0.2	1	0.0167
35.0	0	270	88	0	0.004	0.2	1	0.25
40.0	0	0	0	0	0.004	0.2	2	20:75
41.0	0	0	0	0	0	0	0	0

Table 19.3 Event codes

Event code	Action
0	Continue on
1	Stop for a specified interval given in hours then resume with segment's parameters
2	Stop. Enter solar-tracking battery charging mode until the battery is charged, a specified time is reached, or the simulation time expires, whichever is first

a modular language-based simulation, it may be a separate program unit (or units). Some tasks performed by the driver would be:

1. Keeping the running sums, including the time and the total distance traveled,
2. Reading the road data,
3. Comparing the distance traveled to the distances in the road data file and establishing the new speed (including regeneration braking) and road properties, if necessary,
4. Taking the action required by event codes,
5. Keeping track of the sun's position,
6. Keeping track of the car's latitude and longitude,
7. Reading in solar data,
8. Interpolating to find solar radiation at car's position,
9. Reading in weather data, and
10. Interpolating to find the weather data at the car's position.

In language-based simulations, not all of these tasks are conveniently done in a single program unit. If they were, the unit would be large, complex, and cumbersome to write and debug. Items 5–10 would very likely be done by separate routines,

Table 19.4 Weather parameters

Date	Time	Cloud cover (tenths)	Pressure (mbar)	Temperature (°C)	Wind speed (km/h)	Wind direction (dgr. true)
6/11	10:00	5	1013.25	25	5	180

for example, and the libraries of some program environments may have routines to perform some of the tasks.

Weather Data The barometric pressure (mbar), air temperature (°C), wind speed (km/h), and wind direction (degrees, measured from true north) should be specified. Generally, only hourly values of these parameters are available from the NWS. Values for more closely-spaced times must be interpolated between the hourly values. Table 19.4 shows a sample of weather data at 10 a.m. local time.

The data could be from a single weather station. In this case, the car must be operating the entire time near that station, or the weather system must be taken as uniform over the route.

For a more realistic simulation, hourly typical meteorological (TMY) data from several weather stations in and around the route should be used. The simulation would interpolate in space and time between the values at these stations to find the value at the car's location during each time step.

References

Craparo, J. C., & Thacher, E. F. (1995). A solar-electric vehicle simulation code. *Solar Energy*, 55(3), 221–234.

Duffie, J. A., & Beckman, W. A. (1991). *Solar engineering of thermal processes*. New York: Wiley.

Hu, C., & White, R. M. (1983). *Solar cells from basic to advanced systems*. New York: McGraw-Hill Book Company.

Chapter 20
Rolling Resistance Calculation

20.1 Purpose

This chapter provides a means for estimating the rolling resistance coefficient of a wheel. This information has two uses: it can assist the designer in choosing among candidate wheels and bearings, and it provides rolling resistance coefficients to be used in energy consumption analysis.

Van der Plas (1983) and Kyle (1990) relate the rolling resistance of the small-diameter, narrow tires typically used by solar racing cars and bicycles, to the characteristics of those tires. Bandyopadhyay et al. (1994) present a similar discussion centered on automobile tires. Cenek (1994) and Pillai (1995) are interesting in-depth readings.

20.2 Resistance and Tire Characteristics

Casing, Tread, and Tube According to Bandyopadhyay et al. (1994), loss of energy through working of the tire material when it "passes through" the contact patch is the largest contributor to rolling resistance, causing 90–95 % energy loss. If the casing, tread, and inner tube are made of more elastic, resilient materials the energy loss will be less. Latex-based rubber is better than butyl rubber (Kyle 1990).

Dimensions Larger tire cross-sectional size and larger wheel diameters produce less deformation at the contact patch, and therefore, less rolling resistance.

Pressure and Temperature A higher tire pressure makes the tire stiffer so that it deforms less thus reducing the rolling resistance. The internal friction of rubber decreases as the rubber's temperature increases, lowering the rolling resistance. Cold tires can have as much as twice the rolling resistance than when they are warmed to typical operating temperatures. Rolling resistance tests should not be run without first warming the tires, if typical values are to be found. However, it is of interest to know the rolling resistance of a solar racer's tires when they are cold.

E. F. Thacher, *A Solar Car Primer*, DOI 10.1007/978-3-319-17494-5_20

The energy consumption of the rolling resistance of a solar racer, when it leaves the starting line in the morning, will be higher than later in the day after the tires have warmed. This should be accounted for in the day's energy management strategy.

20.3 Resistance and Operational Parameters

Wheel Load Reducing the wheel load reduces the rolling resistance in direct proportion because the tire deforms less. However, the rolling resistance coefficient is approximately independent of the wheel load.[1] A stiff suspension, because it requires the tire to deform more, will increase the rolling resistance compared to the one soft enough to follow most of the bumps in the road (MacCready et al. 1990).

Speed Rolling resistance increases with speed because the wheel rotational air drag, the bearing frictional torque, and rate of flexing of the tire are all increased by speed. In high-pressure bicycle tires, the flexing effect is small and the rolling resistance coefficient varies linearly with speed. However, the rolling resistance coefficient of automobile tires varies with the square of the speed.

Pavement The smoother the pavement, the lower the rolling resistance. Smoother pavement causes less tire distortion; very rough pavement can double the rolling resistance (Kyle 1990).

Drive Torque Higher-drive torque increases tire distortion and therefore increases rolling resistance.

20.4 Model

The rolling resistance coefficient, μ, for an individual wheel is defined as in Chap. 2,

$$\mu = \frac{R}{N} \tag{20.1}$$

where R denotes the rolling resistance force on the wheel and N the road surface force on the wheel, normal to the contact patch. In Chap. 2, the coefficient is taken to be a linear function of the vehicle's speed.

$$\mu = \mu_1 + \mu_2 V \tag{20.2}$$

[1] At high loads on automobile bias-ply tires Bandyopadhyay et al. (1994) reported a tendency to increase with load.

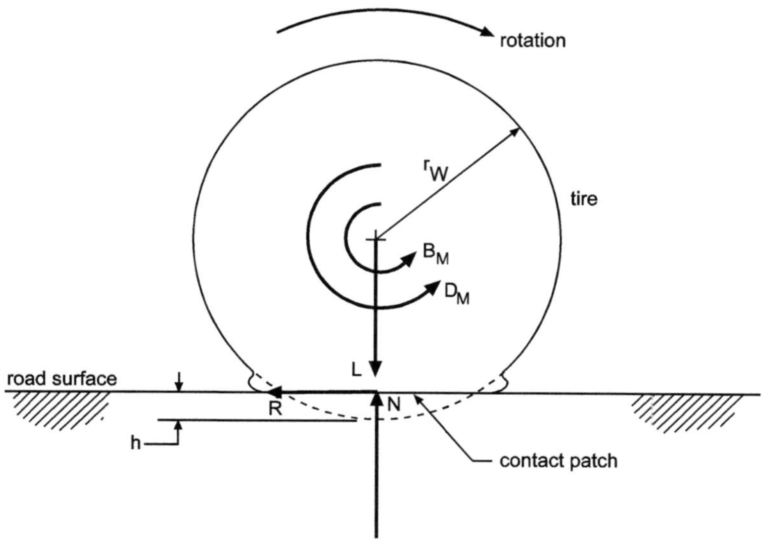

Fig. 20.1 Sinking rate (h)

Both μ and μ_1 are dimensionless and μ_2 has reciprocal speed dimensions. The results of tests on bicycle tires reported in Kyle (1990) show that $\mu_2/\mu_1 \sim 2\%$.

The discussion herein accounts for wheel air drag, bearing friction, tire pressure, wheel diameter, and vehicle weight on the wheel. The model does not directly account for casing, tread, and tube effects. Consequently, it requires a material property (next section) to account for a particular tire's material of construction.

Tire Contribution The formulation used to calculate the rolling resistance of the tire itself is that of Kyle (1990). The sinking rate of a wheel under a vertical load is the amount of flattening of the tire at the center of the contact patch. Figures 20.1 and 2.10 illustrate this idea. The rolling resistance force of the tire is taken to be proportional to the sinking rate.

$$R = Kh \tag{20.3}$$

The constant K is characteristic of the material of the tire and the road surface. Kyle's formulation relates the sinking rate (inches) to the wheel diameter, d_W (inches), wheel road force, N (or L) (lb), and tire inflation pressure, p (psig), as

$$h = \frac{1}{2}\frac{d_0}{d_W}\left(\frac{p}{p_0}\right)^{0.3072}\left(d_W - \sqrt{d_W^2 - \frac{4N}{\pi}\frac{2.456 + 0.251 d_W}{19.58 + 0.5975 p}}\right). \tag{20.4}$$

Kyle used a reference diameter, d_0, of 26 in. and a reference inflation pressure, p_0, of 220 psig. He found that a K-value of 2.47 gave a very good match between the μ measured and predicted for a Continental Olympic racing tire by Eq. (20.1). The model showed the correct variation with wheel diameter and tire pressure. It also correctly predicted that the tire load would have a negligible effect on μ.

Kyle's equation for the sinking rate applies to bicycle-like tires. He found by experiment that these tires have approximately elliptic contact patches and pressure distributions across the contact patch. These observations were built into Eq. (20.4). This equation does not apply to relatively low-pressure automobile radial tires. These tires have approximately square contact patches and pressure distributions that tend to peak near the outer edges of their contact patches.

Air Drag The flow field about a rotating tire installed on a vehicle moving over a road is complex. This chapter considers only the effect of rotating the wheel about its own axis in still air. This drag would be incurred when testing the wheel on some specialized apparatus, such as a rotating drum. It is therefore intended to approximate the aerodynamic resistance to rotation, not the contribution by the wheel to the drag of the vehicle. The rotational air drag is usually small. Walter and Conant (1974) estimate it as 1.5–3 % of the total drag.

The resisting force on the wheel perimeter contributed by air drag torque, τ_D, is

$$R_D = \frac{\tau_D}{r_W}. \tag{20.5}$$

The torque from air drag depends upon the speed, but this dependence is a function of whether the rim is solid or is spoked. The air drag torque on a spoke will be similar to that on a cylinder in cross-flow, but the drag on a wheel with solid rims will be similar to that on a disk. Each will depend on the speed and the dimensions of the spokes or covering disk. Thus

$$R_D = \frac{c_R V}{r_W}. \tag{20.6}$$

where c_R depends on V and the rim geometry.

Suppose the rims are spoked. Kyle idealized the spokes as n_s cylinders with one end attached to the wheel hub. Typically, the spoke diameter, d_s, is small, around 0.1 in. For characteristic wheel dimensions and conditions, even at a vehicle speed of 100 mph, turbulent flow will not occur on one of these spokes. Instead it will experience laminar separated flow over nearly its entire length. A cylinder in this flow has a drag coefficient of 1.2. This model treats each spoke as if it were isolated. In reality each spoke creates a wake. If the spokes were all in the same plane (which they are not) the effect would be to induce turbulent flow about each spoke in regions where the separation between them is at least four spoke diameters, according to the data of Biermann and Herrnstein (1933). This implies a drag coefficient of about 0.3. Closer to the wheel hub the spokes are closer together and near the hub even cross each other, complicating the flow. However, this region is small and the spoke Reynolds numbers associated with it are small. This implies a small contribution to the drag.

Ignore these complications and use Kyle's model, realizing that it may overestimate the drag. This model gives c_R for spoked wheels as

$$c_R = \frac{1}{8}\frac{n_S d_S \left(r_2^4 - r_1^4\right)}{r_W^2}\rho c_D V \tag{20.7}$$

where r_1 and r_2 are the inner and outer spoke radii, respectively, and c_D is 1.2.

Assume the drag on a wheel with solid rims to be approximated by that on a solid disk rotating in free air. This is conservative. The tire actually rotates in a partial housing, but data in Schlichting (1968) for completely-housed rotating disks show that the drag on the free disk is higher. Based on the free-air result

$$c_R = \frac{1}{2} r_W^3 \rho V c_M \tag{20.8}$$

where c_M, the moment coefficient, is

$$c_M = \begin{cases} \dfrac{3.87}{R_{e_W}^{1/2}}, & R_{e_W} \le 3(10^5) \\ \dfrac{0.146}{R_{e_W}^{1/5}}, & R_{e_W} > 3(10^5) \end{cases} \tag{20.9}$$

for laminar and turbulent conditions, respectively. The Reynolds number is defined as

$$R_{e_W} = \frac{V r_W}{\nu}. \tag{20.10}$$

The symbol ν, m²/sec, is the kinematic viscosity, the ratio of the air's viscosity to its density.

Bearing Friction The torque required to overcome friction in the bearings also constitutes a resistance to rotation of the wheels. The rolling resistance contributed by bearing the friction torque, τ_B, is

$$R_B = \frac{\tau_B}{r_W}. \tag{20.11}$$

Kay (1988) tested four bicycle wheel ball bearings at a constant load of 105 lb and over a rotational speed, n_r, (rpm), range of 200–400 rpm. His results show a generally linear dependence on rotational speed. Morelli et al. (1981) tested a particular ball bearing over a 225–1250 rpm range and at loads of 367, 723, and 942 lb. Their results also showed an approximately linear dependence on speed at or below 400 rpm, but a higher-order dependence on speed over their entire speed range, except at the lowest wheel load. At this load, the torque was relatively insensitive to speed but showed a slow increase up to the maximum speed tested. The torque increased nonlinearly with the wheel load. This behavior is represented as follows,

$$\tau_B \approx (a_0 + a_1 n_r + a_2 n_r^{\,2})\left(\frac{L}{L_1}\right)^{f(n_r)} \tag{20.12}$$

where $f(n_r)$ means that the exponent has, in general, a dependence on the rotational speed and L_1 is a reference load.

The foregoing remarks on the data of both papers suggest that at wheel loads characteristic of solar racers, the bearing torque may be approximated as a linear function of speed at a constant load. Also, for L/L_1 ratios of less than 2, the load ratio exponent is approximately independent of n_r and has a value of 2. This approximation was derived by an analysis of the data of Morelli et al. (1981). So, $a_2 \approx 0$, and $f(n_r) \approx 2$. Equation (20.12) may be put in terms of the car speed, V (m/sec). Neglecting contact-patch flattening of the tire, n_r equals $\dfrac{30}{\pi r_W}V$ and putting $c = \dfrac{30}{\pi r_W}$ gives

$$\tau_B \approx (a_0 + a_1 c V)\left(\frac{L}{L_1}\right)^2. \tag{20.13}$$

Total Resistance The total rolling resistance is the sum of the three resistances discussed above

$$R = Kh + R_D + R_B. \tag{20.14}$$

Dividing Eq. (20.14) by the wheel vertical load, separating it into static and speed-dependent parts, and comparing the result to Eq. (20.2) gives these approximations,

$$\mu_1 \approx \frac{Kh + \dfrac{a_0}{r_W}\left(\dfrac{L}{L_1}\right)^2}{L}, \qquad \mu_2 \approx \frac{c a_1\left(\dfrac{L}{L_1}\right)^2 + c_R}{r_W L}. \tag{20.15}$$

20.5 Discussion of Errors

The aerodynamic torque resisting the rotation of wheels which have disk-like rims is probably overestimated by the theory presented. Such wheels housed in wheel wells have been shown to have lower drag than wheels rotating in free air, as the present theory supposes.

It has already been mentioned that the theory for the static rolling resistance coefficient does not apply to passenger car tires. Also, the rolling resistance coefficient of such tires depends more nearly upon the square of the speed at highway speeds.

References

Bandyopadhyay, S., Chandra, A. K., & Mukhopadhyay, R. (1994). *Overview of tyre rolling loss progress in rubber and plastics technology* (Vol. 10, no. 1, pp. 20–53). Shrewsbury England: Rapra Technology Ltd.

Bierman, D., & Herrnstein, Jr., W.H., (1933). Interference between struts in combinations, *NACA TR 468.*

Cenek, P. D. (1994). In B. T. Kulakowski (Ed.), *Rolling resistance characteristics of new zealand road surfaces, vehicle-road interaction, ASTM STP 1225* (pp. 248–262). Philadelphia: American Society for Testing and Materials.

Kay, R. (1988). The new ball bearings. *Bike Technology,* 10–13.

Kyle, C. R. (1990). Lecture 3–3: The sunraycer: wheels, tires, and brakes. In P. MacCready, et al. (Eds.), *Sunraycer case history*. Warrendale: Society of Automotive Engineers.

MacCready, P., et al. (1990). *Sunraycer case history*. Warrendale: Society of Automotive Engineers.

Morelli, A., Nuccio, P., & Visconti, A. (1981) Automobile Aerodynamic Drag on the Road Compared With Wind Tunnel Tests, SAE paper 810186, presented International Congress and Exposition Cobo Hall, Detroit Michigan, February 23–27, 1981.

Pillai, P. S. (1995). Total tire energy loss comparison by the whole tire hysteresis and the rolling resistance methods. *Tire Science and Technology, TSTCA, 23*(4), 256–265.

Schlichting, H. (1968). *Boundary layer theory* (6th ed.). McGraw-Hill Book Company: New York.

Van der Plas, R. (1983). Rolling resistance of bicycle tires. *Bike Tech*, Vol. 2, 6–12.

Walter, J. D., & Conant, F. S. (1974). Energy losses in tyres. *Tyre Science and Technology, 2*(4), 235.

Chapter 21
Stability Calculations

21.1 Purpose

This chapter shows how to investigate certain critical stability limits of a solar car as a function of important design parameters, such as the location of the car's center of gravity (CG). The aim is to present calculations that may be done by hand on a digital calculator or on an electronic spreadsheet. The calculations are therefore based on simplified models. The results still capture the relative importance of the main parameters. However, the errors thus introduced are not always conservative. These errors are qualitatively discussed in each section. For convenience, the notation for moments and forces has been changed from that of Chap. 2.

21.2 Stability

The remarks in this section introduce the concept of stability and provide the justification for the more detailed discussions that follow.

Directional Stability A gust of wind from the side will apply a side force to the car causing motion transverse to the direction of travel. This force will in general not be applied at the CG. Consequently, it will also cause a yawing moment about the CG that will cause the car to turn. The behavior of the car under this sort of disturbance depends upon its *directional stability*.

The driver will react to steer the car back to its intended motion. However, this reaction will not be immediate because the driver's response will lag behind and because there will be a lag in the response of the car to steering inputs. We focus on the motion of the car in the interval between the application of the force and the beginning of the car's response to the driver's correction. This lag time is variable. However, Emmelmann (1987) reported tests that suggest 0.8 s is a representative value. We shall adopt this interval herein. During the lag time, the steering angle will be constant.

E. F. Thacher, *A Solar Car Primer*, DOI 10.1007/978-3-319-17494-5_21

The weight distribution on the front and rear wheels has an important influence on the directional stability. For example, to be directionally stable, a four-wheeled car should have its CG in the front half of its wheelbase. (This rule will be explained later.) Then, the initial transverse and yaw motions caused by a momentary force acting transverse to the car's motion, like the gust in the preceding paragraph, quickly decay. This decay makes the car controllable, and therefore it may be returned to its intended motion.

Now consider the effect of shifting the CG toward the rear wheels. At some more rearward location, oscillations in the gust-induced transverse and directional motions appear. At first, these die away as before; the car is still controllable. However, at a certain weight distribution the car becomes directionally unstable. The gust now causes large and possibly increasing oscillatory directional and transverse motions during the lag time. The driver may not be able to return the car to its intended motion before entering an emergency.

Instead of a momentary application, the gust could be suddenly applied and continue relatively steadily, as when the car emerges from a tunnel on a windy day. The continued presence of the side force exacerbates the deviation of an unstable car from its intended motion. The transient response of a stable car dies out, as described before. But the driver's subsequent steering corrections must counter the continuing side force before some emergency occurs, such as departure from the traffic lane.

Roll Over A side gust also creates a rolling moment about the direction of motion that will tip the car over if it exceeds the stabilizing moment of the weight. Rollover also occurs during turns when the rolling moment of the inertial force transverse to the motion overcomes the stabilizing moment of the weight.

Skid Skidding occurs when the longitudinal inertial force from braking or accelerating or the transverse inertial force from turning exceeds the tire adhesion limit.

21.3 Results Presented

This chapter presents, for three- and four-wheeled cars:

1. Rules for locating the car's CG such that the car will be directionally stable in the sense explained above with respect to a rapidly applied side force
2. A method to estimate the wind gust speed and direction that will cause a directionally stable car to leave its driving lane before corrective steering can take effect
3. A method for estimating the wind gust speed and direction that will cause rollover
4. A method for investigating the resistance to rollover in a circular turn at steady speed when braking and when accelerating

The preceding items require knowledge of the location of the CG and of the polar moment of inertia of the car about that location. Therefore, methods for estimating this information are presented.

21.4 Tires and Turning

This section contains background material that will be used later and may not be familiar to all readers.

Geometry Figure 21.1 shows the geometry of the three cars to be considered: a four-wheeled car, a three-wheeled car with two wheels in front, and a three-wheeled car with two wheels in back. The perpendicular distance, L, between the front and rear axles is called the *wheelbase*. The perpendicular distance, T, between the center planes of the wheels is called the *track*. The tipping axes are drawn anticipating a left turn, resulting therefore in tipping forces in the direction of the negative y-axis. Figure 21.2 shows the *steering angles*, δ_1 and δ_2, of the front tires of the four-wheeled car in a circular turn. These are the angles between the direction of the car's wheelbase and the vertical plane through the center of each wheel's contact patch. The drawing shows, because the inside wheel is closer to the turning center, that the steering system must turn the inside wheel through a larger angle to keep its axis of rotation pointing, approximately, at the turning center.[1]

Cornering The tire axis is pointing only approximately at the turn center because the tires must develop cornering forces on their contact patches, pointing inward parallel to the axis of each wheel, to make the car turn. The faster the car is going, the greater the cornering forces must be to balance the inertial force tending to move the CG of the car away from the turning center.

Figure 21.2 shows how this cornering force is created by tire 1. The sidewise force from the vehicle's rate of change of momentum causes the driver to increase the steering angle past the point where the tire axis points at 0 by an angle called the *slip angle*, α_1. Notice that the slip angle is the angle between the vertical plane through the center of the wheel's contact patch and the direction of travel, which is along the turning circle. The cornering force is proportional to the slip angle.

$$F_1 = C_\alpha \alpha_1 \tag{21.1}$$

The proportionality factor, C_α, is called the *cornering force coefficient* or the *cornering stiffness*. The cornering stiffness is influenced by the tire construction, the inflation pressure, the load on the tire, and the slip angle. But at slip angles below about 5, the cornering stiffness is independent of the slip angle. A bias-ply tire at an inflation pressure of 32 psig has a cornering stiffness of about 150 lbf/degree, for example.

Figure 21.2 also shows that the cornering force is not applied at the center of the contact patch but is offset along the patch by an amount called the *pneumatic trail*, P. This creates a moment about the vertical axis of the tire called the *self-aligning torque*. It is so called because the pneumatic trail is always positioned such that the

[1] This arrangement, called *Ackerman steering*, minimizes scrubbing of the tires during turns. From the geometry if follows that (Wong 1978) $\cot \delta_1 - \cot \delta_2 = T/L$.

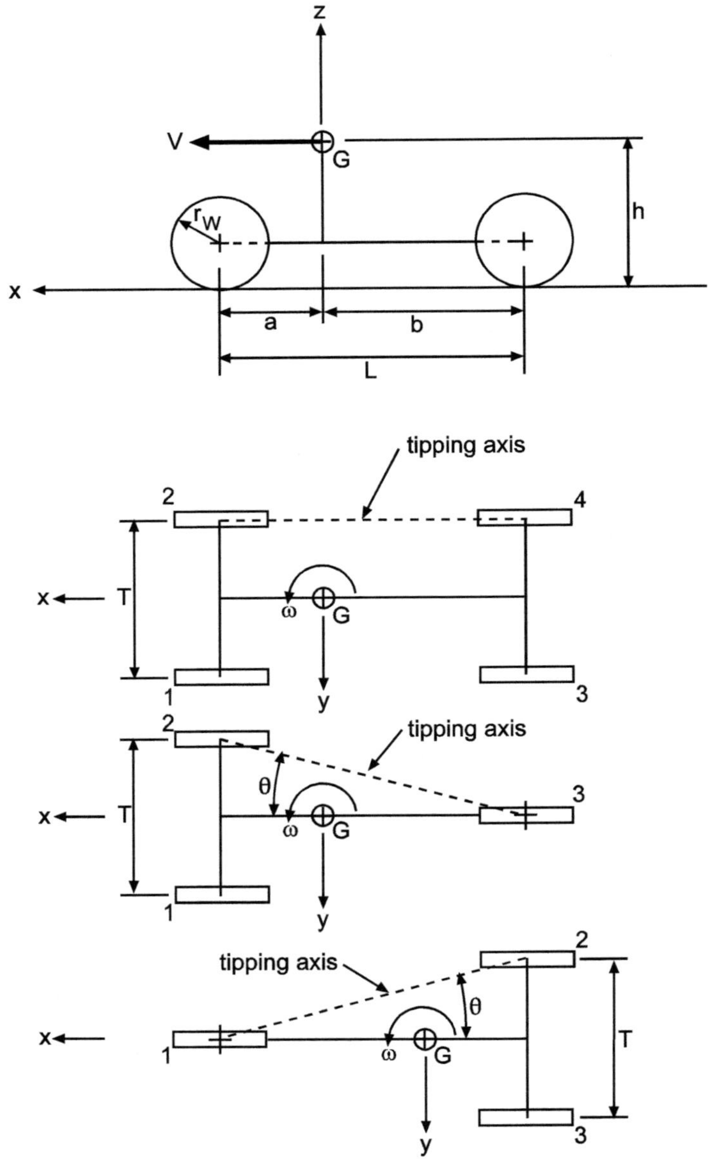

Fig. 21.1 Vehicle geometry

torque tends to rotate the vertical plane of the tire into alignment with the direction of motion. Because this would reduce the cornering force, the driver must balance out the self-aligning torque through the steering system.

$$\tau_{F_1} = F_1 P \tag{21.2}$$

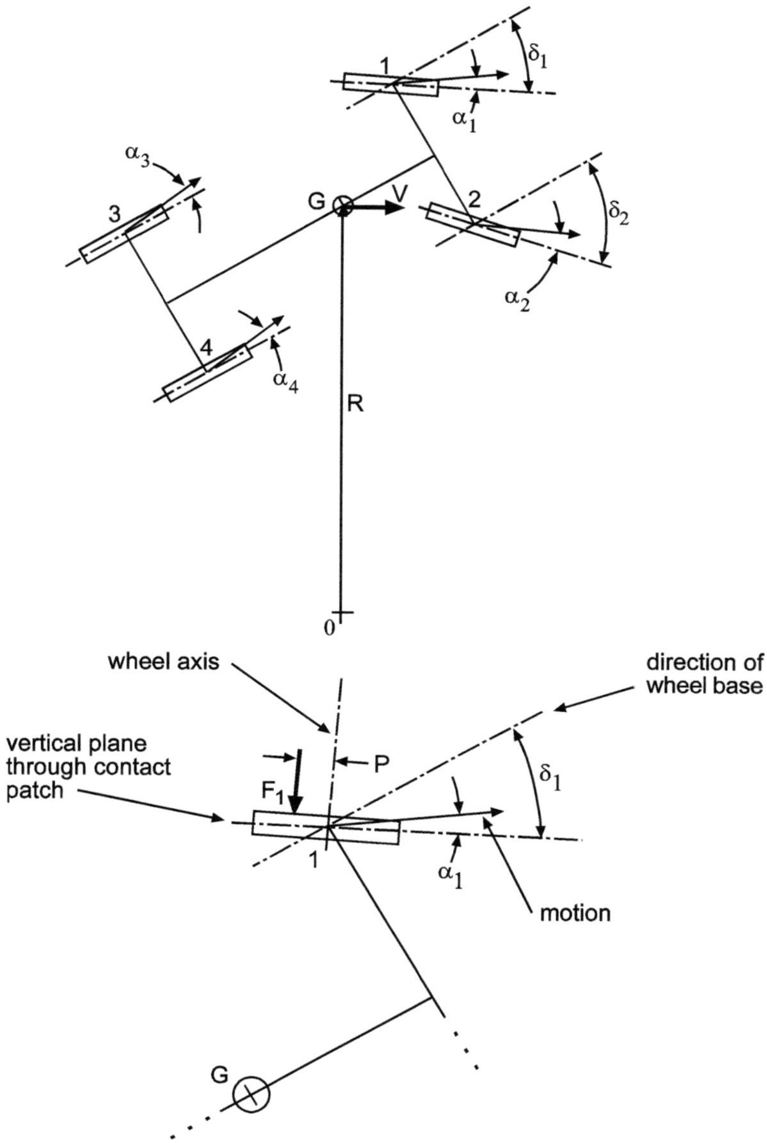

Fig. 21.2 Turning geometry

The self-aligning torque is a function of slip angle and, therefore, also of the quantities that influence the slip angle. At small angles, it is approximately directly proportional to the slip angle. But at larger angles, it varies nonlinearly with the slip angle and may pass through a maximum (Steeds 1960).

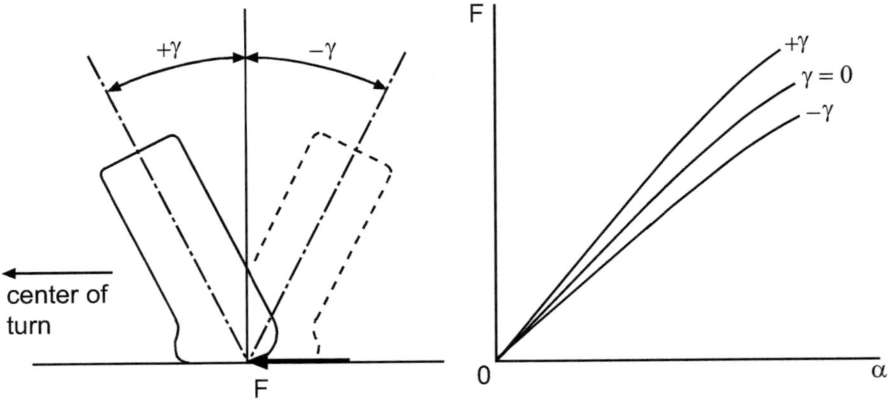

Fig. 21.3 Effect of camber

Camber The cornering force can be increased by causing the wheel's plane of rotation to tilt toward the turn center, as shown in Fig. 21.3. The tilt is called *camber*, and the angle of tilt is called the *camber angle*. The camber angle is positive when the camber is toward the turn center, that is, when the cornering force is increased.

21.5 Skid Limits

Skidding begins when the side force in a turn, or the longitudinal force when braking or accelerating, on the contact patch exceeds the static limit for the existing combination of road surface and tire. Van Valkenburgh et al. (1982) give average values of 0.75 g when cornering and 0.85 g when braking or accelerating for passenger car tires. Figure 21.11 shows the tire traction limit for any combination of cornering and longitudinal acceleration. The curve was assumed to be an ellipse[2] with the cornering and longitudinal traction limits as its minor and major axes, respectively.

Discussion of Error The g-limits for braking and acceleration for the narrower, possibly smooth-surfaced, higher-pressure tires used on student-built solar cars may not be those given above. Consult the tire manufacturer for the g-limits for the tires used on such cars.

The g-limits were based on many tests, but the authors did not characterize the test surfaces. It is reasonable to assume that these surfaces were dry concrete or asphalt.

[2] This shape was suggested by Van Valkenburgh et al. (1982) as the "theoretical" skid boundary. The test results of Van Valkenburgh for several four- and three-wheeled cars all appear elliptical.

21.6 CG Location Rules

A model will now be presented from which rules for locating the CG may be extracted. It will also be used to investigate the response of the vehicle to wind gusts from the side.

Path Geometry Figure 21.4 shows the curved path of the CG of a car and its condition at two times. It also shows the "global" coordinates X and Y which are fixed to the earth. The x- and y-axes are fixed to the CG of the car, as in Fig. 21.1. The *course angle*, C_V, measured from the Y-axis (usually pointing true north), is the actual direction of motion of the CG. This is not the direction in which the car is pointing, that is, the direction of the x-axis, except at the beginning, at zero time. The steering angles are zero, so the slip angle, α, is now the angle between the x-axis and the direction of motion. V_x and V_y are the components of V, the car's velocity in the x- and y-directions.

Simplifications The model incorporates the turning geometry of Fig. 21.2, but with the steering angles the same for each front wheel, and the linear cornering force model (constant C_α) of Eq. (21.1). It further requires a small steer angle and accounts only for transverse forces, that is, in the y-direction. The speed of the car (V_x) in the x-direction remains constant, and the motion of the car is assumed not to affect the weight distribution on the wheels. The simplified equations of motion resulting from these assumptions have been used[3] to investigate lateral stability.

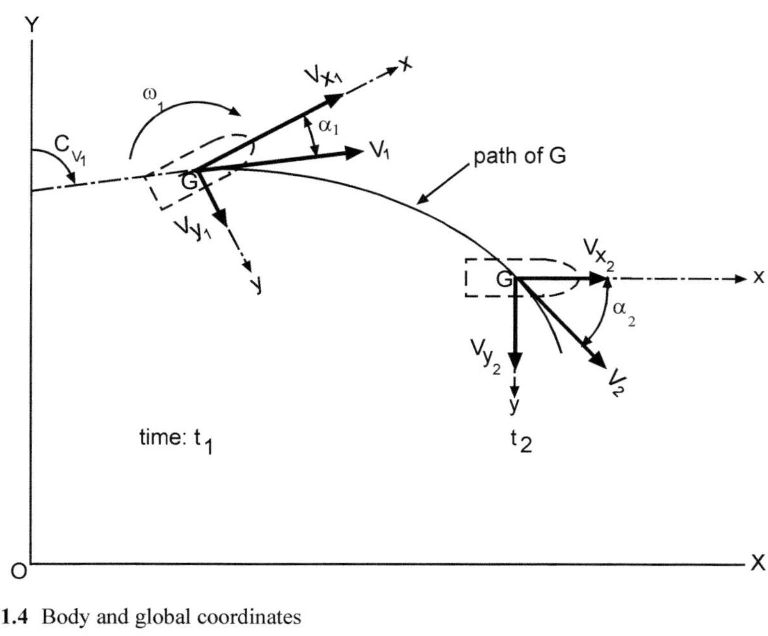

Fig. 21.4 Body and global coordinates

[3] For example, see Wong (1978). Steeds (1960) also discusses more sophisticated models. Sorgatz (1975) reported a detailed model which agreed well with test data, but which also implied that the simplifications of the earlier models were justified. Baker (1991a, b) presented studies for both steady and unsteady wind forces.

They are

$$
\begin{aligned}
Ma_y + c_1 V_y + c_2 \omega &= F_y(t) \\
I_z \alpha_z + d_1 \omega + d_2 V_y &= M_z(t),
\end{aligned}
\tag{21.3}
$$

where $F_y(t)$ is the net side force in the y-direction (Y in Fig. 2.1), $M_z(t)$ is the net yawing moment about the z-axis through the CG (Y_M in Fig. 2.1), a_y is the acceleration in the y-direction, ω is the rate at which the car's course angle is changing, I_z is the car's polar moment of inertia, M is the car's mass, and α_z is the angular acceleration, that is, the rate at which ω is changing, and

$$c_1 = \frac{n_F C_{\alpha F} + n_R C_{\alpha R}}{V_x} \tag{21.3a}$$

$$c_2 = MV_x + \frac{n_F a C_{\alpha F} - n_R b C_{\alpha R}}{V_x} \tag{21.3b}$$

$$d_1 = \frac{n_F a^2 C_{\alpha F} + n_R b^2 C_{\alpha R}}{V_x} \tag{21.3c}$$

$$d_2 = \frac{n_F a C_{\alpha F} - n_R b C_{\alpha R}}{V_x}. \tag{21.3d}$$

If a side force and a yawing moment are suddenly applied to a vehicle modeled as above, Eq. (21.3a–d) predicts that for directional stability to hold at any forward speed,

$$L + \frac{V_x^2}{g} K > 0. \tag{21.4}$$

K, the *understeer gradient*, is

$$K = \left(\frac{W_F}{C_{\alpha F}} - \frac{W_R}{C_{\alpha R}} \right). \tag{21.5}$$

W_F and W_R are the weight on a front and rear wheel, respectively, and $C_\alpha F$ and $C_\alpha R$ are the corresponding cornering stiffnesses.

Because $L > 0$, if $K \geq 0$, the car will be directionally stable at all forward speeds. This requirement means that

$$\frac{W_F}{C_{\alpha F}} \geq \frac{W_R}{C_{\alpha R}}. \tag{21.6}$$

But if there are n_F front wheels and n_R rear wheels,

$$W_F = \frac{Wb}{L n_F}, \qquad W_R = \frac{Wa}{L n_R}. \tag{21.7}$$

Using Eq. (21.7) transforms the inequality Eq. (21.6) into

$$b \geq \frac{n_F \, C_{\alpha F}}{n_R \, C_{\alpha R}} a. \tag{21.8}$$

We assume[4] that for each of the tires (subscript "T")

$$C_{\alpha T} = (A - BW_T)W_T. \tag{21.9}$$

A and B are constants derived from tire data and W_T is the weight on a tire. Using this equation (and $L = a + b$) in the inequality Eq. (21.8) gives

$$b \geq \frac{L}{\frac{n_R}{n_F} + 1}. \tag{21.10}$$

Table 21.1 shows what the inequality Eq. (21.10) requires for the cars in Fig. 21.1.

According to Table 21.1 the CG of the four-wheeler must be at or within the front half of the wheelbase, as previously asserted.

Suppose that $K < 0$. This could be arranged in a two in front–two in rear (2F-2R) car, for example, by shifting weight toward the rear wheels such that b becomes sufficiently less than $L/2$. The inequality Eq. (21.4) shows that as the forward speed, V_x, increases from zero, the car will be directionally stable until

$$L + \frac{V_x^2}{g} K = 0. \tag{21.11}$$

The speed that satisfies Eq. (21.11) is called the *critical speed.* Solving Eq. (21.11) for this speed gives

$$V_{\mathrm{CRIT}} = \sqrt{-\frac{Lg}{K}}. \tag{21.12}$$

Above V_{CRIT} the car will be directionally unstable.

It can be shown[5] that the steering angle of the car model under consideration is given (in radians) by

Table 21.1 CG rules

Car	2F-2R	2F-1R	1F-2R
n_R/n_F	1	1/2	2
$b \geq$	$L/2$	$2\,L/3$	$L/3$

2F-2R two in front–two in rear, *2F-1R* two in front–one in rear, *1F-2R* one in front–two in rear

[4] Following Huston et al. (1982).

[5] See Gillespie (1992), for instance.

$$\delta = \frac{L}{R} + \frac{a_y}{g} K \tag{21.13}$$

for a circular turn of radius R. Thus, the understeer gradient, K, is the slope of the graph of steering angle versus A_y/g. If $K>0$, the steering angle must increase with the transverse acceleration to generate the required cornering forces, that is, it must increase with the square of the speed. This characteristic is called *understeer*. If $K=0$, the steering angle is independent of the acceleration or speed. This is called *neutral steer*. And if $K<0$, the required steering angle decreases as the speed and acceleration increase. This is called *oversteer*.

Notice that following the CG limit rules of Table 21.1 will give understeer when the "greater than" is followed ($K>0$) and neutral steer when the "equality" is applied ($K=0$). Violating those rules produces oversteer ($K<0$). Thus, oversteer corresponds to directional instability under side force loading when at or above the critical speed.

Discussion of Error The model captures the relative importance of main parameters. It does not account for the effect of the suspension, the weight shifts induced by the motion, nor the effect of vertical tire load on the cornering stiffness. Thus, it is best adapted to describing motion at low slip angles and turning at low to moderate a_y/g. Suppose the vehicle were designed for only slight understeer, that is, K were positive but small. Then, it would be possible for the weight shift in a turn to precipitate oversteer. But this would not be revealed by the model used in this section. This effect was reported by Van Valkenburgh et al. (1982) in tests of three-wheelers. The plots of steering angle versus a_y/g show this shift to be relatively rapid. Thus, it would be a safety hazard, possibly inducing rollover, if the driver were not expecting it.

21.7 Side Gust

If the yaw angle of the relative wind is not zero, there will be a transverse aerodynamic force component distributed over the side of the car. This side force will induce motion in the y-direction, rolling about the x-axis through the CG and yawing about the z-axis through the CG. The rolling moment about the x-axis may be large enough to tip over the car.

It is convenient to think of this distributed force as an isolated force applied at a point on the car. The location of this point is such that the moments induced about the x-, y-, and z-axes through the CG by the isolated force are the same as those induced about those axes by the actual, distributed force.

Lateral Deviation Traffic lanes on US highways are about 3.66 m wide. If a 2-m-wide solar car is in the center of the lane, then there is a 0.83-m margin on either side. The question is: will the car travel transversely at least this distance during the 0.8 s lag time?

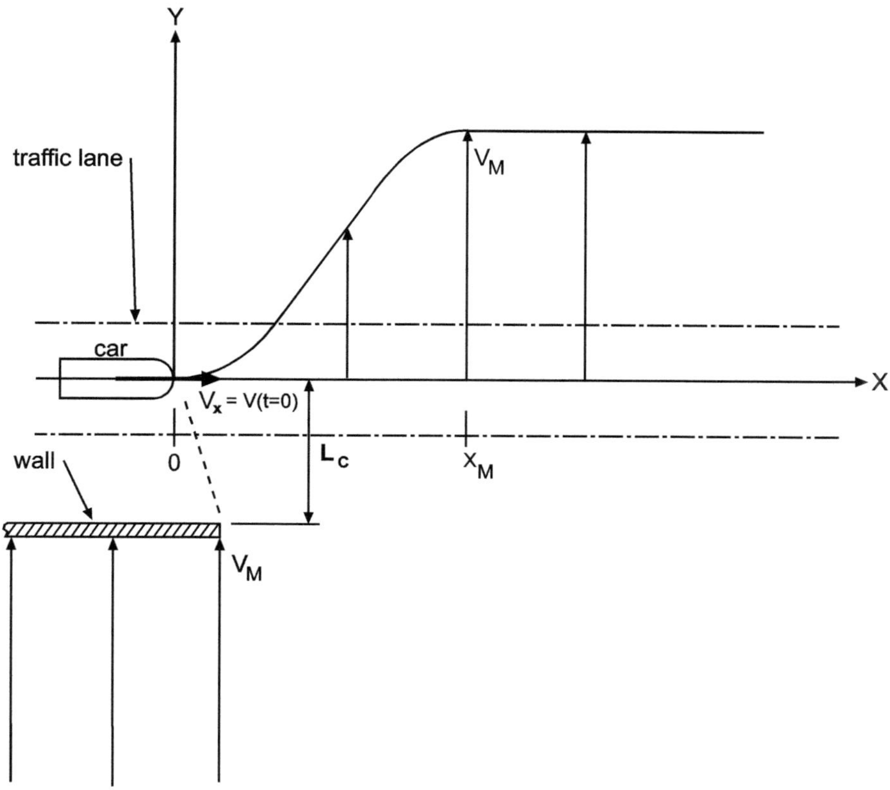

Fig. 21.5 Side gust scenario

Figure 21.5 illustrates the situation imagined. The car is moving at steady speed in a straight line. On the windward side, a wall prevents the wind, blowing steadily at V_M m/s, from striking the car until the car enters the partial jet formed at the end of the wall. The wind speed in the jet is described by[6]

$$V_W(X) = \begin{cases} \dfrac{V_M}{2}\left[1 - cos\left(\pi \dfrac{X}{X_M}\right)\right] & X \leq X_M \\ V_M \quad X > X_M \end{cases}. \tag{21.14}$$

The *mixing length*, X_M, the distance from the beginning of the jet to its core, is proportional to the distance of the car from the wall, L_C. The parameters X_M and V_M should be varied over a range to explore the car's behavior in different gust shapes. Note that vertical variation of V_W has been assumed to be small over the car's height.

[6] Based on the cosine gust shape used by Hucho and Emmelmann (1978).

The lateral deviation is calculated using Eq. (21.3), described in the previous section. We will use the little tap method mentioned in Chap. 2. We imagine that the taps occur at short, equal time intervals, called Δt. We select it such that it is related to the car's length, L_{car}, by

$$\Delta t = \frac{L_{car}}{n_L V_x} = \frac{\Delta u}{V_x}, \tag{21.15}$$

where the number of length segments, n_L, is an integer. The coordinate *"u"* is the distance measured rearward from the car's nose. If we want to use the "little tap" method to predict the transverse velocity after the $(n+1)$th interval from the transverse and angular velocities at the end of the previous interval, n, we would write

$$a_y^{(n)} = \frac{\Delta t}{M}(F_y^{(n)} - c_1\omega^{(n)} - c_2 V_y^{(n)}), \tag{21.16a}$$

$$V_y^{(n+1)} = V_y^{(n)} + a_y^{(n)}\Delta t. \tag{21.16b}$$

The parentheses around the time superscripts distinguish them from exponents. And for ω

$$\alpha^{(n)} = \frac{1}{I_z}(M_z - d_1\omega^{(n)} - d_2 V_y^{(n)}), \tag{21.17a}$$

$$\omega^{(n+1)} = \omega^{(n)} + \alpha^{(n)}\Delta t. \tag{21.17b}$$

The new course angle, using the average of the predicted and previous angular velocities for better accuracy, is

$$C_V^{(n+1)} = C_V^{(n)} + \frac{\Delta t}{2}(\omega^{(n+1)} + \omega^{(n)}). \tag{21.18}$$

The new global position of the CG is

$$X^{(n+1)} = X^{(n)} + V^{(n)} sin(C_V^{(n)})\Delta t.$$

$$Y^{(n+1)} = Y^{(n)} + V^{(n)} cos(C_V^{(n)})\Delta t. \tag{21.19}$$

The speed, V, is given at any time by

$$V = \sqrt{V_x^2 + V_y^2}. \tag{21.19a}$$

At the beginning of the first time interval (or "time step"), the time t, F_y, M_z, V_y, and ω are all zero and C_V, X, and Y have their initial values. The car penetrates a distance Δu in Δt s into the wind, and F_y and M_z are found (as described later) for $t=\Delta t$. The car is "moved" transversely and rotated by this first tap. This process is repeated for $t=2\Delta t$, $3\Delta t$, etc., until at least 0.8 s has elapsed.

An example of the results of this calculation may be found in Chap. 9.

Wind Overturn The aerodynamic rolling moment can be strong enough to overturn a small, relatively light vehicle like a solar racing car, as wind gusts during the World Solar Challenge in Australia have done. We assume the car to be traveling approximately in a straight line at a constant speed. A side force is applied to each of the three configurations in Fig. 21.1 by summing moments about the associated tipping axis. The maximum tolerable wind gust is that which just causes the contact patch forces of the upwind wheels to be zero.

For the four-wheeled car of Fig. 21.1, a moment balance about the tipping axis (situation similar to Fig. 21.10) gives

$$-(h_A F_y)_{\max} + W\frac{T}{2} = 0.$$

Therefore, the limiting tipping moment is

$$(h_A F_y)_{MAX} = \frac{TW}{2} \equiv (M_{T4})_{MAX} \tag{21.20}$$

For a three-wheeled car with one wheel in front (angle θ is defined in Fig. 21.1)

$$-(h_A F_y)_{\max} cos\theta + Wasin\theta = 0.$$

$$(h_A F_y)_{\max} = \frac{TWa}{2L} \equiv (M_{T4})_{\max}\frac{a}{L}. \tag{21.21}$$

And similarly for a three-wheeler with two wheels in front

$$(h_A F_y)_{\max} = (M_{T4})_{\max}\frac{b}{L}. \tag{21.22}$$

The product $h_A F_y$ is limiting, not F_y alone, because h_A may be a variable. The next two sections will make this clear. Note that for the three-wheeled cars, the limit depends on a/L or b/L, both of which are less than one. Therefore, the limiting side force is smaller for these cases, if T and W are the same.

Side Force and Its Moments Figure 21.6 shows the silhouette of the side view of a car. The wind force on this two-dimensional view will represent that on the three-dimensional car.[7] The x-axis extends from the first intercept of the car's nose with the gust. Let u represent the distance from the nose of the car along the x-axis to a point on the car; u cannot exceed L_C.

[7] The approach taken by Hucho and Emmelmann (1973).

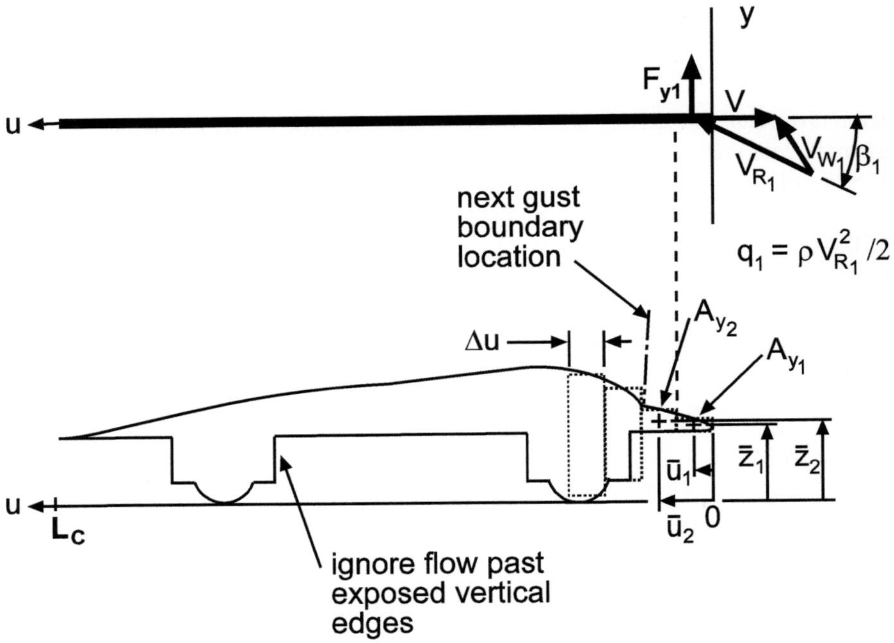

Fig. 21.6 Side force area segments

Let the area of the side view be A_y. The shape of this area may be approximated by dividing it into vertical rectangular area elements, each of length Δu and of area A_{yj}, where $j = 1, 2, 3,\ldots, n_L$. The area elements enter the gust of Fig. 21.5 in sequence, starting with A_{y1}. The two labeled elements enter the gust at Δt and $2\Delta t$, respectively, for example. The next two elements are also shown to illustrate the shape errors introduced at the fairings and wheels by the approximation.

The force perpendicular to each area element is estimated by assuming the flow over a segment to be mostly in the y- and z-directions. That is, the car is assumed to be slender compared to its length. The normal force will be taken as caused mostly by the separation of this two-dimensional flow at the upper and lower boundaries of the rectangular segments.

Because it is separation-driven, the normal force will be approximately independent of the Reynolds number (like the disc in Fig. 2.6). But it will be dependent on the yaw angle, β. This dependence will be through a *normal force coefficient*, $c_N(\beta)$, and the dynamic pressure, $q(\beta)$, of the relative wind (defined in Chap. 2). Equation (21.23) shows how this works. The *"j"* subscript refers to a particular area element.

According to Fig. 21.5, the wind speed will be highest near the nose and lowest at the end of the section in the wind. Thus, as it penetrates the gust, the car will turn downwind. This will change the yaw angle at every location the wind hits. Therefore, the side force on each area element must be recalculated at every time step.

The total side force on the car at any time $t=n\Delta t$, where $n=0, 1, 2,\ldots$, will be the sum of the forces on the elements:

$$F_y^{(n)} = \sum_{j=1}^{n_t} F_{y_j}^{(n)} = \sum_{j=1}^{n_t} (c_N A_y q)_j^{(n)}, \tag{21.23}$$

where n_t is the number of elements in the gust at time, t. How to find c_N will be explained later.

The yawing moment about the CG caused by F_y may be estimated at a particular time as

$$M_z^{(n)} = \sum_{j=1}^{n_t} F_{y_j}^{(n)} (u_{\mathrm{CG}} - \bar{u}_j), \tag{21.24}$$

where positive moments are counterclockwise viewed from the top, as shown in Fig. 2.1. If the forces are assumed to act at the centroid of each segment (located at $(\bar{u}_j, \bar{z}_j)$), applying Eq. (21.23) and simplifying gives the u-coordinate of the total force, $F_y^{(n)}$, at a particular time as

$$\bar{u}_y^{(n)} = \frac{1}{F_y^{(n)}} \sum_{j=1}^{n_t} F_{y_j}^{(n)} \bar{u}_j. \tag{21.25}$$

And taking moments about the x-axis gives the z-coordinate as

$$\bar{z}_y^{(n)} = h_A^{(n)} = \frac{1}{F_y^{(n)}} \sum_{j=1}^{n_t} F_{y_j}^{(n)} \bar{z}_j \tag{21.26}$$

at a particular time.

The rolling moment about the tipping axis from the side force would be at a particular time

$$M_t^{(n)} = \cos\theta \sum_{j=1}^{n_t} F_{y_j}^{(n)} \bar{z}_j. \tag{21.27}$$

A rolling (and pitching) moment may also be caused by upward lift generated by flow from the side. This could occur if the car is tilted up in the direction of the relative wind. This might happen if the car were traversing a banked turn, or if the wind were blowing up a slope toward the road. If lift coefficient data for the car's cross-sectional shape can be found, a lift estimate for 90° yaw could be done using the segment approach above. This would be the limiting case. However, finding the rolling moment would require estimating the pressure distribution about each segment so the point of application of the lift could be calculated.

Element Geometry To find the area elements and the coordinates of their centroids, make a careful, to-scale, side-view drawing of the car on crosshatched paper. Measuring from the nose rearward, find the z-coordinates of the upper and lower edges of the shape at each u-coordinate. Be sure to capture the beginning and ending of each major feature, such as a wheel fairing. Curved surfaces need more samples than flat surfaces. The set of measurements can be interpolated to estimate the z-coordinates of the upper and lower edges at any distance from the nose.

The center of the first area element in Fig. 21.6, A_{y1}, is at $(\overline{u}_1, \overline{z}_1)$, where

$$\overline{u}_1 = \frac{\Delta u}{2} \quad \overline{z}_1 = \frac{z_{L1} + z_{U1}}{2},$$

and $z_L 1$ and $z_U 1$, the coordinates of the lower and upper edges, respectively, are interpolated at $\overline{u}_1$ from the measurements previously taken. The area of element 1 is

$$A_{y_1} = \Delta u (z_{U1} - z_{L1}).$$

The geometry of each element may be calculated in the way just illustrated and used for all subsequent calculations that have the same time step.

Calculation Outline To begin a calculation, we set the car's true course as east. This gives a $C_V^{(0)}$ of 90° clockwise from true north (now the positive y-direction). We also choose the ambient temperature and pressure and calculate the density from the ideal gas law, Eq. (21.3) of Chap. 2. The wind heading, C_W, we will set as north, or 0°; this is a constant. The car's speed, V_X, the wind jet core speed, V_M, and the mixing length, X_M, must also be chosen; these are also constant. Equation (21.14) then gives V_{W1} at $\overline{u}_1$. After one time step, we find the relative wind speed at the centroid of area element 1 using Eq. (12.14), the angles of Fig. 21.6, V=88.5 kph, and V_W=50 kph,

$$V_{R1}^{(1)} = \sqrt{13.89^2 + 24.58^2 - 2 \times 13.89 \times 24.58 \times cos(-90^\circ)} = 28.23 \frac{m}{sec}.$$

Note that at any time the course angle is the same for all area elements. Equation (12.15) gives the yaw angle as

$$\beta_1^{(1)} = sin^{-1}\left[\frac{13.89}{28.23} sin(-90^\circ)\right] = -7.87^\circ.$$

Next, the dynamic pressure, using V_{R1} and assuming ρ=1.20 kg/m³, is

$$q_1^{(1)} = \frac{1}{2} \times 1.20 \times 28.23^2 = 478.16 \frac{N}{m^2}.$$

Using $\beta_1^{(1)}$ in Eq. (21.29), the side force on A_{y1} may now be calculated using Eq. (21.23) and supposing A_{y1}=0.1 m².

$$F_{y1}^{(1)} = 1.98 \times sin^2(-7.86^\circ) \times 0.1 \times 478.16 = 1.77\ N$$

Now find the two moments, supposing u_{CG}=2.6 m and $\bar{u}_1$ =0.25 m:

$$M_{t1}^{(1)} = (2.6 - 0.25) \times 1.77 = 4.16\ N{\cdot}m.$$

And the rolling moment for four wheels and putting $\bar{z}_1$ =0.5 m:

$$M_{z1}^{(1)} = 0.5 \times 1.77 = 0.89\ N{\cdot}m,$$

where

$$cos\theta = \begin{cases} \dfrac{L}{\sqrt{\dfrac{T^2}{4} + L^2}} & 3\ wheels \\ 1 & 4\ wheels \end{cases} . \tag{21.28}$$

Finally, if the constants defined by Eq. (21.3a), Eq. (21.3b), Eq. (21.3c), and Eq. (21.3d) have been calculated, we can substitute the force and moment into Eqs. (21.16)–(21.18) and "move" the car. The elapsed time is now Δt. Then Eq. (21.18) and Eq. (21.19) are used to get a new course angle $C_V^{(1)}$ and the coordinates of the CG, $(X^{(1)}, Y^{(1)})$.

Take another time step. The wind is now blowing on both A_{y1} and A_{y2}. So, the previous steps must be repeated for both area elements. When the summations are made to calculate the total side force and its associated moments, n_t=2 is put in Eq. (21.23), Eq. (21.24), and Eq. (21.27). Then, find $C_V^{(2)}$ and $(X^{(2)}, Y^{(2)})$ as for step 1.

Adding one area element for each new time step continues until the entire car is in the gust. Time continues at least until the specified response delay has elapsed.

The product $h_A F_y$ may be plotted and compared to the appropriate rollover limit, Eq. (21.20), Eq. (21.21), or Eq. (21.22) to learn if rollover occurs. This event may then be related to the car's speed and the gust parameters. From the course angle and y-coordinate of the CG and the layout of the car, the time at which the most *"at risk"* point on the car leaves the traffic lane can be determined.

Normal Force Coefficient Hoerner (1965) cites experimental results for the normal force on a two-dimensional cylinder with its long axis inclined at a yaw angle to the flow to show that

$$c_N(\beta) = c_N(0) sin^2 \beta. \tag{21.29}$$

Equation (21.29) expresses the experimental observation that in cases such as the cylinder, the velocity component normal to the inclined axis determines the flow pattern and pressure forces. In the absence of experimental data for a two-dimensional plate with its long axis yawed to the flow, we will use Eq. (21.29), but with

$$c_N(0) = 1.98. \tag{21.30}$$

This value was given in Hoerner (1965) for a two-dimensional plate normal to the flow. Note that the force used in obtaining $c_N(0)$ was the force per unit length of the plate. The length is the dimension along which the flow (and force) is uniform. The coefficient is therefore defined as

$$c_N = \frac{(F_y \text{ per unit length})}{hq}.$$

Equation (21.23) then gives the force on an area element of length Δu and height $z_U - z_L = h$, or $A_{yj} = h(\Delta u)$.

Discussion of Error The word "conservative" will characterize assumptions that cause overestimation of F_y. Neglected flow in the x-direction that goes around the ends of the car, or around vertical edges such as those of wheel fairings, helps to increase the pressure on the downstream side. This flow therefore reduces the normal force. Consequently, its neglect is conservative. On the other hand, by collapsing the car to a plate, the model substitutes friction over the plate for friction over the top and bottom of the car. The former is parallel to the plate and so makes no contribution to F_y. This substitution is not conservative. Also, the plate data strictly apply to a plate in free air; ground effect is not present. The ground effect of the yawed flow very likely produces lift on the car, probably downward and thus resisting roll. Lift creation would induce drag, that is, add to the side force, through vortex formation, as explained in Chap. 2. The increase in the side force thus induced on the car probably is not large compared to the side force from separation. Separation on the two-dimensional plate model is even more severe.

The mixing length, X_M, was assumed to be constant. This simplifies the calculation, but it is not the case. The car moves transversely under the influence of the side force. This causes X_M to change because it depends on the distance of the car from the object creating the gust (the wall, in the case of Fig. 21.5). The error, which may be conservative or not depending on the car's motion, is small because the motion of the car compared to X_M is small.

The car follows a curved path. Hence, the wind penetration distance in a time step is shorter than Δu. However, because the time interval of the calculation will probably always be on the order of seconds, the error in using Δu will be small.

Bundorf et al. (1963) simulated an aerodynamic side force by applying the thrust from a hydrogen peroxide rocket motor to the side of a station wagon. The motor was placed at one of four locations on the side of the car and operated for about 3 s. The resultant lateral acceleration along the y-axis fixed to the vehicle and the yaw angular velocity about the z-axis were measured. Representative data were plotted in the paper as functions of time.

Because careful measurements of all the relevant parameters of the test vehicle were reported in the paper, it was possible to calculate the parameters c_1, c_2, d_1, and d_2 used in Eq. (21.3). Then, these equations were used to simulate the motion of a rigid-frame vehicle subjected to the rocket thrust. The rocket was located at "position 2," approximately at the intersection of the hood and the windshield.

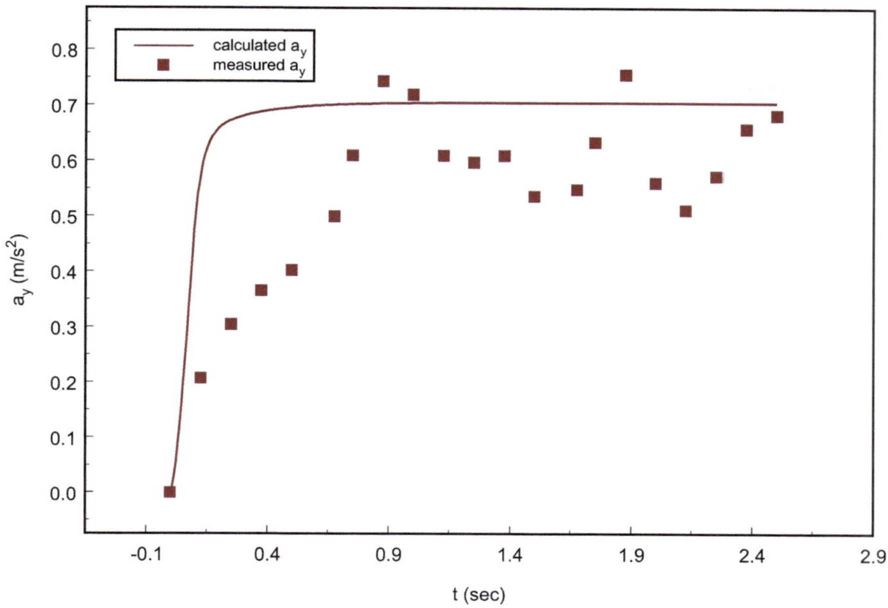

Fig. 21.7 Comparison to Bundorf et al. (1963)

The resulting lateral acceleration was plotted in Fig. 21.7 along with points taken at 1/8-s intervals from the graph of the measured lateral acceleration in Bundorf et al. (1963). Notice that Eq. (21.3) predicts a more rapid motion change and, on average, a greater lateral acceleration. Also, there are oscillations in the data not predicted by the model. The higher-frequency oscillations were caused by unfiltered road noise. The lower-frequency oscillations (which were also captured by the more complicated model used by the authors) were the response of the sprung mass of the car. This complicated response was of course not included in the current model. However, it is apparent that the simplified equations of motion, Eq. (21.3), presented herein are generally conservative at yaw angles near 90 °.

The assumption that the two-dimensional normal coefficient model of Eqs. (21.29) and (21.30) is conservative at yaw angles below 90 ° was tested by measurement in a small wind tunnel. The normal force coefficients of three 1/18-scale car models were measured at different yaw angles and wind speeds. The coefficients were based on the silhouette area of the model cars. The wind speed range was 10–40 mph. This produced a Reynolds number range of $1.58(10^4)$–$7.38(10^4)$, based on the average height of each model, given by $\bar{h} = A_y / L_{\text{car}}$, where A_y is the silhouette area. The upper end of this range is below the onset of turbulence at roughly 10^5, as shown in Fig. 2.6. Figure 21.8a, b, and c show the three scale models, which are realistic. Figure 21.9a shows the comparison with Eq. (21.29) for 20–40 mph and Fig. 21.9b shows the comparison at 10 mph. The data for these figures were adapted from data in Wong et al. (2002).

Fig. 21.8 **a** VW Beetle scale model (Wong et al. 2002). **b** Chevrolet Impala scale model (Wong et al. 2002). **c** Mercedes SUV scale model (Wong et al. 2002)

Also shown in each figure is Eq. (21.29). Figure 21.8a shows that this model is conservative for all cases above a yaw angle of about 45°. Below this angle, the model lies about in the middle of the data. The dashed curve shows the result of arbitrarily selecting an exponent of 3/2 for $\sin\beta$. This empirical model is conservative over a greater range of yaw angles.

The measured results vary widely at 10 mph, as Fig. 21.8b shows, except at 90°. The variation reflects the changing interaction of details of the models' shapes with the flow at different yaw angles. This also occurred at higher Reynolds numbers. However, c_N was more sensitive to yaw at the lowest Reynolds number. Perhaps less of the model cars' shapes was already imbedded in separated flow, as they would be at higher Reynolds numbers, but was tripped into separation by yaw.

At 10 mph, the c_N-models both predict coefficients that lie roughly in the middle of the data and are not conservative. However, the side force at this wind speed is small and this lack of conservatism will not contribute much error when calculating the motion of the solar car.

21.8 Turning Model

The results below relate the design parameters of the car to the combination of speed, turning radius, acceleration, or deceleration, causing rollover. The car will be represented as one of the rigid frames shown in Fig. 21.1.

The inertial force, Ma_y, caused by turning acts at the CG and points outward relative to the instantaneous turning center. This is the coordinate direction *"y"* shown

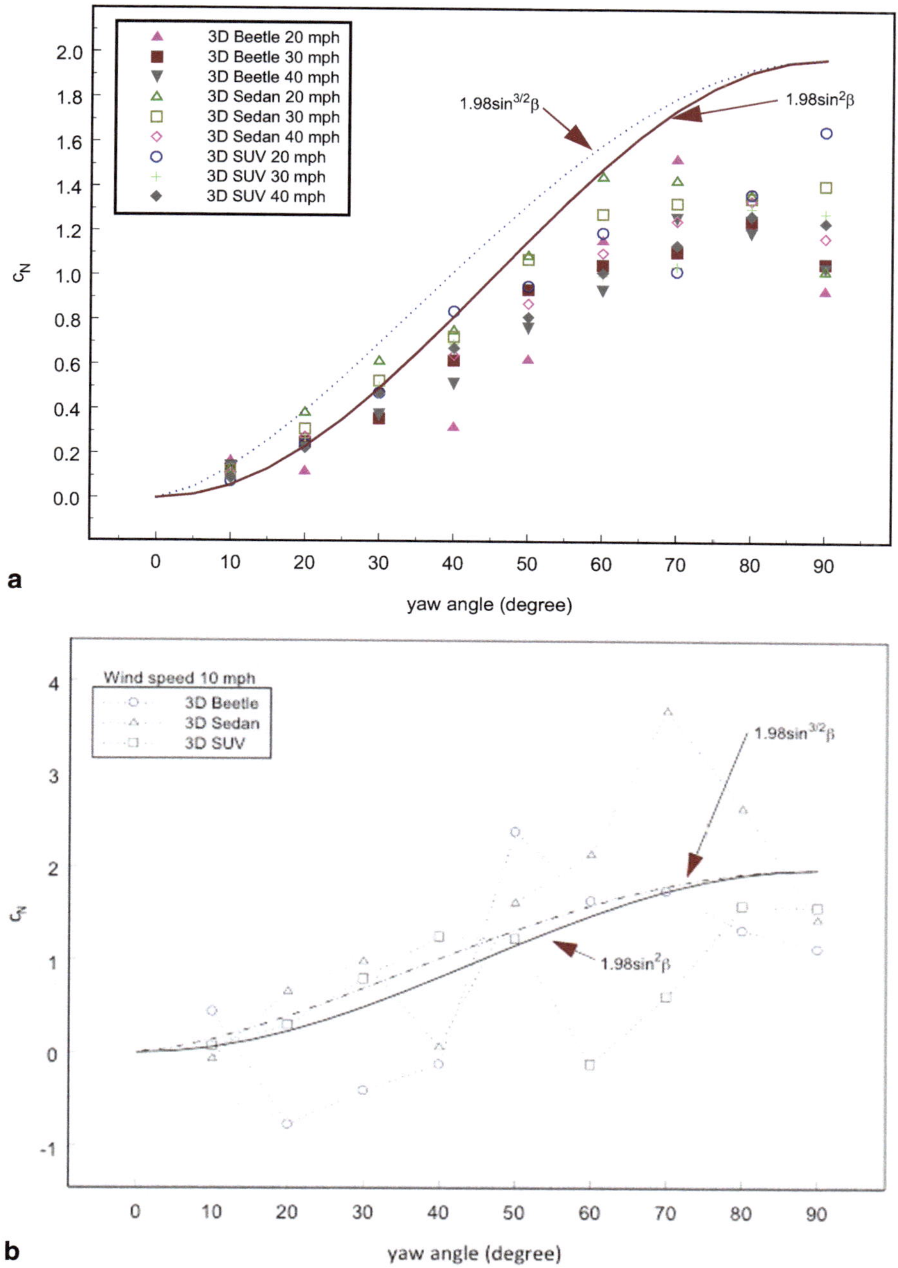

Fig. 21.9 **a** Comparison at 20–40 mph. **b** Comparison at 10 mph

Fig. 21.10 Rollover forces

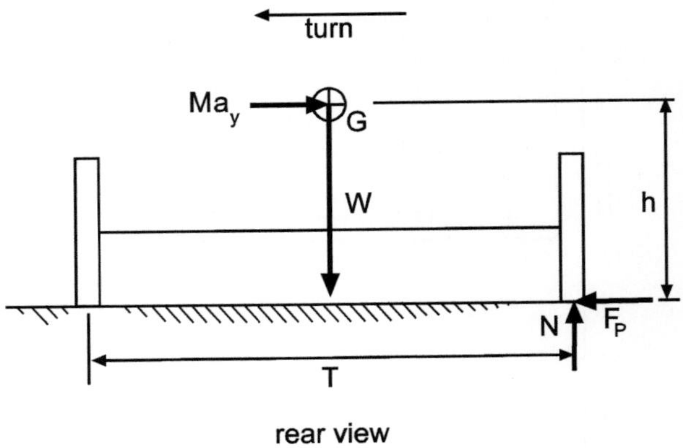

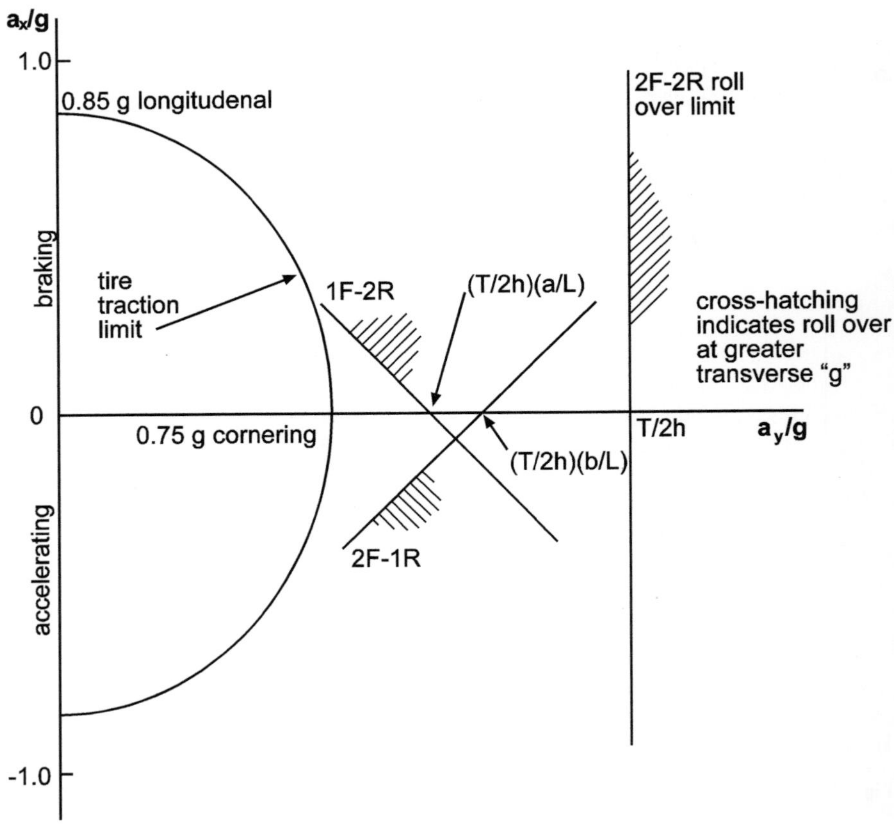

Fig. 21.11 Skid and rollover limits

in Fig. 21.1. The inertial force, Ma_x caused by braking points in the instantaneous direction of motion. This is the coordinate direction "x" shown in the figure. That of acceleration points opposite to the direction of motion, or $-x$ in the figure.

Four Wheels Consider the four-wheeled car of Fig. 21.1. Figure 21.10 shows a rear view of this car in a left turn. The car may be braking (a_x positive) or accelerating (a_x negative). The contact patch forces on the inside wheels will then be zero at the instant roll begins. A moment balance about the tipping axis at that instant gives

$$Ma_y h - W\frac{T}{2} = 0.$$

Dividing through by the weight and expressing the relation as an inequality gives

$$\frac{a_y}{g} \geq \frac{T}{2h}. \tag{21.31}$$

The inequality symbol means that rollover will occur when a_y/g equals or exceeds the value on the right-hand side. Notice that the roll stability of a four-wheeled vehicle is independent of an acceleration in the forward or rear direction because the associated inertial force is parallel to the tipping axis.

Single Wheel in Back A moment balance about the tipping axis of the second vehicle of Fig. 21.1 predicts rollover if

$$\frac{a_y}{g} \geq \frac{T}{2L}\left(\frac{b}{h} + \frac{a_x}{g}\right). \tag{21.32}$$

The roll stability depends on a_x because a braking (a_x>0) or an accelerating (a_x<0) inertial force is not parallel to the tipping axis and therefore creates a moment about it. The rollover limit is increased by braking.

Single Wheel in Front A moment balance about the tipping axis of the last vehicle of Fig. 21.1 predicts rollover if

$$\frac{a_y}{g} \geq \frac{T}{2L}\left(\frac{a}{h} - \frac{a_x}{g}\right). \tag{21.33}$$

The rollover limit is increased by acceleration, rather than by braking as in the previous case.

Figure 21.11 shows qualitatively each of the above inequalities. It also shows the skid ellipse discussed earlier. The "forbidden" regions to the right of the limit lines are indicated by crosshatching.

Three-Wheel Roll Caution It is possible in the three-wheel cases for the roll limit to lie inside the skid ellipse: The car could roll before skidding. To prevent this, the roll intercepts must satisfy

$$\frac{T}{2h}\frac{a}{L}, \quad \frac{T}{2h}\frac{b}{L} > +\sqrt{\frac{A^2T^2}{4L^2}+B^2} \tag{21.34}$$

where A and B are the longitudinal and cornering skid limits, respectively. For $A = 0.85\ g$ and $B = 0.75\ g$

$$+\sqrt{\frac{A^2T^2}{4L^2}+B^2} = +\sqrt{0.723\frac{T^2}{4L^2}+0.563}. \tag{21.35}$$

For the same $T/2h$, the roll limits for both three-wheelers are less than a 2F-2R car. To give the same or better roll limit, $T/2h$ for the three-wheelers must satisfy

$$\frac{T}{2h} \geq \left(\frac{T}{2h}\right)_{\text{2F-2R}} \begin{cases} \dfrac{L}{a} & \text{1F-2R} \\ \dfrac{L}{b} & \text{2F-1R} \end{cases}. \tag{21.36}$$

Limiting Speed If the turn is circular and at a steady speed, V, the lateral acceleration is related to the speed and radius, R, by

$$a_y = \frac{V^2}{R}. \tag{21.37}$$

This means, in the case of a four-wheeled car, that the car will roll if

$$V \geq \sqrt{\frac{TgR}{2h}}. \tag{21.38}$$

The limiting speeds for the three-wheeled cars may be found by substituting Eq. (21.37) into Eq. (21.32) or Eq. (21.33).

Discussion of Error The rigid frame model leads to an overestimate of the rollover stability of the vehicle. The suspension's flexing and tire distortion allow the CG to move toward the tipping axis as the body rolls, thus reducing the restoring moment arm upon which the weight acts. These effects may not be large in solar cars, which usually have relatively stiff suspensions and high-pressure tires.

Nevertheless, the operators of the vehicle should be aware of this lack of conservatism and drive conservatively.

21.9 Center of Gravity

The vehicle must be designed to be stable. Hence, a calculation of the center of gravity (CG) location must be done repeatedly during the design process to assess the effect on stability of configuration changes. This section presents an approximate method for this calculation.

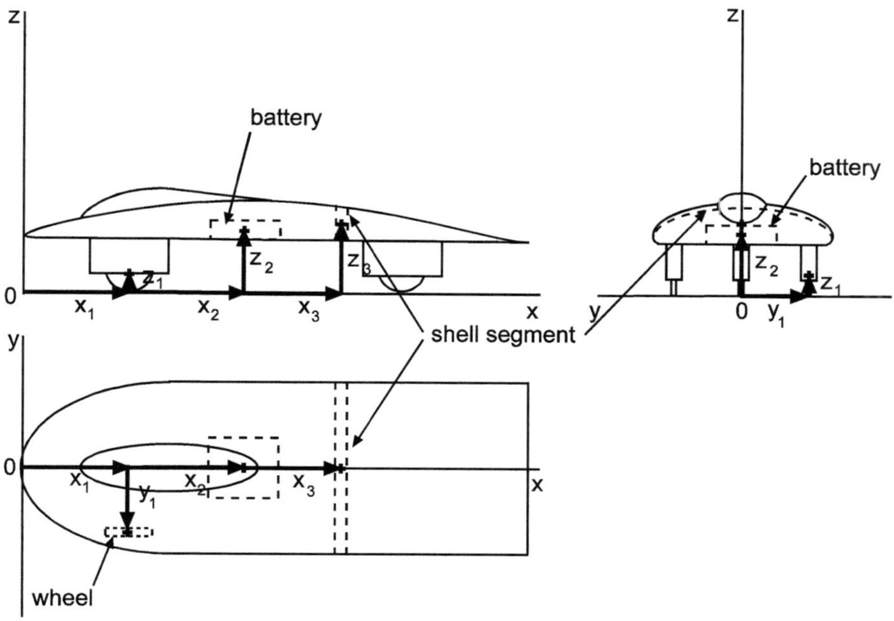

Fig. 21.12 Center of gravity estimation

Figure 21.12 represents a solar car design drawn to scale on crosshatched paper (hatching not shown). The drawing may be done freehand. Freehand drawings can be done relatively quickly and changed just as quickly. The crosshatching allows dimensions and areas to be estimated conveniently.

The location of the CG may be estimated by dividing the car up into pieces. The pieces are chosen such that the location of the CG of each one is known or easily calculated. (The driver should be one of the pieces. The CG shift caused by the size range of the drivers should be calculated.) The mass of the car is the sum of the masses of the pieces. Suppose there are *"m"* of them; the mass of the car would be

$$M = \sum_{i=1}^{m} M_i. \tag{21.39}$$

The CG of each piece has three coordinates. The coordinates of the car's CG are found from these coordinates and the masses of the pieces.

$$\bar{x} = \frac{1}{M}\sum_{j=1}^{m} M_j x_j, \quad \bar{y} = \frac{1}{M}\sum_{j=1}^{m} M_j y_j, \quad \bar{z} = \frac{1}{M}\sum_{j=1}^{m} M_j z_j \tag{21.40}$$

Adopt a convention for the sign of each piece's coordinates that accounts for its moment about the appropriate axis. The arrows in the drawing represent the coordinates of the CG of the left-front wheel, the battery pack, and a segment of the body

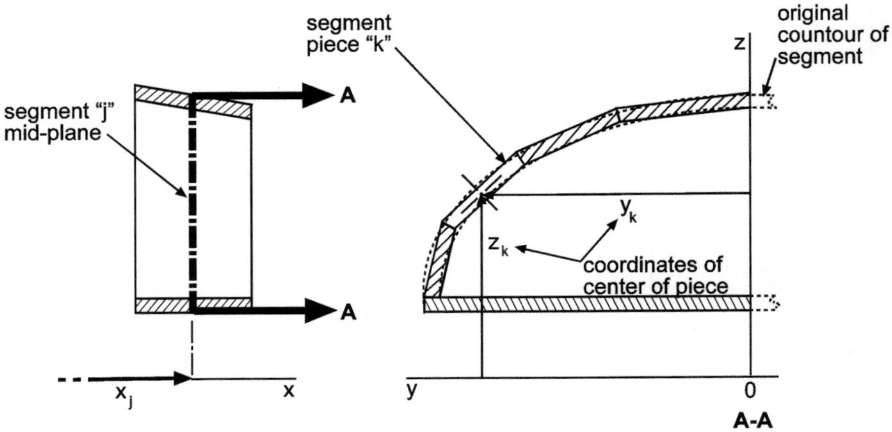

Fig. 21.13 Shell segment

shell measured from an origin on the centerline of the car below the nose and in the horizontal plane of the wheel contact patches.

We assume the wheel to be symmetric about its rotational axis and with respect to the vertical plane perpendicular to that axis and bisecting the wheel. Hence, its CG is located at the point where the axis of rotation pierces this bisecting plane. The moment of the wheel's mass about the x-axis is M_1y_1, taking positive moments to be clockwise when viewed in the positive axis direction, and about the y-axis is M_1x_1. Imagine the car experiencing a centrifugal acceleration in the y-direction, as when turning. The wheel mass would then have a moment $-M_1z_1$ about the x-axis.

Let us assume that the distribution of mass in the battery box is uniform. The CG would be located at the centroid of the box, the point where its three planes of symmetry intersect. The battery box mass has no moment about the x-axis because the y-coordinate of its CG is zero. The other two moments would be calculated similarly to those of the wheel.

The CG of the body shell, or of the canopy, is more difficult to find because the shapes of these objects are in general only symmetric with respect to the plane $y=0$. Approximate coordinates can be found by dividing the shell into segments. One is shown in Fig. 21.13. The segments should be short. Then, the error caused by approximating the curved shell with flat surfaces when calculating properties will be small. The shorter the segments, the smaller this error will be. On the other hand, the shape may be such that large portions of it are uniform along the car's length and may be treated as one segment. Look for such simplifications. Shortening the segments increases the labor in the calculation.

If a segment is of a known cross-sectional shape, such as a rectangle, and has a uniform or at least a mass distribution symmetric with respect to its centroid, its CG will be at its centroid. The segments that have nonstandard cross-sectional shapes must themselves be subdivided. These pieces should be small so that they can be modeled by shapes the volumes of which can be calculated. A subdivided segment is shown in Fig. 21.13.

The x-coordinate of the segment's CG is taken to be the x-coordinate of the mid-plane of the segment, as shown in the figure. Because of the segment's symmetry about the plane, $y=0$, the y-coordinate of the CG is zero. The z-coordinate remains to be found. The mass of the segment is

$$M_j = 2\sum_{k=1}^{n} \Delta M_k. \tag{21.41}$$

The symbol ΔM represents the mass of a piece. The sum is over the subdivided half of the segment and so must be multiplied by 2. The letter "n" represents the number of pieces. For example, the z-coordinate of the segment's CG is

$$z_j = \frac{2}{M_j}\sum_{k=1}^{n} \Delta M_k z_k. \tag{21.42}$$

The z_k are measured in the mid-plane $x=x_j$ and up to the center-plane of the piece. The coordinates of the shell's CG are obtained from the segment CG coordinates using equations Eq. (21.40). The $\bar{y}$-coordinate of the body shell CG is zero because of symmetry about the z-axis.

If the construction of the shell is not uniform through its thickness, this can be accounted for by creating additional segments and segment pieces to isolate the nonuniformities. Then, they are treated as described above.

The forgoing calculations, while not difficult, are tedious and should be automated (as much as possible) in an electronic spreadsheet.

Discussion of Error The method agrees reasonably well with the measured location of the CG (see Chap. 12) in actual solar cars. The error in calculating the contribution of odd-shaped members like shell segments can be reduced by increasing the number of pieces into which they are subdivided. Fortunately, the masses that dominate the calculation are the battery, the motor and controller, the driver, and the wheels. The method works well overall if these components are done carefully.

A loaded wheel is actually not symmetric with respect to its axis of rotation because of the flattening of its tire in the contact patch. Ignoring this causes the CG height to be slightly overestimated because the car is actually closer to the ground.

The mass distribution of the battery box is not uniform. However, each battery module is usually much more massive than the air spaces or other lightweight components of the box. The effect of these components on the box's CG is therefore relatively small.

21.10 Moment of Inertia

The moment of inertia, $I_{Z'}$ of the car about the vertical axis through its CG must be known in order to estimate the motion of the car caused by a side gust (see Eq. (21.3)). It may be estimated by a method similar to that just described for finding the CG.

Fig. 21.14 Calculating I_Z

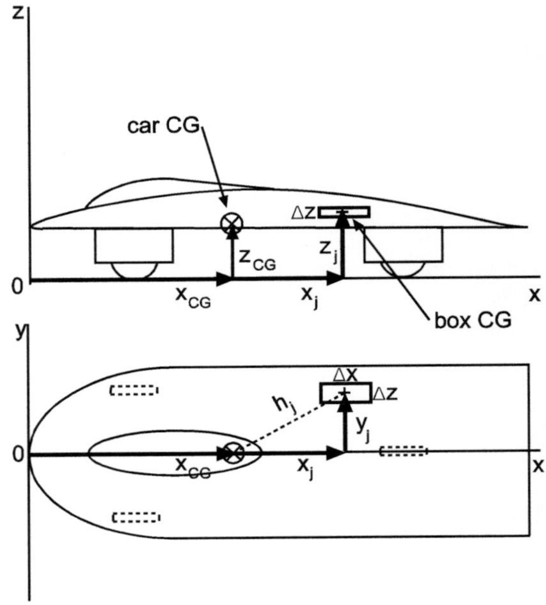

The components used to find CG calculation may also be used for this calculation. We assume each component may be idealized as a well-known shape (i.e., a cylinder, a disc, etc.) of uniform density. The moment of inertia of each such shape about the vertical axis through its CG will be found in standard mathematical tables. We denote this moment as I_j. Figure 21.14 shows the mass M_j of a component (shown as a box for convenience) in a solar car (drawn as before, on crosshatched paper).

The parallel axis theorem then gives the component's moment about the car's CG as

$$I_{Z_j} = I_j + M_j h_j^2. \tag{21.43}$$

The distance h_j is the distance in the x–y plane from the component's CG to the car's CG. From the figure,

$$h_j^2 = (x_j - x_{CG})^2 + (y_j - y_{CG})^2. \tag{21.44}$$

The I_Z of the car is the sum of all the I_{Z_j}:

$$I_Z = \sum_{j=1}^{n} I_{Z_j}. \tag{21.45}$$

A shell or canopy element would be divided into pieces as shown in Fig. 21.13. We assume that the pieces are small enough so that the moment of the jth element about its CG is

$$I_j \approx 2\sum_{k=1}^{n_j} y_k^2 \Delta M_k, \tag{21.46}$$

where n_j is the number of pieces in the element. We have again idealized the shell segment piece as a point mass and taken advantage of the symmetry of the segments with respect to the z-axis. The y_k is the y-coordinate of the CG of the kth piece.

If some of the pieces can be modeled as well-known shapes, their I_j's can be found from a mathematical table. The bottom of the segment in Fig. 21.13 could be modeled as a box, for example. These cases would be added to the summation of Eq. (21.46). Each such piece would have a moment of the form

$$I_k = M_k r_k^2, \tag{21.47}$$

where r_k denotes the *radius of gyration* about the piece's CG. A point mass M_k at radius r_k would have the same I_k about the z-axis through the segment's CG as the actual piece. The moment of the box-shaped object in Fig. 21.13 (assumed to be of uniform density) about the vertical axis through its CG is tabulated in standard reference tables as

$$I_k = M_k \left[\frac{(\Delta x)^2 + (\Delta y)^2}{12} \right]_k, \tag{21.48}$$

where Δx and Δy are the piece's lengths in the x- and y-directions, respectively. The radius of gyration would then be

$$r_k = \left[\frac{(\Delta x)^2 + (\Delta y)^2}{12} \right]_k^{\frac{1}{2}}. \tag{21.50}$$

Discussion of Error This method has not been tested by the author against actual measurements, which are difficult. It suffers from the same kinds of errors that afflict the CG calculation previously discussed. And, as in the CG case, the error is reduced by increasing the number of pieces into which the car is divided and paying close attention to the masses that dominate the calculation.

References

Baker, C. J. (1991a). Ground vehicles in high cross winds, Part 1: Steady aerodynamic forces. *Journal of Fluids and Structures, 5,* 69–90.

Baker, C. J. (1991b). Ground vehicles in high cross winds, Part 2: Unsteady aerodynamic forces. *Journal of Fluids and Structures, 5,* 91–111.

Bundorf, R. T., Pollock, D. E., & Hardin, M. C. (1963). Vehicle handling response to aerodynamic inputs, paper 716B, June, 1963 SAE International Summer Meeting.

Chang, C. -N., & Ding, D. H. (1994). Theoretical stability analyses of the cornering behavior of three- and four-wheel vehicles. *International Journal of Vehicle Design, 15*(3/4/5), 301–316.

Emmelman, H.-J. (1987) Driving stability in side winds, In W. -H. Hucho (Ed.), *Aerodynamics of road vehicles*, (pp. 214–235). Boston: Butterworths.
Gillespie, T. D. (1992). *Fundamentals of vehicle dynamics*. Warrendale: Society of Automotive Engineers, Inc.
Hoerner, S. F. (1965). *Fluid-dynamic drag*. Bakersfield: Hoerner Fluid Dynamics.
Hucho, W.-H., & Emmelmann, H. J. (1973). Theoretical prediction of the aerodynamic derivatives of a vehicle in cross wind gusts, *Proc. Automotive Engineering Congress, Detroit, Michigan, January 1973*, SAE paper 730232.
Huston, J. C., Graves, B. J., & Johnson, D. B. (1982). Three-wheeled vehicle dynamics, SAE paper 820139.
Rocard, Y., (1954). *L'instabilité en Mécanique*. Paris: Masson et Cie.
Sorgatz, U. (1975). Simulation of directional behavior of road vehicles. *Vehicle Systems Dynamics, 5*, 47–66. (Swets and Zeitlinger BV, Amsterdam, August).
Steeds, W. (1960). *Mechanics of Road Vehicles*. London: Iliffe and Sons, Ltd.
Van Valkenburgh, P. G., Klein, R. H., & Kanianthra, J. (1982). Three-wheel passenger vehicle stability and handling, SAE 820140, *SAE Transactions*, vol. 90.
Wong, J. Y. (1978). *Theory of ground vehicles*. New York: Wiley.
Wong, K., Andrews, J., Callan, P., Stewart, M., Baran, O., & Rudisill, S. (2002). Test theory for side gust force and moments on vehicle, Integrated Design-ME445/446 Final Report, Clarkson University MAE Department, Potsdam, NY.

Chapter 22
Structural Load Estimation

22.1 Purpose

This chapter suggests ways to establish design structural load sets for the cruise condition (including remarks on transport of the solar car by trailer) and emergency maneuvers. These load sets can be used by the designers to calculate the loads in the members of the suspension, steering, and frame of the car. The chapter also provides some guidance for the design of the structure under collision loadings.

22.2 Safety and Structural Design

Rules and Inspections The race regulations supply rules for protective structures to enclose the driver and for some other structural elements of the car. The pre-race structural inspection (together with the required structural report) ascertains the car's compliance with the rules. The inspection is professional and strict. Each inspector will interpret omissions in the rules or broadly written rules, conservatively. Therefore, the inspection may seem to the competitors to go beyond the literal requirements of the rules. However, inspectors are like umpires: they do not change their calls. So, to avoid serious problems with inspections, teams should adopt the viewpoint of the inspectors in their design work. Race experience helps here, of course, as does putting questions to the race headquarters in advance.

Design for Safety The sizing and construction of parts can only be done when their loads, and the loads' points of application, have been specified. This is "safety" reduced to practice. The generation of these loads involves judgments about the limiting conditions the structure must withstand, and therefore about the interactions of the solar racing car with its environment that cause these conditions.[1]

[1] Chap. 8, Solar Racer: Specification, points out the need to study the route over which and the conditions in which the car will operate.

E. F. Thacher, *A Solar Car Primer*, DOI 10.1007/978-3-319-17494-5_22

The objectives of the structural design should not merely be to obey race rules and pass inspection, important as these are. Each design team must decide for itself how robust the structural design of its car should be to be safe, because "safe" is a relative term. The loads calculated as suggested herein should be regarded as starting-points; they may not be conservative (i.e., large) enough to satisfy the criteria set up by the design team. However, increased structural strength usually involves either a weight increase or a cost increase (sometimes both). A 2-m wide by 6-m long by 1.6-m high armored tank will resist collapse during a collision very well, but it probably would not be a competitive solar racing car. A titanium space frame will be light and strong, but it will also be expensive and require special welding skills.

22.3 Cruise Condition Loads

Design Objectives The aim of the design under cruise conditions and emergency maneuvers should be to endure the specified vibration, bump, braking, and maneuvering loads without structural damage (although tires may go flat) and continue to race.

Fatigue Cruise condition (defined in Chap. 2) loads are applied only in the x- and z-directions. The average loads applied to the car under these conditions will be less than those in the emergency maneuver and collision categories, and thus not limiting if thought of as static loads. However, the constant vibration about these average values imparted to the car by normal driving over the period beginning with the first testing of the car, extending through the race, and including any post-race driving, causes fatigue. The effects of fatigue are cumulative and may therefore cause failure near the end of the design life of the vehicle, even under the cruise condition loads. The average load, fluctuating load, and number of cycles are a matter of judgment.

Trailering In the author's experience, transporting the car has caused the most structural failures. Why? The solar car is moving at highway speeds and the thus-aggravated bouncing of the trailer interacts with the solar car's suspension and structure. The motion of parts of the car may be out of phase with the trailer's motion. For example, the portion of the body shell behind the rear wheels may behave in this way. It will wave up and down like a cantilevered beam with the trailer bed motion fed through its fixed end. As a consequence, it may fail near its junction with the rest of the shell.[2] Severe bouncing and vibration may go on for many hours, perhaps more than eight, as the team strives to reach its destination. This causes fatigue failure.

Take a hard look at the car-trailer system and deduce the failure modes that could result. By "hard look" it is meant that do not assume that just because the car is

[2] Cracking of the shell of Clarkson's 1999 solar car occurred in exactly this way.

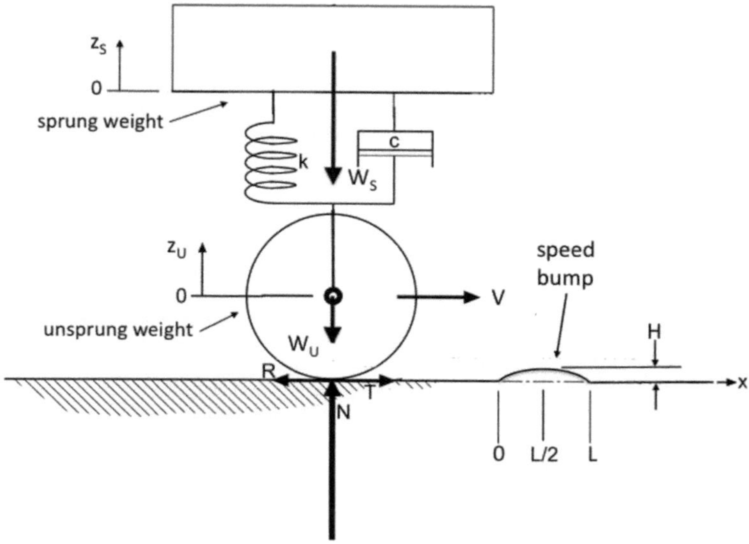

Fig. 22.1 Bump scenario

strapped into the trailer that this is sufficient projection. Individuals charged with the trailer system may not be experienced enough to do this properly. And often that system is an after-thought built at the last moment. This situation is a recipe for disaster. Listen to the experience of those who have done it before and always think of your *racing system*, not just the solar car.

To design the car for these trailer failure modes, to some extent at least, is prudent. But it is better to look at possible failure modes and cradle the car effectively against them and of course to drive more gently when transporting. As a rule, the better-prepared teams will travel to the race at a deliberate pace, instead of a last-minute rush, and trailer less during the race. Therefore, their cars will not be exposed to the stress of transport as much as those of other teams. Transport damage is inversely proportional to preparation.

Bump Also included in the cruise condition is a load that occurs sporadically, but is typical: driving over a bump.[3] The bump loads are calculated for the suspension and wheel with the highest static load on its contact patch. This wheel traverses a bump of specified height and length, as shown in Fig. 22.1.

$$h(x) = 4H\left[\frac{x}{L} - \left(\frac{x}{L}\right)^2\right]. \tag{22.1}$$

The bump in Fig. 22.1 is modeled as a parabola by Eq. (22.1), where *h(x)* is the height of the bump above the normal road surface at a distance *x* from the beginning

[3] This event has been studied extensively. See, for example, Oluwole(2012) and Mohammadzadeh and Haidar (2009).

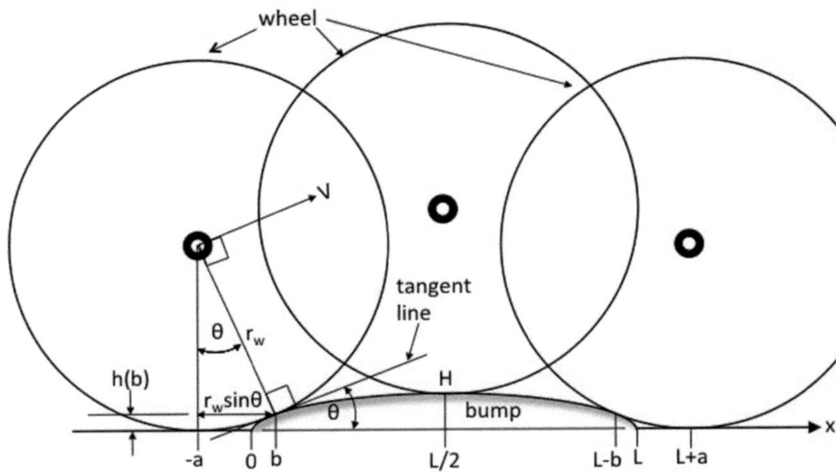

Fig. 22.2 Bump traverse

of the bump, H is the maximum height of the bump, and L is the length of the bump. When the wheel is at x=$-a$, it contacts the bump at height $h(b)$. Fig. 22.2 shows the contact geometry.

Conservatism, that is, the normal force (N, in Fig. 22.1) will be overestimated, is introduced by assuming that the point at which the suspension is attached to the car body does not move up while traversing the bump because $M_S >> M_U$, that the tire is rigid, and that the spring-damper mechanism must absorb the full vertical motion of the wheel. The effect of the tractive force, T, the rolling resistance, R, and the drag on the wheel have been neglected.

The goal is to find an expression for the normal force as a function of the bump parameters, the suspension damping coefficient, c, and the spring constant, k. Just before rolling up the bump, a force balance in the z-direction gives the normal force (subscript "0") as

$$N_0 = W_S + W_U = \frac{Wf}{L_W}, \quad f = \begin{cases} \dfrac{a}{n_R}, & \textit{rear wheels} \\ \dfrac{b}{n_F}, & \textit{front wheels} \end{cases}. \tag{22.2}$$

W_S is the "sprung" weight on the suspension-wheel assembly. As the figure shows, it is the portion of the car's weight above the spring and damper. W_U, the "unsprung" weight, includes the wheel, brake, wheel motor, etc., anything below the suspension.

Equation (22.3) gives the vertical force on the wheel after entering the bump (neglecting the vertical components of drag, tractive force, and rolling resistance).

$$N(x) = M_U a_Z(x) + cV_z(x) + kh(x) + N_0, \tag{22.3}$$

where $a_z(x)$ and $V_z(x)$ are the vertical acceleration and velocity, respectively. Figure 22.1 shows that

$$z = h(x). \tag{22.4}$$

Assume that the tire does not slip while on the bump. Therefore, the velocity V is steady and tangent to the bump. Since the x-position of the tire on the bump is a function of time, $x(t)$, so is $h(x)$. The velocity components in the z- and x-directions are

$$V_z(x) = V\,sin\,\theta(x) = \dot{h}(x), \tag{22.5a}$$

$$V_x(x) = V cos\,\theta(x). \tag{22.5b}$$

where, the overdot indicates the rate of change with time.

Two overdots means acceleration. The tangent angle, θ, is given by the inverse tangent of the slope of the bump

$$\theta(x) = tan^{-1}\left[\frac{4H}{L}\left(1-\frac{2x}{L}\right)\right]. \tag{22.6}$$

The acceleration is

$$a_z(x) = \dot{V}_z(x) = V cos\theta\,(x)\dot{\theta}(x), \tag{22.7}$$

where

$$\dot{\theta}(x) = \frac{4HV^2}{L^2}\frac{sin2\theta(x)}{1+\left[\frac{4H}{L}\left(1-\frac{2x}{L}\right)\right]^2}. \tag{22.8}$$

22.4 Example 22.1

Suppose the Shark encounters a speed bump on its right front side. The bump is approximately parabolic and is 1.0 ft (0.3048 m) long with a maximum height of 2.25 in. (5.715 cm). The Shark's wheel radius is 10 in. or 25.4 cm. If the suspension spring constant and damping coefficients are 18,000 N/m and 370 N sec/m, respectively, what will be the maximum vertical force on the wheel at a speed of 5 mph (2.2352 m/s)?

Solution Equation (22.2) gives N_0 as 988.6 N, 98.3 % of which is sprung weight. Equating Eq. (22.1) and h(x) at $x=b$

$$h(b) = r_w\left(1-cos\theta(b)\right), \tag{22.9}$$

and solving by trial-and-error gives

$$\frac{b}{L} = 0.15.$$

Therefore, b=4.572 cm. From Eq. (22.6),

$$\theta(b) = tan^{-1}\left[\frac{4 \times 0.05715m}{0.3048m}(1 - 2 \times 0.15))\right] = 27.7^{\circ}.$$

The position of the wheel when contact is first made with the bump is

$$a = -\left(r_w cos\theta(b) - b\right) = -\left(25.4cm \times cos27.7^{\circ} - 4.572cm\right) = -17.92cm.$$

The vertical force then shifts abruptly from the pavement to the bump.

Setting up the problem on a spreadsheet and calculating N at intervals of 0.015 m, gave the maximum vertical force as 2125.1 N at x=0.105 m. The maximum occurs between x=b and x=L/2 because the spring force increases with x, while the damper force, proportional to V_z, decreases as the maximum bump height is approached.

Discussion of Errors The normal force calculation is conservative (overestimates the force) because the car body is held stationary. If it were allowed to move up, the spring force would be reduced. Additionally, if tire deflection were allowed, this deflection would reduce the force transmitted to the suspension.

The method presupposes that the *H/L* of the bump is small enough, on the order of a typical artificial speed bump, to allow the tire to roll up. The method will not work for an abrupt bump, like a curb.

The drag, rolling resistance, and tractive force act approximately in the x-direction, while on the bump. Therefore, their influence in the z-direction will be small compared to that of the bump.

22.5 Emergency Maneuvers

Straight-Ahead Stop Each solar car is required to demonstrate its ability to stop from an initial speed V_1 with an average deceleration, a_R. The minimum acceptable magnitudes of each are given in the race regulations (sample: Chap. 16, Sect. 16.6.12.1). The equation developed below estimates the total, constant, braking torque, τ_{BR} (N·m), necessary to do this. Drag and rolling resistance were neglected, therefore the result overestimates the actual torque needed. The calculation allows for the reaction time of the driver-braking system[4] after the need to stop is perceived

[4] The reaction time could be that used by Emmelman (1987), 0.8 s. Even through this includes the delay in the steering system it is at least a place to start having some basis in experience.

by the driver. The magnitude of the deceleration required when a delay of Δt_D seconds is included in the total time interval to stop, Δt_S, is

$$a_B = \frac{V_1}{\Delta t_S - \Delta t_D}.$$

The allowed total stopping time is V_1/a_R. Substituting this in the deceleration equation gives

$$a_B = \frac{a_R}{1 - \dfrac{\Delta t_D a_R}{V_1}}.$$

Multiplying by the effective mass, M_e, and r_W, the wheel radius, and converting a_R to kph/sec gives the equation for the torque as

$$\tau_B = 0.278 \frac{M_e a_R r_W}{1 - \Delta t_D \dfrac{a_R}{V_1}}. \tag{22.10}$$

The braking torque is the total torque developed by all the braked wheels.

22.6 Example G.2

Suppose M_e is 330 kg, V_1 is 50 kph, and Δt_D is 0.8 s. Find the required deceleration and total braking force and torque necessary to satisfy Chap. 16, Sect. 16.6.12.1.

Solution The required total stopping time is

$$\Delta t_S = \frac{50 kph}{17 \dfrac{kph}{sec}} = 2.94 sec.$$

Allowing for 0.8 s of reaction time, the deceleration must be

$$a_B \geq \frac{50 kph}{(2.94 - 0.8) sec} = 23.35 \frac{kph}{sec}.$$

The total braking force is at least

$$F_B \geq 23.25 \frac{kph}{sec} \times 330 kg \times \frac{1000 \dfrac{m}{km}}{3600 \dfrac{sec}{h}} = 2140.6 N.$$

Assume a wheel radius of 10 in. or 25.4 cm. The brakes must supply a total torque of at least

$$\tau_B \geq 2140.6N \times 0.254m = 543.7N \cdot m.$$

Equation (22.10) gives a conservative estimate of the torque required to decelerate the car to meet the race rule requirement. However, the *achievable* braking force is a function of the friction coefficient between the tire and the road surface. (Note that Chap. 16, Sect. 16.6.12.1 requires the braking test be done on a wetted surface.) This *braking coefficient* is defined as the ratio of the braking force to the normal force. For a particular tire

$$\mu_B = \frac{F_X}{N}. \tag{22.11}$$

The achievable total braking force must of course at least equal that required by Eq. (22.10).

The braking coefficient is a strong function of the relative motion, or slip, between the tire and the road. In the cruise condition, the distortion and recovery of the tire as it deforms and rolls through the contact patch and over road surface roughness gives rise to the rolling resistance. When the brakes are applied, the additional braking force, F_X, applied at the contact patch depends upon both the energy absorbed by the flexing of the contact patch caused by the surface texture of the road and the adhesion of the tire to the road. The adhesion dominates the braking force. The wheel begins to decelerate and a difference between the vehicle's and the tire's speed develops. This difference, normalized by the speed, is called the (longitudinal) *slip*.[5]

$$S = 1 - \frac{\omega r_W^*}{V}.$$

The braking force, at first, increases rapidly with slip. The adhesion causes stretching of the tire material, exerting a retarding force on the wheel, just as when a rubber band is stretched. As the slip increases, the stretch also increases, and so does the braking force. This behavior is shown in Fig. 22.3. However, the figure shows a maximum in the braking coefficient, μ_{BM}, occurring at about 25% slip on a dry surface and about 17% slip on a wet surface. At this point, the adhesive bond to the road is overcome by the stretching of the "rubber band" caused by slip. The smallest coefficient at a particular speed, μ_{BS}, occurs at 100% slip; that is, when the brakes have locked and the wheel cannot rotate.

That part of μ_B caused by roughness hysteresis is also shown in the figure. Its dependence on slip is much less than that of the adhesive force and it is not affected by wetness. Like rolling resistance, roughness hysteresis arises from the flexing of

[5] Lateral slip occurs because of side forces when cornering (see Chap. 21).

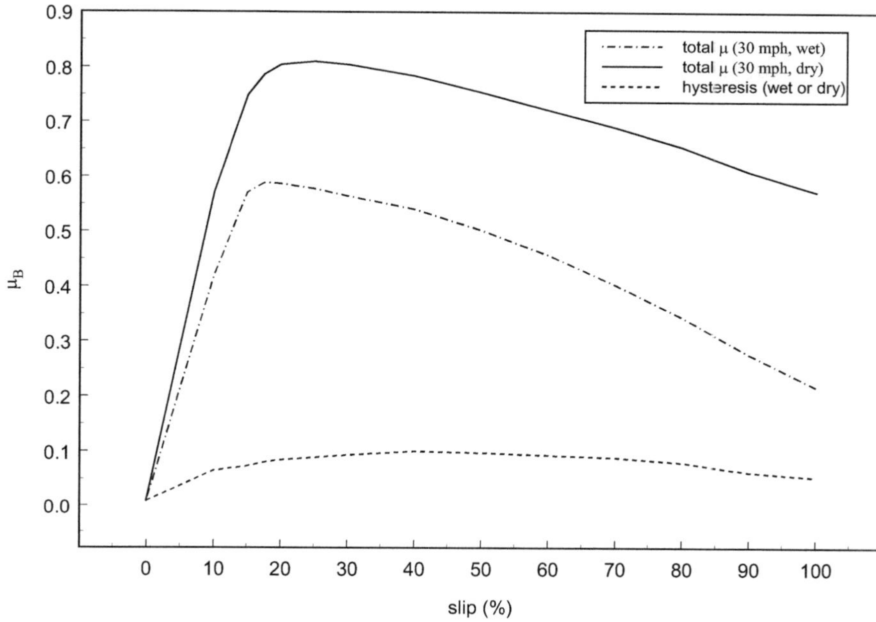

Fig. 22.3 Braking coefficient. (Source: Meyer and Kummer 1962)

small portions of the tire as the contact patch passes over the texture of the road surface. Energy is stored in the small deformations of the tire material and then partially recovered when the tire rolls past the small deformation. The portion not recovered is lost as heat. Hence, the term "hysteresis."

Compare this to the flattening of the contact patch characterized by the "sinking rate" described in Chap. 2 and Chap. 20. Adhesion and hysteresis must be present; else the tire would not exert a tractive force on the car. But the rolling resistance is dominated by the pressure difference caused by the unequal deformation of the tire between the leading and trailing portions of the contact patch.

Like the rolling resistance coefficient, μ_{BP} and μ_{BS} characterize a particular road–tire combination. Besides the construction of the tire and the road surface, the combination includes the vehicle's speed, inflation pressure, and vertical load on the tire. Both coefficients are reduced as speed increases, are insensitive to inflation pressure on dry roads but increase with inflation pressure on wet roads, and are reduced by increasing vertical load. They are reduced roughly 0.01 by a 10 % increase in load near the tire's rated load (Gillespie 1992). Note the different effect of inflation pressure on rolling resistance.

Effective Mass Equation (22.10) includes the effective mass. The kinetic energy of the car, neglecting the inertia of the rotating parts of the transmission and the sinking rate (both of which will be small in solar racing cars), is

$$KE_{CAR} = \frac{1}{2}\left(MV^2 + n_W I_W \omega_W^2 + I_W \omega_W^2\right).$$

Where M and V are the car's mass and speed, respectively, n_W is the number of wheels, I_W the moment of inertia of a wheel about its axis, ω_W is the angular velocity of the wheels, and I_M and ω_M are the moment of inertia of the motor about its axis and its angular velocity, respectively. The angular velocity of the wheels is $2V/d_W$, where d_W is the wheel diameter, and that of the motor is $2n_G V/d_W$, where n_G is the gear ratio of the transmission. Substituting for the angular velocities and collecting terms on $V^2/2$ gives

$$KE_{CAR} = \frac{1}{2}\left(M + \frac{4n_W I_W}{d_W^2} + \frac{4n_G^2 I_M}{d_W^2}\right)V^2 = \frac{1}{2}M_e V^2. \tag{22.12}$$

The term in parentheses is the effective mass.

Wheel Vertical Load The wheel reactions are calculated considering only the inertial effects. By race rules all wheels must be braked. A moment balance about the rear contact patch gives the total reaction, on the front wheels during braking as

$$W_F = \left(1 - \frac{a}{L}\right)W + F_X\frac{h}{L} = \left(1 - \frac{a}{L}\right)W + \frac{\tau_B h}{r_W^* L}. \tag{22.13}$$

where, F_X is the total braking force. A moment balance about the front contact patch gives the total reaction on the rear wheels when braking as

$$W_R = \frac{a}{L}W - F_X\frac{h}{L} = \frac{a}{L}W - \frac{\tau_B h}{r_W^* L}. \tag{22.14}$$

Recognizing that the braking coefficients may be different because of different vertical loads

$$F_X = \mu_{BF}W_F + \mu_{BR}W_R. \tag{22.15}$$

Using (22.13) and (22.14)

$$F_X = \frac{\mu_{BF} - (\mu_{BF} - \mu_{BR})\frac{a}{L}}{1 - (\mu_{BF} - \mu_{BR})\frac{h}{L}}W, \tag{22.16}$$

which reduces to

$$F_X = \mu_B W \le \mu_{BP} W, \tag{22.17}$$

if the front and rear braking coefficients are sufficiently close.

Braking While Turning The wheel carrying the maximum load will be the outside front wheel, in the case of four-wheeled cars or three-wheeled cars with two wheels in front. This wheel must carry the combined effect of the roll and braking weight shift, the cornering force, and the rolling resistance force. The left rear wheel of a 1F–2R three-wheeler carries the transverse inertial load, but reduced by the forward weight shift from braking.

As the discussion of roll over in Chap. 21 implies, the interaction of lateral and longitudinal slip when braking and cornering near the traction limits reduces both the cornering and braking forces. This is shown by the lower g-limits on the skid ellipse, Fig. 21.11, for points where both braking and transverse acceleration exist.

Discussion of Errors Neglect of the deceleration caused by drag and rolling resistance makes the average braking torque estimate in Eq. (22.10) conservative. Not reducing the wheel radius by the sinking rate (defined in Chap. 20) increases the conservatism.

Neglect of rotating gears and shafts causes Eq. (22.12) to underestimate M_e.

22.7 Collisions

Design Objective The objective when designing the vehicle's structure for collision is to protect the driver. That is, to reduce the magnitude of accelerations the driver experiences and to prevent her or him from being struck by pieces of the car set in motion by the collision. Note that the objective is not to prevent failure such that the car can remain in competition (as in the cruise and emergency maneuver categories), but to have the structure fail in such a way that the driver is protected.

Nature of Collisions We consider only events that have the potential to cause major damage, such as striking or being struck by another vehicle, or striking a nearly rigid object, such as a wall, and at an appreciable relative speed–40 kph or above perhaps. What distinguishes the loads arising from these events from the loads in the previous two categories? The intervals over which the cruising condition and emergency maneuver loads are applied are long compared to the time required to propagate the load-caused deflections to the structure away from the point of application. This structure (the suspension A-arms, say) has time to react to the load and contribute to the strength.[6] The interval over which the collision load is applied is short compared to the speed with which the deformation caused by the impact can propagate through the structure. Therefore, the reinforcing members that could provide strength to resist the deformation cannot contribute and the structure tends to fail locally. Also, loads in the cruise and emergency maneuver categories are applied at locations designed to receive them. It is likely that collision loads will be applied first to portions of the solar car's structure which are not intended to carry large forces or deformations, such as the shell supporting the solar cell array. Damage is therefore more probable.

Methods A means of accomplishing collision energy management is by controlled deformation. Race rules mandate this, requiring a certain minimum crush space so

[6] This is not true in the special case of fatigue, which causes local failure.

that collision energy is absorbed by the work necessary to deform structure through the crush space. Also, the average deceleration rate is reduced, and therefore the force exerted on the driver, because of the lengthened time required to bring the vehicle to rest. A complementary strategy is deflection: configuring the roll cage of the solar car so that portions of the car's structure set in motion by the collision are directed away from the driver. Deflection is also mandated by the race rules. Norton (1983) reported a small, three-wheeled vehicle incorporating deformation and deflection in the design of its roll cage.

Each registered team is required to submit a structural report, the detailed requirements for which do contain minimum loads for the following cases: roll over, front impact, side impact, and rear impact.[7] These loads are typically expressed in "g" units, previously defined. Thus a 3-g side impact load on a car weighing 2000 N at 1 g would be equivalent to a force of 6000 N.

Analysis The best design tools are codes that have impact analysis capabilities and consequently provide a good picture of how the structure behaves during a collision. However, these codes are expensive, large, run best on husky work stations, and require some training to be used effectively. Analysis may also be done with codes intended for static problems, and by hand calculations. The structure should be modeled conservatively, so that a safety margin exists, especially when using static codes or hand calculations. Safety margins provide extra strength to compensate for possible errors in the judgments about how the structure fails. These judgments must be made to accomplish simpler analyses.

Proof Tests In addition to analysis, the behavior of solar cars in collisions can be evaluated by crash-testing driverless cars with a dummy inside to simulate the driver, as is done by commercial car companies. This has been done in simplified ways by solar car teams using just a chassis on wheels. It must be done sufficiently in advance of the race, perhaps during the late spring or summer of the year preceding the race, to allow time to incorporate the lessons learned.

At the other end of the scale is testing by applying large (therefore conservative) static loads, non-destructively. To support its hand calculations in a structural report for a race, the Clarkson University Solar Knights submitted a picture of the composite shell of their car undergoing negligible deflection while supported at its ends between two chairs and loaded by six men standing near its center.

References

Emmelman, H.-J. (1987). Driving stability in side winds. In W.-H. Hucho (Eds.), *Aerodynamics of road vehicles* (pp. 214–235). Boston: Butterworths.

Gillespie, T. D. (1992). *Fundamentals of vehicle dynamics*. Warrendale: SAE

Goldsmith, W. (1960). *Impact*. London: Edward Arnold Publishers, Ltd.

[7] In addition to documenting the protection of the driver from the roll over and impact cases mentioned, the reports must include analyses of the tires, wheels, braking, steering, and suspension systems.

Meyer, W. E., & Kummer, H. W., (1962). Mechanism of force transmission between tire and road. SAE 620407.

Mohammadzadeh, A., & Haidar, S. (2009). On the analysis and design of vehicle suspension systems going over speed bumps.Paper AC 2009–1466, American Society for Engineering Education.

Norton, D. G. (1983). The design of a three-wheeled, high performance sports/commuter vehicle. SAE 831182, SAE West Coast International Meeting, British Columbia, August 8–11, 1983.

Oluwole, O. O. (2012). Matlab® and Simulink use in response analysis of automobile suspension system in design. *International Journal of Traffic and Transportation Engineering, 1*(2), 19–31.

Van Valkenburgh, P. G., Klein, R. H., & Kanianthra, J. (1982). Three-wheel passenger vehicle stability and handling, SAE 820140. *SAE Transactions, 90*, 604–627. doi: 10.4271/820140.

Chapter 23
Nomenclature

A

a	Acceleration (m/s^2); distance from front axle to CG (m)
A	An area (m^2), with a subscript to indicate A_D, profile area; Ampere, unit of electric current

B

b	Camber (m); a distance, for example, from the rear axle to the CG or the maximum width of a wheel; a constant
B	Magnetic field strength (Weber/m^2)
BM	Wheel bearing torque (N·m)

C

c	A coefficient with subscript to indicate c_D, overall drag coefficient or c_V, ventilation drag coefficient; chord length (m); a constant
C	Battery charge storage capacity (Ah)
CG	Center of gravity (also termed "center of mass")
C_V	True course angle of vehicle (degrees)
C_W	True course angle of wind (degrees)
C_α	Cornering stiffness of tire (N/degree)

D

d	Diameter (m); a constant
D	Total drag force (N); subscripts indicate component of drag, for example, D_V, ventilation system drag
D_M	Wheel rotational drag moment (N·m)
D_W	True direction from which the wind is blowing (degrees)

E

e	Surface roughness (m); base of natural logarithms

E. F. Thacher, *A Solar Car Primer,* DOI 10.1007/978-3-319-17494-5_23

E Energy (kW.h or J), with subscript to indicate storage location, for example, E_B, battery

F

f Frequency (Hz)

F Force (N); a subscript indicates direction, for example, F_X, a force in the x-direction

G

g Acceleration of gravity (m/s^2)

G Solar irradiance at an instant (W/m^2) on a horizontal surface or with a subscript to indicate a particular surface, for example, G_T, a surface tilted with respect to the local horizontal plane, or source

H

h Height (m); sinking rate of a tire (m); head loss of a fluid (m); an hour

H Daily total radiation on a horizontal surface (kJ/m^2) or with a subscript to indicate a particular surface, for example, H_T, a surface tilted with respect to the local horizontal plane, or source.

I

I Electric current (A); hourly total radiation on a horizontal surface (kJ/m^2) or with a subscript to indicate a particular surface, for example, I_T, a surface tilted with respect to the local horizontal plane; moment of inertia about an axis indicated by a subscript ($kg.m^2$), for example, I_Z, the vertical axis through a point, usually the CG.

J

J Joule, a unit of energy

K

k Hourly average clearness index; proportionality constant, usually with a subscript to indicate k_S, shaft torque constant of a motor (N·m/A)

K Extinction coefficient (m^{-1}); fluid head loss coefficient; Kelvin, a unit of absolute temperature; with overbar, monthly average clearness index

L

l Length (m)

L Lift force (N); vertical load (N); longitude (degrees); distance with a subscript to indicate L_W, length of wheel base (m)

L_a	Latitude (degrees)
L_o	Longitude (degrees)
λ	General symbol for reference length in Reynolds number, also called the "characteristic length"

M

$\dot{m}$	Mass flow rate (kg/s)
M	Mass (kg), with subscript to indicate location or kind, for example, M_e, effective mass of vehicle

N

n	Index of refraction; quantity, for example, n_G, transmission gear ratio
N	Rotational speed (rpm); normal force (N)
N	Newton, unit of force

P

p	Static pressure (N/m^2) with subscript to indicate source, for example, p_A, atmospheric pressure (in Chap. 2, p without a subscript indicates atmospheric pressure)
P	Power (W); perimeter (m)
P_M	Pitching moment about y-axis through vehicle's CG (N·m)

Q

q	Dynamic pressure (N/m^2)
Q	Volumetric flow rate (m^3/s); electric charge (A.h or Coulomb)
Q	Rate at which energy is transferred (W)

R

r	Radius (m), usually with a subscript to indicate r_w, wheel radius
r_w*	Wheel radius (m) reduced by sinking rate in contact patch
R	Total rolling resistance force (N); electric resistance (ohm); radiation tilt correction factor, including a subscript to indicate what radiation component, for example, R_B, beam radiation
R_A	Gas constant for air (kJ/kg K)
Re	Reynolds number
R_M	Rolling moment about x-axis through vehicle's CG (N.m)

S

s	Second
S	Distance along a road (km); slope of a line; tire rotational slip (fraction or %)

T

t Time
T Temperature (°C or K); tractive force (N); track width of vehicle (m)

U

u Horizontal distance measured rearward from the nose of a car (m)
U Overall heat transfer coefficient (W/m² K)

V

V Vehicle velocity (m/s)
V_W Wind velocity (m/s)
V_R Relative wind velocity (m/s)
V_D Drag speed (m/s)

W

W Weight (N); width (m); Watt, unit of power
W_W Weight on a wheel (N)

X

x Coordinate direction, usually the direction of motion; with subscript, a particular distance in the x-direction

Y

y Coordinate direction coplanar with x and perpendicular to it; with subscript, a particular distance in the y-direction
Y_M Yawing moment about z-axis through vehicle's CG (N·m)
Y Side force in y-direction (N)

Z

z Coordinate direction perpendicular to the plane containing coordinates x and y; with subscript, a particular distance in the z-direction

GREEK LETTERS

Note: angles may be measured in degrees or radians, as appropriate

α Angle of inclination; absorptivity; slip angle; angular acceleration (degrees or radians/s^2)
β Yaw angle; tilt angle of plate from horizontal
γ Azimuth of sun measured from south; camber angle
δ Solar declination angle; steering angle
Δ Indicates a change in a quantity, for example, ΔE, a change in energy

η	Efficiency (usually %) and with subscript to indicate η_G, transmission efficiency
θ	Angle of incidence
θ_Z	Zenith angle
μ	Dynamic viscosity (N·m/m^2)
μ_{BP}	Peak braking coefficient (nondimensional)
μ_{BS}	Smallest braking coefficient (nondimensional)
μ_1	Static rolling resistance coefficient
μ_2	Dynamic rolling resistance coefficient (s/m)
π	Pitch angle (+up)
ρ	Density (kg/m^3)
ν	Kinematic viscosity (m^2/s)
τ	Torque (N·m)
ω	Local hour angle; angular speed (degrees or radians/second)

Index

E. F. Thacher, *A Solar Car Primer*, DOI 10.1007/978-3-319-17494-5

MIX
Papier aus verantwortungsvollen Quellen
Paper from responsible sources
FSC® C105338

If you have any concerns about our products,
you can contact us on
ProductSafety@springernature.com

In case Publisher is established outside the EU,
the EU authorized representative is:
Springer Nature Customer Service Center GmbH
Europaplatz 3, 69115 Heidelberg, Germany

Printed by Libri Plureos GmbH
in Hamburg, Germany